Perturbation Signals for System Identification

Prentice Hall International Series in Acoustics, Speech and Signal Processing

Signals and Systems: An Introduction
L. Balmer

Signal Processing, Image Processing and Pattern Recognition
S. Banks

Randomized Signal Processing
I. Bilkinskis and A. Mikelsons

Digital Image Processing
J. Teuber

Restoration of Lost Samples in Digital Signals
R. Veldhuis

Prentice Hall International Series in Acoustics, Speech and Signal Processing

Perturbation Signals for System Identification

Editor Keith Godfrey

Prentice Hall

New York London Toronto Sydney Tokyo Singapore

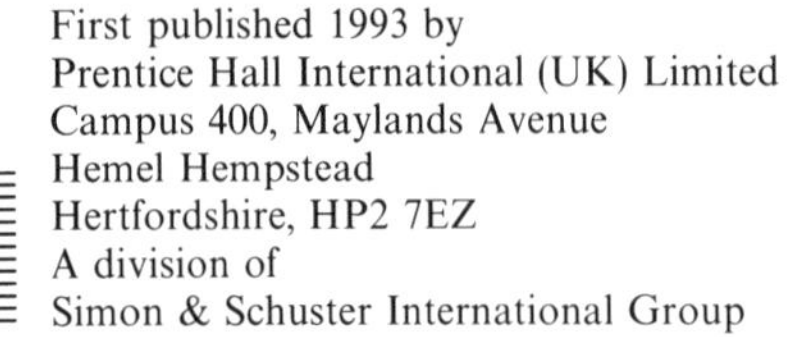

First published 1993 by
Prentice Hall International (UK) Limited
Campus 400, Maylands Avenue
Hemel Hempstead
Hertfordshire, HP2 7EZ
A division of
Simon & Schuster International Group

Typeset in 10/12pt Times Mathematics 569
by International Scientific Communications, Amesbury, Wilts, UK

Printed and bound in Great Britain at
the University Press, Cambridge

Library of Congress Cataloging-in-Publication Data

Library of Congress Cataloging-in-publication data is available from the publisher

British Library Cataloguing in Publication Data

A catalogue record for this book is available from the British Library

ISBN 0-13-656414-3 (hbk)

1 2 3 4 5 97 96 95 94 93

Contents

List of Contributors ix

Preface xi

Nomenclature xviii

1 Introduction to Perturbation Signals for Time-domain System Identification 1
Keith Godfrey
1.1 The time-domain relationship between the input and output of a linear system: the convolution integral 1
1.2 Correlation functions 9
1.3 Correlation functions and process dynamics 19
1.4 Pseudo-random binary signals and related signals 23
1.5 Signals derived from pseudo-random binary signals 35
1.6 Multi-level pseudo-random signals 39
1.7 Perturbation signals for non-linear system identification 49
1.8 Non-periodic signals 55
1.9 Optimal test signals 56
1.10 Pulse transfer function models and persistently exciting signals 58

2 Introduction to Perturbation Signals for Frequency-domain System Identification 60
Keith Godfrey
2.1 Continuous (analog) signal processing in the frequency domain 60
2.2 Discrete signal processing in the frequency domain 70
2.3 Power spectrum of uncorrelated signals 92
2.4 Multi-frequency signals 93
2.5 Linear system identification in the frequency domain 107
2.6 Non-linear system identification in the frequency domain 120

3 Design of Broadband Excitation Signals 126
Johan Schoukens, Patrick Guillaume and Rik Pintelon
3.1 Introduction 126
3.2 Optimization of excitation signals for non-parametric frequency-response measurements 127
3.3 Optimization of excitation signals for parametric measurements 142
3.4 Conclusion 152
Appendix 3.1 Uncertainty on frequency response measurements 153
Appendix 3.2 Crest factor minimization algorithms 155
Appendix 3.3 Some examples of phases 158
Notes 159

4 Periodic Test Signals – Properties and Use 161
Adriaan van den Bos
4.1 Introduction 161
4.2 Discrete-interval multi-frequency signals 162
4.3 Synthesis of discrete-interval multi-frequency signals 163
4.4 Estimation of Fourier coefficients 167
4.5 Use of estimated Fourier coefficients 171
4.6 Conclusions 174

5 Periodic and Non-periodic, Binary and Multi-level Pseudo-random Signals 176
Michael Darnell
5.1 Introduction 176
5.2 Periodic and non-periodic correlation function evaluation 177
5.3 Periodic binary and multi-level pseudo-random signals 181
5.4 Non-periodic binary pseudo-random signals 187
5.5 Non-periodic multi-level pseudo-random signals 194
5.6 System identification: application considerations 201
5.7 Conclusions 208

6 Generation and Applications of Binary Multi-frequency Signals 209
Sandra L. Harris
6.1 Introduction 209
6.2 Signals utilized 210
6.3 Applications to bench- and pilot-scale equipment 214
6.4 Applications using simulations 220
6.5 Conclusions 222
6.6 Nomenclature 222

7 Multi-frequency Binary Sequence Identification of Simulated Control Systems 224
Ian A. Henderson and Joseph McGhee
7.1 Introduction 224
7.2 Identification coding theory and multi-frequency binary sequences 225
7.3 Data measurement using MBS 233
7.4 MBS using phase-shift-keyed modulation 236
7.5 Simulation using SYDLAB 243
7.6 MBS identification monitoring and diagnosis 248
7.7 Conclusions 253

8 Application of Multi-frequency Binary Signals for Identification of Electric Resistance Furnaces 255
Dominik Sankowski, Joseph McGhee, Ian A. Henderson, Jacek Kucharski and Piotr Urbanek
8.1 Introduction 255
8.2 System block diagram and dynamic model 256
8.3 Rotated MBS and switch-on transient 257
8.4 Removal or minimization of MBS switch-on transient 260
8.5 Frequency-response estimation by MBS test signal 261
8.6 Bias error of frequency-response estimates due to the trend 264
8.7 Experimental work 268
8.8 Conclusions 276

9 Temperature Sensor Identification with Multi-frequency Binary Sequences 277
Joseph McGhee, Ian A. Henderson and Lydia M. Jackowska-Strumillo
9.1 Introduction 277
9.2 Dynamic behaviour of contact temperature sensors 278
9.3 Design and application of multi-frequency binary sequences 283
9.4 Data measurement for temperature sensor testing and identification 287
9.5 Conclusions 295

10 Design and Application of Test Signals for Helicopter Model Validation in the Frequency Domain 298
Ron Patton, Martin Miles and Paul Taylor
10.1 Introduction 298
10.2 Frequency-domain testing 300
10.3 Spectral analysis 304
10.4 Frequency-domain testing of a helicopter in flight 311
10.5 The scaled helicopter model 316
10.6 Test signal comparison 318
10.7 Conclusions 320

11 Design of Multi-level Pseudo-random Signals for System Identification 321
H. Anthony Barker
11.1 Introduction 321
11.2 Finite fields 322
11.3 Pseudo-random signals 329
11.4 Autocorrelation functions 332
11.5 Harmonic content 336
11.6 Application 343

12 Design and Application of Non-binary Low-peak-factor Signals for System-dynamic Measurement 348
David Rees and David L. Jones
12.1 Introduction 348
12.2 Non-binary multi-frequency test waveforms 349
12.3 Testing systems using the prime multi-frequency signal 354
12.4 Testing systems with saturation-type non-linearity 362
12.5 Conclusions 368
Appendix 371

13 Multi-frequency Signals for Plant Identification 373
Walter Ditmar and Ray Pettitt
13.1 Introduction 373
13.2 Signal processing 374
13.3 Test-signal properties 377
13.4 Synthesis of pseudo-random Gaussian noise (PRGN) 381
13.5 Results 385
13.6 Conclusions 394
Appendix 395

14 Application of Multi-frequency Test Signals to an Industrial Water Boiler 397
Haydn R. Porch
14.1 Introduction 397
14.2 Excitation signal generation 398
14.3 Plant testing 406
14.4 Frequency-response analysis 408
14.5 Time-domain analysis 410
14.6 Model structure selection 411
14.7 Identification application 413
14.8 Conclusions 421

References 422

Index 436

Contributors

H. Anthony Barker
Department of Electrical and Electronic Engineering,
University College of Swansea, Swansea, SA2 8PP, UK

Adrian van den Bos
Department of Applied Physics
Delft University of Technology
PO Box 5046, 2600 GA, Delft, The Netherlands

Michael Darnell
Department of Electronic Engineering
University of Hull, Hull, HU6 7RX, UK

Walter Ditmar
School of Systems Engineering
University of Portsmouth
Anglesea Road, Portsmouth, PO1 3DJ, UK

Keith Godfrey
Department of Engineering
University of Warwick, Coventry, CV4 7AL, UK

Sandra Harris
Department of Chemical Engineering
Clarkson University, Potsdam, NY13699, USA

Ian Henderson and Joseph McGhee
Industrial Control Centre
University of Strathclyde,
50 George St, Glasgow, G1 1QE, UK

Lidia M. Jacowska-Strumillo
Institute of Textile Engineering
Lodz Technical University
ul Zeromskiego, 116, 90–924 Lodz, Poland

Ron Patton and Martin Miles
Department of Electronics
University of York, Heslington, York, YO1 5DD, UK

Ray Pettitt
Department of Electrical and Electronic Engineering
South Bank University
Borough Road, London, SE1 0AA, UK

Haydn Porch
British Gas plc
Midlands Research Station, Solihull, B29 2JW, UK

David Rees and David L. Jones
Department of Electronics and Information Technology
University of Glamorgan, Pontypridd, CF37 1DL, UK

Dominik Sankowski, Jacek Kucharski and Piotr Urbanek
Lodz Technical University
Al Politechniki 11, 90–924 Lodz, Poland

Johan Schoukens, Patrick Guillaume and Rik Pintelon
Vrije Universiteit Brussel,
Department ELEC, Pleinlaan 2, 1050 Brussels, Belgium

Paul Taylor
Westland Helicopters Ltd
Yeovil, BA20 2YB, UK

Preface

Keith Godfrey

Why a book on perturbation signals?

When measuring the dynamics of a system, it is usually advisable to apply a perturbation signal at the input and to measure the response of the system to the signal. The input and output signals are then processed to give the required estimate of the dynamics of the system. The description of the dynamics may be *non-parametric* – for example, an impulse response, a step response or a frequency response – or it may be *parametric* – for example, a sum of exponentials fit to a step response, a transfer function (continuous or discrete) or a model based on the known properties (e.g. physical, chemical, physiological) of the system under investigation.

In much of the literature on system identification comparatively little attention has been paid to perturbation signal design, other than that the signals should be *persistently exciting*. In the case of linear systems this means effectively that the signal should adequately span the bandwidth of the system being identified. One of the main reasons for this lack of attention has been the emphasis in the literature on identification techniques for parametric models, particularly discrete transfer function (z-transform) models of linear, time-invariant systems, most often with only one input perturbed at a time. Under these circumstances there is not usually a great deal to choose between different perturbation signal designs. However, this is quite an idealized situation in many practical applications and, in practice, many questions of signal design do in fact arise.

Among the most important questions are the following:

1. If the system is noisy and there is only a limited time available for system identification it may be necessary to increase the perturbation signal amplitude to obtain estimates with acceptable accuracy. This brings with it the problem that the assumption of linearity may no longer be valid. Thus, are there signals which

enable the effects of the non-linearities to be minimized, so obtaining a good linear model of the system?
2. Is it possible to design perturbation signals which are particularly suitable for obtaining a characterization of aspects of the non-linear behaviour itself?
3. When identifying a multi-input system, is it possible to obtain several uncorrelated signals, so making it easier to separate out the various input–output relationships?
4. If a system has drift (i.e. a low-frequency disturbance) is it possible to minimize the effects of drift on the system model by suitable signal design?
5. Are there signals and techniques which allow a system to be identified in a short time? This is a particularly important question in many industrial applications.

It is not always easy to find the answers in the literature – hence the need for this book.

The book is quite specifically about the *design and application of signals* for system identification in the time domain and in the frequency domain. It is therefore *complementary* to books on system identification and should be used in conjunction with them. There have been several outstanding books on identification over the last few years, among the most notable being those by Norton (1986), Ljung (1987), Söderström and Stoica (1989), and Schoukens and Pintelon (1991), the last-named placing more emphasis than the others on frequency-domain system identification.

Quite a lot of the material in the book is to do with the identification of non-parametric models. It is important to stress that if a signal is well designed for identifying a non-parametric model of a system, then it is also well designed for identifying a parametric model of the same system. It is hoped that after reading this book the reader will appreciate that, *in practice*, perturbation signal design *is* important in identifying parametric models.

Recent developments in system identification

Over a period of several years a considerable head of steam has been built up in the control engineering area in processing sampled input and output records of a system to obtain a discrete transfer function of the system. The work has been eased by the development by Lennart Ljung of the System Identification Toolbox for use with MATLAB. More recently, there has been increasing interest in techniques based on signal processing in the frequency domain. There are two main reasons for this.

First, until recently, most commercially available digital spectrum analyzers produced only (non-parametric) frequency responses, but some manufacturers have introduced analyzers capable of fitting (parametric) Laplace transfer function models to the frequency responses. The underlying signal processing is via the Fast Fourier Transform, and it is essential to use perturbation signals well designed for use with FFT algorithms. Many analyzers have associated signal generators which avoid some of the pitfalls of frequency-response estimates obtained using the FFT, but even so, some of the signals in these generators are not always particularly well designed and better signals could be incorporated if the relevant work were better known.

Second, an identification Toolset, 'Frequency Domain System Identification', developed by Istvan Kollar of the Technical University of Budapest, has recently been accepted for use with MATLAB. This Toolset, which is complementary to that of Ljung, allows the user to design a wide range of perturbation signals, both binary and non-binary, and to analyze data from system tests to produce frequency responses and corresponding transfer function models.

Overview of the book

The main problem in locating relevant work on perturbation signal design is that it is very widely scattered around the literature. No single journal provides a natural 'home' for papers on this topic.

Considering time-domain system identification first, much of the relevant work appeared in the 1960s and 1970s and stemmed from pseudo-random signals (i.e. deterministic signals with properties similar to those of random signals) based on shift register sequences (Golomb, 1967). A selection of the early work on binary signals for time-domain identification appeared in a monograph by Hoffmann de Visme (1971), and a more recent book on the design and application of such signals is that by Yarmolik and Demidenko (1988), in which the emphasis is on work carried out in the (former) USSR.

The ease of generation, using shift register circuits, of signals based on maximum-length binary sequences has resulted in them being used routinely in a wide range of applications. Such signals are available on many pieces of commercially available hardware. Other types of signal, both binary and non-binary, are less easy to generate using shift register circuitry and so have been much less widely applied. Nevertheless, such signals can be especially useful for multi-input systems or for non-linear systems (see Chapter 1) and, with modern electronic circuitry, can easily be stored in a read-only memory. Also in the time domain, the use of non-periodic signals and non-periodic correlation has been investigated in the area of communications over the last few years, and a number of very interesting signal designs have emerged. This work is applicable much more widely and is reviewed in Chapter 5.

Turning now to frequency-domain identification, early work on signal design dates from the 1960s (unpublished work going back further than that), but most of the relevant papers are much more recent. However, they appear in journals ranging from *Nuclear Science and Engineering* to the *International Journal of Control*, from *Industrial Engineering and Chemistry* to *International Shipbuilding Progress* and in (as least) six different *IEEE Transactions*. Not surprisingly, this work is not as well known as it should be. It is one of the objectives of this book to change this situation by bringing the relevant material together in one location.

In the context of this book it should be emphasized that 'system' is taken very broadly. For example the system may be an industrial process which is being identified with a view to design of suitable controls, or it may be a measuring device whose dynamic characteristics need to be known for it to be incorporated into a larger system to maximum effect. In the latter area, considerable interest has been generated during special sessions on identification methods and perturbation signal design at the annual IEEE Instrumentation and Measurement Technology conferences in recent years. Application areas are very widespread including, for example, work on the design of perturbation signals for identification of respiratory mechanics.

In the early part of the 1990s there were two initiatives to make developments in perturbation signal design known to a wider audience. In the first, David Rees (University of Glamorgan) organized a one-day Colloquium on 'Multifrequency Testing for System Identification' at the Institution of Electrical Engineers in London in June 1990, while in the second, I organized three Special Sessions on 'Perturbation Signals for System Identification' at the IEE International Conference 'Control 91' held in Edinburgh in March 1991. These two initiatives were proposed independently of each other, but when we heard what each other was doing, David and I pooled our ideas to make a very successful Colloquium and three most interesting Conference sessions.

The book emerged from the papers presented in the Special Sessions at 'Control 91'. Twelve of the chapters are considerably revised and extended versions of these papers. The conference proceedings are IEE Publication No. 332 and we are grateful to the Director of Publications of the IEE for permission to proceed with the present book. Two further chapters (6 and 9) have been added so that the book gives a state-of-the-art picture of the design and application of perturbation signals.

Introduction to the book chapters

The book essentially divides into three parts, with Chapters 1 to 5 being overviews of various aspects of signal design and applications, Chapters 6 to 9 considering in more detail the design and application of binary signals, and Chapters 10 to 14 fulfilling a similar role for non-binary signals.

Chapters 1 and 2 are tutorial in nature and describe the fundamentals of signal design for system identification in the time domain (Chapter 1) and the frequency domain (Chapter 2). In preparing them, I have been conscious that a number of texts on system identification are not particularly easy to follow, while from teaching the subject to undergraduates, I am well aware that many students find 'random signals' a difficult topic and need guiding through the material very carefully.

In Chapter 3, Schoukens, Guillaume and Pintelon (Vrije Universiteit, Brussels) consider in detail the design of broadband perturbation signals, carefully distinguishing between the requirements for non-parametric measurements and for parametric modelling. They review many types of signals which have been used in a wide variety of applications, including, for instance, impact testing in mechanical engineering. The emphasis in Chapter 4, by van den Bos (Delft University of Technology), is on system parameter estimation procedures based on Fourier coefficients of the steady-state response measured using multi-frequency perturbation signals. These estimates are simplified through the observations being linear in the Fourier coefficients. Van den Bos shows that the estimation of system parameters from Fourier coefficient estimates can be kept simple by using closed-form instrumental variable or least squares estimators.

Darnell (University of Hull) considers periodic and non-periodic, binary and multi-level pseudo-random signals in Chapter 5. The inclusion in this chapter of the material on non-periodic signals, which are preceded and followed by zeros, is an interesting innovation because until now, they have found little application to system identification, largely because the known signals have been too short. Recently a number of developments have been taking place aimed at producing longer non-periodic signals for communications applications. It may well be that such signals have considerable potential in system identification as well.

Chapter 6 is by Harris (Clarkson University, Potsdam, New York State), one of the pioneers of design and application of binary multi-frequency signals. Author of a number of excellent applications papers, she describes in this chapter several new designs for binary perturbation signals and some new applications and also considers the use of such signals as a perturbation in adaptive algorithms. Many binary multi-frequency signals are of length 128, 256 or 512 (to tie in with data lengths required for processing using Fast Fourier Transform Algorithms), but several designs for shorter signals have been devised by Henderson and McGhee (University of Strathclyde, Glasgow) and are discussed in Chapter 7. Their range of designs includes signals with dominant even harmonics and with dominant odd harmonics, and, using designs based on phase-shift keying, they are able to obtain signals with harmonics close together. Such signals have considerable potential in the identification of lightly damped systems.

The next two chapters describe contrasting applications of binary multi-frequency signals. In Chapter 8, Sankowski (Lodz Technical University, Poland), McGhee, Henderson (University of Strathclyde), Kucharski and Urbanek (Lodz) discuss application to electric resistance furnaces and describe techniques for minimizing switch-on transient and for minimizing error due to drift (i.e. low-frequency disturbance). The potential for application to measuring devices is discussed in Chapter 9 by McGhee and Henderson (Strathclyde) and Jackowska-Strumillo (Lodz), with particular emphasis on temperature sensor identification.

Chapter 10, by Patton, Miles (University of York) and Taylor (Westland Helicopters) acts as a bridge between the applications using binary signals and those using non-binary signals because it discusses the design and application of both types of signal in the context of helicopter model validation in the frequency domain. Clearly, safety considerations are paramount in this particular application, and it is essential that any perturbation signal generation should be safe, well controlled and repeatable.

The next three chapters consider the design of different types of non-binary perturbation signals. In Chapter 11, Barker (University College, Swansea) describes how to design signals based on non-binary pseudo-random sequences. The main factor which has resulted in these signals not finding very wide application up to now is the complexity of the underlying theory, but in his chapter Barker provides a clear explanation of this theory, with the use of sum and product tables to define the field algebra. The reader is also guided to papers describing the identification of non-linear system characteristics using such signals. Rees and Jones (University of Glamorgan) describe in Chapter 12 the design and application of signals containing only prime harmonics (excluding harmonics 1 and 2). They show that when these are applied to non-linear systems a substantial proportion of the higher harmonics generated are at non-prime harmonics, so that if analysis is confined to the prime harmonics only, distortion due to non-linearities is very much reduced. Further types of multi-frequency signals are considered in Chapter 13, by Ditmar (University of Portsmouth) and Pettitt (South Bank University, London), who describe pseudo-random Gaussian noise signals in particular. These signals have an approximately Gaussian probability density function and can result in system response signals of lower amplitude compared with other designs as the system bandwidth is reduced.

The book concludes with a description by Porch (British Gas plc) in Chapter 14 of some features of the Plant System Identification (PSI) package developed at the Midlands Research Station of British Gas plc and marketed by the Industrial Control Centre at the University of Strathclyde, and of the application of multi-frequency test signals incorporated into the package to an industrial water boiler. Fitting of both Laplace transform and z-transform models is discussed and the techniques used for model validation are described.

The fact that this is a multi-author text brings the very considerable advantage of a wide range of designs and applications written by the authors themselves. The disadvantage of many edited multi-author books, that they do not hang together particularly well, has been largely avoided in this book by authors being aware of the presentations at the 'Control 91' conference before preparing their book chapters, by being asked to use the same nomenclature (which follows this preface) and by my requesting authors to revise their chapters as they came in to avoid excessive repetition. There is some repetition between chapters but this is only a small amount, and it serves to reinforce the fact that the book is about a common theme rather than fourteen isolated chapters.

Readership and prerequisites

The book will be of interest to engineers and scientists in industry and academia who wish to measure the dynamics of any form of system. The system could range from a chemical process to a lung, from a nuclear reactor to a measuring device. The book will also be of interest to the designers of system dynamics hardware (including digital spectrum analyzers) and of software for system identification.

Recognizing that some texts on system identification are quite advanced and perhaps prove rather difficult for those unfamiliar with the area, this book starts with two chapters written in a tutorial style to introduce the rest of it. These chapters have been tried out on a number of electronics and electrical engineering students at the outset of their final year at the University of Warwick and have been revised as a result of their comments. Before reading the chapters these students had taken introductory courses in linear system dynamics and in signal processing, both analog and digital. Those readers who have not covered these topics or who need to brush up on them should first study the excellent introductory text on *Signals and Systems* by Balmer (1991).

Nomenclature

General

$j = \sqrt{-1}$
Either ϕ or θ = phase angle
V = amplitude (i.e. $\pm V$) of a binary signal
$\oplus$ = modulo 2 addition
p = prime
q = power of prime
GF(q) = Galois field of q elements

Continuous signals and systems

Time-domain signal processing

$u(t)$ or $x(t)$ = system input
$y(t)$ = system output
$n(t)$ = noise signal
T = period of periodic signal
Δt = clock pulse interval of a discrete interval signal
N = number of clock pulse intervals in one period of a periodic discrete interval signal (for which $T = N\Delta t$)
Either $g(t)$ or $h(t)$ = unit impulse response of a system
$R_{xx}(\tau)$ = autocorrelation function of a signal $x(t)$

$$= \lim_{T\to\infty} \frac{1}{2T} \int_{-T}^{T} x(t)x(t+\tau)\,dt \text{ for a non-periodic signal}$$

$$= \frac{1}{T} \int_{0}^{T} x(t)\,.\,x(t+\tau)\,dt \text{ for a periodic signal}$$

$R_{xy}(\tau)$ = cross-correlation function between $x(t)$ and $y(t)$

so that $R_{xy}(\tau) = \int_0^{T_s} h(\lambda)R_{xx}(\tau - \lambda)\, d\lambda$, for a noise-free process,

where T_s = system settling time.

$$\text{Peak factor} = \frac{x_{max} - x_{min}}{2\sqrt{2}x_{rms}}$$

Frequency-domain signal processing

ω = radian frequency; f = cyclic frequency (Hz).
$X(j\omega)$ = Fourier transform of $x(t)$

$$= \int_{-\infty}^{\infty} x(t)\, e^{-j\omega t}\, dt \text{ for a non-periodic signal}$$

$x(t)$ = Inverse Fourier transform of $X(j\omega)$

$$= \frac{1}{2\pi}\int_{-\infty}^{\infty} X(j\omega)\, e^{-j\omega t}\, d\omega$$

$G(j\omega)$ or $H(j\omega)$ = frequency response of a system

= Fourier transform of $g(t)$ (or $h(t)$).

Use the phrase 'frequency response' for a non-parametric model, i.e. the Fourier transform of the impulse response, and reserve 'transfer function' for parametric models.

$S_{xx}(\omega)$ = power-spectral density of $x(t)$

$$= \int_{-\infty}^{\infty} R_{xx}(\tau)\, e^{-j\omega\tau}\, d\tau \text{ for a non-periodic signal}$$

$S_{xy}(j\omega)$ = cross-power-spectral density between $x(t)$ and $y(t)$

$$= \int_{-\infty}^{\infty} R_{xy}(\tau)\, e^{-j\omega\tau}\, d\tau \text{ for non-periodic signals}$$

Either $\gamma_{xy}^2(\omega)$ or $\Gamma_{xy}^2(\omega)$ = (squared) coherence function

$$= |S_{xy}(j\omega)|^2/[S_{xx}(\omega)S_{yy}(\omega)]$$

For a periodic signal, with period T,

$S_{xx}(\omega)$ = power spectrum of $x(t)$

$$= \frac{1}{T}\int_0^T R_{xx}(\tau)\cos\omega\tau\, d\tau$$

This is a line spectrum, with values only at $f = (k/T)$ Hz (k an integer), and the units are now power content of the signal at those particular frequencies.

Discrete signals and systems

Δ = sampling interval.

$\{u_i\}$ = sequence of values of the signal $u(t)$, sampled at a regular interval Δ.

$$R_{xx}(k) = \frac{1}{N}\sum_{i=0}^{N-1} x_i x_{i+k}$$

$$R_{xy}(k) = \frac{1}{N}\sum_{i=0}^{N-1} x_i\, y_{i+k}$$

where N is the number of sampled values in the sequence. (These definitions do require some modification if dealing with limited amounts of data.)
Discrete Fourier Transform (DFT):

$$X_k = \frac{1}{\mathrm{N}}\sum_{i=0}^{N-1} x_i \exp\left[-\mathrm{j}\frac{2\pi ik}{N}\right]$$

i.e. a sequence of values of the DFT at frequencies $f = (k/N)$ Hz, k an integer.
(NB: Note the $1/N$ in this expression, which has become the most usual location for it in recent texts.)

Inverse DFT $x_i = \sum_{k=0}^{N-1} X_k \exp\left[\mathrm{j}\frac{2\pi ik}{N}\right]$

Discrete Power Spectrum is the DFT of the autocorrelation sequence:

$$S_{xx}(k) = \frac{1}{N}\sum_{i=0}^{N-1} R_{xx}(i) \exp\left[-\mathrm{j}\frac{2\pi ik}{N}\right]$$

As before, this is a sequence of values at frequencies $f = (k/N)$ Hz.
(Similarly for Discrete Cross-Power Spectrum.)

1

Introduction to Perturbation Signals for Time-domain System Identification

Keith Godfrey

1.1 The time-domain relationship between the input and output of a linear system: the convolution integral

The form of system to be considered in most of this book is shown in Figure 1.1. The input signal $u(t)$ produces a system response signal $y(t)$. In many experimental situations, $u(t)$ and $y(t)$ are perturbations around steady-state input and output levels, and in these situations, the terms 'input' and 'output' will be taken to refer to deviations from the steady operating levels. There may be noise in the system and/or the measuring device and this is most commonly represented by a noise signal $n(t)$ added to the system output $y(t)$ to give the *measurable* output signal $z(t)$, so that

$$z(t) = y(t) + n(t) \tag{1.1}$$

The system is assumed to be linear, so that the principle of superposition applies. This means that if $u_1(t)$ produces a response $y_1(t)$, $u_2(t)$ produces a response $y_2(t)$ and so on, then the sum $[u_1(t) + u_2(t) + \ldots]$ produces a response $[y_1(t) + y_2(t) + \ldots]$. The assumption of linearity is very important in the context of this book, because for it to be a reasonable assumption it is usually necessary that the deviations from the operating level of the system caused by the applied input should not be large.

The system response $y(t)$ is given by an integral of weighted inputs which have occurred in the past. The past values of input are multiplied by a function $h(t)$, called

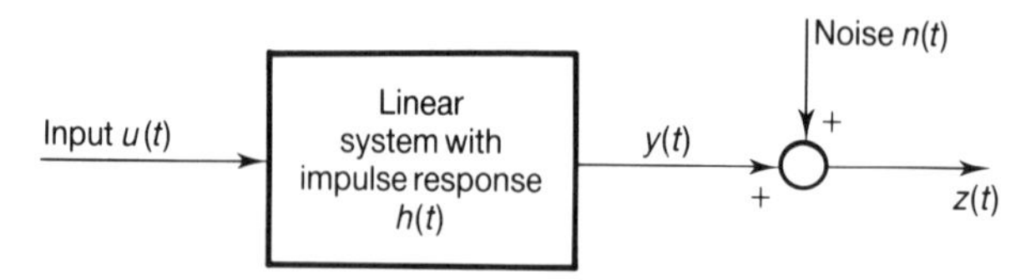

Figure 1.1 Block diagram of a system.

the *weighting function* of the system and the mathematical equation describing the relationship is

$$y(t) = \int_0^\infty h(\lambda)u(t-\lambda)\,\mathrm{d}\lambda \tag{1.2}$$

where λ is a time variable. If the system is stationary, i.e. if its dynamic characteristics do not change with time, then the weighting function is also equal to the *unit impulse response* of the system. This is the response $y(t)$ to an input signal $u(t) = \delta(t)$, where $\delta(t)$, the Dirac delta function centred at $t = 0$, is a function of height which tends to infinity and duration which tends to zero such that (height $\times$ duration) = 1.

Equation (1.2) is called a *convolution integral* and it is illustrated graphically in Figure 1.2. A typical weighting function $h(\lambda)$ is shown in Figure 1.2(a), and an input $u(\lambda)$ consisting, for ease of drawing, of a series of rectangles is shown in Figure 1.2(b). For convolution, $u(\lambda)$ is rotated about the $\lambda = 0$ axis to give $u(-\lambda)$ as in Figure 1.2(c), and this is then shifted by an amount t to give $u(t-\lambda)$, as in Figure 1.2(d); note the position of the marker asterisk in each of these operations. The weighting function $h(\lambda)$ is then multiplied by $u(t-\lambda)$, as in Figure 1.2(e), and this product is then integrated over values of λ from zero to infinity. The procedure is repeated for each value of t for which values of $y(t)$ are required.

In a practical situation, inputs which have occurred at times greater than T_S in the past have negligible effect on the present output of the system so that equation (1.2) may be modified to

$$y(t) = \int_0^{T_S} h(\lambda)u(t-\lambda)\,\mathrm{d}\lambda \tag{1.3}$$

The time T_S is called the *settling time* of the system. Combining equations (1.1) and (1.3), the *measurable* output is given by

$$z(t) = \int_0^{T_S} h(\lambda)u(t-\lambda)\,\mathrm{d}\lambda + n(t) \tag{1.4}$$

If we are concerned with sampled signals, so that $u(t)$, $y(t)$, $z(t)$ and $n(t)$ are all sampled at regular intervals Δ to give corresponding sequences with elements u_r, y_r, z_r and n_r, then the convolution integral (1.2) becomes a convolution sum:

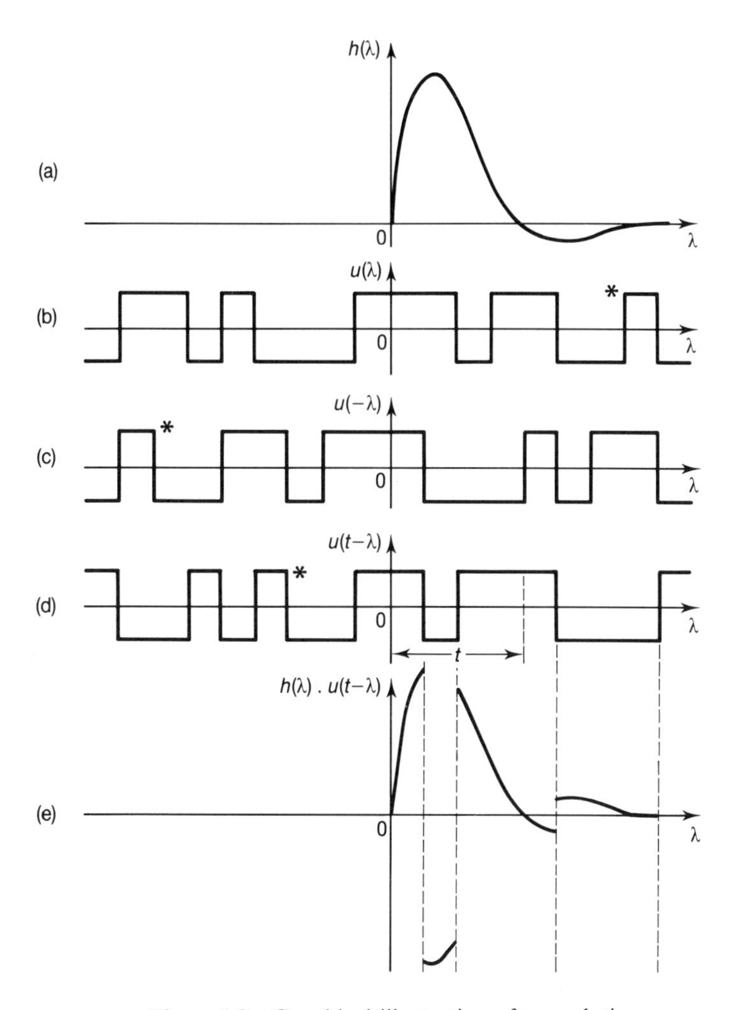

Figure 1.2 Graphical illustration of convolution.

$$y_r = \sum_{i=0}^{\infty} h_i u_{r-i} \tag{1.5}$$

where h_i is the weighting sequence of the system, i.e. $h(t)$ sampled at intervals Δ. As before, in practice it will only be necessary to consider inputs up to a time T_S in the past and denoting $L.\Delta = T_S$, the equivalent of equation (1.4) becomes

$$z_r = \sum_{i=0}^{L} h_i u_{r-i} + n_r \tag{1.6}$$

In theory, equation (1.4) or its discrete equivalent, equation (1.6), could be solved to give an estimate of the system weighting function from almost any input–output records. In practice, deconvolution can give rise to computational problems in the presence of noise unless a simple form of input is used. Two particularly simple forms of input are a step function and a pulse function approximating to a Dirac delta function, and some examples of the application of these will be considered in the remainder of this section.

Application Examples 1.1 and 1.2 below are of step responses of full-scale industrial processes, the first providing reasonable information about the process dynamics and the second providing virtually no useful information. In both cases, with the input signal being a step function, use is being made of the linear system relationship that if $u(t)$ is a unit step, then the process response $y(t)$, i.e. the unit step response, is the (time) integral of the unit impulse response $h(t)$.

APPLICATION EXAMPLE 1.1 STEP RESPONSES FROM A STEAM REFORMING PROCESS

The process in this case was a steam reformer for producing town gas from naphtha and steam. The plant had to respond rapidly to changes in consumer demand, and the requirement was for some mathematical models for the design of feedforward control to minimize changes in product quality in the face of throughput changes (Godfrey and Shackcloth, 1970).

A reformer is an endothermic reactor, and the heat required to sustain the chemical reaction taking place within the reformer tubes was provided in this case by means of burners heating the outside of the tubes. The reformer outlet temperature was a reasonable guide to product quality. A throughput increase resulted in a decrease of outlet temperature and the feedforward strategy was to increase the fuel flow to the burners to compensate for this.

Open-loop responses of outlet temperature to a step increase in burner flow and to a step increase in throughput are shown in Figures 1.3(a) and (b), respectively. The open circles represent the response data, i.e. the *non-parametric* model of the step response. The responses were also modelled by single exponentials with a time delay. Step responses of these are shown as the filled-in circles and it may be seen that these simple *parametric* models provided good approximations to the measured responses. The parametric models were used in the design of the feedforward algorithm. Feedforward is essentially an open-loop strategy, which makes its use without some form of feedback inadvisable in most industrial situations. In this application, feedback from outlet temperature to burner fuel flow was retained, to compensate both for factors other than those being manipulated in the feedforward strategy and for differences between the assumed and actual process dynamics.

A typical feedforward run, as throughput was changed in a ramp from 75% to 43% of full throughput, is shown in Figure 1.4. The result, which was typical of many such runs, was most satisfactory, with the outlet temperature being kept to within ± 3°C of the set point. This compared with variations of some ± 20°C, when the feedforward strategy was implemented manually.

The responses shown in Figure 1.3 are unusually noise-free for an industrial process. In the construction of the plant, instrument cables were well shielded and

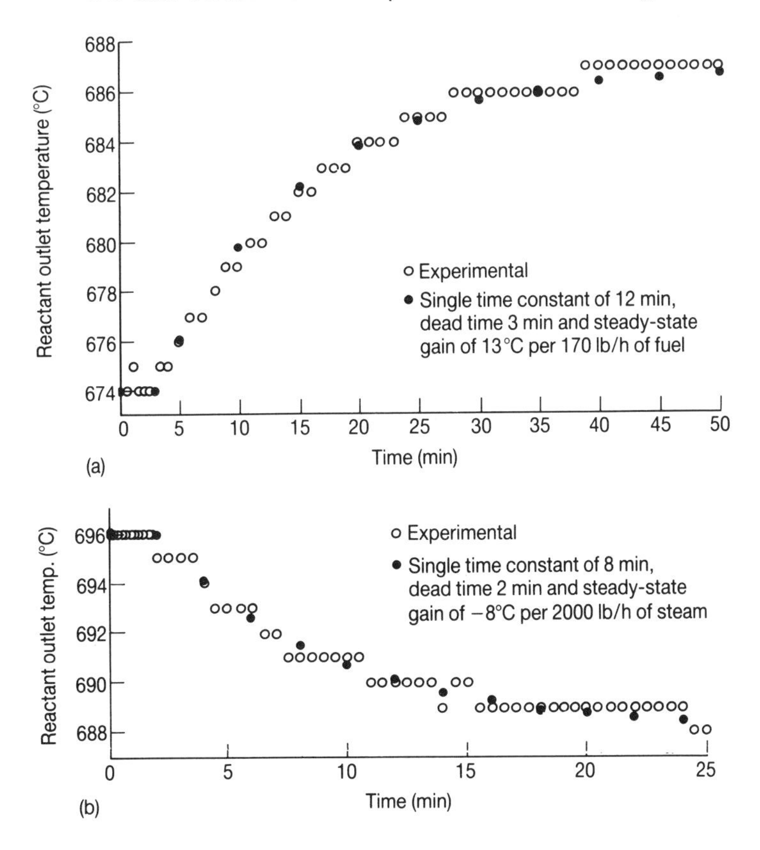

Figure 1.3 Open-loop responses of reformer outlet temperature. (a) To an increase in burner fuel flow; (b) to an increase in steam flow (naphtha in ratio).

were kept as distant as possible from the power cables. The relatively small extra cost which this entailed during the plant-construction phase paid handsome dividends in the simplicity and effectiveness of the control system design once the plant was operational.

The simple parametric models fitted to the step-response data corresponded to (Laplace) transfer functions of the form

$$H(s) = K_p \exp(-sT_D)/(1 + sT_1) \tag{1.7}$$

where K_p is the steady-state gain of the process (negative in the case of the throughput changes), T_D is the time delay and T_1 is the time constant. As pointed out earlier, these simple models proved very effective in the feedforward design. However, the amplitude quantization in Figure 1.3 was 1 °C, and this coarseness would make the accurate estimation of any second time constant virtually out of the question. A further test was carried out with the response to a step change in burner fuel flow measured with the outlet temperature quantized to $\frac{1}{3}$°C; this is shown in Figure 1.5.

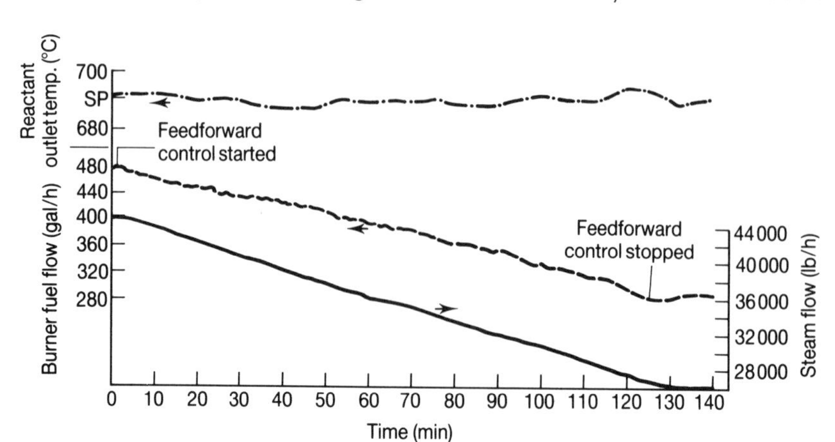

Figure 1.4 Demonstration of the feedforward/feedback control scheme.

It may be seen from Figure 1.5 that it would still have been difficult to obtain a reasonably accurate estimate of any shorter time constant. The apparent dead time was slightly reduced using the finer quantization interval, as would be expected. If there were, in fact, a shorter time constant present, so that the transfer function was of the form

$$H(s) = K_p \exp(-sT_D)/[(1 + sT_a)(1 + sT_b)] \tag{1.8}$$

then it would often be difficult to estimate T_a and T_b accurately from actual step responses from a full-scale process, and the resulting fit would often be of the form of equation (1.7) with $T_1 = T_a + T_b$. This results in the loss of the s^2 term in the denominator of equation (1.8). Since frequency-response information is obtained from

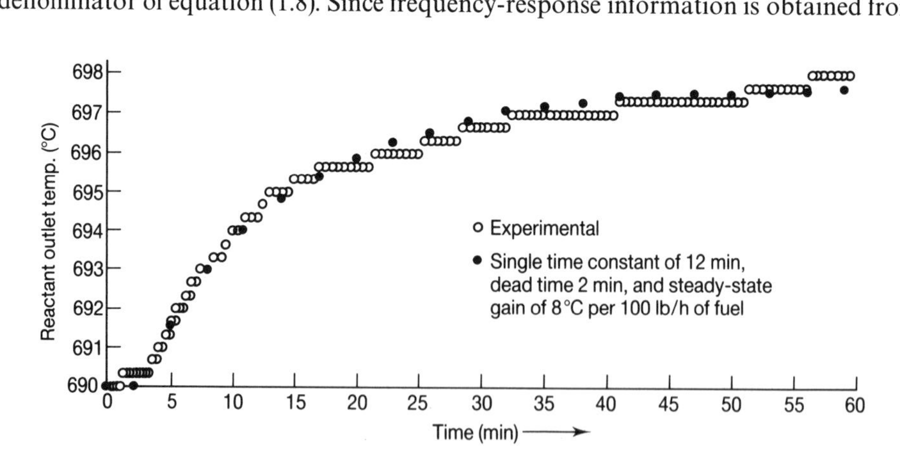

Figure 1.5 Open-loop step response of reformer outlet temperature to a step increase in burner fuel flow with measurement at $(\frac{1}{3})$°C intervals.

a Laplace transfer function by substitution of $j\omega$ for s, it is evident that a step response does not usually convey high-frequency information with any great accuracy.

APPLICATION EXAMPLE 1.2 STEP RESPONSE FROM AN OIL REFINERY DISTILLATION COLUMN

An example from an industrial process in which virtually nothing could be deduced from a step response was from an oil refinery distillation column, in which the response of a particular column temperature to step changes in feed heater temperature set point was measured (Godfrey, 1969b). Typical responses to upward and downward step changes are shown in Figure 1.6. The magnitudes of the step changes were at the maximum level permitted on safety grounds by the plant manager. It is clear that it would be difficult to obtain any meaningful estimate of the process dynamics from the response. Further results designed to get round this problem by perturbing the reboiler set point with a pseudo-random binary signal will be considered in Application Example 1.4.

The second very simple form of input which removes the need for deconvolution is a short-duration pulse of area (amplitude $\times$ time) K_1, in which case $z(t)$ in equation (1.4) is given approximately by

$$z(t) \simeq K_1 h(t) + n(t) \tag{1.9}$$

where the approximation becomes more accurate as the pulse duration becomes shorter, i.e. as $u(t) \to K_1 \delta(t)$. This approach is widely used in acoustic measurements and in mechanical testing (hammer excitation) where special-purpose equipment is available. The problem with using this approach for the identification of industrial processes is that, in many cases, it is necessary for the pulse amplitude to become large in order that the $K_1 h(t)$ term should predominate over the $n(t)$ term while maintaining a short pulse duration. Alternatively, it may be necessary to average many such responses. This is a time-consuming procedure and, in practice, it is not

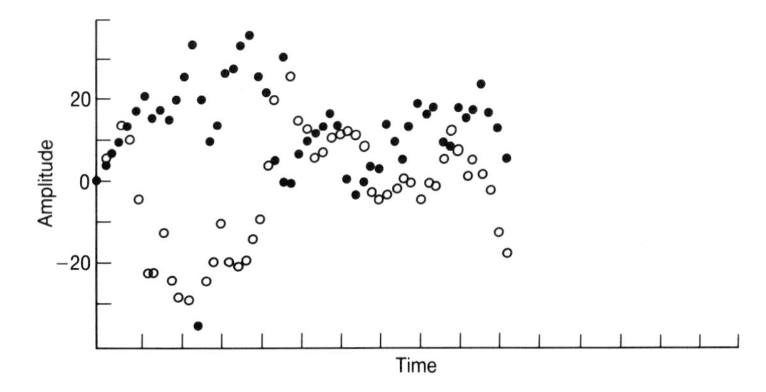

Figure 1.6 Responses of column temperature to upward and downward step changes of feed heater temperature set point. ● Upward response, ○ downward response.

always easy to keep a system stationary (in dynamical terms) over a long period of time. For these reasons, testing with short-duration pulses has found comparatively little application to industrial processes.

By contrast, in biomedical work a rapidly administered dose (either intravenous or oral) is by far the most frequently encountered form of input function. The measured response $z(t)$ to this approximately impulsive input is then often fitted by a sum of exponentials (Godfrey, 1983). Many such applications are characterized by good-quality measurements over a wide range of amplitudes and with comparatively little noise $n(t)$. Most of the measurements are discrete in time, and with a well-designed sampling schedule there are short intervals between measurements when the response $z(t)$ is changing rapidly and wider intervals when it is not. Even so, it is rarely possible to fit more than three exponentials to such responses and even to fit three, measurements usually have to be available for a long time, starting very soon after the impulsive input.

Just occasionally, it is possible to fit four exponentials. For example, Foster *et al.* (1979) fitted the following curve to the plasma response to a rapidly administered intravenous injection of ^{69m}Zn in human beings:

$$z(t) = 0.790e^{-176t} + 0.175e^{-73.4t} + 0.022e^{-5.87t} + 0.013e^{-0.053t}$$

where the amplitude has been normalized so that $z(t) = 1$ when $t = 0^+$ and time t is in days. The data consisted of almost noise-free sampled measurements from a few minutes after the injection up to five days after, inter-sample intervals were a few minutes in the early part of the response, increasing to a day between four and five days after. Although the amplitudes of the third and fourth terms are small, there is no particularly close pole-zero cancellation in the Laplace transform of $z(t)$:

$$Z(s) = \frac{(s + 92.66)(s + 9.475)(s + 1.171)}{(s + 176)(s + 73.4)(s + 5.87)(s + 0.053)}$$

(which is, of course, proportional to the transfer function of the process because of the impulsive form of input). However, by the end of the measurement period ($t = 5$), the fourth term in $z(t)$ has decreased to only $e^{-0.265} = 0.767$ of its initial value so that, even with this long response record, good initial estimates would be required to permit its accurate estimation using a non-linear least-squares algorithm.

In this section it has been seen that use of the input–output relationship of equation (1.4) to estimate system dynamics can give rise to problems in many practical situations. Unless the input $u(t)$ is of a comparatively simple form, solution of equation (1.4) for $h(t)$ requires a deconvolution procedure, and this can often result in computational problems with real data. If $u(t)$ is kept simple (for example, of step or impulsive form) the noise term $n(t)$ can easily mask the system response, so that to obtain a reasonable estimate of the weighting function can require either a large-amplitude perturbation (which may well violate the assumption of linearity) or the averaging of several responses, which can become time consuming and which may not even be possible in an industrial situation. In terms of industrial applications, the steam reformer step responses of Figures 1.3 and 1.5 are very good in terms of signal-to-noise ratio, while the distillation-column response of Figure 1.6 is unusually poor; the usual situation is something in between the two. At best, the step responses prove perfectly satisfactory for the purpose for which they were obtained (e.g. design of feedforward control in the steam reformer example). More generally, a step response will, in most cases, prove helpful in the design of further experiments to obtain improved characterization of the process dynamics.

To obtain such improved characterization, alternative methods of signal processing, using different forms of perturbation signal, are needed; these should enhance the system component at the expense of the noise component. In the time domain, improved results can often be obtained through the use of correlation methods; frequency-domain methods will be discussed in Chapter 2.

1.2 Correlation functions

1.2.1 Definitions

In this section extensive use will be made of the *mean* or *expected value* of a signal; for a signal $x(t)$ this will be denoted by $E[x(t)]$ or $\mu_x(t)$.

The *autocorrelation function* $R_{xx}(t_1, t_2)$ is defined by

$$R_{xx}(t_1, t_2) = E[x(t_1)x(t_2)] \tag{1.10}$$

while the closely related *autocovariance function* $C_{xx}(t_1, t_2)$ is defined by

$$C_{xx}(t_1, t_2) = E\{[x(t_1) - \mu_x(t_1)] \,.\, [x(t_2) - \mu_x(t_2)]\} \tag{1.11}$$

Note that

$$R_{xx}(t_1, t_2) = C_{xx}(t_1, t_2) + \mu_x(t_1) \,.\, \mu_x(t_2) \tag{1.12}$$

A signal is said to be *strictly stationary* if its probability characteristics remain invariant under a shift in the time origin. Such a requirement can only rarely be assumed completely, but the above expressions can be simplified if the signal is assumed to be *wide-sense stationary* or *weakly stationary*, which only requires that its mean value does not vary with time. The autocorrelation function then depends only on the time difference $|t_2 - t_1|$ and setting this equal to τ gives

$$R_{xx}(\tau) = E[x(t) \,.\, x(t + \tau)] \tag{1.13}$$

For almost all stationary signals of practical interest, it is found that the mathematical expectation operator is equivalent, under fairly general conditions, to an average performed on the signal over an infinite time interval, i.e. the ensemble averages and time averages are equivalent. This is known as the ergodic hypothesis. For a stationary, ergodic signal $x(t)$,

$$E[x(t)] = \mu_x = \lim_{T \to \infty} \frac{1}{2T} \int_{-T}^{T} x(t)\, \mathrm{d}t \tag{1.14}$$

$$R_{xx}(\tau) = \lim_{T \to \infty} \frac{1}{2T} \int_{-T}^{T} x(t) . x(t + \tau)\, \mathrm{d}t \tag{1.15}$$

$$C_{xx}(\tau) = \lim_{T \to \infty} \frac{1}{2T} \int_{-T}^{T} [x(t) - \mu_x][x(t + \tau) - \mu_x]\, \mathrm{d}t \tag{1.16}$$

If $x(t)$ is a periodic signal with period T, then these infinite time averages are equivalent to averages over a time kT, where k is an integer ($\geqslant 1$). For example, for such a periodic signal,

$$R_{xx}(\tau) = \frac{1}{T} \int_{0}^{T} x(t) . x(t + \tau)\, \mathrm{d}t \tag{1.17}$$

Similar expressions to those above are obtained for the *cross-correlation function*, $E[x(t) . y(t + \tau)]$ between two stationary, ergodic signals $x(t)$ and $y(t)$:

$$R_{xy}(\tau) = \lim_{T \to \infty} \frac{1}{2T} \int_{-T}^{T} x(t) . y(t + \tau)\, \mathrm{d}t \tag{1.18}$$

in general and

$$R_{xy}(\tau) = \frac{1}{T} \int_{0}^{T} x(t) . y(t + \tau)\, \mathrm{d}t \tag{1.19}$$

if both $x(t)$ and $y(t)$ are periodic with period T.

The physical interpretation of a cross-correlation function of the form of equation (1.18) is that very long portions of the signals $x(t)$ and $y(t)$ are taken, and $y(t)$ is shifted by an amount τ with respect to $x(t)$. The signals $x(t)$ and $y(t + \tau)$ are then multiplied together and the integral of this multiplication is determined. The result is then divided by the time over which the integration is carried out. The procedure is illustrated in Figure 1.7.

1.2.2 Properties of autocorrelation functions of stationary signals

PROPERTY A

From equation (1.15),

$$R_{xx}(0) = E[x^2(t)] \tag{1.20}$$

i.e. the autocorrelation function of a signal at delay $\tau = 0$ is equal to the mean squared value of the signal. Similarly, from equation (1.16),

$$C_{xx}(0) = E[(x(t) - \mu_x)^2] = \sigma_x^2 \tag{1.21}$$

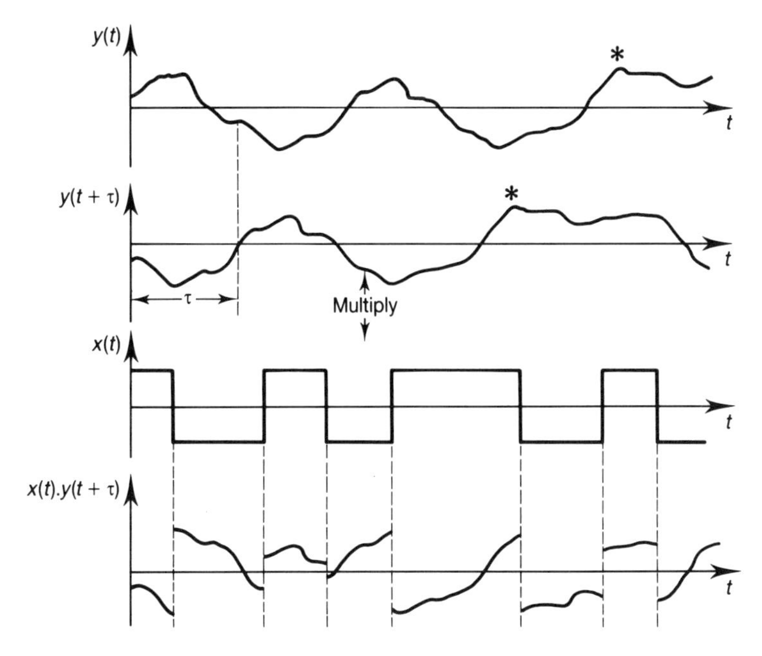

Figure 1.7 Graphical illustration of cross-correlation.

i.e. the autocovariance function of a signal at delay $\tau = 0$ is equal to the variance of the signal.

PROPERTY B

For a stationary process,

$$R_{xx}(\tau) = R_{xx}(-\tau) \tag{1.22}$$

i.e. the autocorrelation function is an even function of τ (similarly for the autocovariance function).

PROPERTY C

$$R_{xx}(0) \geqslant |R_{xx}(\tau)| \tag{1.23}$$

i.e. the autocorrelation function for $\tau = 0$ is larger than or equal to the magnitude of the autocorrelation function for any other value of τ. To show this, we know that:

$$E\{[x(t) \pm x(t+\tau)]^2\} \geqslant 0$$

which, on rearranging, gives

$$E[x^2(t)] + E[x^2(t+\tau)] \geqslant \pm 2E[x(t) . x(t+\tau)]$$

so that

$$2R_{xx}(0) \geqslant \pm 2R_{xx}(\tau)$$

PROPERTY D
If $z(t) = x(t) + y(t)$, then $R_{zz}(\tau) = R_{xx}(\tau) + R_{xy}(\tau) + R_{yx}(\tau) + R_{yy}(\tau)$, which, if $x(t)$ and $y(t)$ are uncorrelated, becomes

$$R_{zz}(\tau) = R_{xx}(\tau) + R_{yy}(\tau) \tag{1.24}$$

Note that, in general, these properties do not apply to cross-correlation functions. Property B becomes $R_{xy}(\tau) = R_{yx}(-\tau)$, so that the cross-correlation function is not (in general) an even function of τ; also, the largest magnitude point on a cross-correlation function can occur for any value of τ.

1.2.3 Analytical calculation of some autocorrelation functions

EXAMPLE 1A AUTOCORRELATION FUNCTION OF A SINE WAVE
Consider the sine wave given by

$$x(t) = \text{V} \sin(\omega t + \varphi)$$

Let us first determine the autocorrelation function using the general expression of equation (1.15). Thus:

$$\begin{aligned}
R_{xx}(\tau) &= \lim_{T\to\infty} \frac{1}{2T} \int_{-T}^{T} V^2 \sin(\omega t + \varphi) . \sin(\omega t + \omega\tau + \varphi)\, \mathrm{d}t \\
&= \lim_{T\to\infty} \frac{1}{2T} \int_{-T}^{T} \tfrac{1}{2} V^2 \left[\cos \omega\tau - \cos(2\omega t + \omega\tau + 2\varphi)\right] \mathrm{d}t \\
&= \tfrac{1}{2} V^2 \lim_{T\to\infty} \frac{1}{2T} \left[t \cos \omega\tau - \frac{\sin(2\omega t + \omega\tau + 2\varphi)}{2\omega} \right]_{-T}^{T}
\end{aligned}$$

As the limit $T \to \infty$ is taken, neither $\sin(\infty)$ nor $\sin(-\infty)$ is determinate, but neither is less than -1 nor greater than $+1$. Thus the second term in the bracket lies between $-(1/\omega)$ and $+(1/\omega)$ and it disappears when the division by $2T$ (with $T \to \infty$) is made. The $2T$ cancels out in the first term, leaving

$$R_{xx}(\tau) = \tfrac{1}{2} V^2 \cos \omega\tau$$

Note that in the autocorrelation function the phase information is lost.

Let us also confirm the expression for $R_{xx}(\tau)$ by using equation (1.17) for periodic functions; in this case, $T = 2\pi/\omega$. Hence:

$$\begin{aligned}
R_{xx}(\tau) &= \frac{\omega}{2\pi} \int_{0}^{2\pi/\omega} V^2 [\sin(\omega t + \varphi) . \sin(\omega t + \omega\tau + \varphi)]\, \mathrm{d}t \\
&= \frac{\omega}{2\pi} \cdot \frac{V^2}{2} \int_{0}^{2\pi/\omega} [\cos \omega\tau - \cos(2\omega t + \omega\tau + 2\varphi)]\, \mathrm{d}t
\end{aligned}$$

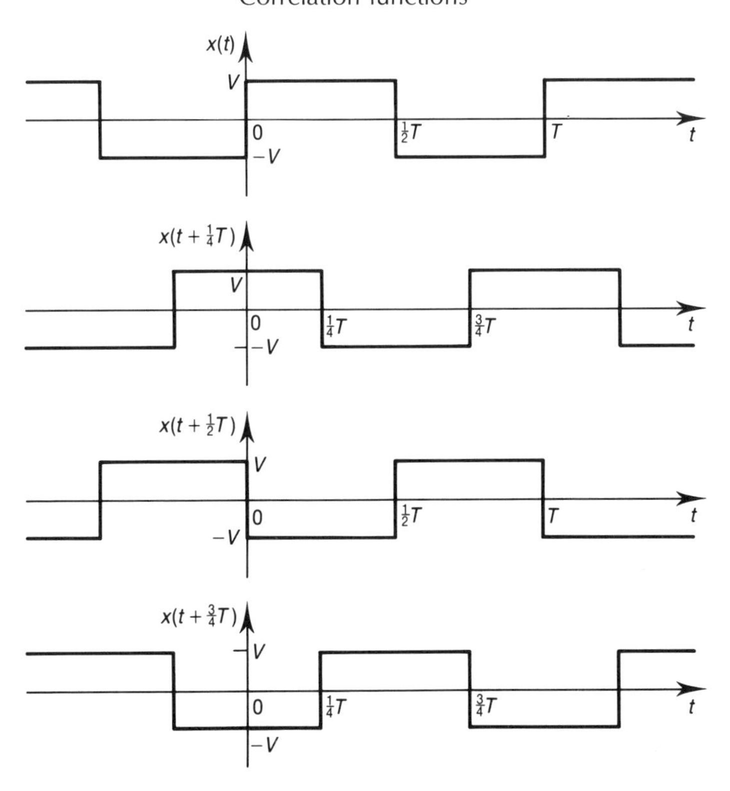

Figure 1.8 Squarewave and its shifted versions.

$$= \frac{\omega}{2\pi} \cdot \frac{V^2}{2} [t \cos \omega\tau]_0^{2\pi/\omega}$$

$$- \frac{\omega}{2\pi} \cdot \frac{V^2}{2} \left[\frac{\sin(2\omega t + \omega\tau + 2\varphi)}{2\omega}\right]_0^{2\pi/\omega}$$

$$= \tfrac{1}{2}V^2 \cos \omega\tau, \text{ as before}$$

Note that the autocorrelation function has the same period as the original signal and that $R_{xx}(\tau)$ does not die away as τ increases. This property gives rise to one of the main uses of autocorrelation – the detection of a periodic component in noise. If $z(t) = x(t) + n(t)$, where $x(t)$ is a periodic signal and $n(t)$ is a non-periodic noise signal uncorrelated with $x(t)$, then from Property D,

$$R_{zz}(\tau) = R_{xx}(\tau) + R_{nn}(\tau)$$

Thus if the two components have zero mean, $R_{nn}(\tau)$ dies away to zero as τ becomes large, leaving only the periodic term $R_{xx}(\tau)$. The noise component could, of course, be reduced by filtering if the bandwidth of the noise were well separated from the

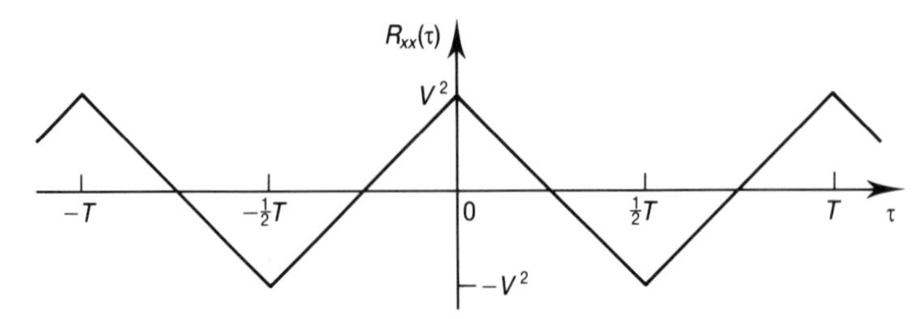

Figure 1.9 Autocorrelation function of the squarewave of Figure 1.8.

frequency of the periodic component. If the frequencies are similar, any filtering done to reduce the noise term will also affect the periodic component, whereas autocorrelation techniques still prove effective.

EXAMPLE 1B AUTOCORRELATION FUNCTION OF A SQUAREWAVE

Consider now a squarewave $x(t)$ with levels $\pm V$ and period T, as shown in Figure 1.8. Recall that the autocorrelation function is the average value of the product of the waveform $x(t)$ and a shifted version $x(t+\tau)$. In Figure 1.8 the shifted versions of the waveform are shown for values of τ equal to $\frac{1}{4}T$, $\frac{1}{2}T$ and $\frac{3}{4}T$, and it is clear that

$$\begin{aligned} R_{xx}(\tau) &= V^2, & \tau &= 0 \\ &= 0, & \tau &= \tfrac{1}{4}T \\ &= -V^2, & \tau &= \tfrac{1}{2}T \\ &= 0, & \tau &= \tfrac{3}{4}T \\ &= V^2, & \tau &= T \end{aligned}$$

It is also clear that, for any intermediate value of τ, the autocorrelation function would lie on a straight line joining these values (because $x(t)$ is constant at either $+V$ or $-V$ over each half-period). Hence the complete autocorrelation function is as shown in Figure 1.9.

EXAMPLE 1C AUTOCORRELATION FUNCTION OF A DISCRETE-INTERVAL RANDOM BINARY SIGNAL

Consider now a discrete-interval random binary signal, which assumes values $+V$ or $-V$ for time intervals of duration Δt changing with probability 0.5 at regularly spaced 'event points' 0, Δt, $2\Delta t$, ... The value of $x(t)$ in any particular interval between event points could, for instance, be obtained by tossing a fair coin, with Heads $\rightarrow +V$ and Tails $\rightarrow -V$ or vice versa. Clearly, $\mu_x = 0$, while from Property A, $R_{xx}(0) = V^2$. A typical portion of such a signal is shown in Figure 1.10.

Now consider the product $x(t).x(t+\tau)$ for $\tau > \Delta t$; for such values an event point has definitely occurred in the time interval from t to $t+\tau$, so that, from the definition of the signal, $x(t)$ and $x(t+\tau)$ are independent. Thus for $\tau > \Delta t$,

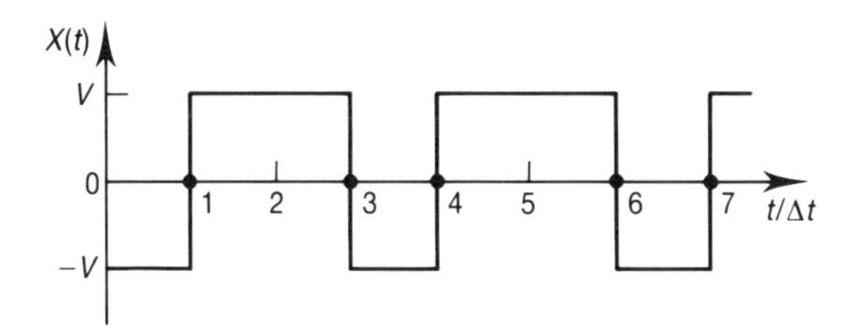

Figure 1.10 Portion of a discrete-interval random binary signal.

$R_{xx}(\tau) = E[x(t).x(t+\tau)] = E[x(t)].E[x(t+\tau)] = \mu_x.\mu_x = 0$. (An alternative way of looking at this is that, for $\tau > \Delta t$, the product $x(t).x(t+\tau)$ is $(+V).(-V)$ with probability 0.25, $(+V).(+V)$ with probability 0.25, $(-V).(+V)$ with probability 0.25 and $(-V)(-V)$ with probability 0.25, thus contributing to a product $+V^2$ with probability 0.5 and $-V^2$ with probability 0.5.)

For $\tau < \Delta t$, the probability of an event point in the time interval from t to $t+\tau$ is $\tau/\Delta t$, in which case $E[x(t).x(t+\tau)] = 0$. The probability of no event point in the same time interval is $1 - (\tau/\Delta t)$, in which case $E[x(t).x(t+\tau)] = V^2$. Thus for $\tau < \Delta t$,

$$R_{xx}(\tau) = E[x(t).x(t+\tau)] = \left(1 - \frac{\tau}{\Delta t}\right).V^2 + \frac{\tau}{\Delta t}.0$$

$$= \left(1 - \frac{\tau}{\Delta t}\right).V^2$$

(The autocorrelation function thus reduces linearly with τ from V^2 at $\tau = 0$ to zero at $\tau = \Delta t$; the straight line is inevitable because $x(t)$ is constant at either $+V$ or $-V$ between event points, as was the squarewave (Example 1B) over successive half-periods.)

From Property B, $R_{xx}(-\tau) = R_{xx}(\tau)$, so that the complete autocorrelation function is as shown in Figure 1.11.

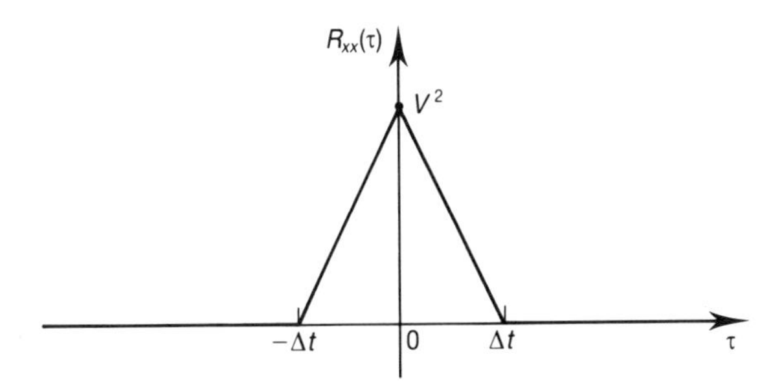

Figure 1.11 Autocorrelation function of a discrete-interval random binary signal.

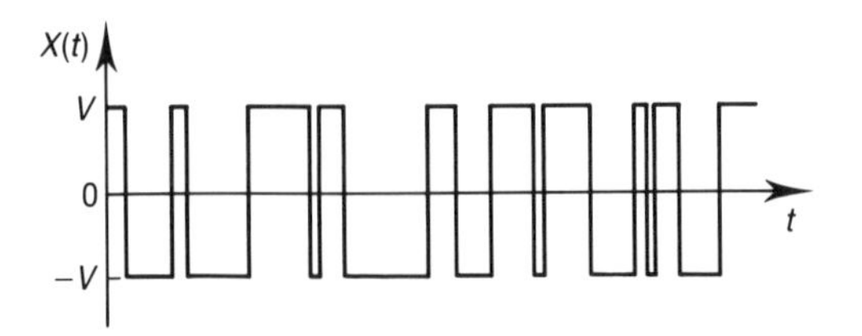

Figure 1.12 Portion of a random binary signal.

EXAMPLE 1D AUTOCORRELATION FUNCTION OF A RANDOM BINARY SIGNAL

Consider now a rather different random binary signal – one in which the transitions from one level to the other can occur at any time. Assume again that the two levels of the signal are $+V$ and $-V$ and let the average number of zero crossings per unit time be v. The zero crossings thus follow a Poisson distribution, with the probability of n zero crossings in a time interval τ being given by:

$$P(n) = \frac{(v\tau)^n}{n!}\exp(-v\tau)$$

A typical portion of such a signal is shown in Figure 1.12.

As before, $R_{xx}(0) = V^2$ from Property A. For a given τ, the product $x(t).x(t+\tau)$ will be $+V^2$ if there have been 0, 2, 4, ..., zero crossings and $-V^2$ if there have been 1, 3, 5, ..., zero crossings. Thus:

$$\begin{aligned} E[x(t).x(t+\tau)] &= V^2\left[\sum_{n\text{ even}}\frac{(v\tau)^n}{n!}\exp(-v\tau) - \sum_{n\text{ odd}}\frac{(v\tau)^n}{n!}\exp(-v\tau)\right] \\ &= V^2\left[1 - v\tau + \frac{(v\tau)^2}{2!} - \frac{(v\tau)^3}{3!} + \dots\right].\exp(-v\tau) \\ &= V^2\exp(-2v\tau) \end{aligned}$$

This is illustrated in Figure 1.13.

Note that, for large values of τ, $x(t)$ and $x(t+\tau)$ become virtually unrelated, so that:

$$\begin{aligned} E[x(t).x(t+\tau)] &\rightarrow E[x(t)].E[x(t+\tau)] \\ &= \mu_x.\mu_x = 0 \end{aligned}$$

This leads to a further property which applies to *any* non-periodic signal $x(t)$.

PROPERTY E

If $x(t)$ is non-periodic, then $R_{xx}(\tau) \rightarrow (\mu_x)^2$ as $\tau \rightarrow \infty$. (This applied, as expected, in Example 1C, for which μ_x was again zero.)

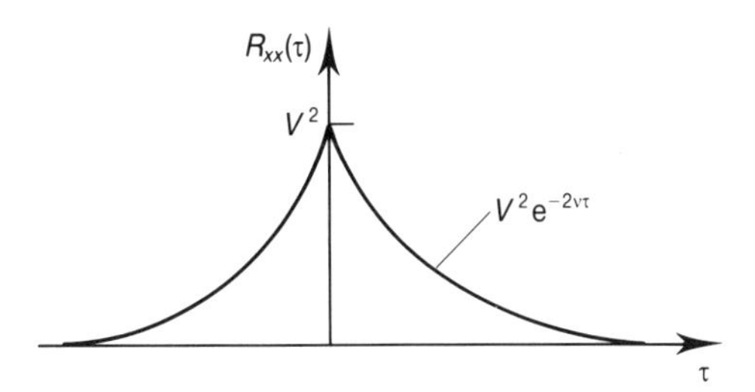

Figure 1.13 Autocorrelation function of a random binary signal.

1.2.4 Computing correlation functions in practice

Recall the (theoretical) expressions (1.15) and (1.18) for auto- and cross-correlation functions:

$$R_{xx}(\tau) = \lim_{T\to\infty} \frac{1}{2T} \int_{-T}^{T} x(t)\,.\,x(t+\tau)\,\mathrm{d}t \tag{1.25}$$

$$R_{xy}(\tau) = \lim_{T\to\infty} \frac{1}{2T} \int_{-T}^{T} x(t)\,.\,y(t+\tau)\,\mathrm{d}t \tag{1.26}$$

The processes involved in this computation are (1) time shift by an amount τ; (2) multiplication; (3) averaging the result over a very long (theoretically infinite) time. In practice, computation of correlation functions via this direct form will require some modification.

First, the first signal is delayed rather than advancing the second signal, so that

$$R_{xx}(\tau) = \lim_{T\to\infty} \frac{1}{2T} \int_{-T}^{T} x(t-\tau)\,.\,x(t)\,\mathrm{d}t \tag{1.27}$$

$$R_{xy}(\tau) = \lim_{T\to\infty} \frac{1}{2T} \int_{-T}^{T} x(t-\tau)\,.\,y(t)\,\mathrm{d}t \tag{1.28}$$

Second, it is necessary to compute the correlation function over a finite time (T_{f}, say) so that approximate correlation functions are obtained:

$$\hat{R}_{xx}(\tau) = \frac{1}{T_{\mathrm{f}}} \int_{0}^{T_{\mathrm{f}}} x(t-\tau)\,.\,x(t)\,\mathrm{d}t \tag{1.29}$$

$$\hat{R}_{xy}(\tau) = \frac{1}{T_{\mathrm{f}}} \int_{0}^{T_{\mathrm{f}}} x(t-\tau)\,.\,y(t)\,\mathrm{d}t \tag{1.30}$$

In some cases only a limited length of data is available, so that there is loss at the end as the time shift τ is increased from zero. Equations (1.29) and (1.30) are then further modified to

$$\hat{R}_{xx}(\tau) = \frac{1}{T_f - \tau} \int_0^{T_f - \tau} x(t - \tau) . x(t) \, dt \tag{1.31}$$

$$\hat{R}_{xy}(\tau) = \frac{1}{T_f - \tau} \int_0^{T_f - \tau} x(t - \tau) . y(t) \, dt \tag{1.32}$$

Clearly, the maximum usable value of τ in this restricted data case is very much less than the length (T_f) of the record.

If the correlation functions are computed digitally (as usual) with a total of N equally spaced samples of $x(t)$ and $y(t)$ within the time T_f, then the expressions corresponding to equations (1.29) and (1.30) for unrestricted data length are

$$\hat{R}_{xx}(k) = \frac{1}{N} \sum_{r=0}^{N-1} x_{r-k} x_r \tag{1.33}$$

$$\hat{R}_{xy}(k) = \frac{1}{N} \sum_{r=0}^{N-1} x_{r-k} \cdot y_r \tag{1.34}$$

while the expressions corresponding to equations (1.31) and (1.32) for restricted data length are

$$\hat{R}_{xx}(k) = \frac{1}{N - k} \sum_{r=0}^{N-k-1} x_{r-k} \cdot x_r \tag{1.35}$$

$$\hat{R}_{xy}(k) = \frac{1}{N - k} \sum_{r=0}^{N-k-1} x_{r-k} \cdot y_r \tag{1.36}$$

If $x(t)$ is a periodic function with period T, then expressions (1.29) and (1.33) are exact autocorrelation functions provided $T_f = kT$, where k is a integer ($\geqslant 1$). Similarly, if both $x(t)$ and $y(t)$ are periodic functions with period T, then expressions (1.30) and (1.34) are exact cross-correlation functions provided $T_f = kT$.

Bendat and Piersol (1966) derived an expression for the variance of a cross-correlation function $R_{xy}(\tau)$ due to taking measurements over a finite time T_f rather than an infinite time, on the assumption that both $x(t)$ and $y(t)$ are zero-mean signals with Gaussian probability density functions:

$$\text{Var}[R_{xy}(\tau)] = \frac{1}{T_f} \int_{-\infty}^{\infty} [R_{xx}(u) R_{yy}(u) + R_{xy}(u + \tau) R_{xy}(\tau - u)] \, du \tag{1.37}$$

The corresponding expression for variance of an autocorrelation function is obtained by putting $y = x$:

$$\text{Var}[R_{xx}(\tau)] = \frac{1}{T_f} \int_{-\infty}^{\infty} [(R_{xx}(u))^2 + R_{xx}(u + \tau) . R_{xx}(u - \tau)] \, du \tag{1.38}$$

so that for $\tau = 0$

$$\text{Var}[R_{xx}(0)] = \frac{4}{T_f} \int_0^{\infty} [R_{xx}(u)]^2 \, du \tag{1.39}$$

Many modern digital signal analyzers compute correlation functions via Fourier transformation of the time-domain data into the frequency domain and then inverse Fourier transformation back into the time domain. This makes use of the computational efficiency of the Fast Fourier Transform, which can be used for both the transformations, and which means that this seemingly complicated approach to computing correlation functions can be more efficient than use of the direct formulae given above. A result of this, though, is that the resulting correlation functions are periodic, regardless of whether the original signals are, and some care is needed interpreting them. For a detailed discussion of the effect of this, see Chapter 11 of Newland (1984).

1.3 Correlation functions and process dynamics

1.3.1 Convolution integral relationships

It has been shown in Section 1.1 that if the simple input–output convolution integral relationship of equation (1.4) is used to estimate the dynamics of a process, the noise term can often prove substantial in practice, so leading to poor parameter-estimation accuracy. A feasible alternative is to use corresponding correlation function relationships. Multiplying equation (1.4) throughout by $u(t-\tau)$,

$$u(t-\tau)\,.\,z(t) = \int_0^{T_s} h(\lambda)u(t-\tau)u(t-\lambda)\,\mathrm{d}\lambda + u(t-\tau)n(t) \tag{1.40}$$

Averaging the result over a long (theoretically infinite) time gives

$$R_{uz}(\tau) = \int_0^{T_s} h(\lambda)\,.\,R_{uu}(\tau-\lambda)\,\mathrm{d}\lambda + R_{un}(\tau) \tag{1.41}$$

which is seen to be another convolution integral relationship. If the input signal $u(t)$ and the noise signal $n(t)$ are uncorrelated, then $R_{un}(\tau) = 0$.

In practice, it is necessary to use finite time correlation functions (equations (1.29) and (1.30)), so that equation (1.41) becomes

$$\hat{R}_{uz}(\tau) = \int_0^{T_s} h(\lambda)\,.\,\hat{R}_{uu}(\tau-\lambda)\,\mathrm{d}\lambda + \hat{R}_{un}(\tau) \tag{1.42}$$

and if $u(t)$ and $n(t)$ are uncorrelated, the effect of $\hat{R}_{un}(\tau)$ is to appear as a 'noise' on $\hat{R}_{uz}(\tau)$, which diminishes in amplitude as the correlation time is increased (tending to zero as the correlation time tends to infinity). It is important to stress this point: using correlation function relationships, the effects of noise diminish as the averaging time is increased, which was not so for the simple input–output relationship of equation (1.4).

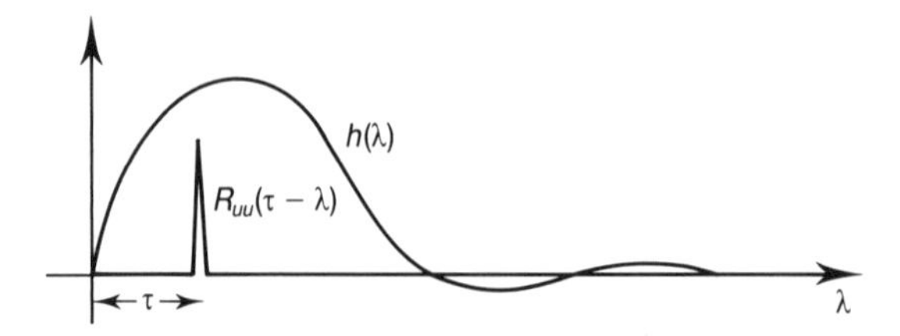

Figure 1.14 Convolution of a weighting function with the autocorrelation function of a discrete-interval random binary signal.

If signals $u(t)$ and $z(t)$ are sampled at regular intervals Δ to give corresponding sequences with elements u_r and z_r, then the convolution integral relationship (1.42) is approximated by a convolution sum:

$$\hat{R}_{uz}(k) = \sum_{i=0}^{L} h_i \hat{R}_{uu}(k - i) + \hat{R}_{un}(k) \tag{1.43}$$

where $L\Delta$ is equal to the system-settling time T_s and h_i is the weighting sequence of the system, as before.

To avoid a potentially awkward deconvolution, the objective now is to make the autocorrelation function $R_{uu}(\tau - \lambda)$ of approximately impulsive form. This will be illustrated for a discrete-interval random binary signal.

1.3.2 Illustration of the convolution integral relationship for a discrete-interval random binary signal input

The convolution integral

$$\int_0^{T_s} h(\lambda) R_{uu}(\tau - \lambda)\, \mathrm{d}\lambda$$

is illustrated in Figure 1.14 for a discrete-interval random binary signal. If the clock pulse interval Δt of the signal is sufficiently small, then

$$R_{uu}(\tau - \lambda) \simeq V^2 \Delta t\ \delta(\tau - \lambda) \tag{1.44}$$

where $\delta(\tau - \lambda)$ is a delta function centred at $\tau = \lambda$. Equation (1.41) then simplifies to

$$R_{uz}(\tau) \simeq V^2 \Delta t\, h(\tau) \tag{1.45}$$

so that the weighting function is directly proportional to the input–output cross-correlation function.

If the signals are sampled, and the sampling interval Δ is equal to the clock pulse interval Δt of the input signal, then

$$R_{uu}(k - i) = V^2, \qquad k = i \tag{1.46a}$$

$$R_{uu}(k-i) = 0, \qquad k \neq i \tag{1.46b}$$

This leads to a very simple approximation to the deconvolution, with

$$R_{uz}(k) = V^2 . h_k \tag{1.47}$$

Use of a discrete-interval random binary signal brings about two problems. First, such a signal is not particularly easy to generate, although one is incorporated into the Hewlett Packard HOI-3722A noise generator. Second, it brings the further problem of just how closely the finite time $\hat{R}_{uu}(\tau)$ approaches the infinite time $R_{uu}(\tau)$ of Figures 1.11 and 1.14, or, for sampled signals, the very simple form of equations (1.46a) and (1.46b). This is a different problem from how nearly the finite time $\hat{R}_{un}(\tau) \rightarrow 0$, and it can be avoided by using periodic signals and averaging over an integer number of signal periods.

Before proceeding to that, though, let us examine whether we can, in practice, obtain anything meaningful from normal operating records.

1.3.3 The use of normal operating records – or do we need a perturbation signal?

Data logging of normal operating records can give useful information about the behaviour of a process especially when step disturbances occur. When trying to estimate dynamics, it is very tempting to base models on such records. It is necessary to check first that the normal operating inputs are *persistently exciting* – effectively, that their frequency components adequately span the passband of the system (see Section 1.10 for a further discussion of this). Normal operating records may not be persistently exciting, especially if they are zero (or constant) for lengthy periods of time. Even when this check proves positive, there may be other pitfalls, as the next example shows.

APPLICATION EXAMPLE 1.3 ESTIMATION OF THE DYNAMICS OF A STEELWORKS BLAST FURNACE

Normal operating records from a blast furnace at the Port Talbot Steelworks of British Steel were available over several weeks of furnace operation. From these, a section consisting of 49 casts was selected for analysis on the grounds that it was free from breaks in the records and that casting was done at regular intervals of 3 hours within that period (of just over 6 days).

Two inputs – ore/coke ratio and blast temperature – were examined. Both appeared to be suitable for process-identification purposes because they both varied at frequent intervals, both had autocorrelation functions approximating to delta functions and the cross-correlation between them was approximately zero. Thus, the cross-correlation functions between each input in turn and output variable of interest (percentage silicon in the cast iron) should give values proportional to the two impulse responses.

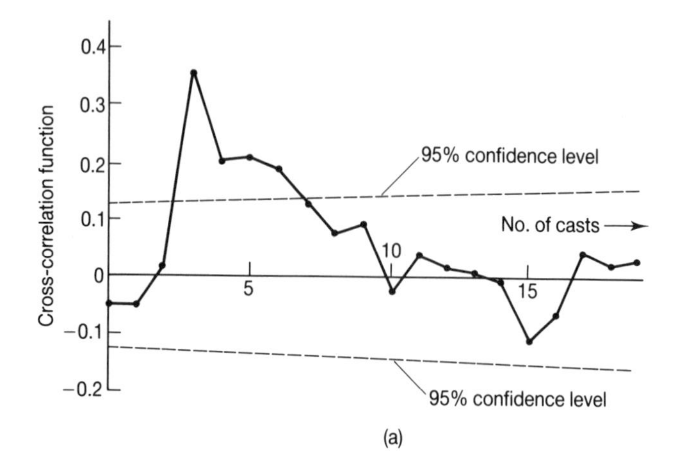

(a)

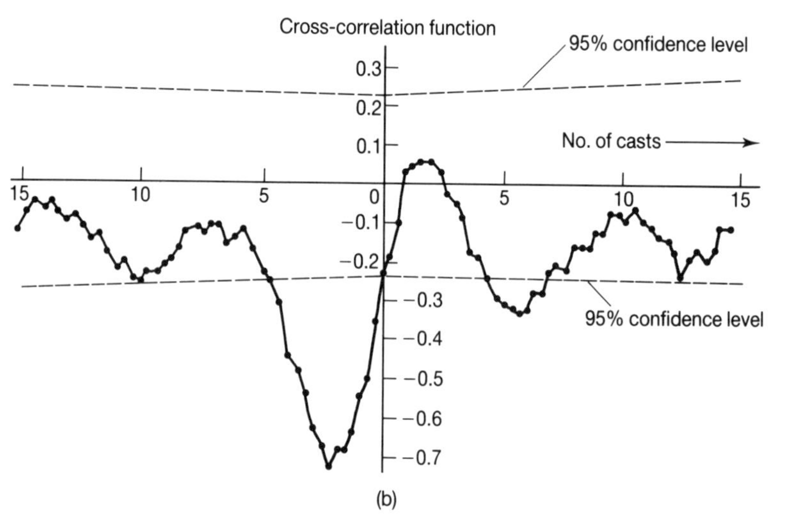

(b)

Figure 1.15 Cross-correlation functions of normal operating records from a steelworks blast furnace. (a) Between ore/coke ratio and percentage silicon in the cast iron; (b) between blast temperature and percentage silicon in the cast iron.

Cross-correlation functions were computed using the restricted data length formula of equation (1.36); these should be more accurately described as cross-covariance functions, because means of all records were removed. The cross-correlation functions are shown in Figure 1.15. Confidence intervals are shown in both Figures 1.15(a) and 1.15(b) and are based on the assumption of discrete Gaussian processes, for which

$$\operatorname{Var}[\hat{R}_{uz}(k)] = \frac{1}{N-k}\,\sigma_u^2\sigma_z^2$$

(Jones, 1986), where σ_u^2 and σ_z^2 are the variances of the input and output, respectively. Clearly, the assumption of Gaussian probability density functions is difficult to check with such short data records, but the confidence limits shown do nevertheless provide a useful guideline, with values considerably above the upper or below the lower limits then being assumed to denote a statistically significant input–output relationship.

The result for the ore/coke ratio input is very satisfactory, and a reasonable idea of the dynamic relationship between this ratio and percentage silicon could be obtained from Figure 1.15(a). The blast temperature result (Figure 1.15(b)) is not satisfactory, with no statistically significant relationship between blast temperature as input and percentage silicon output (i.e. for positive τ). The problem here is that blast temperature was used as a control variable, with manual adjustment of the temperature according to the value of silicon content of the cast. The manual control reduced the blast temperature when the silicon content was high and increased it when it was low, and this shows as a substantial negative cross-correlation for *negative* values of τ, which completely swamps the forward path relationship. Ore/coke ratio is not normally used as a control variable, and no values of cross-correlation for negative τ were outside the confidence intervals for this relationship.

Although it is possible to identify the open-loop dynamics from closed-loop experiments under certain conditions, the procedure would have been difficult to apply in this example because of the variability of the manual feedback.

Even for processes with feedback applied automatically, it is most inadvisable to try to estimate the dynamics of a process from closed-loop experiments using normal operating records (Wellstead, 1977). For instance, if a non-parametric representation of the dynamics is required in the form of a frequency-response function (see Chapter 2), it is quite possible to estimate the inverse of the frequency response of the feedback path rather than that of the forward path. Problems of this nature can be overcome by use of a suitable externally applied reference signal.

1.4 Pseudo-random binary signals and related signals

One of the most useful types of periodic signal for process identification is a pseudo-random binary signal (PRBS) which has the following properties:

1. The signal has two levels, and it can switch from one level to the other only at certain event points $t = 0, \Delta t, 2\Delta t, \ldots$
2. Whether the signal changes level at any particular event point is *predetermined*, so that the PRBS is deterministic and experiments are repeatable. (This is in contrast to the discrete-interval random binary signal.)
3. The PRBS is periodic with period $T = N\Delta t$, where N is an odd integer.
4. In any one period, there are $\frac{1}{2}(N + 1)$ intervals when the signal is at one level and $\frac{1}{2}(N - 1)$ intervals when it is at the other.
5. If the two levels are $\pm V$, the autocorrelation function of a PRBS is as shown in Figure 1.16.

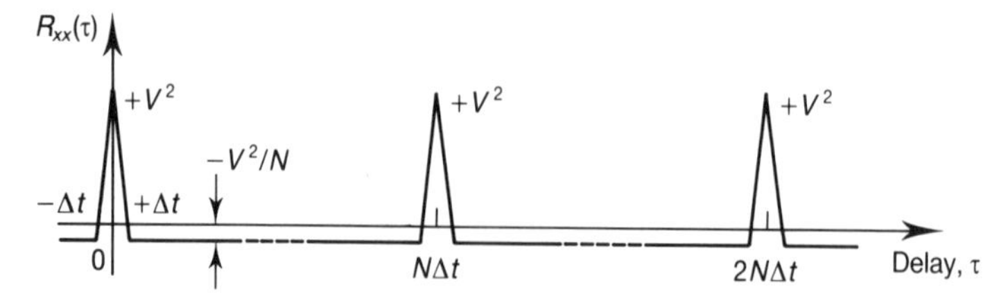

Figure 1.16 Autocorrelation function of a PRBS.

A PRBS is based on a pseudo-random binary sequence of length N. The most commonly used signals are based on maximum-length sequences (m-sequences) and the reason for their popularity is that they can readily be generated using feedback shift register circuits.

1.4.1 Pseudo-random binary signals based on maximum-length sequences

Binary m-sequences exist for $N = 2^n - 1$, where n is an integer (>1). They can be generated using an n-stage feedback shift register with feedback to the first stage consisting of the modulo-2 sum of the logic value of the last stage and one or more of the other stages. The binary logic values are taken (as usual) as 1 and 0 and modulo-2 addition is given by

$$1 \oplus 1 = 0 \oplus 0 = 0$$

$$1 \oplus 0 = 0 \oplus 1 = 1$$

Transformation from sequence logic values to signal voltage levels is made either by $1 \rightarrow +V$, $0 \rightarrow -V$ or by $1 \rightarrow -V$, $0 \rightarrow +V$.

A shift register circuit for generating a pseudo-random binary signal based on an m-sequence of length $2^4 - 1 = 15$ is shown in Figure 1.17. The register can be started with any binary number other than 0, 0, 0, 0 (which would obviously give a cycle of length 1). In Table 1.1 the register is started (arbitrarily) with 0, 0, 0, 1 and it is seen from Table 1.1(a) that this number reappears after $2^4 - 1$ clock pulses. Each binary number from 0, 0, 0, 1 to 1, 1, 1, 1 appears as the register contents exactly once during the cycle.This is a general result for all binary m-sequences: each binary number from 1 to $2^n - 1$ appears in the sequence exactly once. If feedback connections from stages $k, l, m, \ldots, n$ result in an m-sequence, then so do feedback connections from stages $n-k, n-l, n-m, \ldots, n$; the second sequence is the first sequence reversed. This is confirmed in Table 1.1(b), which is for feedback from stages 3 and 4.

Table 1.1 Binary maximum-length sequences from a four-stage shift register with feedback connections corresponding to a primitive polynomial (modulo 2)

Number of clock pulses	Shift register stage			
	1	2	3	4
(a) Feedback from stages 1 and 4				
1	0	0	0	1
2	1	0	0	0
3	1	1	0	0
4	1	1	1	0
5	1	1	1	1
6	0	1	1	1
7	1	0	1	1
8	0	1	0	1
9	1	0	1	0
10	1	1	0	1
11	0	1	1	0
12	0	0	1	1
13	1	0	0	1
14	0	1	0	0
15	0	0	1	0
16	0	0	0	1
(b) Feedback from stages 3 and 4				
1	0	0	0	1
2	1	0	0	0
3	0	1	0	0
4	0	0	1	0
5	1	0	0	1
6	1	1	0	0
7	0	1	1	0
8	1	0	1	1
9	0	1	0	1
10	1	0	1	0
11	1	1	0	1
12	1	1	1	0
13	1	1	1	1
14	0	1	1	1
15	0	0	1	1
16	0	0	0	1

Very few of the possible feedback connections for an n-stage shift register result in sequence of the maximum length of $2^n - 1$. A shift register with connections resulting in a maximum-length sequence has a *characteristic equation* in the delays, D, in the register which corresponds to a primitive polynomial, modulo 2. This is a polynomial

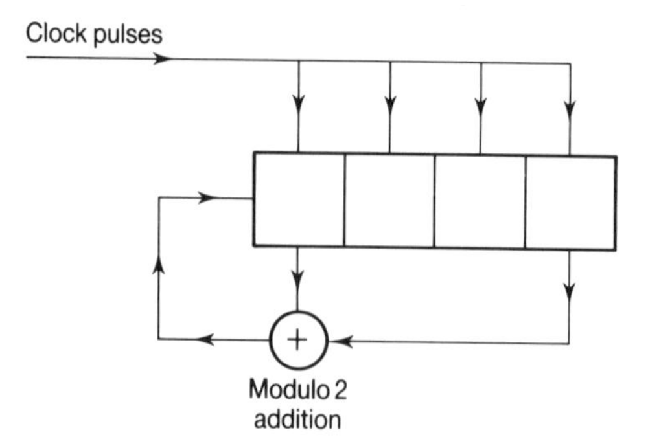

Figure 1.17 Shift register circuit for generating a PRBS based on a 15-digit maximum-length sequence.

which is irreducible (having no factors, modulo 2) and which is not a factor $x^r \oplus 1$ for any $r < 2^n - 1$. A listing of all irreducible polynomials, modulo 2, up to $n = 16$ is provided by Peterson (1961), who also gives several examples for $17 \leqslant n \leqslant 34$; there is a clear indication in the listings of those polynomials which are primitive. Watson (1962) gives one example of a primitive polynomial, modulo 2, for $2 \leqslant n \leqslant 100$ and for $n = 107$ and 127.

If the polynomial $c_n x_n + c_{n-1}x^{n-1} + \ldots + c_1 x + c_0$ is primitive, modulo 2, then the corresponding characteristic equation is in the delays, D, introduced by the shift register (Briggs *et al.*, 1965):

$$c_n D^n + c_{n-1} D^{n-1} + \ldots + c_1 D + c_0 = 0, \textit{ modulo } 2 \tag{1.48}$$

The logic connections back to the first stage of the register are given by

$$c_0 X = -c_1 DX - \ldots - c_{n-1} D^{n-1} X - c_n D^n X, \text{ modulo } 2 \tag{1.49}$$

where X is the input sequence to the register, DX is the sequence at the output of the first stage of the register and so on, so that $D^n X$ is the sequence at the output of the last stage of the n-stage register. Since modulo 2 subtraction is the same as modulo 2 addition, the feedback configuration is given by

$$c_0 X = c_1 DX + \ldots + c_{n-1} D^{n-1} X + c_n D^n X, \text{ modulo } 2 \tag{1.50}$$

For binary m-sequences, $c_0 = c_n = 1$, and the remaining coefficients $c_1, \ldots, c_{n-1}$ are either 1 or 0. The characteristic equation for the shift register shown in Figure 1.17 is thus:

$$D^4 + D + 1 = 0, \text{ modulo } 2$$

and the corresponding primitive polynomial is $x^4 + x + 1$, modulo 2.

An example of one possible feedback configuration for each n in the range $2 \leqslant n \leqslant 100$ and for $n = 127$ is given in Table 1.2; the sources are, for $n \leqslant 34$, Peterson (1961) and for $n > 34$, Watson (1962).

Table 1.2 Feedback configurations for the generation of binary m-sequences; one example for each n from 2 ($N = 3$) to 100 ($N = 1.268 \times 10^{30}$) and for $n = 127$ ($N = 1.701 \times 10^{38}$)

n	Feedback from stages	n	Feedback from stages
2	1, 2	52	3, 52
3	2, 3	53	1, 2, 6, 53
4	3, 4	54	2, 3, 4, 5, 6, 54
5	3, 5	55	1, 2, 6, 55
6	5, 6	56	2, 4, 7, 56
7	4, 7	57	2, 3, 5, 57
8	4, 5, 6, 8	58	1, 5, 6, 58
9	5, 9	59	1, 3, 4, 5, 6, 59
10	7, 10	60	1, 60
11	9, 11	61	1, 2, 5, 61
12	6, 8, 11, 12	62	3, 5, 6, 62
13	9, 10, 12, 13	63	1, 63
14	4, 8, 13, 14	64	1, 3, 4, 64
15	14, 15	65	1, 3, 4, 65
16	4, 13, 15, 16	66	2, 3, 5, 6, 8, 66
17	14, 17	67	1, 2, 5, 67
18	11, 18	68	1, 5, 7, 68
19	14, 17, 18, 19	69	2, 5, 6, 69
20	17, 20	70	1, 3, 5, 70
21	19, 21	71	1, 3, 5, 71
22	21, 22	72	1, 2, 3, 4, 6, 72
23	18, 23	73	2, 3, 4, 73
24	17, 22, 23, 24	74	3, 4, 7, 74
25	22, 25	75	1, 3, 6, 75
26	20, 24, 25, 26	76	2, 4, 5, 76
27	22, 25, 26, 27	77	2, 5, 6, 77
28	25, 28	78	1, 2, 7, 78
29	27, 29	79	2, 3, 4, 79
30	7, 28, 29, 30	80	1, 2, 3, 5, 7, 80
31	28, 31	81	4, 81
32	10, 30, 31, 32	82	1, 4, 6, 7, 8, 82
33	20, 33	83	2, 4, 7, 83
34	7, 32, 33, 34	84	1, 3, 5, 7, 8, 84
35	2, 35	85	1, 2, 8, 85
36	1, 2, 4, 5, 6, 36	86	2, 5, 6, 86
37	1, 2, 3, 4, 5, 37	87	1, 5, 7, 87
38	1, 5, 6, 38	88	1, 3, 4, 5, 8, 88
39	4, 39	89	3, 5, 6, 89
40	3, 4, 5, 40	90	2, 3, 5, 90
41	3, 41	91	2, 3, 5, 6, 7, 91
42	1, 2, 3, 4, 5, 42	92	2, 5, 6, 92
43	3, 4, 6, 43	93	2, 93
44	2, 5, 6, 44	94	1, 5, 6, 94
45	1, 3, 4, 45	95	1, 2, 4, 5, 6, 95
46	1, 2, 3, 5, 8, 46	96	2, 3, 4, 6, 7, 96
47	5, 47	97	6, 97
48	1, 2, 4, 5, 7, 48	98	1, 2, 3, 4, 7, 98
49	4, 5, 6, 49	99	4, 5, 7, 99
50	2, 3, 4, 50	100	2, 7, 8, 100
51	1, 3, 6, 51	127	1, 127

The two m-sequences listed in Table 1.1 turn out to be the only two binary m-sequences of length 15. The sequences resulting from the only other possible two-stage feedback (stages 2 and 4) are listed in Table 1.3(a), from which it may be seen that there are three possible sequences, depending on the starting state of the register (assumed not to be four zeros), two of length 6 and one of length 3. The polynomial corresponding to this feedback, $x^4 + x^2 + 1$, is not irreducible, modulo 2, but is equal to $(x^2 + x + 1)(x^2 + x + 1)$. There is one further polynomial for $n = 4$ which *is* irreducible modulo 2 and this is $x^4 + x^3 + x^2 + x + 1$. This corresponds to feedback from all four stages, and the (non-zero) sequences resulting from this configuration are listed in Table 1.3(b). It is seen that there are three possible sequences each of length 5. The problem is that while this polynomial is irreducible, it is not primitive, being equal to $(x^5 + 1)/(x + 1)$, modulo 2.

Table 1.3 How not to generate a binary maximum-length sequence from a four-stage shift register. (Polynomial corresponding to (a) is not irreducible; polynomial corresponding to (b) is irreducible but not primitive)

Number of clock pulses	Shift register stage			
	1	2	3	4
(a) Feedback from stages 2 and 4				
1	0	0	0	1
2	1	0	0	0
3	0	1	0	0
4	1	0	1	0
5	0	1	0	1
6	0	0	1	0
7	0	0	0	1
1	0	0	1	1
2	1	0	0	1
3	1	1	0	0
4	1	1	1	0
5	1	1	1	1
6	0	1	1	1
7	0	0	1	1
1	0	1	1	0
2	1	0	1	1
3	1	1	0	1
4	0	1	1	0

Table 1.3 *Continued*

Number of clock pulses	Shift register stage			
	1	2	3	4
(b) Feedback from stages 1, 2, 3 and 4				
1	0	0	0	1
2	1	0	0	0
3	1	1	0	0
4	0	1	1	0
5	0	0	1	1
6	0	0	0	1
1	0	0	1	0
2	1	0	0	1
3	0	1	0	0
4	1	0	1	0
5	0	1	0	1
6	0	0	1	0
1	0	1	1	1
2	1	0	1	1
3	1	1	0	1
4	1	1	1	0
5	1	1	1	1
6	0	1	1	1

None of the six sequences listed in Table 1.3 is useful for system identification, thus illustrating the importance of correct choice of the feedback configuration.

1.4.2 Pseudo-random binary signals based on other classes of sequences

It is sometimes thought that all pseudo-random binary signals have to be based on m-sequences, but there are other classes of binary sequences on which they can be based (Everett, 1966), although these cannot (in general) be as easily generated using feedback shift registers. The most interesting alternative class of sequences is quadratic residue codes, for which there are many more possible sequence lengths N than for m-sequences.

Quadratic residue codes exist for $N = 4k - 1$, where k is an integer and N is prime, i.e. for $N = 3, 7, 11, 19, 23, 31, 43, 47, 59, 67, 71, 79, 83, 103, 107, 127, \ldots$ The sequence $\{x_r\}$, $r = 1, 2, \ldots, N$ is formed from the rule that digit $x_r = 1$ if r is a square, modulo N and $x_r = 0$ otherwise; x_n can be either 1 or 0. (Note that r^2 modulo N

simply means $r^2 - aN$, where a is the largest integer such that aN is less than or equal to r^2.)

The rule is applied for $N = 23$ in Table 1.4, where the numbers in the fourth column are squares, modulo 23. It follows that the quadratic residue code of length 23 has 1's in positions 1, 2, 3, 4, 6, 8, 9, 12, 13, 16 and 18 and 0's in positions 5, 7, 10, 11, 14, 15, 17, 19, 20, 21 and 22. The Nth digit can be either 0 or 1. It may be seen from Table 1.4 that the same numbers are obtained in the fourth column from $r = \frac{1}{2}(N + 1)$ onwards as those from $r = 1$ up to $r = \frac{1}{2}(N - 1)$, but in the reverse order. This is true for all quadratic residue codes, so that it is only necessary to compute the squares, modulo N, up to $r = \frac{1}{2}(N - 1)$. The corresponding PRBS is obtained as before by transformation to voltage levels either by $1 \to +V$, $0 \to -V$ or by $1 \to -V$, $0 \to +V$, and the signal possesses exactly the same properties as one based on an m-sequence.

Table 1.4 Computation of the quadratic residue code of length 23

r	r^2	Multiple of 23 immediately below r^2	r^2mod23	x_r
1	1	0	1	1
2	4	0	4	1
3	9	0	9	1
4	16	0	16	1
5	25	23	2	0
6	36	23	13	1
7	49	46	3	0
8	64	46	18	1
9	81	69	12	1
10	100	92	8	0
11	121	115	6	0
12	144	138	6	1
13	169	161	8	1
14	196	184	12	0
15	225	207	18	0
16	256	253	3	1
17	289	276	13	0
18	324	322	2	1
19	361	345	16	0
20	400	391	9	0
21	441	437	4	0
22	484	483	1	0
23	529	(529)	0	0 or 1

Other classes of binary sequences on which pseudo-random binary signals can be based are Twin Prime Sequences, which exist for $N = k(k + 2)$, where both k and $k + 2$ are prime (i.e. $N = 15, 35, \ldots$) and Hall Sequences, which exist for $N = 4k^2 + 27$, where N is prime (i.e. $N = 31, 43, 127, \ldots$) (Everett, 1966).

1.4.3 Process identification using a PRBS

The convolution of process weighting function $h(\lambda)$ with the autocorrelation function of a PRBS is illustrated in Figure 1.18. It is assumed that the period $N\Delta t$ of the signal is greater than the settling time T_s of the process. It may be seen that the autocorrelation function consists of a triangle of width $2\Delta t$ and height $V^2(1 + 1/N)$, so giving an area $V^2(1 + 1/N)\Delta t$, together with a d.c. level of $-V^2/N$. If the clock pulse interval Δt is sufficiently small (i.e. if $h(\lambda)$ does not vary appreciably over the width of the autocorrelation function triangle), then

$$\hat{R}_{uz}(\tau) = V^2\left(1 + \frac{1}{N}\right)\Delta t\ h(\tau) - \frac{V^2}{N}\int_0^{T_s} h(\lambda)\ \mathrm{d}\lambda + \hat{R}_{un}(\tau) \tag{1.51}$$

The second term on the right-hand side is a d.c. level, the integral

$\int_0^{T_s} h(\lambda)\ \mathrm{d}\lambda$ being the steady-state gain of the process.

Use of a PRBS has overcome the two problems discussed in Section 1.3.2 for testing with a discrete-interval random binary signal. First, the PRBS is easy to generate and second, if the correlation averaging time is an exact number of PRBS periods, then the finite time autocorrelation function of the PRBS is exactly the same as that for infinite time averaging, $R_{uu}(\tau)$. The PRBS has introduced a problem which did not exist previously – that of a d.c. bias on $\hat{R}_{uz}(\tau)$, but in nearly all practical applications this can readily be taken into account. We are, of course, still left with the problem of how closely $\hat{R}_{un}(\tau)$ approaches zero. Its effect is that the weighting function estimates become scattered on either side of the true weighting function. The scatter can be reduced by increasing the correlation time and, in practice, this often has to be a matter of trial and error, averaging over successively longer times until (hopefully!) an acceptable estimate of the process dynamics is obtained.

Cross-correlation reduces the effect of noise, and the influence of $\hat{R}_{un}(\tau)$ on the input–output cross-correlation results is less serious than that of $n(t)$ if the 'direct' relationship of equation (1.4) is used. To illustrate this point, let us return to the

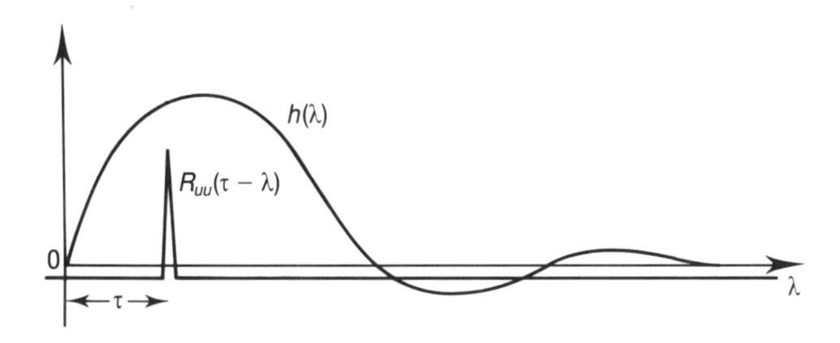

Figure 1.18 Convolution of a weighting function with the autocorrelation function of a PRBS.

experiments on an oil refinery distillation column and recall from Application Example 1.2 that no meaningful estimate of the process dynamics could have been obtained from the step responses of Figure 1.6.

APPLICATION EXAMPLE 1.4 CROSS-CORRELATION FUNCTIONS FROM AN OIL REFINERY DISTILLATION COLUMN WITH A PRBS PERTURBATION SIGNAL

A 127-digit PRBS with amplitude $\pm 2.8\,^{\circ}\mathrm{C}$ about the steady level was used to perturb feed heater temperature, and recordings of column temperature response over several signal periods were obtained. While the experiment was in progress, a number of control changes were made, but it was possible to obtain one signal period of data relatively unaffected by control changes. The cross-correlation function between feed heater temperature and the column temperature of Application Example 1.2 for this period of data is shown in Figure 1.19. A much better idea is obtained of the dynamics than from the corresponding step response (to steps of size $\pm 5.6\,^{\circ}\mathrm{C}$) of Figure 1.6.

The cross-correlation function points in Figure 1.19 are the filled-in circles, but the input–output data were also used to compute z-transfer functions of the form

$$H(z) = \left(\frac{z^{-k}(b_1 z^{-1} + b_2 z^{-2} + \ldots + b_n z^{-n})}{1 + a_1 z^{-1} + a_2 z^{-2} + \ldots + a_n z^{-n}} \right) \tag{1.52}$$

The pulse response of the model of order $n = 2$ is shown as the squares in Figure 1.19. For any specified n, the best fit of k (where the initial time delay is k sampling intervals) and the parameters $b_1, b_2, \ldots, b_n, a_1, a_2, \ldots, a_n$ was computed using a Generalized Least Squares procedure (Clarke, 1967). Fitting models of this type is discussed further in Section 1.10.

Hazlerigg and Noton (1965) discuss the error in estimating the weighting function in the presence of a wide bandwidth random noise disturbance $n(t)$. They showed that the RMS error in the weighting function reduces by a factor approximately equal to the square root of the total averaging time: this is the inaccuracy resulting from $\hat{R}_{un}(\tau)$ not being zero over a finite averaging time in equation (1.48). They also showed that the error depends on Δt, with error decreasing as Δt is increased. With Δt chosen too small the process response signal $y(t)$ has little variability in it due to the input $u(t)$; effectively, the input signal switches between levels before $y(t)$ has had much of

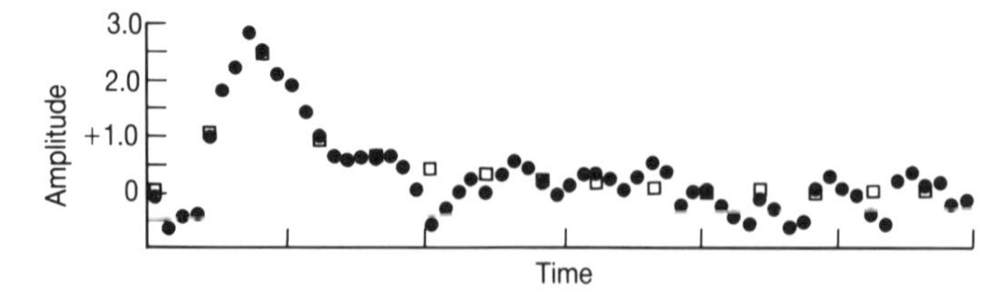

Figure 1.19 Cross-correlation function between feed heater temperature set point and column temperature.

a chance to respond. On the basis of this alone, therefore, Δt should be made large, but we then start to get into difficulties over the simplification of the general convolution integral of equation (1.41) to the particular PRBS input form of equation (1.48), since the latter assumes that the process weighting function does not vary appreciably over the width of the autocorrelation function triangle.

This is quite a tricky compromise in the design of a suitable PRBS for a particular application. The choice can often be helped by having reasonable step response data before using a PRBS. The author, in fact, initially used too small a clock pulse interval in the distillation column modelling described in Application Example 1.4. The result was far too small a variation of the column temperature. This wasted considerable experimentation time and in order to get the improved results which are shown in Figure 1.19, a more reasonable clock pulse interval was chosen from the step response of one of the other column temperatures.

Once a value for Δt has been selected, there is nothing to be gained in time-domain identification from using a signal of period $N\Delta t$ very much greater than the process-settling time T_s. If a long experimentation time is possible, it is better to average results over several shorter signal periods rather than one long period, not least because if anything goes wrong during the experiment there may still be usable periods of data, of at least one signal period duration using the former scheme. When first applying the PRBS it is necessary to wait for one process-settling time T_s for the process to achieve a 'dynamic steady state'. In practice, it is usual to wait for one complete signal period $N\Delta t$ before commencing correlation.

In general, the point in the PRBS at which the correlation is commenced does not matter. However, Barker (1967) showed that if there is a drift component $a_1 t$ in the process response $y(t)$, then there is a particular starting location where the effect of this component can be removed (the small d.c. component of the PRBS also needs to be removed). For all $N = 2^n - 1 > 3$ there is more than one m-sequence on which a PRBS can be based; for example, there are eighteen for $N = 127$. In general, there is nothing to choose between them, but Barker (1967) showed that if there is a quadratic drift component $a_2 t^2$ in the output $y(t)$, then it does matter which signal is used, and it is possible by suitable choice of signal to minimize the error due to this term. Barker lists the optimal m-sequence (in the sense of minimizing this error) for $N = 3, 7, 15, 31, 63, 127$ and 255, and gives the starting point for the removal of error due to the $a_1 t$ component.

The off-peak autocorrelation function of a PRBS can be made zero by altering the levels in the signal from $\pm V$. Hoffmann de Visme (1971) shows that if in the sequence there are $\frac{1}{2}(N + 1)$ digits with level α and $\frac{1}{2}(N - 1)$ digits with level $\rho.\alpha$, then for the off-peak autocorrelation to be zero, the ratio ρ is given by

$$\rho = [-(N + 1) - 2\sqrt{N + 1}]/(N - 3) \tag{1.53}$$

Thus, for $N = 15$, $\rho = -2$, so that there are eight digits at level α and seven digits at level -2α. The on-peak autocorrelation function is thus

$$\frac{1}{15}\left[8\alpha^2 + 7(2\alpha)^2\right] = \frac{36\alpha^2}{15}$$

and if we make this equal to V^2 as before, $\alpha = 0.6455V$. Thus the two levels in the signal are $+0.6455V$ and $-1.291V$. The drawback to this is that the magnitude of the d.c. level in the signal is greater than before: it is now

$$\frac{1}{15}\left[8 \times 0.6455V - 7 \times 1.291V\right] = -0.2582V$$

compared with $V/15 = 0.0667V$ in the original signal with eight intervals $+V$ and seven intervals $-V$.

Summarizing the design of a PRBS:

1. The amplitude will often be imposed on safety or linearity considerations in a practical situation.
2. The clock pulse interval Δt needs to be chosen with great care: too small a value results in too little variability in the response, while too large a value leads to the simple deconvolution form of equation (1.48) not being valid. A guideline often used is to set Δt to approximately 0.4/BW, where BW is the bandwidth of the system under test.
3. The signal period $N\Delta t$ should be greater than the process-settling time T_s, but not very much greater. If a long averaging time is possible, it is better on experimental grounds to average over several short signal periods than over one long signal period.
4. If there is drift of the form $a_1 t$ and/or $a_2 t^2$ in the output $y(t)$, the PRBS should be based on the particular m-sequence listed by Barker (1967) to minimize the effect of the $a_2 t^2$ term and the correlation should be started at the point indicated (with the small d.c. component in the PRBS removed) to remove the effect of the $a_1 t$ term.

There was much research on the theory and applications of pseudo-random binary signals in the 1960s and early 1970s, and a review paper published in 1969 (Godfrey, 1969a) considered a number of applications to full-scale industrial processes in the chemical, steel, petroleum, paper, electricity supply and nuclear power industries. Subsequent use of these signals has become widespread and routine, largely due to the availability of commercial signal generators. Examples are the Hewlett Packard HOI-3722A Noise Generator which can generate pseudo-random binary signals based on m-sequences for all $N = 2^n - 1$ between 15 and 1 048 575 inclusive ($4 \leqslant n \leqslant 20$), and the Solartron JM1861 Pseudo-Random Signal Generator, which can generate similar signals for $n = 5, 6, 7, 9, 10, 15, 17$ and 20. As we will see in the next section, both these generators include facilities for generating other signals related to pseudo-random binary signals.

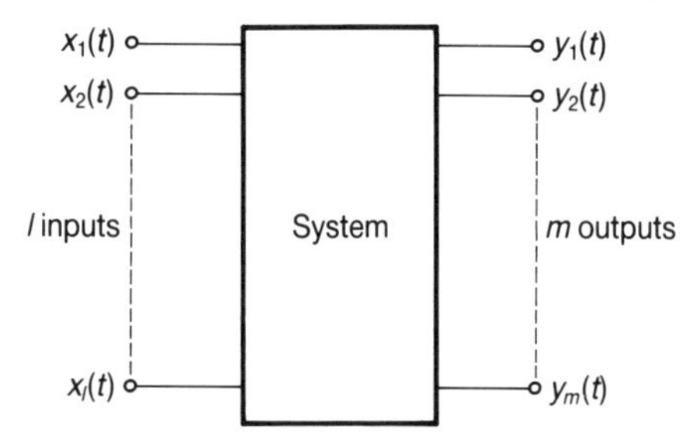

Figure 1.20 Multivariable system.

1.5 Signals derived from pseudo-random binary signals

1.5.1 Uncorrelated signals for multi-input system identification

Consider the l-input, m-output process shown in Figure 1.20. The cross-correlation function between input p and output q, over an integration time T_f, is given by

$$
\begin{aligned}
R_{u_p y_q}(\tau) &= \frac{1}{T_f}\int_0^{T_f} u_p(t) \,.\, y_q(t+\tau)\,\mathrm{d}t \\
&= \int_0^{T_f} h_{pq}(\lambda) R_{u_p u_p}(\tau-\lambda)\,\mathrm{d}\lambda \\
&\quad + \sum_{i \neq p} \int_0^{T_f} h_{iq}(\lambda) R_{u_i u_p}(\tau-\lambda)\,\mathrm{d}\lambda
\end{aligned}
\tag{1.54}
$$

where $i, p = 1, 2, \ldots, l$; $q = 1, 2, \ldots, m$, and $h_{iq}(\lambda)$ is the weighting function between input i and output q. Evaluation of equation (1.54) is greatly simplified if all the cross-correlation functions between the input signals are zero, in which case

$$
R_{u_p y_q}(\tau) = \int_0^{T_f} h_{pq}(\lambda) R_{u_p u_p}(\tau-\lambda)\,\mathrm{d}\lambda \tag{1.55}
$$

which is exactly the same form as for a single-input experiment. If it is also possible to make $R_{u_p u_p}(\tau-\lambda)$ of impulsive form, then the deconvolution would be simple as in the single-input case.

What is being sought therefore is a set of signals with the following properties:

1. The cross-correlation functions $R_{u_i u_j}(\tau)$, $i, j = 1, 2, \ldots, p$; $i \neq j$ are negligible (preferably zero) for τ less than the maximum settling time T_{smax} of the process;
2. The autocorrelation functions $R_{u_i u_i}(\tau)$, $i = 1, 2, \ldots, p$ are small, except for the spike around $\tau = 0$, for $\tau < T_{smax}$.

Completely uncorrelated signals of the same length do exist (see, for example, MacWilliams, 1967), but they do not possess autocorrelation properties which are suitable for system identification. Briggs and Godfrey (1976) showed that it is not possible to obtain independent estimates of the weighting functions of multi-input, single-output systems unless the common period of the perturbation signals is at least as long as the sum of the settling times of the input–output relationships. Thus, it is impossible to obtain uncorrelated signals of the same period with autocorrelation functions of delta function form. It is nevertheless possible to obtain sets of uncorrelated signals of *different* periods with suitable autocorrelation functions.

Consider first a two-input system, with one input perturbed by a PRBS $u_1(t)$ based on a sequence of length N; recall that N must be odd. An *inverse-repeat* binary signal (Simpson, 1966) is based on a sequence obtained by inverting every other digit in the pseudo-random binary sequence, i.e. by adding, modulo 2, the sequence 0 1 0 1 0 1... to the first sequence. The second signal $u_2(t)$ is then obtained by transformation to voltage levels as before. This is illustrated in Table 1.5 for the sequence at the fourth stage of the register in Table 1.1(a). Since N must be odd, the period of the inverse-repeat signal $u_2(t)$ is $2N\Delta t$ (i.e. twice that of $u_1(t)$) and its second half is the negative of the first half, as indicated by the intentional gap in Table 1.5.

Thus over the common period $2N\Delta t$, the cross-correlation function $R_{u_1u_2}(\tau) = 0$; in other words, the two signals are uncorrelated. The question remaining is whether the autocorrelation function $R_{u_2u_2}(\tau)$ is of impulse-like form.

To determine this, let us return to sampled data form and consider the sequence $\{x_r\}$ obtained by sampling the PRBS $u_1(t)$ every clock pulse interval Δt. We know that

$$R_{xx}(k) = V^2 \qquad k = 0 \bmod N \tag{1.56a}$$

$$= -V^2/N, \quad k \neq 0 \bmod N \tag{1.56b}$$

where $R_{xx}(k)$ has been obtained by averaging over N or any integer multiple of N. Denote by $\{d_r\}$ the sequence obtained by sampling the inverse-repeat signal $u_2(t)$ every Δt, so that

$$d_r = (-1)^{r+1} x_r \tag{1.57}$$

if the digits $r = 2, 4, 6, \ldots, 2N$ in $\{x_r\}$ are inverted to obtain $\{d_r\}$, as in Table 1.5. The autocorrelation function $R_{dd}(k)$, defined by averaging over $2N$ or any integer multiple of $2N$, is thus given by:

$$R_{dd}(k) = \frac{1}{2N} \sum_{r=1}^{2N} d_r d_{r+k} \tag{1.58}$$

so that from equation (1.57) we obtain:

$$R_{dd}(k) = \frac{1}{2N} \sum_{r=1}^{2N} (-1)^{r+1} x_r . (-1)^{r+1+k} x_{r+k}$$

$$= (-1)^k R_{xx}(k) \tag{1.59}$$

Table 1.5 Pseudo-random binary signal of period $15\Delta t$ and corresponding inverse-repeat signal of period $30\Delta t$

Number of clock pulses	m-sequence	PRBS $u_1(t)$	Inverse-repeat sequence	Inverse-repeat signal $u_2(t)$
1	1	$+V$	1	$+V$
2	0	$-V$	1	$+V$
3	0	$-V$	0	$-V$
4	0	$-V$	1	$+V$
5	1	$+V$	1	$+V$
6	1	$+V$	0	$-V$
7	1	$+V$	1	$+V$
8	1	$+V$	0	$-V$
9	0	$-V$	0	$-V$
10	1	$+V$	0	$-V$
11	0	$-V$	0	$-V$
12	1	$+V$	0	$-V$
13	1	$+V$	1	$+V$
14	0	$-V$	1	$+V$
15	0	$-V$	0	$-V$
16	1	$+V$	0	$-V$
17	0	$-V$	0	$-V$
18	0	$-V$	1	$+V$
19	0	$-V$	0	$-V$
20	1	$+V$	0	$-V$
21	1	$+V$	1	$+V$
22	1	$+V$	0	$-V$
23	1	$+V$	1	$+V$
24	0	$-V$	1	$+V$
25	1	$+V$	1	$+V$
26	0	$-V$	1	$+V$
27	1	$+V$	1	$+V$
28	1	$+V$	0	$-V$
29	0	$-V$	0	$-V$
30	0	$-V$	1	$+V$

Hence, from equations (1.56a) and (1.56b)

$$R_{dd}(k) = V^2,\ k = 0 \bmod 2N \tag{1.60a}$$

$$= -V^2,\ k = N \bmod 2N \tag{1.60b}$$

$$= -V^2/N,\ k = 2, 4, 6 \ldots \neq 0 \bmod 2N \tag{1.60c}$$

$$= +V^2/N,\ k = 1, 3, 5, \ldots \neq N \bmod 2N \tag{1.60d}$$

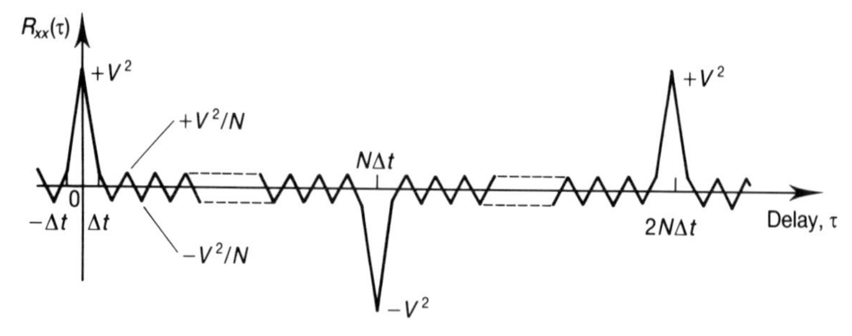

Figure 1.21 Autocorrelation function of an inverse-repeat binary signal of period $2N\Delta t$.

The autocorrelation function of the corresponding inverse-repeat signal is shown in Figure 1.21.

The Solartron JM1861 signal generator can generate inverse-repeat signals based on sequences of length $2(2^n - 1)$ for the same values of n as those listed at the end of Section 1.4.3.

In practice, the sawtooth 'ripple' on the autocorrelation function becomes small as N gets larger and provided the two process settling times are less than $N\Delta t$, the measured cross-correlation functions for values of τ less than $N\Delta t$ will yield the two process-weighting functions in exactly the same way as for a single-input experiment.

The argument can be followed through to provide further uncorrelated signals. If we return to sequences of logic levels 1 and 0, we have seen that the inverse-repeat signal was based on a sequence obtained by adding, modulo 2, 0 1 0 1 0 1 ... to the original sequence, so providing two signals which are uncorrelated over their common period $2N\Delta t$. A third uncorrelated signal is based on a sequence obtained by adding, modulo 2, 0 0 1 1 0 0 1 1... to the original sequence (i.e. by inverting digits 3, 4, 7, 8,... in the $\pm V$ sequence $\{x_r\}$). The three signals are then uncorrelated over their common period $4N\Delta t$. The next signal is based on a sequence obtained by adding, modulo 2, 0 0 0 0 1 1 1 1 0 0 0 0 1 1 1 1... to the original sequence, so providing four signals uncorrelated over their common period $8N\Delta t$, all having impulsive autocorrelation functions in the range up to $\tau = N\Delta t$ (Briggs and Godfrey, 1966).

This is a very convenient way of generating uncorrelated signals, and the resulting signals also have advantages in suppressing the effects of non-linear terms, as we will see in Section 1.7.1. However, there is a drawback from the process-identification viewpoint, and this is that the minimum correlation period is doubled for each successive input. If the system-setting times are very unequal in a particular application it may be better to resort to delayed versions of the same PRBS as the perturbation signals, with the delays chosen so that weighting function terms do not appear on top of one another in the resulting input–output cross-correlation functions. Such an experiment would have to be designed with considerable care, and initial step-response information would be almost essential for proper design of the signal and the delays.

1.5.2 Signals with Gaussian properties

A signal having a probability density function (PDF) which is approximately Gaussian can be obtained easily from a PRBS, by low-pass filtering it. Considerable care needs to be taken to obtain a reasonably Gaussian PDF in this way, and there are three conditions which must be met:

1. The PRBS needs to be fairly long, and it is usually suggested that a reasonably Gaussian PDF cannot be obtained for $N < 4095$ ($n < 12$).
2. The filter cut-off frequency, f_{CO}, must not be too high, otherwise recognizable features of the original signal are retained and the PDF is not smooth. The maximum value of f_{CO} for which the PDF can reasonably be regarded as Gaussian depends, to some extent, on the order of the low-pass filter. For a fourth-order filter, from observation on a signal analyzer the PDF is definitely non-Gaussian if $f_{CO} = 0.2/\Delta t$, but it is reasonably Gaussian if $f_{CO} = 0.1/\Delta t$.
3. If f_{CO} is too low, the PDF becomes skewed (Tomlinson and Galvin, 1974). Again, for a fourth-order filter, the skewing begins to be noticeable on a signal analyzer if $f_{CO} = 0.02/\Delta t$ and is very significant if $f_{CO} = 0.01/\Delta t$.

The Hewlett Packard HOI-3722A noise generator includes a Gaussian signal feature by low-pass filtering a PRBS with a fourth-order filter with cut-off frequency $f_{CO} = 0.05/\Delta t$ Hz, so avoiding the problems associated with (2) and (3) above. The user is warned not to expect a Gaussian PDF unless $N \geqslant 4095$. We will see in Chapter 2 that the (half-power) bandwidth of a PRBS is $0.443/\Delta t$, so the bandwidth of the Gaussian signal ($0.05/\Delta t$) is much less than that of the PRBS from which it is obtained.

The skewing referred to in (3) can be avoided by low-pass filtering an inverse-repeat signal rather than a PRBS (Godfrey, 1975). A true Gaussian signal has the property that it is uncorrelated with its squared value (both with mean levels removed); this property is possessed by a suitably filtered inverse-repeat signal but not by a filtered PRBS.

The skewing can also be avoided by filtering a discrete-interval random binary signal. As noted above, this type of signal is much less easily generated than a PRBS, but the HOI-3722A can generate such a signal.

1.6 Multi-level pseudo-random signals

Most multi-level pseudo-random signals considered in the literature have been based on multi-level maximum-length sequences, and in this chapter this will be the only type to be considered. Multi-level m-sequences exist for the number of levels, q, equal to a prime or a power of a prime $p(>1)$, i.e. for $q = 2, 3, 4, 5, 7, 8, 9, 11, 13, \ldots$ (Zierler, 1959). The length of such a sequence $\{x_r\}$ is $q^n - 1$, where n is an integer.

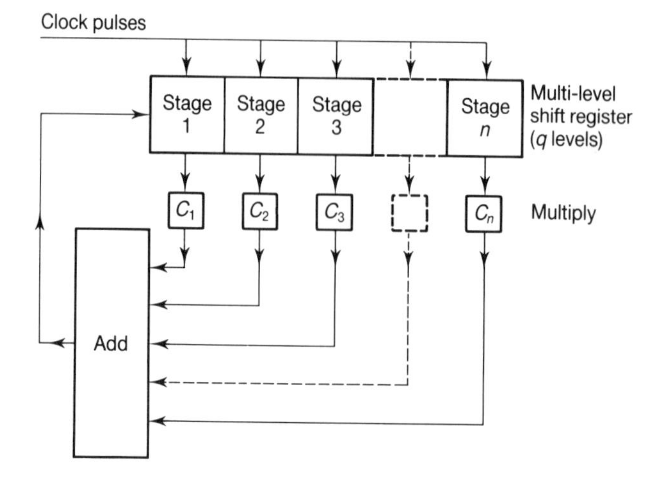

Figure 1.22 Generation of a q-level m-sequence.

One of the main reasons for considering signals based on m-sequences with $q > 2$ is that they can provide a better estimate than binary signals of the linear dynamics of a process with non-linearities and they can also be of use in the identification of the non-linear characteristics themselves. Some of these aspects are discussed in Section 1.7.

1.6.1 Multi-level maximum-length sequences with q prime

When q is prime, the digits of the sequence $\{x_r\}$ are the integers $0, 1, \ldots, (q - 1)$ and the sequence can be thought of as being generated by a q-level shift register with feedback to the first stage consisting of the modulo q sum of the outputs of the other stages multiplied by coefficients $c_1, \ldots, c_n$ which are also the integers $0, 1, \ldots, (q - 1)$. This is illustrated in Figure 1.22.

For q the power of a prime, the addition and multiplication are no longer modulo q and they need redefining. The case $q = 4$ is described separately in Section 1.6.3 and the cases $q = 4$, 8 and 9 are considered by Barker in Chapter 11.

As with the binary case, the feedback configuration corresponds to a primitive polynomial, modulo q, but rather more care is needed than in the binary case when deriving the corresponding feedback configuration. If the polynomial $a_n x^n + a_{n-1} x^{n-1} + \ldots + a_1 x + a_0$ is primitive, modulo q, then, in a similar manner to the binary case (Section 1.4.1), the logic connections to the first stage are given by

$$a_0 X = -a_1 DX - \ldots - a_{n-1} D^{n-1} X - a_n D^n X, \text{ modulo } q \tag{1.61}$$

where $a_0 = 1$ and the remaining coefficients $a_1, \ldots, a_n$ have integer values in the range from 0 to $q - 1$. However, for $q > 2$, modulo q subtraction is not the same as modulo q addition so that the feedback has to be revised to

$$X = c_1 DX + \ldots + c_{n-1} D^{n-1} X + c_n D^n X, \text{ modulo } q \quad (1.62)$$

where

$$c_r = (q - a_r), \; r = 1, 2, \ldots, n \quad (1.63)$$

For example, $2x^4 + 2x^3 + 1$ is primitive modulo 3, and the feedback to the input of the corresponding four-stage, three-level shift register is given by the modulo 3 sum:

$$X = D^3 X + D^4 X$$

i.e. the modulo 3 sum of the output of the third and fourth stages. The resulting sequence has logic values 0, 1 and 2 and is of period $3^4 - 1 = 80$. Similarly, the polynomial $3x^4 + 4x^3 + 4x^2$ is primitive, modulo 5, and the feedback to the input of the corresponding four-stage, five-level shift register is given by the modulo 5 sum:

$$X = D^2 X + D^3 X + 2D^4 X$$

i.e. the modulo 5 sum of the output of stages 2 and 3 and twice the output of stage 4. The resulting sequence has logic values 0, 1, 2, 3 and 4 and is of period $5^4 - 1 = 624$.

Thus, for q prime, an m-sequence of length $q^n - 1$ can be produced from an n-stage, q-level shift register with appropriate feedback and with its initial state any combination of logic values 0, 1, 2, ..., $(q - 1)$, except n zeros. Each combination of length n of these logic values except n zeros appears as the state of the register exactly once during the sequence. In the sequence, there are q^{n-1} of each of the logic values 1, 2, ..., $(q - 1)$ and $q^{n-1} - 1$ of the logic value 0. A further property is that the sequence can be divided into $(q^n - 1)/(q - 1)$ segments, such that for k an integer,

$$\{x_{r + k(q^n - 1)/(q - 1)}\} = \{g x_r\} \quad (1.64)$$

where g is one of the elements 1, 2, ..., $(q - 1)$.

Church (1935) lists all irreducible polynomials, modulo 3 for $2 \leqslant n \leqslant 7$, modulo 5 for $2 \leqslant n \leqslant 5$ and modulo 7 for $2 \leqslant n \leqslant 4$, indicating clearly which of the polynomials are primitive.

Two examples of feedback connections for generating three-level m-sequences for each n in the range $2 \leqslant n \leqslant 7$ (sequence lengths $N = 3^n - 1 = 8, 26, 80, 242, 728$ and 2186) are given in Table 1.6, and it should be recalled that the feedback is

Table 1.6 Two examples of feedback connections for the generation of three-level m-sequences of each length between 8 and 2186 (inclusive)

n	N	Feedback coefficients						
		c_1	c_2	c_3	c_4	c_5	c_6	c_7
2	8	2	1	–	–	–	–	–
2	8	1	1	–	–	–	–	–
3	26	0	1	2	–	–	–	–
3	26	1	2	2	–	–	–	–
4	80	0	0	2	1	–	–	–
4	80	0	0	1	1	–	–	–
5	242	0	0	0	1	2	–	–
5	242	0	0	1	2	2	–	–
6	728	0	0	0	0	2	1	–
6	728	0	0	0	0	1	1	–
7	2186	0	0	0	0	2	1	2
7	2186	0	0	0	0	1	0	2

according to the multiplication indicated in Table 1.6 followed by modulo 3 *addition*, as in Figure 1.21. An illustration of the generation of a 26-digit, three-level m-sequence using the first of the two possible feedback connections listed in Table 1.6 for $n = 3$ is given in Table 1.7. As expected, there are nine 2's, nine 1's and eight 0's in the sequence. The gap in the middle of Table 1.7 is intentional, emphasizing that the second half of the sequence can be obtained from the first half by interchanging logic values 1 and 2; this corresponds to $g = 2$ in equation (1.64). The signal levels shown in the right-hand column of Table 1.7 will be considered in more detail in Section 1.6.2.

Table 1.7 Sequence and corresponding signal from a three-stage, three-level register with feedback given by $X = D^2X + 2D^3X$, modulo 3

Number of clock pulses	Shift register stage			Corresponding signal
	1	2	3	
1	0	0	1	$+V$
2	2	0	0	0
3	0	2	0	0
4	2	0	2	$-V$
5	1	2	0	0
6	2	1	2	$-V$
7	2	2	1	$+V$
8	1	2	2	$-V$
9	0	1	2	$-V$
10	2	0	1	$+V$
11	2	2	0	0
12	2	2	2	$-V$
13	0	2	2	$-V$

Table 1.7 *Continued*

Number of clock pulses	Shift register stage 1	2	3	Corresponding signal
14	0	0	2	$-V$
15	1	0	0	0
16	0	1	0	0
17	1	0	1	$+V$
18	2	1	0	0
19	1	2	1	$+V$
20	1	1	2	$-V$
21	2	1	1	$+V$
22	0	2	1	$+V$
23	1	0	2	$-V$
24	1	1	0	0
25	1	1	1	$+V$
26	0	1	1	$+V$
27	0	0	1	$+V$

Further feedback connections (with modulo q addition) for generating q-level m-sequences are given in Table 1.8, for $q = 5$ and $n = 2$, 3, 4 and 5 and for $q = 7$ and $n = 2$, 3 and 4, derived from the primitive polynomials listing of Church (1935), and for $n = 3$ and $q = 11$, 13, 19, 23, 29 and 31 (all prime numbers), from Table 1 of Everett (1966).

Table 1.8 Feedback configurations for some q-level m-sequences of length $N = q^n - 1$

			Feedback coefficients				
q	n	N	c_1	c_2	c_3	c_4	c_5
5	2	24	1	3	–	–	–
5	3	124	0	1	2	–	–
5	4	624	0	1	1	2	–
5	5	3 124	0	0	0	1	2
7	2	48	1	4	–	–	–
7	3	342	0	1	5	–	–
7	4	2 400	0	1	1	4	–
11	3	1 330	0	10	7	–	–
13	3	2 196	0	12	7	–	–
17	3	4 912	0	16	14	–	–
19	3	6 858	0	18	15	–	–
23	3	12 166	0	22	20	–	–
29	3	24 388	0	28	18	–	–
31	3	29 790	0	30	17	–	–

The five-level m-sequence of length 24 corresponding to the entry in Table 1.8 for $n = 2$, $q = 5$ is shown in Table 1.9. The sequence has been started (arbitrarily) from the register state 0, 1 and the gaps after 6, 12 and 18 clock pulses are intentional, emphasizing that each segment of $(q^n - 1)/(q - 1) = 6$ can be obtained from the preceding segment by multiplication modulo 5 by an integer between 1 and $(q - 1)$, as indicated by equation (1.64). In Table 1.9 the multiplication factor is 2.

Table 1.9 Sequence and corresponding inverse-repeat signals from a two-stage, five-level register, with feedback given by $X = DX + 3D^2X$, modulo 5

Number of clock pulses	Shift register stage		Inverse-repeat signals	
	1	2		
1	0	1	$+V$	$+V$
2	3	0	0	0
3	3	3	$-2V$	$+2V$
4	2	3	$-2V$	$+2V$
5	1	2	$+2V$	$-2V$
6	2	1	$+V$	$+V$
7	0	2	$+2V$	$-2V$
8	1	0	0	0
9	1	1	$+V$	$+V$
10	4	1	$+V$	$+V$
11	2	4	$-V$	$-V$
12	4	2	$+2V$	$-2V$
13	0	4	$-V$	$-V$
14	2	0	0	0
15	2	2	$+2V$	$-2V$
16	3	2	$+2V$	$-2V$
17	4	3	$-2V$	$+2V$
18	3	4	$-V$	$-V$
19	0	3	$-2V$	$+2V$
20	4	0	0	0
21	4	4	$-V$	$-V$
22	1	4	$-V$	$-V$
23	3	1	$+V$	$+V$
24	1	3	$-2V$	$+2V$
25	0	1	$+V$	$+V$

1.6.2 Pseudo-random signals based on multi-level maximum-length sequences

The main application for pseudo-random signals based on q-level m-sequences with q a prime ($\geqslant 3$) has been with signal levels chosen to give zero mean. For a three-level

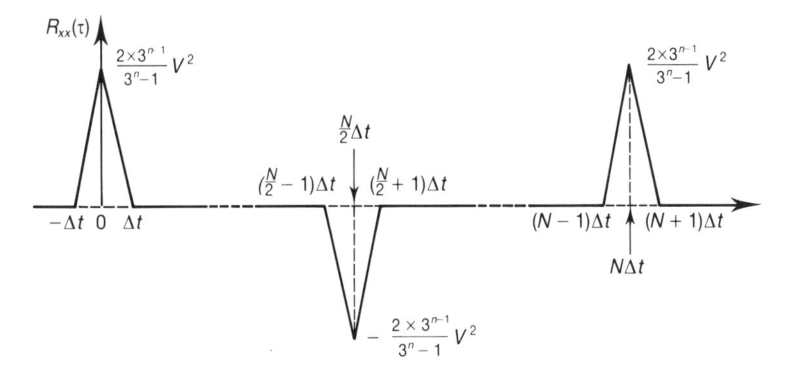

Figure 1.23 Autocorrelation function of a signal based on a three-level m-sequence.

signal this can be done by conversion from logic values to signal levels either by $0 \to 0$, $1 \to +V$, $2 \to -V$ or by $0 \to 0$, $1 \to -V$, $2 \to +V$. The signal levels shown in the right-hand column of Table 1.7 are from the first of these two choices for conversion. Properties of signals based on three-level m-sequences with either of these two conversions can be listed in a manner similar to those for pseudo-random binary signals given in Section 1.4:

1. The signal has three levels, 0, $+V$ and $-V$ and may switch from one level to another only at certain event points $t = 0, \Delta t, 2\Delta t, \ldots$
2. The signal is deterministic and experiments are repeatable.
3. The signal is periodic with period $T = N\Delta t$, where $N = 3^n - 1$, n an integer.
4. In any one period, there are 3^{n-1} intervals when the signal has level $+V$, 3^{n-1} intervals when it has level $-V$ and $3^{n-1} - 1$ intervals when it has level 0.
5. The second half is the negative of the first half.
6. The autocorrelation function of the signal is as shown in Figure 1.23; note the similarity in shape to that of a binary inverse-repeat signal (Figure 1.20).

Because of the inverse-repeat property of the signal, its odd-order autocorrelation functions are zero. Thus, in continuous time,

$$R_{xx \cdots x}(\tau_1, \tau_2, \ldots, \tau_{v-1}) \equiv \frac{1}{T}\int_0^T x(t).x(t+\tau_1).x(t+\tau_2)\ldots x(t+\tau_{v-1})\,dt$$
$$= 0 \text{ for odd } v \qquad (1.65)$$

Similarly, for sampled signals, with the signal $x(t)$ sampled once per clock pulse interval Δt to obtain a sequence $\{x_r\}$, $r = 0, 1, 2, \ldots, (N-1)$,

$$R_{xx\ldots x}(k_1, k_2, \ldots, k_{v-1}) \equiv \frac{1}{N}\sum_{r=0}^{N-1} x_r.x_{r+k_1}.x_{r+k_2}.\ldots.x_{r+k_{v-1}}$$
$$= 0 \text{ for odd } v \qquad (1.66)$$

For $v = 1$, this simply gives that the signal mean is zero, but the result for $v = 3$, 5, 7,..., is of use in non-linear system identification, as we will see in Section 1.7. Equations (1.65) and (1.66) also apply to the binary inverse-repeat signals considered in Section 1.5.1, with integration over the appropriate period $T = 2N\Delta t$ or summation over the appropriate sequence length $r = 0$ to $2N - 1$.

For signals based on q-level m-sequences with q a prime >3, there are several choices for the conversion from sequence logic value to signal level which will result in a zero-mean signal. The logic value 0 should correspond to the level zero in the signal, but the other logic values can then be assigned arbitrarily to symmetric voltage values. However, only some of these will result in signals with an inverse-repeat property, which is usually the property being sought for identification of the characteristics of processes with non-linearities.

To demonstrate this, consider the case $q = 5$, for which the logic values are 0, 1, 2, 3 and 4. A zero-mean signal can be achieved by assigning $0 \rightarrow 0$ and then logic values $\{1,2,3,4\}$ to signal levels $\{-2V, -V, +V, +2V\}$ in any order. We know that the m-sequence can be split into four segments, each of length $(5^n - 1)/4$, with each segment being an integer multiple, modulo 5, of the preceding segment. To achieve an inverse-repeat property, the four segments of the signal must be either in the ratio 1, 2, -1, -2, or 1, -2, -1, 2; any other possibilities are simply shifted versions of these two. The two signals resulting from this are shown in Table 1.9 for the sequence from stage 2 of the shift register.

The properties of the zero-mean signal with inverse-repeat property are then similar to those of a signal based on a three-level m-sequence, with the number 5 replacing the number 3 where appropriate. For example, property (4) above is revised to read that each of the levels $-2V$, $-V$, $+V$ and $+2V$ occurs during 5^{n-1} of the intervals, while level zero occurs during $5^{n-1} - 1$ of the intervals. The autocorrelation function is similar in shape to that of Figure 1.23, with the peaks having value $\pm 2V^2 \times 5^n/(5^n - 1) = \pm 2.08333V^2$ for $n = 2$ and $\pm 2.016129V^2$ for $n = 3$ (Chang, 1966).

An application of an inverse-repeat signal based on a five-level m-sequence to an industrial process was described by Chang *et al.* (1968). The process was a four-stand cold-rolling mill at the Port Talbot Works of British Steel (the same steelworks as in Application Example 1.3). Impulse responses were determined from cross-correlation functions and were used in turn to estimate the coefficients of frequency responses $H(j\omega)$.

1.6.3 Multi-level m-sequences with q the power of a prime

When the number of levels, q, is the power of a prime, the logic values in the sequence can still be taken as 0, 1, 2, ..., $(q - 1)$, but the arithmetic operations are no longer modulo q. As an example, $q = 4$ will be described here; the cases $q = 4$, 8 and 9 are considered further by Barker in Chapter 11. Four-level m-sequences can be thought of as being generated by four-level feedback shift register circuits and, as before, feedback is according to

$$X = c_1DX + c_2D^2X + \dots c_nD^nX$$

However, unlike the case for q prime, the addition and multiplication are not modulo q but are according to the operations shown in Table 1.10.

Table 1.10 Addition and multiplication operations for the generation of four-level m-sequences

	0	1	2	3		0	1	2	3
0	0	1	2	3	0	0	0	0	0
1	1	0	3	2	1	0	1	2	3
2	2	3	0	1	2	0	2	3	1
3	3	2	1	0	3	0	3	1	2
	Addition					Multiplication			

Examples of feedback polynomials which result (using this arithmetic) in four-level m-sequences of length $4^n - 1$ have been given by Balza *et al.* (1967) for $n = 3$, 4, 5 and 6 and one example for each of these values together with a suitable feedback polynomial for $n = 2$ is listed in Table 1.11. The sequence of length 15

Table 1.11 Feedback configurations for some four-level m-sequences

		Feedback coefficients					
n	N	c_1	c_2	c_3	c_4	c_5	c_6
2	15	1	2	–	–	–	–
3	63	1	1	2	–	–	–
4	255	1	2	2	2	–	–
5	1023	1	2	3	0	3	–
6	4095	0	0	0	1	1	2

generated using the listed polynomial $X = DX + 2D^2X$ (with arithmetic as given in Table 1.10) is shown in Table 1.12. As expected, there are four 1's, four 2's, four 3's and three 0's in the sequence, which has $(q - 1) = 3$ segments, each of length 5. Each successive segment is multiplied by $g = 2$ to obtain the next segment, recalling from Table 1.10 that, in the required arithmetic, $2 \times 1 = 2$, $2 \times 2 = 3$ and $2 \times 3 = 1$.

With the division of each sequence into three segments, the corresponding signals cannot be made of inverse-repeat form. Briggs and Godfrey (1968) show that if the transformation from logic values to signal levels in a four-level m-sequence of length $N = 4^n - 1$ is made according to $0 \rightarrow a_0$, $1 \rightarrow a_1$, $2 \rightarrow a_2$ and $3 \rightarrow a_3$, then the mean level of the signal is given by

$$\bar{x} = \frac{1}{4}(a_0 + a_1 + a_2 + a_3) + \frac{1}{4(4^n - 1)}(a_0 + a_1 + a_2 + a_3 - 4a_0) \tag{1.67}$$

They further show that if the signal is sampled once per clock pulse interval Δt to give a sequence $\{x_r\}$, then the normalized autocorrelation function of $\{x_r\}$ is given by

$$R_{xx}(0) = \frac{1}{4}(a_0^2 + a_1^2 + a_2^2 + a_3^2) + \frac{1}{4(4^n - 1)}(a_0^2 + a_1^2 + a_2^2 + a_3^2 - 4a_0^2) \tag{1.68a}$$

Table 1.12 Sequence from a two-stage, four-level register with feedback given by $X = DX + 2D^2X$, with addition and multiplication as in Table 1.10

Number of clock pulses	Shift register stage	
	1	2
1	0	1
2	2	0
3	2	2
4	1	2
5	2	1
6	0	2
7	3	0
8	3	3
9	2	3
10	3	2
11	0	3
12	1	0
13	1	1
14	3	1
15	1	3
16	0	1

$$R_{xx}\left(\frac{N}{3}\right) = R_{xx}\left(\frac{2N}{3}\right)$$
$$= \frac{1}{4}(a_0^2 + a_1a_2 + a_2a_3 + a_3a_1)$$
$$+ \frac{1}{4(4^n - 1)}(a_0^2 + a_1a_2 + a_2a_3 + a_3a_1 - 4a_0^2) \tag{1.68b}$$

and, for other values of k,

$$R_{xx}(k) = \frac{1}{16}(a_0 + a_1 + a_2 + a_3)^2 + \frac{1}{16(4^n - 1)}[(a_0 + a_1 + a_2 + a_3)^2 - 16a_0^2] \tag{1.68c}$$

For the logic value to signal level transformation $0 \to -1.5V$, $1 \to -0.5V$, $2 \to +0.5V$, $3 \to +1.5V$ and for $n = 3$ ($N = 63$), $\bar{x} = 0.0238V$, $R_{xx}(0) = 1.2341V^2$, $R_{xx}(21) = R_{xx}(42) = 0.4722V^2$ and $R_{xx}(k) = -0.0357V^2$ ($k \neq 0$, 21 or 42, modulo 63), while for the transformation $0 \to -0.5V$, $1 \to +0.5V$, $2 \to +1.5V$, $3 \to -1.5V$ (again with $n = 3$), $\bar{x} = 0.0079V$, $R_{xx}(0) = 1.2659V^2$, $R_{xx}(21) = R_{xx}(42) = -0.5119V^2$ and $R_{xx}(k) = -0.0040V^2$. The autocorrelation functions of the corresponding signals are shown in Figure 1.24.

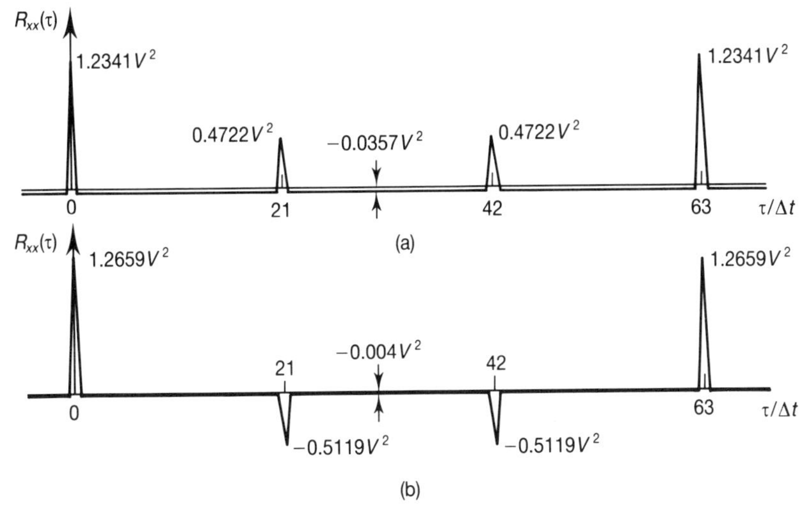

Figure 1.24 Autocorrelation functions of signals based on a four-level m-sequence of length 63. (a) Logic value to signal level transformation $\{0,1,2,3\} \rightarrow \{-1.5V, -0.5V, +0.5V, +1.5V\}$; (b) logic value to signal level transformation $\{0,1,2,3\} \rightarrow \{-0.5V, +0.5V, +1.5V, -1.5V\}$.

1.7 Perturbation signals for non-linear system identification

We have seen in Section 1.1 that the input $u(t)$ and output $y(t)$ of a *linear* system are related by a convolution integral (equation (1.2)). One of the most general mathematical representations of the corresponding relationship for a *non-linear* system is the Volterra functional series expansion:

$$
\begin{aligned}
y(t) = &\int_0^\infty h(\lambda)u(t-\lambda)\,\mathrm{d}\lambda \\
&+ \int_0^\infty \int_0^\infty h_2(\lambda_1, \lambda_2)u(t-\lambda_1)u(t-\lambda_2)\,\mathrm{d}\lambda_1\,\mathrm{d}\lambda_2 \\
&+ \dots \\
&+ \int_0^\infty \int_0^\infty \dots \int_0^\infty h_n(\lambda_1, \lambda_2, \dots, \lambda_n)u(t-\lambda_1)u(t-\lambda_2)\dots \\
&\qquad u(t-\lambda_n)\,\mathrm{d}\lambda_1\,\mathrm{d}\lambda_2\dots\mathrm{d}\lambda_n \\
&+ \dots
\end{aligned}
\tag{1.69}
$$

The function $h_n(\lambda_1, \lambda_2, \dots \lambda_n)$ is called the Volterra kernel of order n; the kernel of order 1 is the (linear) system weighting function. Most non-linear systems can be described by this form of expansion, although processes which give rise to sub-harmonic generation are excluded (Simpson and Power, 1972). For many systems of practical interest, the effect of the higher-order kernels on the output $y(t)$ decreases quite rapidly with increasing n.

1.7.1 Identification of the linear approximant to a non-linear system

In many experimental situations involving non-linear systems the desired result is simply as accurate a characterization as possible of the linear dynamics of the system. In the time domain this is of the linear weighting function $h(t)$ or a corresponding parametric model. In this case the use of a signal with an inverse-repeat property is desirable, with either an inverse-repeat binary signal (Section 1.5.1) or a three-level pseudo-random signal with levels $-V$, 0 and $+V$ (Section 1.6.2). This is because the odd-order autocorrelation functions (equation (1.64)) of an inverse-repeat signal are zero, so that if $u(t)$ and $y(t)$ in equation (1.68) are cross-correlated over an integer number of periods of the perturbation signal, then the contributions from the even-order kernels ($n = 2, 4, 6, \ldots$) disappear, leaving only the linear term ($n = 1$) and terms involving kernels of order 3, 5, 7, ... As pointed out above, in many practical situations the effects of the higher-order kernels on the output diminish quite rapidly with increasing n, so that, with the effects of the second-order kernel in particular removed, a better estimate of the linear kernel $h(t)$ is obtained.

This will be illustrated for one of the most widely occurring departures from linearity – that when the dynamics of the system are different according to whether the variable under investigation is increasing or decreasing. This situation occurs on a large scale in industry. One example is a temperature loop operating above ambient temperature where, owing to heat losses, an increase in the variable by a given amplitude can take longer than a decrease in the same amplitude (the opposite can be true for a very well-insulated thermal process – for instance, a metal-heating furnace). Another example is a steam drum, where pressure can fall quite slowly when the flow out of the drum is increased due to flashing off of steam. When the flow out decreases, flashing off does not occur and pressure can rise more rapidly as a consequence.

The identification of direction-dependent non-linear systems was studied by Godfrey and Briggs (1972), who obtained theoretical results for a system with first-order dynamics and simulation results for systems with first- and higher-order dynamics. Three types of discrete-interval binary signal were considered, based on the following sequences:

1. A 31-digit maximum-length sequence, obtained from a five-stage shift register with feedback from stages 3 and 5;
2. A 31-digit quadratic residue code, with $+1$ in the 31st position;
3. A 62-digit inverse-repeat sequence obtained by inverting every other digit of the sequence in (1).

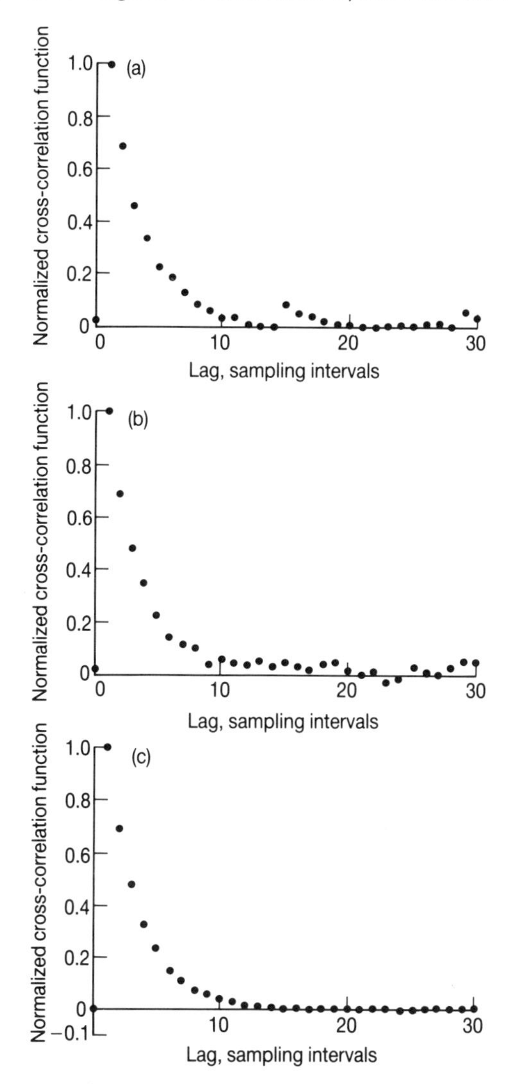

Figure 1.25 Cross-correlation functions for a process with direction-dependent dynamic responses. Perturbation signals based on (a) 31-digit m-sequence; (b) 31-digit quadratic residue code; (c) 62-digit inverse repeat sequence.

Input–output cross-correlation functions using these three types of signal as input to a first-order system with time constant $2\Delta t$ in the positive direction and $4\Delta t$ in the negative direction (where Δt is the clock pulse interval of the signal) are shown in Figure 1.25.

The theory given by Godfrey and Briggs (1972) predicts that, for the m-sequence signal, there will be noticeable discontinuities in the cross-correlation function starting at time delays $15\Delta t$ and $29\Delta t$ and a smaller discontinuity starting at delay $6\Delta t$, and this is confirmed in Figure 1.25(a). For the signal based on the Quadratic Residue code, the effects of the departure from linearity are more distributed over the cross-correlation function (Figure 1.25(b)) and are very similar in appearance to noise.

For the inverse-repeat signal, only the first half of the cross-correlation function is shown in Figure 1.25(c) because the second half is the negative of the first half; also, the small-amplitude sawtooth pattern resulting from the sawtooth off-peak pattern in the autocorrelation function (Figure 1.21) has been removed. It may be seen that the cross-correlation function is quite smooth. Since the contributions of the even-order non-linearities have been removed, this suggests that the contributions from odd-order kernels of order 3 and above are so small as to be negligible. If the cross-correlation function of Figure 1.25(c) is interpreted as a linear system weighting function, then it is found that the dynamics lie between those of first-order systems with time constants $2\Delta t$ and $4\Delta t$ (Godfrey and Briggs, 1972).

The cross-correlation function for the m-sequence-based input signal (Figure 1.25(a)) is somewhat unexpected, with its sharp discontinuities. (These in fact start at values of delay which depend on the particular m-sequence which is used.) The discontinuities could lead to quite misleading conclusions, as the final Application Example in this chapter demonstrates.

APPLICATION EXAMPLE 1.5 GAS TURBINE ENGINE DYNAMICS

For a complex process, the additional peaks in the input–output cross-correlation function when testing with a PRBS based on an m-sequence can easily be attributed at first sight to reflections or 'echoes' of the dynamics around the process itself. Such an effect has been observed in dynamic testing of gas turbine engines (Moore, 1970; Godfrey and Moore, 1974). Little sense could be made from step responses of the maximum permitted amplitude and even to obtain reasonable results from cross-correlation functions it was necessary to correlate over several periods of the perturbation signal, which was based on a 511-digit m-sequence with feedback shift register connections from stages 5 and 9. The cross-correlation function between fuel flow input and compressor speed response on a Pegasus Vertical Take-Off Engine is shown in Figure 1.26.

The additional peaks in the cross-correlation function could easily be attributed to echoes of the main dynamics, attenuated and delayed, but in fact, the additional peaks occur in positions expected from the theory of Godfrey and Briggs (1972) for a process with direction-dependent dynamic responses. The main source of the departure from linearity was thought to be the fuel flow activator rather than the main engine dynamics; it was felt that, with the small amplitude of perturbation signals used, the engine dynamics should be linear. It was not possible to conduct tests on an isolated fuel flow activator for the Pegasus engine, but it was possible to test a similar activator from an Olympus 593 (Concorde) engine. The cross-correlation function (Figure 9 of Godfrey and Moore, 1974) again showed extra peaks in the

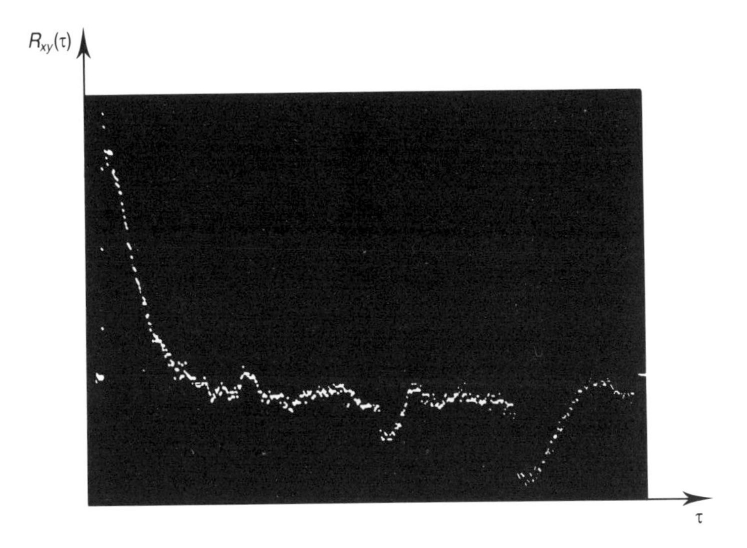

Figure 1.26 Cross-correlation function between fuel flow and compressor speed on the Pegasus Vertical Take-Off engine. Perturbation signal based on a 511-digit m-sequence with feedback shift register connections from stages 5 and 9.

positions expected from the theory of Godfrey and Briggs (1972), and this confirmed that the activator is primarily responsible for the bilinear (direction-dependent) dynamic behaviour. The main peak of the cross-correlation function of Figure 1.26 is also affected by the departure from linearity, with the dynamics lying between the individual dynamic responses in the upward and downward directions.

1.7.2 Identification of non-linear characteristics of a system

In Section 1.7.1 methods have been described for minimizing the effects of non-linearities on an input–output cross-correlation function, so allowing as good an estimate as possible to be made of the system kernel of order 1 (i.e. the linear weighting function). This leads to the interesting question of whether it is possible to identify aspects of the non-linearities themselves using a single perturbation signal (as distinct from several perturbation signals of differing amplitude).

In a Volterra series expansion it becomes extremely difficult to identify kernels of order three or more, but time-domain techniques for identifying the second-order kernel $h_2(\lambda_1, \lambda_2)$ have been developed (Barker *et al.*, 1972; Barker and Davy, 1978). The technique involves using discrete-interval signals sampled once per clock pulse interval. The output sequence $\{y_r\}$ is cross-correlated not with the input sequence $\{u_{r-k}\}$ but with the sequence given by the product $\{u_{r-I}u_{r-I-J}\}$.

For the method to prove tractable, the input signal must be of inverse-repeat form because then the effects of all odd-order kernels, including the linear kernel $h(\lambda)$, are zero due to the odd-order autocorrelation functions of the input (equation (1.66)) being zero. This leaves only those terms in the cross-correlation function between the input product sequence and the output which involve even-order kernels. It is assumed that the effects of kernels of order 4 and above are negligible compared with those of the kernels of order 2 so that, in this way, it is possible, subject to a further restriction, to obtain an estimate of the second-order kernel sequence which has elements $h_2(I, I + J)$.

The further restriction is that the diagonal elements (i.e. those with $J = 0$) cannot be estimated independently using a binary inverse-repeat signal (Section 1.5.1) and it is necessary to use inverse-repeat signals with three or more levels. Barker *et al.* (1972) considered the use of such signals based on three- and five-level m-sequences, while Barker and Davy (1978) discussed the use of the three-level signals in more detail.

If we consider *linear* system identification using a signal based on a three-level m-sequence, then because of the anti-symmetric property of the signal, the signal period has to be more than twice the settling time of the system to obtain independent estimates of the weighting function; also, there is nothing to choose between the different m-sequences of the same length. Neither of these is true when identifying the second-order kernel in the way discussed above. First, independent estimates of the kernel are only available for delays considerably less than $\frac{1}{2}N$, so that the signal length has to be a good deal longer than twice the settling time of $h_2(\lambda_1, \lambda_2)$ in either of its two (time) dimensions λ_1 and λ_2. Second, for a given sequence length N, there is a preferred sequence which is better than any of the others in the sense that it allows the second-order kernel values to be estimated from the cross-correlation function between $\{u_{r-I}u_{r-I-J}\}$ and $\{y_r\}$ with a minimum of calculation over the greatest number of delays. A table listing the preferred three-level m-sequences with N $(= 3^n - 1)$ in the range 8 to 6560 is given by Barker and Davy (1978).

In some senses, this latter point is similar to the preferred binary m-sequences for linear system identification for a system with an additive quadratic drift term at the output, discussed in Section 1.4.3. When the whole system is linear, there is nothing to choose between different m-sequences of the same length, but when there is a non-linearity present, one particular sequence gives the best results.

Using an approach similar to that above, Parker and Moore (1980) showed that it is possible to estimate the non-linear characteristics of a cascaded linear–non-linear–linear system (with the non-linear part being a static non-linearity) and they obtained the gain characteristic of an electrohydraulic servomechanism (Parker and Moore, 1982) using a signal based on a three-level m-sequence. The same approach can be used for bilinear systems, and Baheti *et al.* (1980) developed an algorithm to estimate parameters of a class of discrete-time bilinear systems using second-order correlations. Again, using pseudo-random ternary signals, they applied their algorithm to the one-delayed group kinetic equations for neutron density in a nuclear reactor, one of which contains a multiplicative term involving neutron density and reactivity.

The use of signals based on three-level m-sequences in the simultaneous estimation of the first and second derivatives of a cost function (performance index) expressed as a quadratic function of an input parameter followed by linear dynamics has been discussed by Clarke and Godfrey (1966). The technique involves cross-correlating the system output with the signal $u(t)$ to obtain an estimate of the first derivative and with the square of the signal $u^2(t)$ (with its d.c. bias removed) to obtain an estimate of the second derivative. The method exploits the property of such a signal that the cross-correlation function between $u(t)$ and $u^2(t)$ is zero, and it enables the minimum of a cost function to be reached in a single step in a noise-free process. A simulation study of the performance of the scheme (1) on non-quadratic cost functions, (2) in the presence of noise and (3) when the period of the signal has been set rather too low for the system dynamics (Clarke and Godfrey, 1967) showed that, provided a suitable lower bound is set on the second derivative estimate, the scheme will work well under a wide variety of conditions.

1.8 Non-periodic signals

Most system-identification applications have been based around the use of periodic perturbation signals, with signal processing over an integer number of signal periods. Using this approach, it is necessary to wait for one system-settling time before starting the signal processing, for the system to get into a dynamic 'steady state'; in practice, it is usual to wait for one complete signal period. For time-domain signal processing this would not be necessary if suitable non-periodic signals, consisting of a sequence of non-zero values embedded in an all-zero stream of digits, could be found, with the signal processing being done using non-periodic correlation.

Because the signal consists of a burst of non-zero values surrounded by zeros, the definition of non-periodic correlation has to be slightly different from any of those in Section 1.2.4. The autocorrelation function is given by

$$R_{uu}(k) = \frac{1}{N} \sum_{r=0}^{N-k-1} u_r u_{r+k} \tag{1.70}$$

while the corresponding input–output cross-correlation function is given by

$$R_{uy}(k) = \frac{1}{N} \sum_{r=0}^{N-k-1} u_r y_{r+k} \tag{1.71}$$

In both of these equations N is the number of non-zero digits in the sequence; the normalization by dividing by N is optional. For system identification, the objective is to obtain an autocorrelation function of approximately impulsive form, so that, as before, the input–output cross-correlation function can readily be interpreted in terms of the system-weighting sequence h_k, without having to resort to complicated deconvolution.

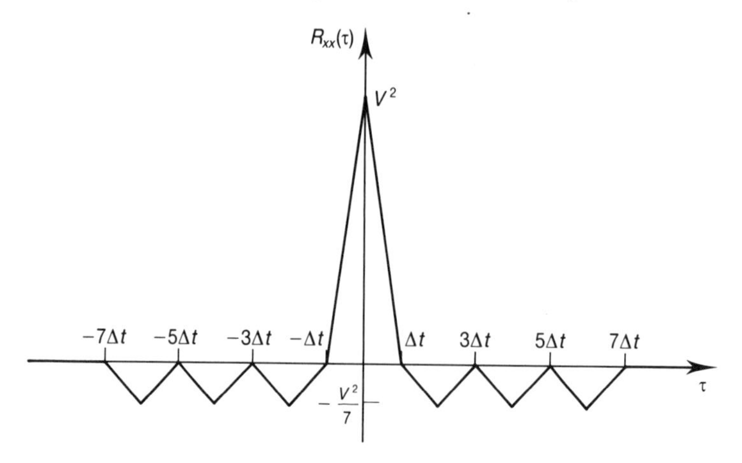

Figure 1.27 Non-periodic autocorrelation function of a signal based on the Barker sequence of length 7.

The best-known set of sequences having approximately impulsive non-periodic correlation are Barker sequences (Barker, 1953), which are binary. If the levels are taken as $\pm V$, then the on-peak autocorrelation ($k = 0$ in equation (1.69)) is V^2, while the off-peak values are either zero or $\pm V^2/N$. The non-periodic autocorrelation function of the Barker sequence with $N = 7$ is determined in Table 1.13 and the autocorrelation function of the corresponding signal (with clock pulse interval Δt) is shown in Figure 1.27.

The small off-peak autocorrelation values make Barker sequences appear very attractive for system identification but unfortunately, the longest known Barker sequence is for $N = 13$, which is too short to be really useful in most practical applications. Several ideas for longer sequences, both binary and non-binary, with good non-periodic autocorrelation properties have been emerging over the last few years, mainly in the context of radio communication design and signal processing. These ideas are reviewed by Darnell in Chapter 5.

1.9 Optimal test signals

When a good mathematical model of a system is available, there are, potentially, considerable benefits to be gained from good experiment design (Goodwin, 1987), including (in the context of this book) the use of optimal input signals (Mehra, 1974). Goodwin (1971) described a procedure, applicable to both linear and non-linear systems, for designing optimal test signals for a system described by a state-space formulation. The signals are designed to minimize the variances of the parameter

Table 1.13 Non-periodic autocorrelation function of the Barker sequence of length 7

r	-7	-6	-5	-4	-3	-2	-1	0	1	2	3	4	5	6	7	8	9	10	11	12	13		
x_r	0	0	0	0	0	0	0	$+V$	$+V$	$+V$	$-V$	$-V$	$+V$	$-V$	0	0	0	0	0	0	0	k	$R_{xx}(k)$
x_{r+0}	0	0	0	0	0	0	0	$+V$	$+V$	$+V$	$-V$	$-V$	$+V$	$-V$	0	0	0	0	0	0	0	0	$+V^2$
x_{r+1}	0	0	0	0	0	0	$+V$	$+V$	$+V$	$-V$	$-V$	$+V$	$-V$	0	0	0	0	0	0	0	0	1	0
x_{r+2}	0	0	0	0	0	$+V$	$+V$	$+V$	$-V$	$-V$	$+V$	$-V$	0	0	0	0	0	0	0	0	0	2	$-V^2/7$
x_{r+3}	0	0	0	0	$+V$	$+V$	$+V$	$-V$	$-V$	$+V$	$-V$	0	0	0	0	0	0	0	0	0	0	3	0
x_{r+4}	0	0	0	$+V$	$+V$	$+V$	$-V$	$-V$	$+V$	$-V$	0	0	0	0	0	0	0	0	0	0	0	4	$-V^2/7$
x_{r+5}	0	0	$+V$	$+V$	$+V$	$-V$	$-V$	$+V$	$-V$	0	0	0	0	0	0	0	0	0	0	0	0	5	0
x_{r+6}	0	$+V$	$+V$	$+V$	$-V$	$-V$	$+V$	$-V$	0	0	0	0	0	0	0	0	0	0	0	0	0	6	$-V^2/7$
x_{r+7}	$+V$	$+V$	$+V$	$-V$	$-V$	$+V$	$-V$	0	0	0	0	0	0	0	0	0	0	0	0	0	0	7	0
x_{r-1}	0	0	0	0	0	0	0	0	$+V$	$+V$	$+V$	$-V$	$-V$	$+V$	$-V$	0	0	0	0	0	0	-1	0
x_{r-2}	0	0	0	0	0	0	0	0	0	$+V$	$+V$	$+V$	$-V$	$-V$	$+V$	$-V$	0	0	0	0	0	-2	$-V^2/7$
x_{r-3}	0	0	0	0	0	0	0	0	0	0	$+V$	$+V$	$+V$	$-V$	$-V$	$+V$	$-V$	0	0	0	0	-3	0
x_{r-4}	0	0	0	0	0	0	0	0	0	0	0	$+V$	$+V$	$+V$	$-V$	$-V$	$+V$	$-V$	0	0	0	-4	$-V^2/7$
x_{r-5}	0	0	0	0	0	0	0	0	0	0	0	0	$+V$	$+V$	$+V$	$-V$	$-V$	$+V$	$-V$	0	0	-5	0
x_{r-6}	0	0	0	0	0	0	0	0	0	0	0	0	0	$+V$	$+V$	$+V$	$-V$	$-V$	$+V$	$-V$	0	-6	$-V^2/7$
x_{r-7}	0	0	0	0	0	0	0	0	0	0	0	0	0	0	$+V$	$+V$	$+V$	$-V$	$-V$	$+V$	$-V$	-7	0

estimates and the state-space estimates. The inputs can be restricted by inequality constraints on the inputs themselves and/or the states, due to physical, economic or safety limits on the system operation. The procedure requires an initial estimate of the parameters and the initial states, which is obtainable from an experiment with a non-optimal perturbation signal.

Goodwin (1971) applied the procedure to designing optimal inputs for a steam-generating plant, for which a four-state model was available. The inputs were subject to three maximum flow constraints and there were two further constraints on the states resulting from plant-safety considerations. The results were compared with those from an experiment in which the inputs had been adjusted manually to drive the plant over a wide operating range, and a sixfold improvement was obtained in the trace of the parameter covariance matrix.

As with all aspects of optimal experiment design, the quality of results obtained using optimal test inputs depends on the similarity between the behaviour of the assumed model and that of the process itself. In the real world, the time and cost of experimentation have to intervene at some stage in the process of convergence towards the perfect result. The design of optimal test signals is discussed further in Chapter 3.

1.10 Pulse transfer function models and persistently exciting signals

In Application Example 1.4 a z-transfer function of the form of equation (1.52) with $n = 2$ was calculated from the input–output data using a Generalized Least Squares procedure (Clarke, 1967). The method, which can readily be formulated for recursive computation (Hastings-James and Sage, 1969), is one of several different algorithms for estimating the parameters of a z-transfer function (pulse transfer function) model. Under sometimes differing assumptions regarding model structure and noise structure, these include maximum likelihood estimation (Åström and Bohlin, 1965), instrumental variables (Young, 1970), least squares estimation from a cross-correlation function (Isermann and Baur, 1974) and stochastic approximation (Sakrison, 1966; Saridis and Stein, 1968). Comparisons of some of these techniques were made by Isermann *et al.* (1974) and Saridis (1974), and further details of these methods (and others) can be found in textbooks on system identification (Norton, 1986; Ljung, 1987; Söderström and Stoica, 1989).

The requirement for a perturbation signal in such parameter estimation is that it should be *persistently exciting*. Earlier in this chapter it was stated that this means effectively that its frequency content should adequately span the bandwidth of the system being tested. However, this is not a sufficiently rigorous definition of persistently exciting. In the extreme, a two-harmonic signal could be regarded as spanning the bandwidth of a system but not contain sufficient information to allow the parameters of a high-order model to be estimated.

The precise definition needs careful thought (Norton, 1986, Section 8.2.3) and the order of the model to be identified sets the requirement for the number of independent components which must be present in the signal. The definition given by Norton is that a sampled signal $\{u_r\}$ is persistently exciting of order n if the following three conditions apply:

1. The mean $\bar{u}$ should exist, where

$$\bar{u} = \lim_{N \to \infty} \frac{1}{N} \sum_{r=1}^{N} u_r$$

2. The autocovariance function $C_{uu}(k)$ should exist, where

$$C_{uu}(k) = \lim_{N \to \infty} \frac{1}{N} \sum_{r=1}^{N} (u_r - \bar{u})(u_{r+k} - \bar{u})$$

3. The $n \times n$ matrix with element i, j given by $C_{uu}(i - j)$ should be positive definite.

The pseudo-random signals described in this chapter possess these properties provided they are long enough. However, even then, it is sometimes necessary for further conditions to be satisfied, depending on the particular application. For example, as discussed in Section 1.7.2, Barker and his co-workers (1972, 1978) found that a binary inverse-repeat signal was not suitable for estimating the on-diagonal terms of the sequence of second-order kernel values of a Volterra series expansion of a non-linear system, whereas a signal based on a three-level m-sequence was suitable. Barker (1991) concludes that signals based on inverse-repeat binary sequences are not persistently exciting with respect to amplitude for the application. This point emphasizes the need to tailor perturbation signal design to the specific application in question.

2

Introduction to Perturbation Signals for Frequency-domain System Identification

Keith Godfrey

The increasing interest in system identification in the frequency domain has been mainly due to the computational efficiency of the Fast Fourier Transform algorithm. This is a discrete signal-processing algorithm and, as such, needs interpreting with considerable care. In this chapter we will first consider continuous (analog) signal processing in the frequency domain, then move on to discrete Fourier transformation and to a discussion of two types of signal especially designed for frequency-domain system identification. We will then describe linear system relationships in the frequency domain and conclude the chapter with a brief discussion of non-linear system relationships.

2.1 Continuous (analog) signal processing in the frequency domain

In the frequency domain it is necessary to distinguish carefully between non-periodic and periodic signals, and these will be considered in turn.

2.1.1 Non-periodic signals

Frequency-domain signal processing is based around the Fourier transform. For a non-periodic signal $x(t)$, the continuous Fourier transform is given by

$$X(\mathrm{j}\omega) = \int_{-\infty}^{\infty} x(t)\, \mathrm{e}^{-\mathrm{j}\omega t}\, \mathrm{d}t \tag{2.1a}$$

$$= \int_{-\infty}^{\infty} x(t) \cos \omega t \, \mathrm{d}t - \mathrm{j} \int_{-\infty}^{\infty} x(t) \sin \omega t \, \mathrm{d}t \tag{2.1b}$$

The signal $x(t)$ can be obtained from $X(\mathrm{j}\omega)$ by inverse Fourier transformation:

$$x(t) = \frac{1}{2\pi} \int_{-\infty}^{\infty} X(\mathrm{j}\omega)\, \mathrm{e}^{\mathrm{j}\omega t}\, \mathrm{d}\omega \tag{2.2}$$

The exact location of the multiplying factor $1/2\pi$ in the Fourier transform pair is not important, but the location in the inverse transform (equation (2.2)) is that most commonly used in the literature.

Also of considerable interest in frequency-domain system identification is the Fourier transform pair between the *power-spectral density* $S_{xx}(\omega)$ and the autocorrelation function $R_{xx}(\tau)$ of the signal:

$$S_{xx}(\omega) = \int_{-\infty}^{\infty} R_{xx}(\tau)\, \mathrm{e}^{-\mathrm{j}\omega\tau}\, \mathrm{d}\tau \tag{2.3}$$

$$R_{xx}(\tau) = \frac{1}{2\pi} \int_{-\infty}^{\infty} S_{xx}(\omega)\, \mathrm{e}^{\mathrm{j}\omega\tau}\, \mathrm{d}\omega \tag{2.4}$$

Since the autocorrelation function is an even function of τ, $S_{xx}(\omega)$ is a real function of ω and is most easily calculated from

$$S_{xx}(\omega) = 2 \int_{0}^{\infty} R_{xx}(\tau) \cos \omega\tau \, \mathrm{d}\tau \tag{2.5}$$

To determine the units of the power-spectral density we note from equation (2.4) that, for $\tau = 0$,

$$R_{xx}(0) = \frac{1}{2\pi} \int_{-\infty}^{\infty} S_{xx}(\omega)\, \mathrm{d}\omega = \int_{-\infty}^{\infty} S_{xx}(2\pi f)\, \mathrm{d}f \tag{2.6}$$

Recalling from Chapter 1 that $R_{xx}(0)$ is the mean squared value of $x(t)$,

$$E[x^2(t)] = \int_{-\infty}^{\infty} S_{xx}(2\pi f)\, \mathrm{d}f \tag{2.7}$$

Thus, if the units $x(t)$ are volts, then the units S_{xx} are volts^2 per cycle per second (i.e. volts^2/Hz). Integrating S_{xx} between any two cyclic frequencies f_1 and f_2 gives the power in the signal between those two frequencies.

To illustrate the theory, we will return to the two non-periodic signals considered in Examples 1C and 1D in Chapter 1.

EXAMPLE 2A POWER-SPECTRAL DENSITY OF A DISCRETE-INTERVAL RANDOM BINARY SIGNAL

It was found in Example 1C that a discrete-interval random binary signal with voltage levels $\pm V$ and clock pulse interval Δt has the autocorrelation function shown in Figure 1.11, which, for $\tau \geqslant 0$, is described by

$$R_{xx}(\tau) = \left(1 - \frac{\tau}{\Delta t}\right) . V^2, \quad 0 \leqslant \tau < \Delta t$$

$$= 0, \qquad \tau > \Delta t$$

Substituting in equation (2.5), the power-spectral density is given by

$$S_{xx}(\omega) = 2V^2 \int_0^{\Delta t} \left(1 - \frac{\tau}{\Delta t}\right) \cos \omega\tau \, d\tau$$

$$= 2V^2 \left[\frac{\sin \omega\tau}{\omega}\right]_0^{\Delta t} - \frac{2V^2}{\Delta t}\left[\frac{\tau \sin \omega\tau}{\omega}\right]_0^{\Delta t} + \frac{2V^2}{\Delta t}\int_0^{\Delta t} \frac{\sin \omega\tau}{\omega} \, d\tau$$

$$= \frac{2V^2}{\omega^2 \Delta t}[1 - \cos \omega\Delta t]$$

$$= V^2 . \Delta t \frac{\sin^2(\omega\Delta t/2)}{(\omega\Delta t/2)^2}$$

This is shown in Figure 2.1. Note that the power-spectral density is zero at the clock pulse frequency $f = 1/\Delta t$; it is also zero at integer multiples of the clock pulse frequency, i.e. $f = k/\Delta t$, except for $k = 0$.

The expression for $S_{xx}(\omega)$ is for positive and negative frequencies, and from the symmetry it is expected that half the power will be at negative frequencies and half at positive frequencies. To check this, consider the power for positive frequencies, which is given by

$$\int_0^{\infty} S_{xx}(2\pi f) \, df = V^2 \Delta t \int_0^{\infty} \frac{\sin^2(\pi . \Delta t . f)}{(\pi . \Delta t . f)^2} \, df$$

But

$$\int_0^{\infty} \frac{\sin^2 ax}{x^2} \, dx = \frac{a\pi}{2}$$

(Gradshteyn and Ryzhik, 1980, 3.821, No. 9), so that

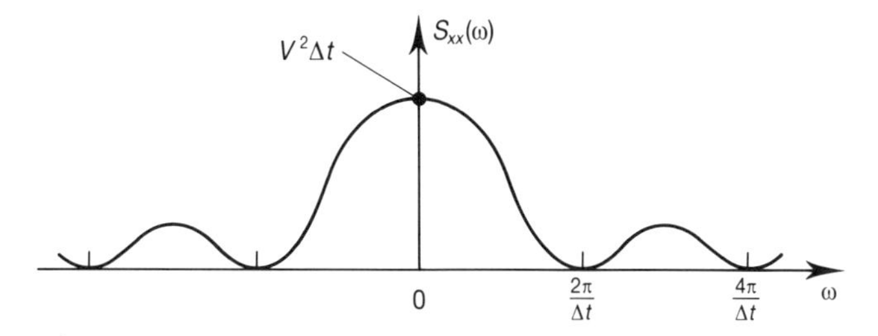

Figure 2.1 Power-spectral density of a discrete-interval random binary signal.

$$\int_0^\infty S_{xx}(2\pi f)\,\mathrm{d}f = V^2 . \Delta t . \frac{\pi}{2} . \frac{1}{\pi . \Delta t}$$

$$= \tfrac{1}{2}V^2, \text{ as expected}$$

For system identification, the assumption in the time domain that the system impulse response $h(t)$ does not vary appreciably over the width of the triangle of the autocorrelation function $R_{xx}(\tau)$ corresponds (approximately) in the frequency domain to the power-spectral density $S_{xx}(\omega)$ not varying appreciably over the bandwidth of the system being tested. The half-power point in the signal is thus of interest and this is reached at a (cyclic) frequency, f, at which

$$\frac{\sin^2(\pi . \Delta t . f)}{(\pi . \Delta t . f)^2} = \frac{1}{2}$$

i.e. at a frequency of approximately $0.443/\Delta t$ Hz.

EXAMPLE 2B POWER-SPECTRAL DENSITY OF A RANDOM BINARY SIGNAL

It was found in Example 1D that the autocorrelation function of a binary signal with voltage levels $\pm V$ and with zero crossings distributed according to a Poisson distribution with an average of v per unit time is of the form shown in Figure 1.13. For $\tau \geqslant 0$

$$R_{xx}(\tau) = V^2\,\mathrm{e}^{-2v\tau}$$

Substituting in equation (2.5), the power-spectral density is given by

$$\begin{aligned} S_{xx}(\omega) &= 2V^2 \int_0^\infty \mathrm{e}^{-2v\tau} \cos \omega\tau \,\mathrm{d}\tau \\ &= 2V^2\, Re\left\{\int_0^\infty \mathrm{e}^{-(2v+\mathrm{j}\omega)\tau}\,\mathrm{d}\tau\right\} \\ &= 2V^2\, Re\left\{\left[-\frac{\mathrm{e}^{-(2v+\mathrm{j}\omega)\tau}}{2v+\mathrm{j}\omega}\right]_0^\infty\right\} \end{aligned}$$

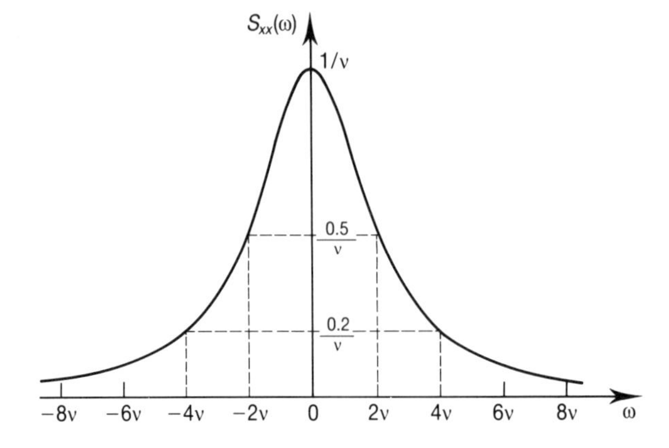

Figure 2.2 Power-spectral density of a random binary signal.

$$= 2V^2\, Re\left\{\frac{1}{2v + j\omega}\right\}$$

$$= \frac{4V^2 v}{4v^2 + \omega^2}$$

This is shown in Figure 2.2.

As in Example 2A, it is expected that half the power will be at negative frequencies and half at positive frequencies. To check this, the power-spectral density is integrated from 0 to infinity to give

$$\int_0^\infty S_{xx}(2\pi f)\, df = 4V^2 v \int_0^\infty \frac{1}{(2\pi f)^2 + 4v^2}\, df$$

$$= \frac{4V^2 v}{4\pi^2} \cdot \frac{\pi}{v}\left[\tan^{-1}\frac{\pi}{v} f\right]_0^\infty$$

$$= \frac{4V^2 v}{4\pi^2} \cdot \frac{\pi}{v} \cdot \frac{\pi}{2} = \tfrac{1}{2}V^2, \text{ as expected}$$

The half-power point in the signal is reached at a (radian) frequency $\omega = 2v$, i.e. at a (cyclic) frequency $f = (v/\pi)$ Hz.

2.1.2 Periodic signals

The Fourier transform pair of equations (2.1a) and (2.2) applies for a *non-periodic* continuous signal. If the signal is periodic, then the Fourier transform has to be replaced by a Fourier series. If the period of the signal is T s then the signal can contain power only at harmonically related cyclic frequencies $f = (k/T)$ Hz, where k is an integer.

The most familiar form of Fourier series is the trigonometric which, for positive and negative frequencies, is given by

$$x(t) = \sum_{k=-\infty}^{\infty} a_k \cos\left(\frac{2\pi kt}{T}\right) + \sum_{k=-\infty}^{\infty} b_k \sin\left(\frac{2\pi kt}{T}\right) \tag{2.8}$$

where

$$a_k = \frac{1}{T}\int_0^T x(t)\cos\left(\frac{2\pi kt}{T}\right) \mathrm{d}t, \quad k = 0, \pm 1, \pm 2, \ldots \tag{2.9a}$$

$$b_k = \frac{1}{T}\int_0^T x(t)\sin\left(\frac{2\pi kt}{T}\right) \mathrm{d}t, \quad k = \pm 1, \pm 2, \ldots \tag{2.9b}$$

(The coefficient b_0 is zero because $\sin 0 = 0$.)

At first sight, this does not appear much like the Fourier transforms of Section 2.1.1, but the comparison is much closer if the Fourier series is expressed in exponential form (see, for example, Balmer, 1991):

$$x(t) = \sum_{k=-\infty}^{\infty} X_k \exp\left(\mathrm{j}\,\frac{2\pi kt}{T}\right) \tag{2.10}$$

$$\begin{aligned} X_k &= \frac{1}{2}(a_k - \mathrm{j}b_k) \\ &= \frac{1}{T}\int_{-T/2}^{T/2} x(t)\exp\left(-\mathrm{j}\,\frac{2\pi kt}{T}\right) \mathrm{d}t, \quad k = 0, \pm 1, \pm 2, \ldots \end{aligned} \tag{2.11}$$

The range of integration must be one period of the signal, but it can start anywhere within that period. In equation (2.11) the range has been taken from $-\frac{1}{2}T$ to $+\frac{1}{2}T$ to increase the similarity with the Fourier transform of equation (2.1a). It is often possible to choose the location of the time origin to simplify the calculations by eliminating either the cosine terms for an odd function of time or the sine terms for an even function of time.

There is a similar Fourier series relationship between the autocorrelation function $R_{xx}(\tau)$ of a periodic signal and its power spectrum $S_{xx}(f = k/T)$, which is the amount of signal power at the frequency $f = (k/T)$ Hz. The spectrum is referred to as a line spectrum and the signal has no power at any other frequencies. If the signal $x(t)$ is a voltage signal, then the units of S_{xx} are volts^2.

Since $R_{xx}(\tau)$ is an even function of τ, only cosine terms need to be taken into account in the Fourier series, so that

$$S_{xx}(f = 0) = \frac{1}{T}\int_0^T R_{xx}(\tau)\,\mathrm{d}\tau = \frac{2}{T}\int_0^{T/2} R_{xx}(\tau)\,\mathrm{d}\tau \tag{2.12}$$

$$S_{xx}\left(f=\frac{k}{T}\right)=\frac{1}{T}\int_0^T R_{xx}(\tau)\cos\left(\frac{2\pi k\tau}{T}\right)\mathrm{d}\tau$$

$$=\frac{2}{T}\int_0^{T/2} R_{xx}(\tau)\cos\left(\frac{2\pi k\tau}{T}\right)\mathrm{d}\tau,\quad k=\pm 1,\pm 2,\ldots \tag{2.13}$$

In the next two examples the power spectrum of the two periodic signals whose autocorrelation functions were found in Examples 1A and 1B will be determined, then the power spectrum of a PRBS will be evaluated.

EXAMPLE 2C POWER SPECTRUM OF A SINE WAVE

This may seem a rather trivial example because the result is so well known, but it will help to consolidate some of the relationships developed in this section. Consider the sine wave:

$$x(t)=V\sin(\omega t+\varphi)$$

which has period $T=2\pi/\omega$. The mean squared value of the signal is $\frac{1}{2}V^2$ and if we are considering both positive and negative frequencies (as we have up to now), then $\frac{1}{4}V^2$ is at the (radian) frequency $-\omega$ and $\frac{1}{4}V^2$ is at (radian) frequency $+\omega$.

It was found in Example 1A that

$$R_{xx}(\tau)=\frac{1}{2}V^2\cos\omega\tau$$

so that, from equation (2.13),

$$S_{xx}\left(f=\frac{k\omega}{2\pi}\right)=\frac{\omega}{2\pi}\int_0^{2\pi/\omega}\frac{1}{2}V^2\cos\omega\tau\,.\cos\left(\frac{2\pi k\tau\omega}{2\pi}\right)\mathrm{d}\tau$$

$$=\frac{\omega V^2}{4\pi}\int_0^{2\pi/\omega}\cos\omega\tau\,.\cos k\omega\tau\,\mathrm{d}\tau$$

$$=0\text{ except for }k=\pm 1 \tag{2.14a}$$

For $k=1$

$$S_{xx}\left(f=\frac{\omega}{2\pi}\right)=\frac{\omega V^2}{4\pi}\int_0^{2\pi/\omega}\frac{1}{2}(1+\cos 2\omega\tau)\,\mathrm{d}\tau$$

$$=\frac{\omega V^2}{8\pi}\left[\tau+\frac{\sin 2\omega\tau}{2\omega}\right]_0^{2\pi/\omega}$$

$$=\frac{1}{4}V^2 \tag{2.14b}$$

Similarly,

$$S_{xx}\left(f=-\frac{\omega}{2\pi}\right)=\frac{1}{4}V^2 \tag{2.14c}$$

EXAMPLE 2D POWER SPECTRUM OF A SQUAREWAVE

Consider now a squarewave with voltage levels $\pm V$ and period T. The autocorrelation function $R_{xx}(\tau)$ is as shown in Figure 1.9, and substituting in equations (2.12) and (2.13) to find the power spectrum, we obtain

$$S_{xx}(f=0)=0$$

$$\begin{aligned}
S_{xx}(f=k/T) &= \frac{2}{T}\int_0^{T/2} V^2\left(1-\frac{4\tau}{T}\right)\cos\left(\frac{2\pi k\tau}{T}\right)\mathrm{d}\tau, \quad k=\pm 1, \pm 2, \pm 3, \ldots \\
&= \frac{2V^2}{T}\cdot\frac{T}{2\pi k}\left\{\left[\sin\left(\frac{2\pi k\tau}{T}\right)\right]_0^{T/2}\right. \\
&\quad \left. -\frac{4\tau}{T}\left[\sin\left(\frac{2\pi k\tau}{T}\right)\right]_0^{T/2} + \frac{4}{T}\int_0^{T/2}\sin\left(\frac{2\pi k\tau}{T}\right)\mathrm{d}\tau\right\} \\
&= \frac{2V^2}{T}\cdot\frac{4}{T}\cdot\left(\frac{T}{2\pi k}\right)^2\left[-\cos\left(\frac{2\pi k\tau}{T}\right)\right]_0^{T/2} \\
&= \frac{2V^2}{(\pi k)^2}(1-\cos\pi k)
\end{aligned}$$

Thus

$$S_{xx}\left(f=\frac{k}{T}\right)=\frac{4V^2}{(\pi k)^2} \quad \text{for } k \text{ odd} \tag{2.15a}$$

$$= 0 \text{ for } k \text{ even} \tag{2.15b}$$

The Fourier series for a squarewave is a familiar one. If the time origin is taken at one of the transitions of $x(t)$ from $-V$ to $+V$ (as in the uppermost waveform of Figure 1.8), then the waveform is an odd function of time and its Fourier series is

$$x(t)=\frac{4V}{\pi}\left(\sin\frac{2\pi t}{T}+\frac{1}{3}\sin\frac{6\pi t}{T}+\frac{1}{5}\sin\frac{10\pi t}{T}+\ldots\right)$$

It is necessary to exercise some care here, because the Fourier series representation of a waveform is traditionally for non-negative frequencies only. Thus the power in harmonic k (where k is odd) is

$$\frac{16V^2}{\pi^2}\cdot\frac{1}{2}\cdot\frac{1}{k^2}=\frac{8V^2}{(\pi k)^2}$$

i.e. the sum of the power spectrum at harmonics $-k$ and $+k$. Note that since

$$1+\frac{1}{9}+\frac{1}{25}+\frac{1}{49}+\ldots=\frac{\pi^2}{8}$$

the total power in the harmonics is

$$\frac{16V^2}{\pi^2} \cdot \frac{1}{2} \cdot \frac{\pi^2}{8} = V^2$$

i.e. the total power in the Fourier series harmonics is the same as that in the waveform itself, as expected.

The power spectrum of a squarewave decreases too rapidly with increasing k for the waveform to be useful in most frequency-domain identification applications.

In the example above it was possible to determine the power spectrum both by finding the power in each harmonic of the Fourier series of the waveform itself and by finding the Fourier series of the autocorrelation function. However, many waveforms have no particularly simple form of Fourier series, but their autocorrelation function is relatively simple in form, so that the power spectrum can be determined via this latter route. This applies to a PRBS, which is the subject of the next example.

EXAMPLE 2E POWER SPECTRUM OF A PRBS

The autocorrelation function $R_{xx}(\tau)$ of a pseudo-random binary signal of period $T = N.\Delta t$ and with voltage levels $\pm V$ is as shown in Figure 1.16. To determine S_{xx}, it is necessary to characterize $R_{xx}(\tau)$ algebraically over the time range from $\tau = 0$ to $\tau = \frac{1}{2}T$. Within this range, the autocorrelation function is most easily described by the sum of a d.c. offset of $-V^2/N$ throughout the range and a triangle of height $V^2(N+1)/N$ starting at $\tau = 0$ and decreasing linearly to zero at $\tau = \Delta t$. Hence within the range from $\tau = 0$ to $\tau = \frac{1}{2}N.\Delta t$,

$$\begin{aligned} R_{xx}(\tau) &= \frac{N+1}{N} V^2 \left(1 - \frac{\tau}{\Delta t}\right) - \frac{V^2}{N}, \qquad 0 \leqslant \tau < \Delta t \\ &= -\frac{V^2}{N}, \qquad \Delta t \leqslant \tau < \frac{1}{2} N.\Delta t \end{aligned}$$

Substituting in equation (2.12),

$$\begin{aligned} S_{xx}(0) &= \frac{2V^2}{N^2 \Delta t} \left\{ \int_0^{\Delta t} (N+1)\left(1 - \frac{\tau}{\Delta t}\right) d\tau - \int_0^{N.\Delta t/2} 1 \, d\tau \right\} \\ &= \frac{2V^2}{N^2.\Delta t} \left[(N+1)\,\Delta t - \frac{1}{2}(N+1).\Delta t - \frac{1}{2} N.\Delta t \right] \\ &= V^2/N^2 \end{aligned} \tag{2.16a}$$

This was expected from the signal itself, which has a d.c. offset of either $+V/N$ or $-V/N$, depending on the way that the transformation has been made from the sequence logic values (0 and 1) to the signal voltage levels ($\pm V$).

The power in the remaining harmonics at frequencies $f = k/T$, $k = \pm 1, \pm 2, \pm 3, \ldots$ is then obtained from equation (2.13):

$$S_{xx}\left(f=\frac{k}{T}\right)=\frac{2V^2}{N^2.\Delta t}\left\{\int_0^{\Delta t}(N+1)\left(1-\frac{\tau}{\Delta t}\right)\cos\left(\frac{2\pi k\tau}{N.\Delta t}\right)\mathrm{d}\tau\right.$$

$$\left.-\int_0^{N.\Delta t/2}\cos\left(\frac{2\pi k\tau}{N.\Delta t}\right)d\tau\right\}$$

$$=\frac{2V^2}{N^2.\Delta t}\left\{(N+1).\frac{N.\Delta t}{2\pi k}\left[\sin\left(\frac{2\pi k\tau}{N.\Delta t}\right)\right]_0^{\Delta t}\right.$$

$$-(N+1).\frac{N.\Delta t}{2\pi k}\left[\frac{\tau}{\Delta t}\sin\left(\frac{2\pi k\tau}{N.\Delta t}\right)\right]_0^{\Delta t}$$

$$+(N+1).\frac{N.\Delta t}{2\pi k}.\frac{1}{\Delta t}\int_0^{\Delta t}\sin\left(\frac{2\pi k\tau}{N.\Delta t}\right)\mathrm{d}\tau$$

$$\left.-\frac{N.\Delta t}{2\pi k}\left[\sin\left(\frac{2\pi k\tau}{N.\Delta t}\right)\right]_0^{N.\Delta t/2}\right\}$$

The first two terms on the right-hand side cancel each other out, while the fourth term is zero at both limits of integration (since k *is* an integer), leaving only the third term:

$$S_{xx}\left(f=\frac{k}{T}\right)=\frac{2V^2}{N^2.\Delta t}.\frac{(N+1)}{\Delta t}.\left(\frac{N.\Delta t}{2\pi k}\right)^2\left[-\cos\left(\frac{2\pi k\tau}{N.\Delta t}\right)\right]_0^{\Delta t}$$

$$=\frac{V^2}{2}.\frac{N+1}{(\pi k)^2}\left[1-\cos\frac{2\pi k}{N}\right]$$

$$=V^2.\frac{N+1}{N^2}\frac{\sin^2(\pi k/N)}{(\pi k/N)^2},\quad k=\pm 1,\ \pm 2,\ \ldots \tag{2.16b}$$

This is illustrated in Figure 2.3 for a PRBS with $N=15$. As N becomes large, $S_{xx}(0)\rightarrow 0$, while the power in the first harmonic ($k=+1$) tends to V^2/N. The harmonic spacing ($1/N.\Delta t$) also gets smaller. It is interesting to note that the power-spectral density of the discrete interval random binary signal considered in Example 2A is, for small values of ω, given by

$$S_{xx}(\omega)\simeq V^2.\Delta t$$

so that the area under the power-spectral density between

$$f=\frac{1}{2N.\Delta t}\quad\text{and}\quad f=\frac{3}{2N.\Delta t}$$

is approximately V^2/N, i.e. the same as that of a PRBS (with large N) at the first harmonic,

$$f=\frac{1}{N.\Delta t}$$

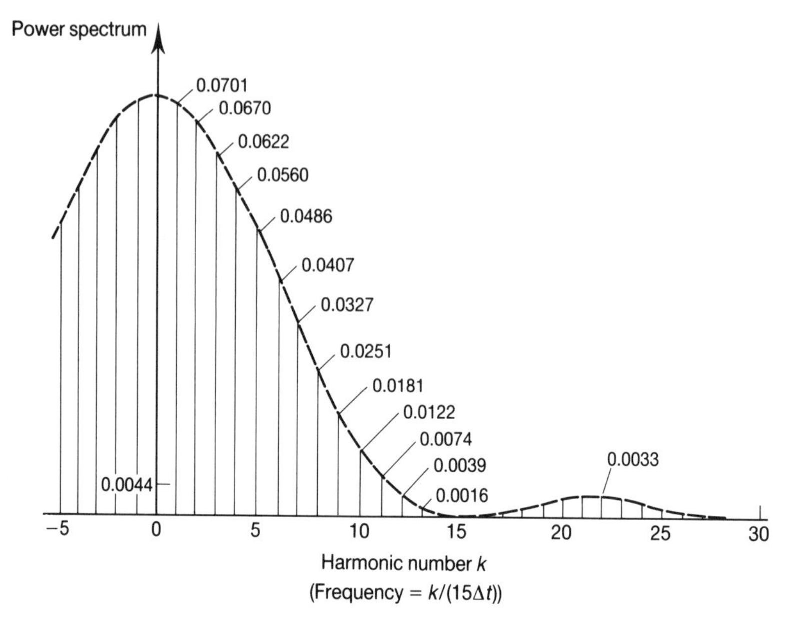

Figure 2.3 Power spectrum of a PRBS with voltage levels $\pm V$ and $N = 15$.

This example has illustrated that the power spectrum of a PRBS is spread over a large number of harmonics, with a comparatively small amount of power in each of them.

2.2 Discrete signal processing in the frequency domain

Most signal processing is now done digitally. Since most signals are continuous in time, the first step in the procedure is to sample the signal $x(t)$ at regular intervals of time, Δ, to form a sequence x_r, $r = 0, 1, \ldots, (N-1)$. The segment, or block, of signal which is digitized is thus of length $T_0 = N.\Delta$.

2.2.1 Discrete Fourier Transform (DFT)

The Discrete Fourier Transform X_k is then obtained from x_r:

$$X_k = \frac{1}{N}\sum_{r=0}^{N-1} x_r \exp\left(-\mathrm{j}\,\frac{2\pi kr}{N}\right) \tag{2.17a}$$

$$= \frac{1}{N}\sum_{r=0}^{N-1} x_r\left(\cos\frac{2\pi kr}{N} - \mathrm{j}\sin\frac{2\pi kr}{N}\right) \tag{2.17b}$$

where $k = 0, 1, \ldots, (N-1)$.

Thus X_k is a sequence of N complex numbers giving an approximation to a sampled version of the continuous Fourier transform $X(j\omega)$ at harmonically related frequencies $k/N.\Delta$ Hz, where k is an integer. Because of the cyclic properties of cosine and sine, the sequence X_k is cyclic with period $1/\Delta$ Hz, referred to as the *repeat frequency* of the DFT. Further, the magnitude of X_k is an even function of frequency around the frequency $1/2\Delta$ Hz, while the phase of X_k is an odd function around this frequency; the frequency $1/2\Delta$ Hz is referred to as the *folding frequency*. The useful information in the DFT is contained in the harmonics $0, 1/N\Delta, 2/N\Delta, \ldots, 1/2\Delta$ Hz; no additional information is obtained by examining higher harmonics. It is important to note that this folding effect does not occur in the continuous Fourier transform $X(j\omega)$.

The time sequence, x_r, can be obtained from X_k by the inverse Discrete Fourier Transform:

$$x_r = \sum_{k=0}^{N-1} X_k \exp\left(j\frac{2\pi kr}{N}\right) \tag{2.18}$$

where $r = 0, 1, \ldots, (N-1)$. Note that if r is continued above $(N-1)$, x_r is a periodic sequence $(x_{r+N} = x_r)$ even if the original continuous time function $x(t)$ is not. The exact location of the factor $1/N$ in the DFT pair (equations (2.17) and (2.18)) is variable in the literature, but there is an increasing trend to place it as shown above. It is, of course, only an amplitude-scaling factor.

The discrete power spectrum $S_{xx}(k)$ of a signal $x(t)$ can be found by taking the DFT of the autocorrelation function $R_{xx}(\tau)$ which has been sampled at regular intervals Δ to produce a sequence $R_{xx}(r)$, $r = 0, 1, \ldots, (N-1)$:

$$S_{xx}(k) = \frac{1}{N}\sum_{r=0}^{N-1} R_{xx}(r)\exp\left(-j\frac{2\pi kr}{N}\right) \tag{2.19a}$$

Since $R_{xx}(r)$ is symmetrical about $r = 0$, $S_{xx}(k)$ is a sequence of real numbers and can be expressed as:

$$S_{xx}(k) = \frac{1}{N}\sum_{r=0}^{N-1} R_{xx}(r)\cos\frac{2\pi kr}{N} \tag{2.19b}$$

The corresponding inverse DFT is:

$$R_{xx}(r) = \sum_{k=0}^{N-1} S_{xx}(k)\exp\left(\frac{2\pi kr}{N}\right) \tag{2.20}$$

which is periodic with period $N\Delta$ even if the original continuous autocorrelation function $R_{xx}(\tau)$ is not.

Equation (2.20) leads to an alternative way of computing discrete power spectra. From Chapter 1 we know that

$$R_{xx}(r) = \frac{1}{N}\sum_{i=0}^{N-1} x_i . x_{i+r} \tag{2.21}$$

But from equation (2.18)

$$x_{i+r} = \sum_{k=0}^{N-1} X_k \exp\left(j \frac{2\pi k(i+r)}{N}\right) \tag{2.22}$$

Hence from equations (2.21) and (2.22),

$$\begin{aligned} R_{xx}(r) &= \frac{1}{N} \sum_{i=0}^{N-1} x_i \left[\sum_{k=0}^{N-1} X_k \exp\left(j \frac{2\pi k i}{N}\right) . \exp\left(j \frac{2\pi k r}{N}\right)\right] \\ &= \sum_{k=0}^{N-1} \left\{ \left[\frac{1}{N} \sum_{i=0}^{N-1} x_i \exp\left(j \frac{2\pi k i}{N}\right) X_k \exp\left(j \frac{2\pi k r}{N}\right)\right]\right\} \\ &= \sum_{k=0}^{N-1} X_k^* X_k \exp\left(j \frac{2\pi k r}{N}\right) \end{aligned} \tag{2.23}$$

where X_k^* denotes the complex conjugate of X_k. Comparing equations (2.20) and (2.23), it is seen that

$$S_{xx}(k) = X_k^* X_k \tag{2.24}$$

This is the approach to computing power spectra used in most modern digital spectrum analyzers.

For a non-periodic signal $x(t)$, the discrete power spectrum $S_{xx}(k)$ provides an estimate of the continuous power-spectral density $S_{xx}(f)$ at the cyclic frequency $f = k/N.\Delta$, subject to a scaling factor. Since the frequency interval in the DFT is $1/N.\Delta$, we are effectively obtaining an integral of the continuous power-spectral density over a frequency interval of $1/N.\Delta$, so that the discrete spectrum has to be multiplied by $N.\Delta = T_0$ to give us the estimate of the power-spectral density. Thus

$$\hat{S}_{xx}\left(f = \frac{k}{N.\Delta}\right) = T_0 . S_{xx}(k) \tag{2.25}$$

For a periodic signal $x(t)$, the scaling factor of T_0 is not needed, with the discrete spectrum providing us with an estimate of the continuous power spectrum, provided the segment (block) length T_0 is an integer multiple of the signal period. If this proviso is not met, then spectral leakage occurs; this will be discussed further in Section 2.2.4.

There are several factors which need to be taken into account when computing the DFT and then interpreting it in terms of a corresponding continuous (analog) quantity, and the most important of these will be considered in Sections 2.2.2–2.2.4.

2.2.2 Averaging

We have seen that if the length of the data segment is $T_0 = N.\Delta$, then frequency information is available at harmonically related cyclic frequencies $f = k/N.\Delta$, where k is an integer. It is very easy to think that the variance of estimates of the true (analog) power spectrum will be improved by increasing the record length, but, somewhat surprisingly, this is not so. The accuracy of a spectral measurement depends on the product of the effective bandwidth of the measurement and the record length (Newland, 1984), so that, as T_0 becomes larger, the spectral bandwidth decreases, while the product remains the same.

The solution is to choose T_0 such that the frequency interval $1/T_0$ is sufficiently small for us to resolve the power spectrum in enough detail for the particular requirement and then to average several segments (blocks) of data, each of length T_0; this procedure does reduce the variance of the spectral estimates.

Modern digital spectrum analyzers all make use of the Fast Fourier Transform (FFT), which is simply the DFT with a particular choice of N (Brigham, 1974, 1988). Because of the properties of sine and cosine, the DFT algorithm can be made very efficient for certain values of N, with the most efficient choice being $N = 2^n$, where n is an integer. A typical value of N in a spectrum analyzer is 1024.

In situations where only a limited amount of data is available it is still possible to benefit from averaging by allowing some overlapping of segments. Obviously, the amount of overlap must not be too great, otherwise the segments are almost identical, but in many situations, overlapping by up to 60% can prove beneficial (Carter *et al.*, 1973). The Advantest TR9403 is an example of a modern commercial digital spectrum analyzer, and this has a facility for 50% overlapping of segments. Overlapping is considered further by Patton, Miles and Taylor in Chapter 10, in the context of helicopter model validation, where limited overall record length often proves a problem.

2.2.3 Aliasing

When sampling a signal at regular intervals Δ, power occurring in the continuous signal at frequencies above the Nyquist (Shannon) frequency $f = 1/2\Delta$ appears at lower frequencies in the DFT. The effect is known as aliasing and occurs unless $x(t)$ is band-limited with a maximum frequency less than $1/2\Delta$.

For this reason, anti-aliasing filters are incorporated into the input channels of modern digital spectrum analyzers. They are low-pass filters with very low (hopefully, negligible) gain at frequencies of $1/2\Delta$ or above. Inevitably they have some effect on frequencies below the folding frequency in the DFT, and for this reason the maximum frequency displayed on an analyzer is rather less than $1/2\Delta$. Taking two modern analyzers as examples, both the Hewlett Packard 5420B and the Advantest TR9403 work with $N = 1024$, so that folding corresponds to frequency point 512. The 5420B displays 256 of these points, while the TR9403 displays 400. In both cases it is considered that the anti-aliasing filter has negligible effect on both the gain and phase of estimates of continuous quantities up to the following values:

$$\left(\frac{256}{1024\Delta} = \frac{0.25}{\Delta} \text{ for the 5420B and } \frac{400}{1024\Delta} = \frac{0.390625}{\Delta} \text{ for the TR9403}\right)$$

An anti-aliasing filter is automatically included when using either of these signal analyzers, so that it is not possible to illustrate the effects of aliasing using them. It is, nevertheless, possible to illustrate the problem theoretically and to do this, we will consider the power spectrum of a pseudo-random binary signal of length 7 clock pulse intervals. We will see in Section 2.2.4 that spectral leakage will occur unless the period $N.\Delta$ of the DFT is an integer multiple of the period of the signal ($7\Delta t$ in this case). We will therefore make the period of the DFT the same as that of the signal and consider the following cases: (1) $N = 7$, $\Delta = \Delta t$; (2) $N = 14$, $\Delta = 0.5\Delta t$; (3) $N = 28$; $\Delta = 0.25\Delta t$. With these choices, the differences between the digital power spectrum and the analog power spectrum are due to aliasing and not due to leakage.

EXAMPLE 2F DISCRETE POWER SPECTRUM OF A PSEUDO-RANDOM BINARY SIGNAL

From Example 2E we know that the (continuous) power spectrum of a PRBS with period $T = N.\Delta t$ and levels $\pm V$ is given by equations (2.16a) and (2.16b).

Values of this power spectrum are listed for k up to 14 in the left-hand column of Table 2.1. Let us now consider in turn the discrete power spectrum for the three cases listed above.

Table 2.1 Continuous and discrete power spectra of a PRBS with voltage levels $\pm V$ and period $7\Delta t$

k	Continuous $S_{xx}(f)$	Discrete $S_{xx}(f)$		
		$N = 7$	$N = 14$	$N = 28$
0	$0.02041V^2$	$0.02041V^2$	$0.02041V^2$	$0.02041V^2$
± 1	$0.15259V^2$	$0.16327V^2$	$0.15518V^2$	$0.15323V^2$
± 2	$0.12387V^2$	$0.16327V^2$	$0.13253V^2$	$0.12597V^2$
± 3	$0.08560V^2$	$0.16327V^2$	$0.09980V^2$	$0.08891V^2$
± 4	$0.04815V^2$	$0.16327V^2$	$0.06347V^2$	$0.05152V^2$
± 5	$0.01982V^2$	$0.16327V^2$	$0.03074V^2$	$0.02204V^2$
± 6	$0.00424V^2$	$0.16327V^2$	$0.00808V^2$	$0.00494V^2$
± 7	0.00000	$0.02041V^2$	0.00000	0.00000
± 8	$0.00238V^2$	$0.16327V^2$	$0.00808V^2$	$0.00315V^2$
± 9	$0.00612V^2$	$0.16327V^2$	$0.03074V^2$	$0.00870V^2$
± 10	$0.00770V^2$	$0.16327V^2$	$0.06347V^2$	$0.01195V^2$
± 11	$0.00637V^2$	$0.16327V^2$	$0.09980V^2$	$0.01089V^2$
± 12	$0.00344V^2$	$0.16327V^2$	$0.13253V^2$	$0.00656V^2$
± 13	$0.00090V^2$	$0.16327V^2$	$0.15518V^2$	$0.00190V^2$
± 14	0.00000	$0.02041V^2$	$0.02041V^2$	0.00000

(i) Sampling Interval $\Delta = \Delta t$: Number of Samples = 7 In the first part of this example the discrete power spectrum will be calculated first by using equation (2.24) and then by using equation (2.19b), to confirm that they give the same result. When calculating a discrete power spectrum, provided the segment length is an integer number of periods of the signal (as it is in this case), the starting point in the signal is immaterial. It is convenient computationally to take the sequence x_r, $r = 0, 1, \ldots, 6$ as:

$$x_r = \{+V, -V, -V, +V, -V, +V, +V\}$$

From equation (2.17b), the corresponding DFT is given by:

$$X_k = \frac{1}{7}\sum_{r=0}^{6} x_r\left(\cos\frac{2\pi kr}{7} - \mathrm{j}\sin\frac{2\pi kr}{7}\right)$$

$$= \frac{V}{7}\left[1 - \cos\frac{2\pi k}{7} - \cos\frac{4\pi k}{7} + \cos\frac{6\pi k}{7} - \cos\frac{8\pi k}{7} + \cos\frac{10\pi k}{7} + \cos\frac{12\pi k}{7}\right.$$

$$\left. - \mathrm{j}\left(-\sin\frac{2\pi k}{7} - \sin\frac{4\pi k}{7} + \sin\frac{6\pi k}{7} - \sin\frac{8\pi k}{7} + \sin\frac{10\pi k}{7} + \sin\frac{12\pi k}{7}\right)\right]$$

This gives

$$X_0 = \frac{V}{7}, \qquad X_0^* = \frac{V}{7}, \qquad X_0^*X_0 = \frac{V^2}{49}$$

$$X_1 = \frac{V}{7}(1 + \mathrm{j}\,2.6456), \quad X_1^* = \frac{V}{7}(1 - \mathrm{j}\,2.6456), \quad X_1^*X_1 = \frac{8V^2}{49}$$

$$X_2 = \frac{V}{7}(1 + \mathrm{j}\,2.6456), \quad X_2^* = \frac{V}{7}(1 - \mathrm{j}\,2.6456), \quad X_2^*X_2 = \frac{8V^2}{49}$$

$$X_3 = \frac{V}{7}(1 - \mathrm{j}\,2.6456), \quad X_3^* = \frac{V}{7}(1 + \mathrm{j}\,2.6456), \quad X_3^*X_3 = \frac{8V^2}{49}$$

The DFT folds around the half-way point, with the magnitude being an even function of frequency and the phase an odd function, and it is readily confirmed that

$$X_4 = \frac{V}{7}(1 + \mathrm{j}\,2.6456)$$

and

$$X_5 = X_6 = \frac{V}{7}(1 - \mathrm{j}\,2.6456)$$

so that

$$X_4^*X_4 = X_5^*X_5 = X_6^*X_6 = \frac{8V^2}{49}$$

The autocorrelation sequence is:

$$R_{xx}(r) = \left\{V^2, -\frac{V^2}{7}, -\frac{V^2}{7}, -\frac{V^2}{7}, -\frac{V^2}{7}, -\frac{V^2}{7}, -\frac{V^2}{7}\right\}$$

so that, from equation (2.19b),

$$S_{xx}(k) = \frac{1}{7}\sum_{r=0}^{6} R_{xx}(r)\cos\frac{2\pi kr}{7}, \quad k = 0, 1, \ldots, 6$$

This gives:

$$S_{xx}(0) = \frac{V^2}{7}\left(1 - \frac{6}{7}\right) = \frac{V^2}{49}$$

$$S_{xx}(1) = \frac{V^2}{7}\left[1 - \frac{2}{7}(0.6235 - 0.2225 - 0.9010)\right] = \frac{8V^2}{49}$$

$$S_{xx}(2) = \frac{V^2}{7}\left[1 - \frac{2}{7}(-0.2225 - 0.9010 + 0.6235)\right] = \frac{8V^2}{49}$$

$$S_{xx}(3) = \frac{V^2}{7}\left[1 - \frac{2}{7}(-0.9010 + 0.6235 - 0.2225)\right] = \frac{8V^2}{49}$$

With the folding around the mid-point, $S_{xx}(4) = S_{xx}(5) = S_{xx}(6) = 8V^2/49$.

These calculations have confirmed that $S_{xx}(k) = X_k^* X_k$ (equation (2.24)), and we also see that the total power (V^2) in the signal is contained in the discrete power spectrum at the harmonics corresponding to $k = 0, \pm 1, \pm 2$ and ± 3. The d.c. value of the discrete spectrum is the same as that of the continuous spectrum, but the rest of the discrete spectrum bears little resemblance to the continuous spectrum. Since the autocorrelation sequence $R_{xx}(r)$ is constant for $r = 1, 2, \ldots, 6$, the discrete power spectrum is a constant for $k = 1, 2, \ldots, 6$. (This is to be expected because the DFT of an impulsive sequence (1 in position $r = 0$ and 0 in positions $r = 1$ to $(N - 1)$) is simply a constant value of $1/N$ for all k from 0 to $(N - 1)$. Here, the only difference is that the constant value of the autocorrelation sequence is non-zero for $r = 0$.)

In the continuous spectrum, only 74.45% of the power lies in the harmonics corresponding to $k = 0, \pm 1, \pm 2$ and ± 3. The remaining 25.55% is at harmonics with $k > 3$ or $k < -3$, and this contributes to the aliasing in the discrete power spectrum.

It is important to emphasize that without an anti-aliasing filter the discrete power spectrum of a PRBS when the sampling interval Δ is made the same as the clock pulse interval Δt does not provide a reasonable estimate of the continuous power spectrum, except at d.c.

(ii) Sampling Interval $\Delta = 0.5\Delta t$: Number of Samples = 14 The situation can be improved by sampling twice per clock pulse interval. The autocorrelation sequence is now

$$R_{xx}(r) = V^2\left\{1, \frac{3}{7}, -\frac{1}{7}, -\frac{1}{7}, -\frac{1}{7}, -\frac{1}{7}, -\frac{1}{7}, -\frac{1}{7}, -\frac{1}{7}, -\frac{1}{7}, -\frac{1}{7}, -\frac{1}{7}, -\frac{1}{7}, \frac{3}{7}\right\}$$

and the corresponding discrete power spectrum is

$$S_{xx}(k) = \frac{V^2}{14}\left\{1 + \frac{6}{7}\cos\left(\frac{2\pi k}{14}\right) - \frac{2}{7}\left[\cos\frac{4\pi k}{14} + \cos\frac{6\pi k}{14} + \cos\frac{8\pi k}{14}\right.\right.$$
$$\left.\left. + \cos\frac{10\pi k}{14} + \cos\frac{12\pi k}{14}\right] - \frac{1}{7}\cos\pi k\right\}$$

This is shown in the fourth column of Table 2.1. All of the signal power is in the discrete power spectrum at harmonics corresponding to $k = 0, \pm 1, \pm 2, \pm 3, \pm 4, \pm 5$ and ± 6. As before, the discrete and continuous values of $S_{xx}(0)$ are identical, and the other values in the range $-6 \leqslant k \leqslant 6$ are in better agreement than in (i). This time, 88.90% of the signal power lies in continuous spectrum harmonics in this range, leaving 11.10% to contribute to aliasing in the discrete power spectrum.

(iii) Sampling Interval $\Delta = 0.25\Delta t$: Number of Samples = 28 If we sample four times per clock pulse interval, the autocorrelation sequence $R_{xx}(r)$ is now V^2 at $r = 0$, $5V^2/7$ at $r = 1$ and 27, $3V^2/7$ at $r = 2$ and 26, $V^2/7$ at $r = 3$ and 25 and $-V^2/7$ for $4 \leqslant r \leqslant 24$. The corresponding discrete power spectrum is shown in the fifth column of Table 2.1. All of the signal power is contained in the discrete power spectrum at harmonics corresponding to k in the range $-13 \leqslant k \leqslant 13$, while 94.28% of the signal power is contained in the continuous power spectrum at these harmonics, leaving only 5.72% to contribute to the aliasing of the discrete spectrum. The agreement between the continuous and discrete power spectrum is improving, and it would continue to do so as the sampling interval is reduced, leaving less of the total signal power to contribute to the discrete spectrum aliasing. The continuous power spectrum and the three discrete power spectra ($N = 7$, 14 and 28) are shown in Figure 2.4.

2.2.4 Spectral leakage and windows

We saw in Section 2.2.2 that the variance of spectral estimates can be reduced by averaging estimates from segments (blocks) of data, each of length $T_0 = N.\Delta$. The finite length of the segments brings about a further problem if the signal being processed is not periodic with period T_0/K, where K is an integer, i.e. its (continuous) power spectrum does not have lines spaced by K/T_0 in the frequency domain.

Consider a segment of data obtained by multiplying the signal $x(t)$ by a rectangular function $w_R(t)$, given by

$$w_R(t) = 1, \qquad -\tfrac{1}{2}T_0 \leqslant t < \tfrac{1}{2}T_0 \tag{2.26a}$$

$$= 0, \qquad \text{for other values of } t \tag{2.26b}$$

as shown in Figure 2.5.

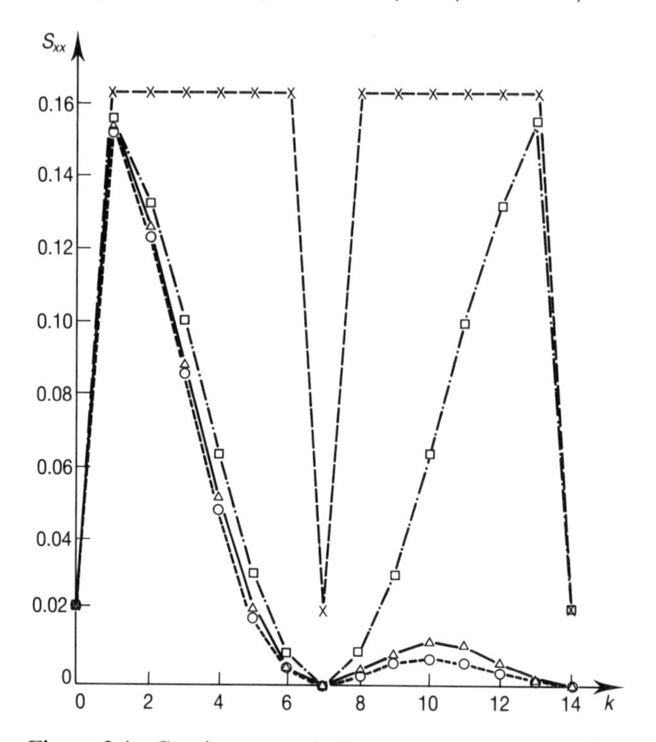

Figure 2.4 Continuous and discrete power spectra of a PRBS with voltage levels ± 1 and period $7\Delta t$. o Continuous power spectrum, x discrete power spectrum, $N = 7$, □ discrete power spectrum, $N = 14$, Δ discrete power spectrum, $N = 28$.

The multiplication $x(t).w_R(t)$ in the time domain becomes the convolution $X(j\omega) * W_R(\omega)$ of the Fourier transforms of $x(t)$ and $w_R(t)$ in the frequency domain. From equations (2.26a) and (2.26b) we obtain

$$W_R(\omega) = 2 \int_0^{T_0/2} \cos \omega t \; dt \tag{2.27}$$

$$= T_0 \, . \, \frac{\sin \omega T_0/2}{\omega T_0/2} \tag{2.28}$$

so that

$$W_R(f) = T_0 \, . \, \frac{\sin \pi f T_0}{\pi f T_0} \tag{2.29}$$

which is shown in Figure 2.6.

Due to the convolution, values of $X(j\omega)$ occurring at frequencies well away from that under consideration may contribute to the Discrete Fourier Transform at that

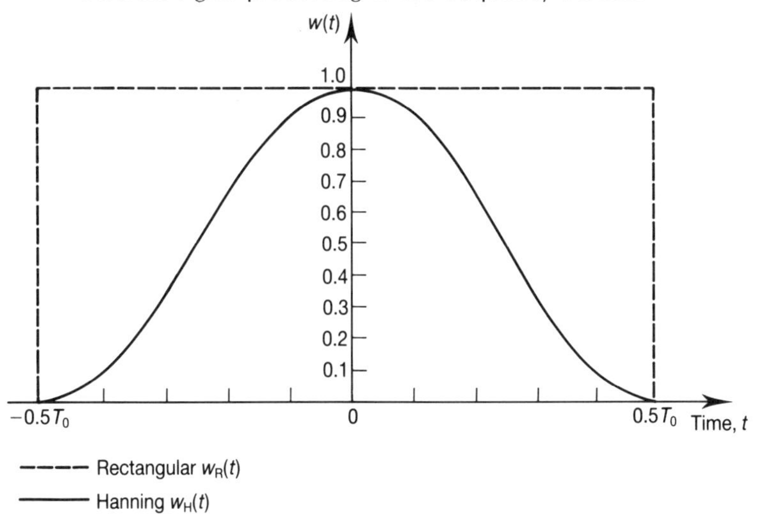

Figure 2.5 Rectangular and Hanning window functions $w_R(t)$ and $w_H(t)$.

frequency. This effect is known as *leakage* and is particularly noticeable for this window because the side lobes of $W_R(\omega)$ are slow to decay with increasing frequency.

By differentiating equation (2.26) it is readily seen that $W_R(\omega)$ has zero slope at (positive) cyclic frequencies f given by

$$\pi f T_0 = \tan \pi f T_0 \tag{2.30}$$

The first zero is thus at a frequency of approximately $1.43/T_0$, at which $W_R(f) = -0.217T_0$; hence $|W_R(f)|$ is only 13.27 dB less than the main peak. It is also seen from equation (2.29) that as f becomes large the side-lobe magnitudes decay by only 6 dB/octave.

It is important to stress that if $x(t)$ is periodic, and its spectral lines are at cyclic frequencies with spacing K/T_0, where K is an integer (i.e. the period of $x(t)$ is T_0/K), then spectral leakage does not occur, because all the spectral lines of the signal occur

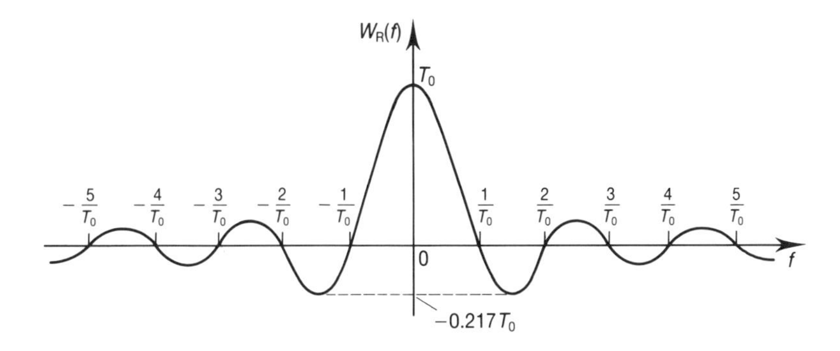

Figure 2.6 Fourier transform $W_R(f)$ of the rectangular window function w_R(t).

at zeros in $W_R(\omega)$. However, if $x(t)$ is non-periodic, or if it is periodic other than T_0/K, then spectral leakage will occur (see Brigham, 1988, Chapter 9, for an illustration of the effect for a sine wave). Its effect can be reduced significantly by multiplying the segment of data by a different time function, whose Fourier transform has lower side lobes than those of the rectangular window. There are numerous possibilities for the choice of window function (see Harris, 1978, for an excellent review).

One of the best-known designs is the Hanning window, given by:

$$w_H(t) = K_H\left(1 + \cos\frac{2\pi t}{T_0}\right), \quad -\frac{1}{2}T_0 \leqslant t < \frac{1}{2}T_0 \tag{2.31a}$$

$$= 0, \text{ for other values of } t \tag{2.31b}$$

where K_H is a constant. If $K_H = 0.5$, $w_H(0) = w_R(0) = 1$, as illustrated in Figure 2.5, and the Fourier transform of $w_H(t)$ is then given by

$$\begin{aligned} W_H(\omega) &= 2\int_0^{T_0/2} 0.5\left(1 + \cos\frac{2\pi t}{T_0}\right)\cos\omega t \, dt \\ &= \int_0^{T_0/2} \cos\omega t \, dt + 0.5\int_0^{T_0/2} \cos\left(\omega + \frac{2\pi}{T_0}\right)t \, dt \\ &\quad + 0.5\int_0^{T/2} \cos\left(\omega - \frac{2\pi}{T_0}\right)t \, dt \end{aligned} \tag{2.32}$$

By comparison with equation (2.27), it is seen that

$$W_H(\omega) = 0.5W_R(\omega) + 0.25W_R\left(\omega + \frac{2\pi}{T_0}\right) + 0.25W_R\left(\omega - \frac{2\pi}{T_0}\right) \tag{2.33}$$

Hence

$$\begin{aligned} W_H(\omega) &= \frac{T_0}{2}\left[\frac{2\sin\omega T_0/2}{\omega} - \frac{\sin\omega T_0/2}{\omega + (2\pi/T_0)} - \frac{\sin\omega T_0/2}{\omega - (2\pi/T_0)}\right] \\ &= \frac{T_0}{2}\frac{\sin\omega T_0/2}{\omega T_0/2}\cdot\frac{4\pi^2}{4\pi^2 - \omega^2 T_0^2} \end{aligned} \tag{2.34}$$

so that

$$W_H(f) = \frac{T_0}{2}\frac{\sin\pi f T_0}{\pi f T_0}\cdot\frac{1}{1 - (fT_0)^2} \tag{2.35}$$

This has considerably improved side-lobe amplitude compared with the rectangular window. It first has zero slope at a (cyclic) frequency f of approximately $2.36/T_0$, and the magnitude of this lobe is some 32 dB less than that of the main peak. As fT_0 becomes large, side-lobe magnitudes fall off at a rate of 18 dB/octave, compared with 6 dB/octave for the rectangular window. The gains $|W_R(f)|$ and $|W_H(f)|$ are shown on a dB scale in Figure 2.7, in which the gain of $W_H(f)$ at $f = 0$ has been normalized to 1 (0 dB) by making $K_H = 1$.

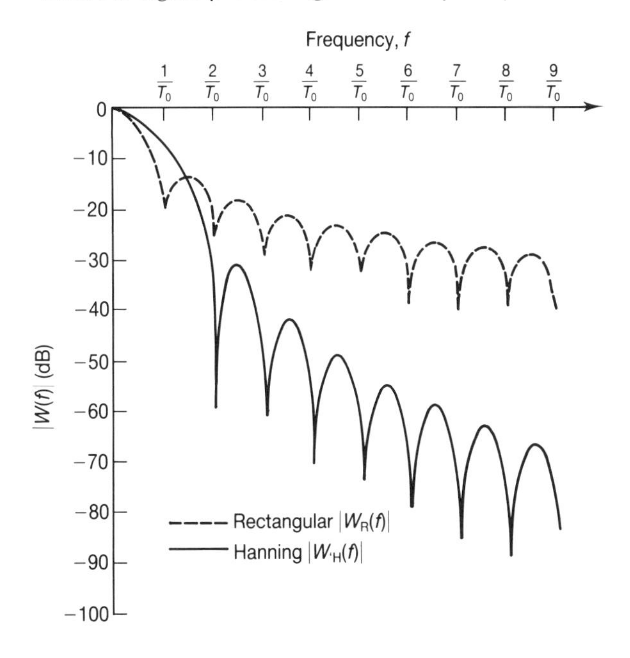

Figure 2.7 Comparison of the magnitude of the Fourier transform of the rectangular and Hanning window functions.

The improved side-lobe characteristic of $W_H(f)$ is obtained at the expense of less good frequency resolution. The first zero of $W_H(f)$ occurs at $f = 2/T_0$, compared with $1/T_0$ for the rectangular window. The value of $W_H(f = 1/T_0)$ is, from equation (2.33), or by applying L'Hopital's rule to equation (2.35), half that at $f = 0$. The main peak of $W_H(f)$ is thus twice as wide as that of $W_R(f)$, so that it is more difficult to resolve two spectral peaks which are close together if a Hanning window is used.

Windows with side lobes considerably lower than −32 dB are available (Harris, 1978), but there is always a conflict between good amplitude resolution and good frequency resolution, so that the choice of which window to use has to depend on the requirements of the particular application.

The next two examples are designed to illustrate some aspects of the choice of window functions. In both examples, three different windows are applied to segments of data from periodic signals already discussed in this chapter – sine waves and pseudo-random binary signals. In both examples, discrete power spectra have been measured using a commercial digital spectrum analyzer (an Advantest TR9403). This automatically includes an anti-aliasing filter, but the user can choose from a range of window functions. The three used in the examples are a rectangular window, a Hanning window (with $K_H = 1$ so that $W_H(0) = W_R(0)$) and a 'flat-pass' window, designed to give excellent amplitude characteristics; some other manufacturers refer to this as a flat-top window. As mentioned earlier, this analyzer uses a 1024-point FFT algorithm and it displays 400 of the 1024 frequency points. In common with

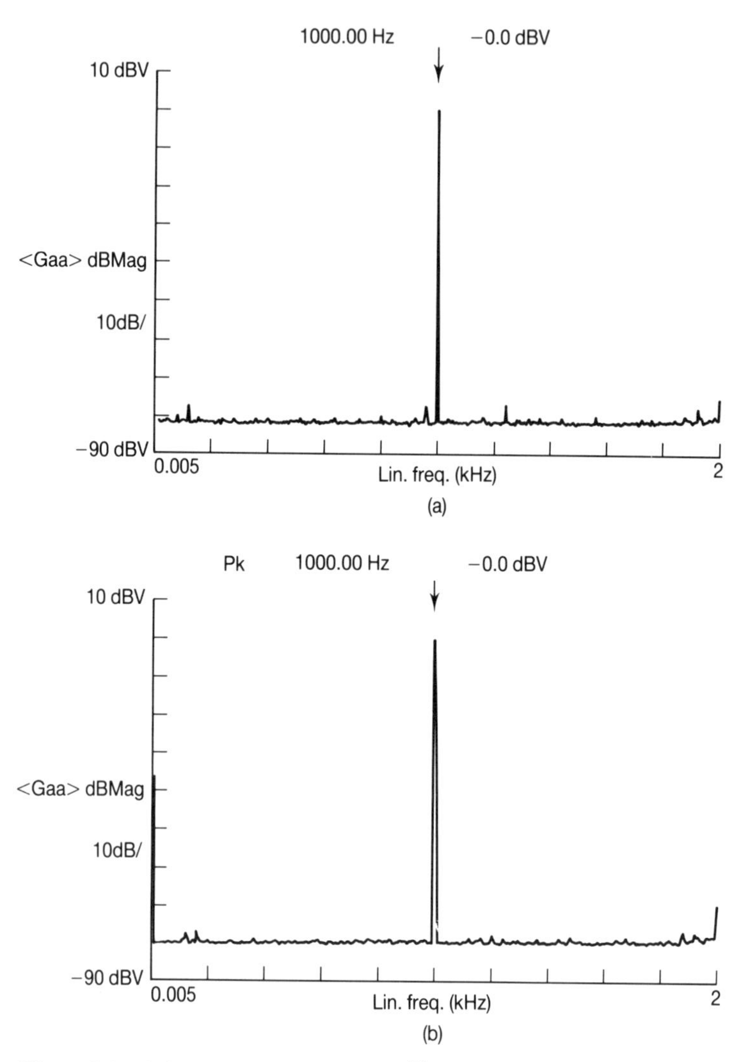

Figure 2.8 (Discrete power spectrum)$^{1/2}$ of a 1 volt RMS sine wave of frequency 1 kHz; dBV vertical scale. (a) Rectangular window; (b) Hanning window.

most commercial spectrum analyzers, values are displayed for positive frequencies only, with components at $-\omega$ being added to those at $+\omega$.

EXAMPLE 2G DISCRETE POWER SPECTRUM OF A SINE WAVE

The analyzer was set to a maximum displayed frequency of 2 kHz, so that discrete spectral lines are at intervals of 2000/400 = 5 Hz. Each segment of data is of length (T_0) equal to $^1/_5$ second = 200 ms, and the sampling interval is 200/1024 = 0.1953125 ms.

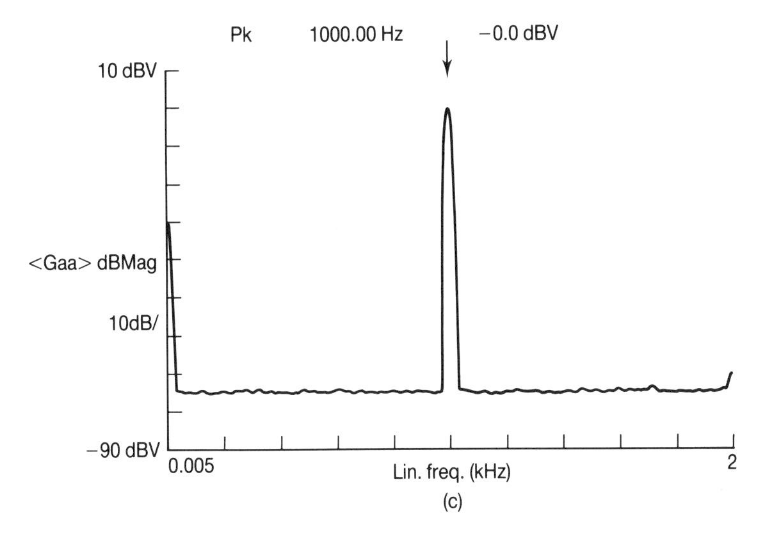

Figure 2.8 (cont.) (c) Flat-pass window.

In both parts of the Example, a sine wave of 1 volt RMS was used, and with the power at negative frequencies being added to that at positive frequencies, the continuous power spectrum is thus 1 volt^2 at the frequency of the sine wave, and zero at all other frequencies.

(i) Comparison of the windows with no leakage In the first part of the Example the frequency of the sine wave was set to 1 kHz, which is the frequency of one of the discrete spectral lines, so that there is no leakage. (The segment length T_0 contains exactly 200 periods of the signal.) The discrete power spectrum using the three windows are shown in Figure 2.8, with the vertical scale being the square root of the discrete power spectrum, on a dBV scale (i.e. $20 \log_{10} \sqrt{S_{xx}(f)}$). With this setting, the continuous spectrum consists of a single value of 0.0 dBV, at $f = 1000$ Hz. Although averaging should not be necessary in this case, the lower end of each scale is -90 dBV, and there is some effect at these low values from laboratory electrical noise; therefore 256 segments of data were averaged, the total length of data being 51.2 s. (Longer averaging resulted in negligible improvement in the variance of the low-level values.) The features to note are the following:

1. *Rectangular window.* The peak is at a single frequency point and is of correct magnitude (0.0 dBV). Off-peak values are around -82 dBV, except for some small spikes, all below -75 dBV. The off-peak value is sometimes referred to as the noise floor of the analyzer with this window.
2. *Hanning window.* The peak is now spread over three frequency points, the value at 1000 Hz again being 0.0 dBV, and those at 995 Hz and 1005 Hz being

−6.0 dBV. Off-peak values are around −79 dBV (the noise floor) and are somewhat smoother than those in (1).

3. *Flat-pass window.* The main peak is now much more spread out, with values stretching from 975 Hz to 1025 Hz, as shown in Table 2.2. The off-peak values are around −74 dBV (the noise floor) and are considerably smoother than those in (2).

Table 2.2 (Discrete power spectrum)$^{0.5}$ of a 1 kHz, 1 volt RMS sine wave, using a flat-pass window function

	(Discrete power spectrum)$^{0.5}$	
f (Hz)	dB	Volts
970, 1030	−74.2	1.90×10^{-4}
975, 1025	−30.8	2.89×10^{-2}
980, 1020	−14.8	1.82×10^{-1}
985, 1015	−6.0	5.00×10^{-1}
990, 1010	−1.8	8.15×10^{-1}
995, 1005	−0.3	9.68×10^{-1}
1000	0.0	1.00×10^{0}

In (2), there is a component at the first discrete harmonic (5 Hz) and this is noticeable up to the fifth discrete harmonic (25 Hz) in (3). This is due to a small d.c. component spreading out into adjacent frequency points (it does not occur in (1), as expected). The decibel scale exaggerates this component, which, in (3), is seen to be 30 dB down on the main peak.

(ii) Comparison of the windows with leakage In the second part of the Example the frequency of the sine wave was set to 997.5 Hz, exactly half-way between the discrete spectral lines at 995 Hz and 1000 Hz. The period of the sine wave is now 1.0025 ms, and each segment of data consists of 199.5 periods of the signal. As before, 256 segments were averaged. Each segment should (hopefully) start at a different point in the signal, so that averaging will change the effect of the leakage. It was found there was negligible change to any of the computed discrete spectra if the number of segments averaged was increased above 256. The discrete power spectrum is shown in Figure 2.9. Values over a range of frequencies on either side of the peak are also shown in Table 2.3.

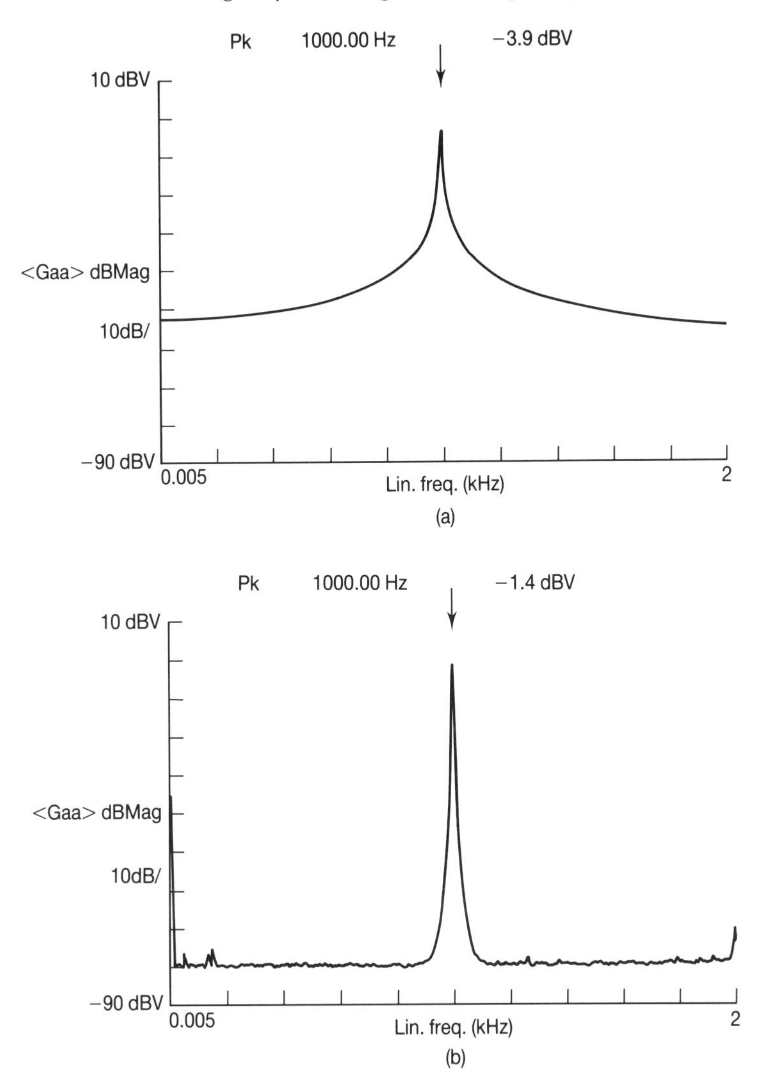

Figure 2.9 (Discrete power spectrum)$^{1/2}$ of the 1 volt RMS sine wave of frequency 997.5 Hz; dBV vertical scale. (a) Rectangular window; (b) Hanning window.

The features to note from Figure 2.9 and Table 2.3 are the following:

1. *Rectangular window.* The discrete spectrum is now very spread out. The peak value (at discrete frequency points 995 Hz and 1000 Hz) is -3.9 dBV, considerably less than the RMS of the signal (0.0 dBV).

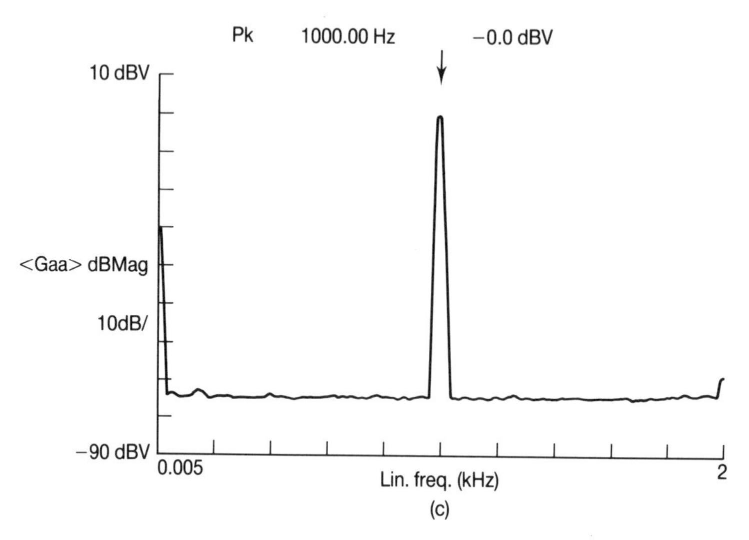

Figure 2.9 (cont.) (c) Flat-pass window.

Table 2.3 (Discrete power spectrum)$^{0.5}$ of a 997.5 Hz, 1 volt RMS sine wave, using three different window functions

	(Discrete power spectrum)$^{0.5}$					
	Rectangular window		Hanning window		Flat-pass window	
f (Hz)	dB	Volts	dB	Volts	dB	Volts
800, 1195	−41.8	8.11×10^{-3}	−79.4	1.07×10^{-4}	−74.1	1.98×10^{-4}
900, 1095	−35.7	1.64×10^{-2}	−78.8	1.15×10^{-4}	−74.5	1.89×10^{-4}
960, 1035	−27.4	4.25×10^{-2}	−62.2	7.73×10^{-4}	−74.1	1.98×10^{-4}
965, 1030	−26.2	4.90×10^{-2}	−58.5	1.19×10^{-3}	−67.8	4.07×10^{-4}
970, 1025	−24.7	5.79×10^{-2}	−54.1	1.97×10^{-3}	−44.2	6.19×10^{-3}
975, 1020	−23.0	7.08×10^{-2}	−48.7	3.67×10^{-3}	−21.6	8.34×10^{-2}
980, 1015	−20.8	9.11×10^{-2}	−41.9	8.07×10^{-3}	−9.7	3.27×10^{-1}
985, 1010	−17.9	1.28×10^{-1}	−32.3	2.42×10^{-2}	−3.4	6.73×10^{-1}
990, 1005	−13.5	2.12×10^{-1}	−15.4	1.71×10^{-1}	−0.8	9.17×10^{-1}
995, 1000	−3.9	6.36×10^{-1}	−1.4	8.49×10^{-1}	0.0	9.97×10^{-1}

2. *Hanning window.* The discrete spectrum is less spread out, being confined to approximately ±100 Hz about the peak. The peak value itself, at −1.4 dBV, (at 995 Hz and 1000 Hz) is still wrong, but is closer to the RMS value of the signal. The values away from the main peak are unchanged at around −79 dBV.

3. *Flat-pass window.* The main peak is much narrower than for the other windows, stretching from 965 Hz to 1030 Hz, so that it is only a little more stretched out than

in part (i) of the Example. The peak value (at 995 Hz and 1000 Hz) is correct to two significant figures at 0.0 dBV. (Measuring on a voltage scale, the peak was found to be 9.97×10^{-1} volts, an error of only 0.3 %.) The values away from the main peak are unchanged at around −74 dBV.

With leakage, the flat-pass window provides the best amplitude resolution, as expected. However, with the width of its main peak, this window can give misleading results in a situation with no leakage, but with a signal with harmonics which are close together. This is illustrated in part (i) of the next Example.

EXAMPLE 2H DISCRETE POWER SPECTRUM OF A PRBS

The analyzer was again set to a maximum displayed frequency of 2 kHz, so that discrete spectral lines are at intervals of 5 Hz, each segment of data is of length 200 ms and the sampling interval $\Delta = 0.1953125$ ms.

The signal in this case was a PRBS with 15 digits and voltage levels ± 1 volt obtained from a Hewlett Packard HOI-3722A noise generator. Since the contributions to the power spectrum at $-\omega$ are added to those at $+\omega$, the continuous power spectrum is, from equations (2.16a) and (2.16b), given by

$$S_{xx}(0) = \frac{V^2}{N^2} = \frac{1}{225}$$

$$S_{xx}\left(f = \frac{k}{15 \, . \, \Delta t}\right) = 2V^2 \frac{N+1}{N^2} \frac{\sin^2(\pi k/15)}{(\pi k/15)^2}, \quad k = 1, 2, 3, \ldots$$

$$= \frac{32}{225} \frac{\sin^2(\pi k/15)}{(\pi k/15)^2}, \quad k = 1, 2, 3, \ldots$$

(i) Comparison of the windows with no leakage In the first part of the Example, Δt was set to 3.33 ms, so that the continuous spectrum lines are at 0, 20, 40, 60, ... Hz which correspond to discrete spectral line locations, so that there is no leakage. (The signal period is 50 ms, so that there are exactly four periods of the signal in each segment of data.) The discrete power spectra using the three windows are shown in Figure 2.10; the vertical scale in this case is the (positive) square root of the power spectrum, to give reasonable accuracy at low values without attenuating high values as in the dB scale. The corresponding continuous spectrum quantity is thus

$$\frac{4\sqrt{2}\,|\sin(\pi k/15)|}{\pi k}, \quad k = 1, 2, 3, \ldots$$

The expected first harmonic value ($k = 1$) is 0.374 volts. The features to note are the following:

1. *Rectangular window*. The discrete spectral lines are distinct, as expected, and the peak value of 0.378 volts at 20 Hz is in close agreement with that expected. (The voltage level from the signal generator was in fact slightly more than 1 volt; there

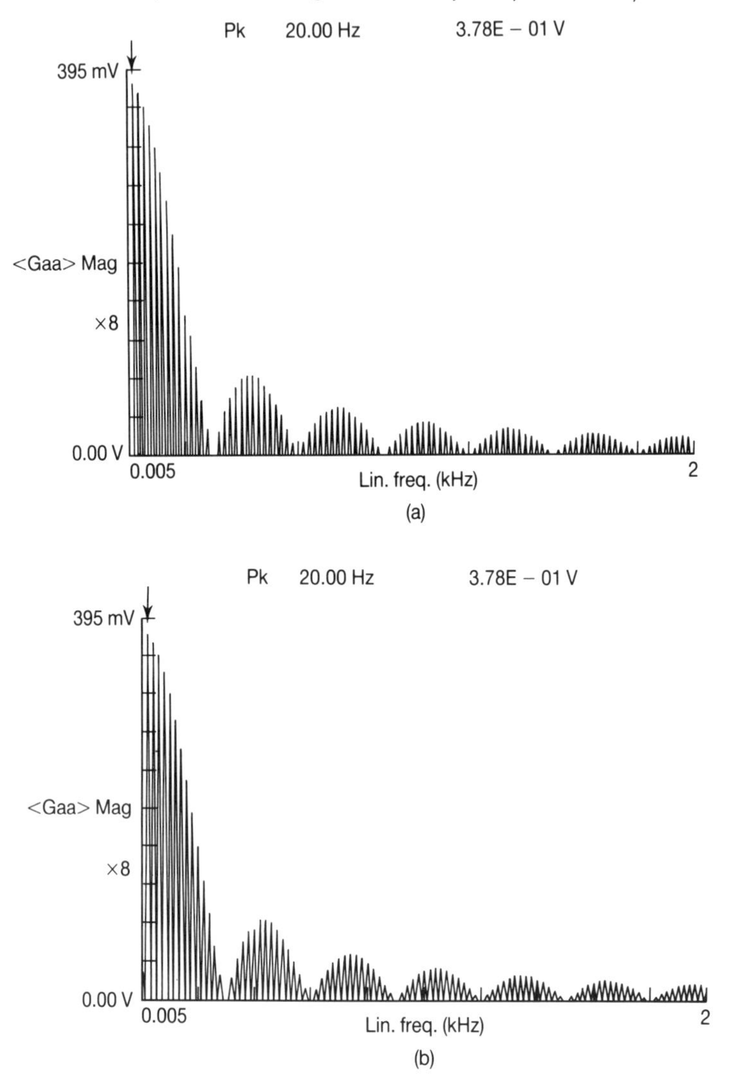

Figure 2.10 (Discrete power spectrum)$^{1/2}$ of a 15-digit PRBS with levels ± 1 volt and clock frequency 300 Hz. (a) Rectangular window, (b) Hanning window.

is no facility on the generator used for (continuous) fine tuning of the voltage level.)

2. *Hanning window*. Each peak is accompanied by a value equal to half its value at the discrete frequency locations on either side of it (as a result of the wider main lobe of $W_H(f)$). For example, the peak of 0.378 volts at 20 Hz is accompanied by values of 0.189 volts at 15 Hz and 25 Hz. Thus the discrete spectral lines are still distinct, but only just.

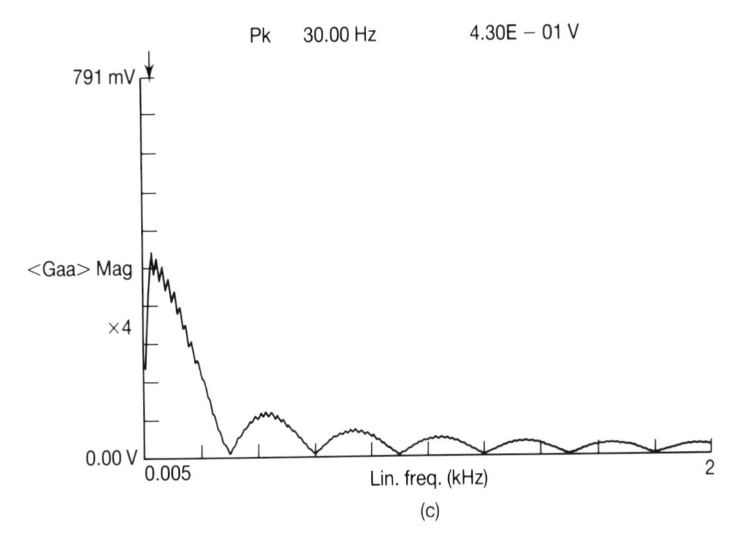

Figure 2.10 (cont.) (c) Flat-pass window.

3. *Flat-pass window.* We saw in the previous Example that the main peak is spread over ± 25 Hz around zero and it is clear from (3) in that Example that the discrete spectral lines have merged into each other. The maximum value (0.430 volts) and its location (30 Hz) give incorrect estimates of the corresponding continuous spectrum quantities, due to the merging.

(ii) Comparison of the windows with leakage In the second part of the Example, Δt was set to 1 ms, so that the continuous spectrum lines are at 0, $66.\dot{6}$, $133.\dot{3}$, 200, $266.\dot{6}$, ... Hz. The lines at 200, 400, 600, ... Hz correspond to discrete spectral line locations, but the others do not, so that there is leakage. It is interesting to compare the first three continuous spectrum harmonics with the discrete spectrum values at the nearest discrete spectrum frequency, and this is done in Table 2.4. The discrete power spectra using the three windows are shown in Figure 2.11. The features to note from Figure 2.11 and Table 2.4 are the following:

Table 2.4 Comparison of continuous and discrete spectrum values of the first three harmonics of a PRBS with 15 digits, $V = 1$ volt and $\Delta t = 1$ ms

Harmonic number	f (Hz)	Continuous $(S_{xx}(f))^{0.5}$	Nearest discrete spectrum frequency point (Hz)	Discrete $(S_{xx}(f))^{0.5}$		
				Rectangular	Hanning	Flat-pass
1	$66\frac{2}{3}$	0.374	65	0.314	0.352	0.376
2	$133\frac{1}{3}$	0.366	135	0.307	0.345	0.368
3	200	0.353	200	0.356	0.356	0.356

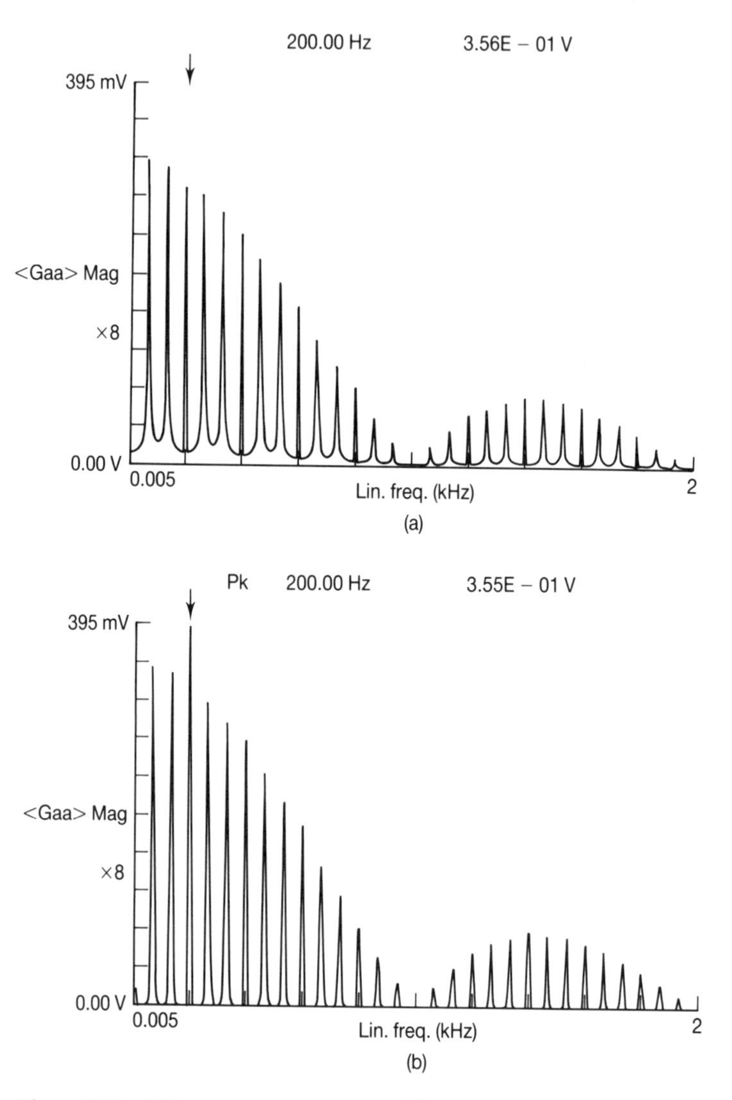

Figure 2.11 (Discrete power spectrum)$^{1/2}$ of a 15-digit PRBS with $V = 1$ volt and clock frequency 1 kHz. (a) Rectangular window; (b) Hanning window.

1. *Rectangular window.* The peaks are narrow where the continuous and discrete spectral lines are the same (200, 400, 600, ... Hz), but are wider elsewhere with substantial leakage. At both the first two harmonics, discrete spectrum values (at the nearest discrete frequency value) are considerably less than continuous spectrum values for $k = 1$ and 2.

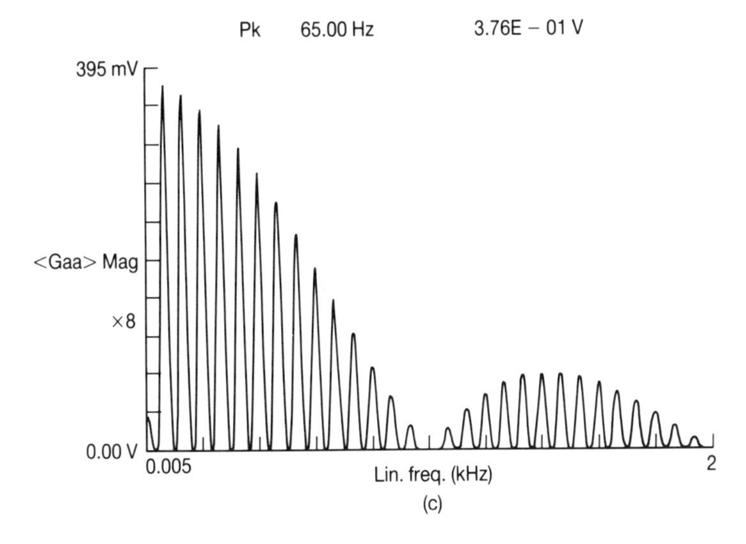

Figure 2.11 (cont.) (c) Flat-pass window.

2. *Hanning window.* The leakage is reduced and the discrete spectrum values are in closer agreement with the continuous spectrum values. The largest value is still at 200 Hz (corresponding to $k = 3$).
3. *Flat-pass window.* With the continuous spectrum line spacing being $66.\dot{6}$ Hz, discrete spectral peaks no longer run into each other as they did in (i), and the spectrum envelope is of the expected shape. Discrete spectrum values are now in good agreement with continuous spectrum values.

Both of these Examples have shown that, if there is no leakage, the rectangular window function is best in the sense that it has the optimum frequency resolution. If there is leakage, a different window function can be used to provide good amplitude accuracy, but only at the expense of frequency resolution. It is clear that leakage should be avoided if at all possible and from the point of view of perturbation signal design for system identification in the frequency domain, it is desirable that the segment length T_0 should be an integer number of periods of the signal. For this reason, there is an increasing trend towards providing signal generators as an integral part of digital spectrum analyzers. We will return to the problem of leakage when considering system identification in Application Example 2.1 (Section 2.5.3).

It should be emphasized that the problems with leakage in Examples 2G and 2H have arisen from the use of computations based on a radix-2 FFT, as used on virtually all commercially available digital spectrum analyzers. A powerful technique for removing the effects of leakage is to use the chirp z-transform (Rabiner *et al.*, 1969), which allows the conversion of a radix-2 FFT into an arbitrary radix DFT using three FFT-type computations. It is then possible to tailor the number of digits,

N, to the particular application, ensuring that an exact number of periods of the perturbation signal is used. Nevertheless, the approach on most digital spectrum analyzers is to use a radix-2 FFT and then provide several data windows to reduce the resulting spectral leakage to an acceptable level in a particular application.

2.3 Power spectrum of uncorrelated signals

In Section 1.5 we saw that, by inverting every other digit in the sequence on which a PRBS is based, it is possible to obtain a signal, called an inverse-repeat binary signal, which is uncorrelated with the PRBS over their common period of $2N.\Delta t$, where N is the number of digits in the PRBS and Δt is the clock pulse interval. The autocorrelation function of the inverse-repeat is shown in Figure 1.21.

Since the common period is $2N.\Delta t$, the spacing between the spectral lines in the continuous power spectrum is $1/(2N.\Delta t)$. The power spectrum of the PRBS is as found in Example 2E, with power at harmonics $f = k/(2N.\Delta t)$ for $k = 0, \pm 2, \pm 4, \ldots$; i.e. the power is at the even harmonics only. Fourier analysis of the autocorrelation function of the inverse-repeat signal gives a power spectrum with power at the odd harmonics only, i.e. for $k = \pm 1, \pm 3, \pm 5, \ldots$ Thus the two signals which are uncorrelated in the time domain do not have power at the same harmonics in the frequency domain. This is a result which we will also encounter in Section 2.4.1 when considering binary multi-frequency signals.

The result extends to the further uncorrelated signals considered in Section 1.5, obtained by inverting alternate pairs of digits, then alternate runs of four digits (and so on) in the sequence on which the PRBS is based. The Fourier analysis must be done over the longest period of the set of uncorrelated signals. For example, the third signal in the set obtained from a PRBS (inverting alternate pairs of digits in the original sequence) has period $4N.\Delta t$, so that the harmonic spacing is $1/(4N.\Delta t)$. The PRBS contains power at harmonics 0, 4, 8, ...; the inverse-repeat signal contains power at harmonics 2, 6, 10, ...; and the third signal contains power only at odd harmonics.

The power spectrum of a PRBS and the corresponding inverse-repeat binary signal are illustrated in the next Example.

EXAMPLE 2I: POWER SPECTRUM OF A PRBS AND THE CORRESPONDING INVERSE-REPEAT BINARY SIGNAL

The two signals were obtained from a Solartron JM1861 Signal Generator. The minimum number of digits in a PRBS from this device is 31, so that exact comparison with the PRBS used in Example 2H is not possible. The clock pulse interval was set to 1 ms and the voltage level to ± 1 volt. As in the previous two examples, we will be interested only in positive frequencies, so that, from equation (2.16b), the expected continuous power spectrum of the PRBS is given by

$$S_{xx}\left(f = \frac{k}{62\Delta t}\right) = \frac{64}{31^2}\frac{\sin^2(\pi k/62)}{(\pi k/62)^2}, \quad k = 2, 4, 6, \ldots$$

Hence the square root of the power spectrum is given by

$$\sqrt{S_{xx}\left(\frac{k}{62\Delta t}\right)} = \frac{16\,|\sin(\pi k/62)|}{\pi k}, \quad k = 2, 4, 6, \ldots$$

The largest value is that for $k = 2$ and is 0.2576 volts. The harmonics are at cyclic frequencies $1000k/62 = (16.129032k)$ Hz, $k = 2, 4, 6, \ldots$, so that the frequency corresponding to $k = 2$ is 32.258065 Hz.

The discrete spectra of the two signals were measured using an Advantest TR9403 digital spectrum analyzer. It is not possible to adjust the discrete frequencies to be the same as those of the continuous spectrum, so that leakage is inevitable; thus to obtain good amplitude resolution, a flat-pass window was used. However, if the maximum displayed frequency is set to 2 kHz, the spectral lines using this window merge, as in part (i) of Example 2H. To ensure separation of the spectral lines, the maximum displayed frequency was therefore set to 1 kHz, giving discrete spectral lines separated by $1000/400 = 2.5$ Hz.

The discrete spectra of the two signals are shown in Figure 2.12. It is seen from Figure 2.12(a) that the PRBS has power only at even harmonics, as expected. The value (0.2576) for $k = 2$ in the continuous spectrum (32.258065 Hz) is very close to the discrete spectrum value (0.257) at the nearest discrete frequency point (32.5 Hz). Again, this is expected, given the excellent amplitude characteristics of the flat-pass window function. It is seen in Figure 2.12(b) that the inverse-repeat binary signal does indeed have power only at the odd harmonics and that the envelope of its power spectrum is very similar in shape to that of the PRBS except that the power in harmonic 31 is much reduced. This also proves to be true of harmonics which are odd multiples of the number of digits in the PRBS (in this case 93, 155, 217, ...) as illustrated in Figure 2.13, which is the same as Figure 2.12(b) except that the clock frequency $1/\Delta t$ has been reduced from 1 kHz to 300 Hz. With the closeness of the harmonics (the spacing in the continuous spectrum is now only 9.6774195 Hz), the discrete spectrum peaks run into each other, as in Figure 2.10(c), so that the maximum value (0.301) of the discrete spectrum is a little larger than it should be (0.2576), but the purpose of this figure is to demonstrate the reduction in power, compared with the envelope, at harmonics 31, 93 and 155.

2.4 Multi-frequency signals

Pseudo-random binary signals have a fixed amount of power in each harmonic, and the amount of power in any given harmonic is rather small. Confining ourselves to positive frequencies only, the maximum amount of power in a PRBS with levels $\pm V$ and period $N.\Delta t$ is only

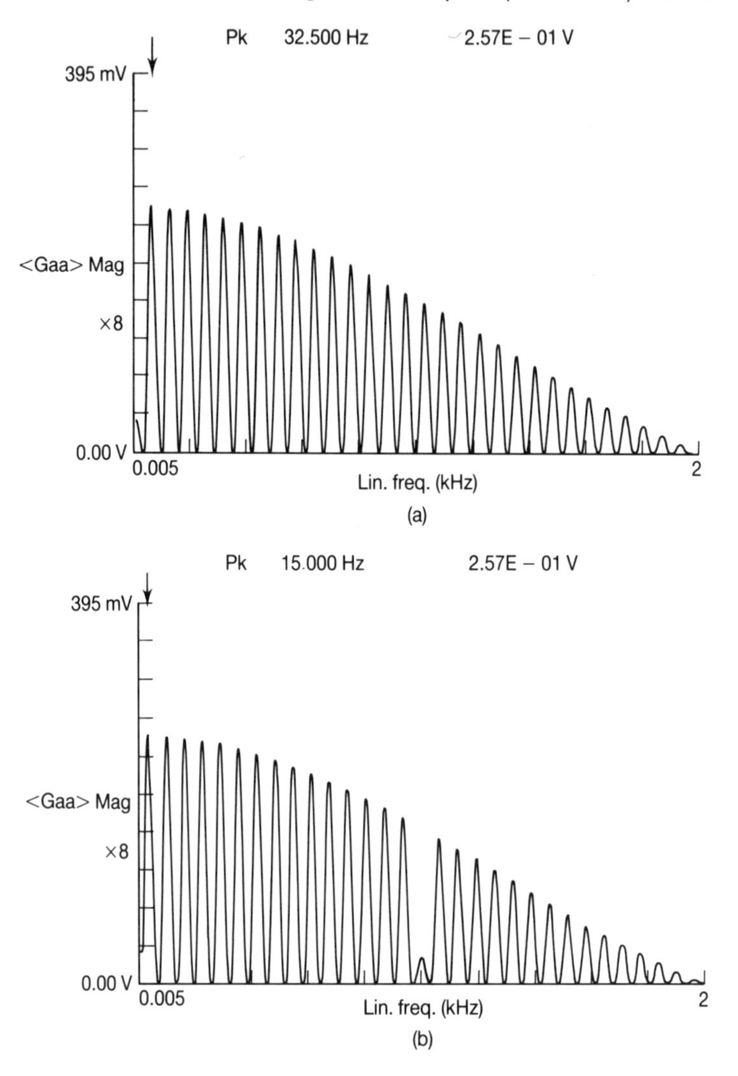

Figure 2.12 (Discrete power spectrum)$^{1/2}$ of (a) a 31-digit PRBS and (b) the corresponding 62-digit inverse-repeat binary signal, both with $V = 1$ volt and clock frequency 1 kHz (data segments multiplied by a flat-pass window function).

$$2V^2 \frac{N+1}{N^2} \frac{\sin^2(\pi/N)}{(\pi/N)^2}$$

at the first harmonic $f = 1/N \, . \, \Delta t$. For example, if $N = 127$, this maximum power level is only $0.01587V^2$, compared with the total signal power of V^2. This can give rise to problems in identification of noisy systems, as we will see in Application

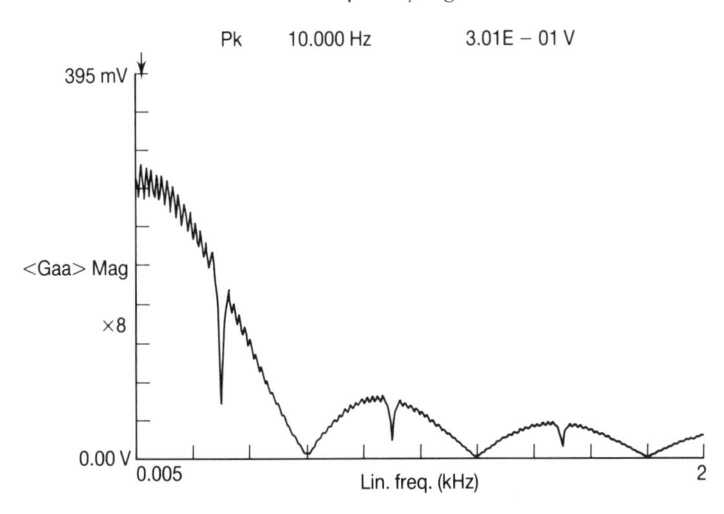

Figure 2.13 (Discrete power spectrum)$^{1/2}$ of a 62-digit inverse-repeat binary signal with $V = 1$ volt and clock frequency 300 Hz. (Data segments multiplied by a flat-pass window function).

Example 2.3 in Section 2.5.3. Further, since N is odd for all pseudo-random binary sequences on which the signals are based (e.g. $N = 2^n - 1$ for a binary m-sequence), the signals are not very suitable for frequency-domain system identification using the FFT, which uses segments of data of length 2^n (where n is an integer) – typically 1024.

This has led to an interest in signal design specifically with the frequency domain in mind. The amount of power in any given harmonic can be specified by the user, and most of the designs contain a larger amount of power in a smaller number of harmonics. There have been two different types of design – binary multi-frequency signals and non-binary, sum-of-harmonics signals – and these will be discussed in the remainder of this section.

2.4.1 Binary multi-frequency signals

These are discrete-interval, periodic signals (period $= N \,.\, \Delta t$) with as much power as possible in certain harmonics specified by the user. Nearly all designs are for N a power of 2, which makes it easy to ensure no spectral leakage using FFT signal processing. The earliest published designs for such signals are given by van den Bos (1967), although he notes that Jensen, of the Danish Technical University in Copenhagen, had designed multi-frequency signals on intuitive grounds and had included them in unpublished notes in 1959.

The pattern of values ($+V$ or $-V$) over each clock pulse interval to achieve the optimal amount of power in the specified harmonics is determined by theory backed up by computer search. Further designs of multi-frequency signals have been published

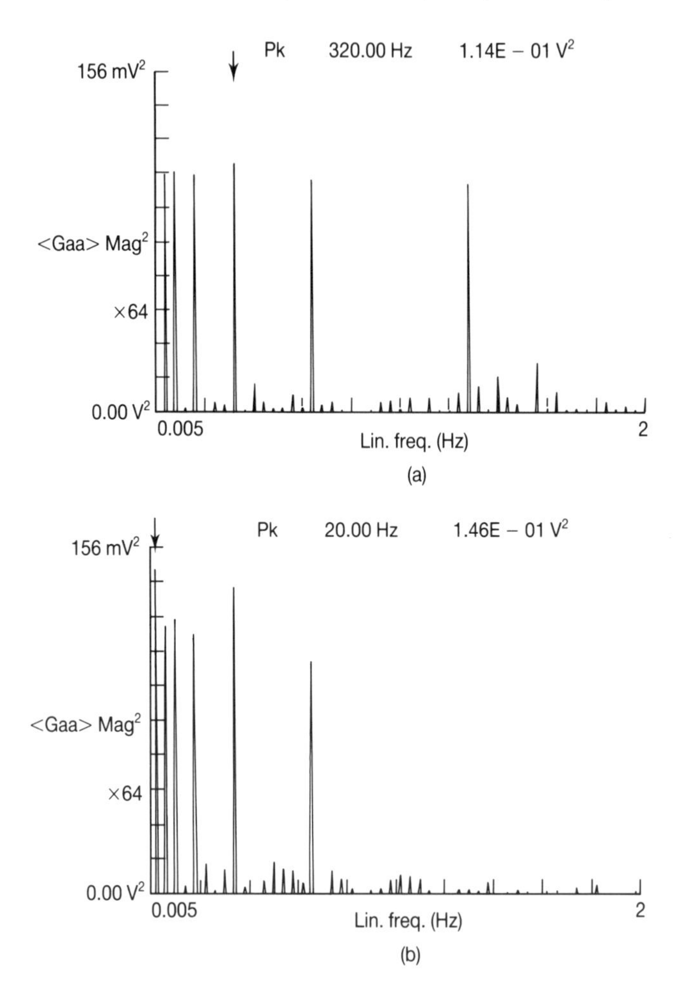

Figure 2.14 Discrete power spectra of four binary multi-frequency signals. (a) 6-harmonic signal (T12) from Paehlike (1980); (b) 6-harmonic signal (No. 2) from Buckner and Kerlin (1972).

by van den Bos (1970), Buckner and Kerlin (1972), Paehlike and Rake (1979), Paehlike (1980), van den Bos and Krol (1979) and Harris and Mellichamp (1980).

Typical designs for multi-frequency signals have $N = 128, 256$ or 512. This allows a reasonable spread of specified harmonics and such signals are useful for the identification of low-order processes. However, Harris and Mellichamp (1980) also give designs with a smaller spread of the selected harmonic range. These signals, referred to as *compact* signals, are suitable for the identification of higher-order systems or closed-loop systems. The idea of compact binary multi-frequency signals has been

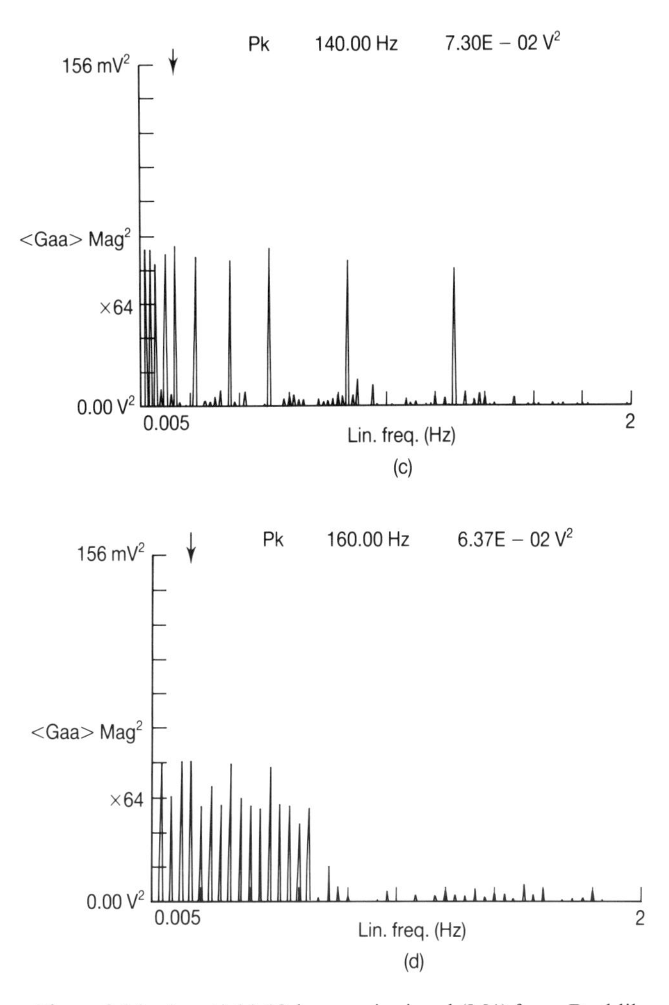

Figure 2.14 (cont.) (c) 10-harmonic signal (M1) from Paehlike (1980); (d) 16-harmonic signal (T16) from Paehlike (1980).

taken further by Henderson and McGhee (1990b) in a number of designs with $N = 16$, 32 or 64. Several of their designs have a number of the specified harmonics very close together, so that the signals can be used to 'zoom in' and highlight the peak of the frequency response of a lightly damped system. The reader is referred to Chapter 7 and to the references therein.

A number of binary multi-frequency signals have been incorporated into a signal generator constructed at the University of Warwick, and the measured discrete power spectra of four of them are shown in Figure 2.14. All four signals have voltage levels ± 1 volt, so that the total power in each signal is 1 volt^2.

The power spectra were measured using an Advantest TR9403 digital spectrum analyzer with the maximum displayed frequency set to 2 kHz, as in most of the preceding examples: with this setting, data segment lengths are 200 ms. Two of the signals have $N = 128$ and the other two have $N = 256$. For convenience of display, the clock pulse frequency of the binary multi-frequency signal generator was set to 5120 Hz, so that the first harmonic of the signal is at 40 Hz for the signals with $N = 128$ and at 20 Hz for the signals with $N = 256$. Each data segment length is an integer multiple of the signal period, so that there is no spectral leakage; because of this, a rectangular window was used to maximize the frequency resolution. Details of the four signals, with the theoretical power in the specified harmonics, are given in Table 2.5. Agreement between theoretical power and measured power (Figure 2.14) was extremely close, as expected.

Table 2.5 Details of the four binary multi-frequency signals considered in Section 2.4.1

Signal A: Signal T12 from Paehlike (1980). $N = 128$.

$2^+\ 2^-\ 3^+\ 1^-\ 2^+\ 3^-\ 1^+\ 2^-\ 2^+\ 2^-\ 14^+\ 2^-\ 6^+\ 11^-\ 2^+\ 1^-\ 2^+\ 14^-\ 1^+\ 12^-\ 6^+\ 1^-\ 6^+\ 2^-$
$6^+\ 3^-\ 1^+\ 2^-\ 1^+\ 3^-\ 7^+\ 2^-\ 2^+\ 1^-$

(The notation denotes 2 clock pulse intervals with the signal at its positive level, then 2 with the signal at its negative level and so on.)

Longest run in the signal = 14 clock pulse intervals.

Specified harmonics and theoretical power in these harmonics:

Harmonic number k	Power (% of total)
1	10.85
2	11.05
4	10.95
8	11.37
16	10.84
32	10.84
	65.90

Percentage deviation in power in the specified harmonics, i.e. [(maximum − minimum)/average] × 100% = 4.83%.

Measured power spectrum shown in Figure 2.14(a).

Signal B: Signal No. 2 from Buckner and Kerlin (1972). $N = 256$.

$12^+\ 11^-\ 4^+\ 13^-\ 1^+\ 13^-\ 4^+\ 26^-\ 6^+\ 3^-\ 4^+\ 2^-\ 6^+\ 3^-\ 12^+\ 8^-$
$12^-\ 11^+\ 4^-\ 13^+\ 1^-\ 13^+\ 4^-\ 26^+\ 6^-\ 3^+\ 4^-\ 2^+\ 6^-\ 3^+\ 12^-\ 8^+$

(NB: Signal is inverse-repeat.)

Longest run in the signal = 26 clock pulse intervals.

Specified harmonics and theoretical power in these harmonics:

Harmonic number k	Power (% of total)
1	14.6
3	12.0
5	12.5
9	11.7
17	13.9
33	10.9
	75.6

Percentage deviation in power in the specified harmonics = 29.4%.

Measured power spectrum shown in Figure 2.14(b).

Signal C: Signal M1 from Paehlike (1980). $N = 256$.

$1^+\ 2^-\ 2^+\ 6^-\ 2^+\ 3^-\ 1^+\ 6^-\ 2^+\ 2^-\ 10^+\ 3^-\ 22^+\ 1^-\ 1^+\ 7^-\ 3^+\ 9^-\ 2^+\ 3^-\ 5^+\ 2^-\ 10^+\ 15^-\ 2^+\ 5^-$
$2^+\ 2^-\ 2^+\ 6^-\ 2^+\ 2^-\ 2^+\ 23^-\ 5^+\ 14^-\ 6^+\ 2^-\ 26^+\ 3^-\ 1^+\ 3^-\ 5^+\ 2^-\ 10^+\ 2^-\ 3^+\ 5^-\ 1^+$

Longest run in the signal = 26 clock pulse intervals.

Specified harmonics and theoretical power in these harmonics:

Harmonic number k	Power (% of total)
1	7.28
2	7.18
3	6.65
5	7.17
7	7.29
11	6.76
18	6.70
26	7.31
42	6.91
64	6.91
	70.16

Percentage deviation in power in the specified harmonics = 9.41%

Measured power spectrum shown in Figure 2.14(c).

Signal D: Signal T18 from Paehlike (1980). $N = 128$.

$2^+\ 13^-\ 6^+\ 1^-\ 12^+\ 5^-\ 5^+\ 10^-\ 5^+\ 15^-\ 4^+\ 5^-\ 5^+\ 6^-\ 4^+\ 5^-\ 11^+\ 3^-\ 1^+\ 1^-\ 9^+$

Longest run in the signal = 15 clock pulse intervals.

Specified harmonics and theoretical power in these harmonics:

Harmonic number k	Power (% of total)	Harmonic number k	Power (% of total)
1	6.21	9	4.62
2	4.59	10	5.10
3	6.25	11	4.22
4	6.36	12	6.00
5	4.22	13	4.40
6	5.18	14	4.35
7	4.32	15	4.25
8	6.12	16	4.32
			80.51

Percentage deviation in power in the specified harmonics = 42.53%.

Measured power spectrum shown in Figure 2.14(d).

The points to note from Figure 2.14 and Table 2.5 are:

1. It is possible to get only a certain percentage of the total signal power into the specified harmonics of binary multi-frequency signals; there is inevitably a proportion of the power at the other (unspecified) harmonics. In many cases, it is also possible to concentrate somewhat more of the power in the specified harmonics if there are rather unequal amounts of power in these harmonics. This is illustrated by comparing signals A and B (Figures 2.14(a) and 2.14(b)). In signal A, the deviation between the maximum and minimum power in these harmonics, expressed as a percentage of the average power in these harmonics, is $100\ (11.37 - 10.84)/10.98 = 4.83\%$, but only 65.9% of the power is in these harmonics. In signal B, 75.6% of the power is in the six specified harmonics, but at the expense of a 29.4% deviation in the power in these harmonics.

   Up to a point, this deviation does not matter, but it is important that it is taken into account when the power in any specified harmonic becomes small. On the basis of total power in the specified harmonics, a squarewave might seem a particularly suitable signal for system identification if the specified harmonics are a fundamental and its odd harmonics. In a squarewave, 93.3% of the power is in harmonics 1, 3 and 5, while 96.6% is in harmonics 1, 3, 5, 7, 9 and 11, but there is a very unequal amount of power in the harmonics. The power in harmonic 5 is only 1/25 of that in the fundamental while that in harmonic 11 is only 1/121, so making this signal unsuitable for system identification in the frequency domain.
2. As the number of specified harmonics is increased, the average amount of power in each specified harmonic decreases, but the total amount of power in the specified harmonics of signals with similar deviations between maximum and minimum power in the specified harmonics increases. This is because of the reduction in the number of harmonics which are not specified. In signal D, more than 80% of the signal power is in the sixteen specified harmonics.
3. Provided the number of specified harmonics is not too large, the power in the specified harmonics is considerably greater than the power in the harmonics of a PRBS of similar length. For instance, the maximum power in any harmonic of a PRBS is in the first harmonic and is, for voltage levels ± 1 volt, 0.01587 volt^2 for $N = 127$ and 0.007873 volt^2 for $N = 255$.
4. A further figure of merit which is important in some applications is the longest run of clock pulse intervals with the signal at the same level ($+V$ or $-V$). If the signal is at one level for too long, the system response may start to deviate from its mean level by an unacceptable amount. For a PRBS based on a maximum length sequence with $N = 2^n - 1$, the longest run of consecutive clock pulse intervals with the signal at the same level is n (e.g. 7 for $N = 127$ and 8 for $N = 255$). The longest runs for the two binary multi-frequency signals with $N = 128$ in Table 2.5 are 14 for signal A and 15 for signal D, while those for the signals with $N = 256$ are 26 for signal B and 26 for signal C. The longest run of most binary multi-frequency signals is longer than that of a PRBS of comparable length.

5. Signal B has $N = 256$ and is inverse-repeat (i.e. its second half is the negative of the first half), in common with the six designs given by Buckner and Kerlin (1972). Thus, it is uncorrelated with signal A over their common period of 256 clock pulse intervals. This idea for generating uncorrelated binary multi-frequency signals was originated by van den Bos (1970). It is, of course, a similar idea to that of a PRBS and an inverse-repeat binary signal (Section 2.3) and, in common with these, the two signals do not have power at the same frequencies. Taking the harmonic spacing as $1/256\Delta t$, the inverse-repeat multi-frequency signal has power only at odd harmonics (1, 3, 5, 9, 17 and 33 for signal B), while the other signal has power only at even harmonics (2, 4, 8, 16, 32 and 64 for signal A).

2.4.2 Low-peak-factor multi-frequency signals

Binary multi-frequency signals contain only a certain percentage of the total signal power in the specified harmonics, but they automatically have a low ratio between peak-to-peak amplitude and total signal power, because they are binary. An alternative approach to multi-frequency signal design is to generate a (non-binary) signal which is the sum of the specified harmonics. This obviously has all the signal power in the specified harmonics, but the main design requirement for such signals is to phase the harmonics in such a way as to avoid one or more large peaks occurring at some time during the signal period, with only a small signal amplitude between the peaks.

This design requirement can be expressed in terms of the *peak factor*, which for a signal $x(t)$ is traditionally defined by

$$\text{Peak factor} = \frac{x_{\max} - x_{\min}}{2\sqrt{2}\, x_{\text{rms}}} \tag{2.36}$$

where $x_{\max}$, $x_{\min}$ and x_{rms} are, respectively, the maximum, minimum and RMS values of $x(t)$. The normalizing factor of $2\sqrt{2}$ in the denominator means that the peak factor of a single sine wave (with zero mean) is 1, while that of a binary signal with equal positive and negative levels ($+V$ and $-V$) is $1/\sqrt{2}$. It is evident that low peak factor is a design objective for a multi-frequency signal.

The weakness with the traditional definition of equation (2.36) is that peak factor can be affected by non-zero mean value or, in the case of a binary signal, by unequal positive and negative levels. For example, for a binary signal with levels $+V$ with probability p and $-V$ with probability $1-p$, $x_{\text{rms}} = V$, so that, provided $0 < p < 1$, the peak factor is $1/\sqrt{2}$. On the other hand, if the levels are $+2V$ with probability p and 0 with probability $1-p$, $x_{\text{rms}} = 2V\sqrt{p}$. The range $x_{\max} - x_{\min}$ is still $2V$, but the peak factor is now $1/2\sqrt{2p}$ and is dependent on p; for instance, if $p = 0.5$, the peak factor is 0.5. If p is fixed, a smaller peak factor would result as the positive level is increased.

This may seem a somewhat trivial point, but it is significant when considering a PRBS, which has levels $+V$ with probability $(N+1)/2N$ and $-V$ with probability $(N-1)/2N$. (Note that these probabilities can be reversed, according to the transformation from sequence logic values to signal voltage levels.) The peak factor of such a signal is $1/\sqrt{2}$. In some applications it is essential to use a zero mean signal, so that the signal mean V/N is subtracted from each level to give levels $(N-1)V/N$ with probability $(N+1)/2N$ and $-(N+1)V/N$ with probability $(N-1)/2N$. For this signal, the mean squared value is given by

$$E[x^2] = \frac{1}{N}\left[\frac{1}{2}(N+1)\frac{(N-1)^2}{N^2}V^2 + \frac{1}{2}(N-1)\frac{(N+1)^2}{N^2}V^2\right]$$

$$= V^2\frac{N^2-1}{N^2}$$

The peak factor is thus

$$\frac{1}{\sqrt{2}} \cdot \frac{N}{\sqrt{N^2-1}}$$

and is slightly larger than previously.

Several alternative definitions of peak factor which overcome these problems are possible, but that given in equation (2.36) is the one most widely used in the literature. It presents no problems with binary signals with equal positive and negative levels, or with sum of harmonics signals with zero mean, which are the types of signals which will be considered in the remainder of this chapter.

A non-binary multi-frequency signal (also called a multi-harmonic signal or a sum-of-harmonics signal) has its components at N_F specified harmonically related cyclic frequencies $f = k_1/T, k_2/T, \ldots$, where T is the period of the signal and $k_1, k_2, \ldots$ are specified integers. It is given by

$$x(t) = \sum_k (2P_k)^{1/2} \cos\left(\frac{2\pi kt}{T} + \varphi_k\right) \tag{2.37}$$

where P_k is the relative power of harmonic k and $k = k_1, k_2 \ldots$ We will consider the design of such signals with total power normalized to 1, i.e. $\Sigma_k P_k = 1$.

To illustrate the design problem, suppose that it is required to design a signal with 20 consecutive harmonics (1 to 20) with equal power in each, so that $P_k = 0.05$ (and so $(2P_k)^{1/2} = 0.316$). If the total signal power is 1 volt^2, then

$$x(t) = \sum_{k=1}^{20} 0.316 \cos\left(\frac{2\pi kt}{T} + \varphi_k\right) \text{ volts}$$

With all phases zero, the signal is a sum of cosines, all of which add up at $t = 0$ to give $x_{max} = 6.32$ volts; this signal is shown in Figure 2.15(a). For the unphased cosine waveform, it is found that $x_{min} = -1.57$ volts, so that the peak factor is, from equation (2.36), $(6.32 + 1.57)/2\sqrt{2} = 2.79$ (since $x_{rms} = 1$ due to the normalization).

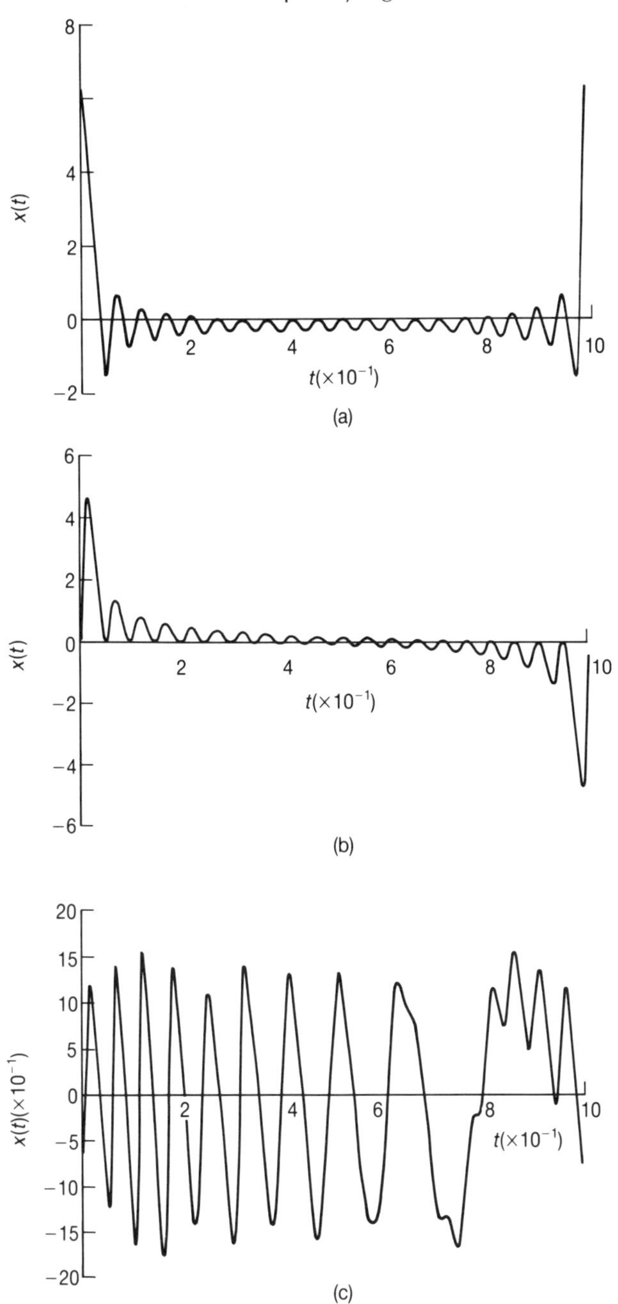

Figure 2.15 Multi-harmonic signals with 20 consecutive harmonics (1 to 20) and with total power 1 volt^2. (a) Unphased cosine waveform; (b) unphased sine waveform; (c) Schroeder-phased waveform.

If all the phases φ_k are fixed at $-90°$, the signal is a sum of sines, as shown in Figure 2.15(b). It is found that $x_{max} = +4.69$ volts and $x_{min} = -4.69$ volts, so that the peak factor is 3.32.

Neither signal is very attractive for use in system identification. The large peaks in each waveform mean that any system non-linearities, particularly saturation non-linearities, could easily be reached, while the small amplitude of both signals throughout most of the rest of their periods could well lead to inaccuracies due to amplitude quantization if the signal is digitized.

Fortunately, a considerable reduction in peak factor can be achieved in most cases by using a simple formula for the phases, devised by Schroeder (1970). He noted that frequency-modulated signals have a low peak factor and that Woodward's theorem states that the spectrum of a high-index frequency-modulated waveform is approximately the probability distribution of its instantaneous frequency. Working this through, it is found that, for harmonic k, the phase should be given by

$$\varphi_k = 2\pi \sum_{i=1}^{k} i\,.\,P_i \tag{2.38}$$

(Flower *et al.*, 1978b).

For a spectrum containing N consecutive harmonics of equal power content, so that $P_i = 1/N$, $i = 1, 2, \ldots, N$, equation (2.38) becomes

$$\begin{aligned}\varphi_k &= \frac{2\pi}{N} \sum_{i=1}^{k} i \\ &= \frac{\pi}{N}(k^2 + k)\end{aligned} \tag{2.39}$$

The term $\pi k/N$ can be dropped because its effect is simply to produce a linear phase in the frequency domain, which is equivalent to a pure time delay in the time domain. Thus for our N consecutive harmonic signal,

$$\varphi_k = \frac{\pi}{N} k^2 \tag{2.40}$$

Other, similar, formulae for the phases have been given in the literature and these are reviewed by van der Ouderaa *et al.* (1988b).

For our example of 20 equally spaced harmonics with equal power ($P_i = 1/20$, $i = 1, 2, \ldots, 20$), substitution in equation (2.38) gives $\varphi_1 = 2\pi/20$, $\varphi_2 = 6\pi/20$, $\varphi_3 = 12\pi/20, \ldots, \varphi_{20} = 420\pi/20$. (Note that we could equally have used equation (2.40).) The resulting signal is shown in Figure 2.15(c). The maximum and minimum values are $+1.57$ volts and -1.76 volts, so that the peak factor is 1.18; this is much lower than for the unphased cosine or sine signals.

Schroeder's formula results in low peak factors in most cases. Two instances where it does not work well are signals with only a small number of harmonics and those where the harmonics are widely separated. It is possible to improve upon the formula by computer optimization of the phases. Van den Bos (1987) developed an iterative procedure to reduce the peak factor, and this has recently been further refined by van der Ouderaa *et al.* (1988b), who give two examples of multi-harmonic signals with a peak factor of 0.993 (i.e. less than that of a single sine wave).

In another recent development, Rees (1990) has designed a 20-harmonic low-peak-factor signal in which the harmonic frequencies are prime number multiples of a fundamental which is itself excluded from the signal. The signal provides total immunity from harmonic distortion due to even power non-linearities and a substantial reduction compared with a PRBS, in harmonic distortion due to odd power non-linearities. This is because some of the distortion appears at frequencies not in the original signal. This will be discussed further in Section 2.6 (see also Chapter 12).

Low-peak-factor multi-frequency signals are beginning to appear in commercially available software packages and hardware. For example, Schroeder-phased signals have been incorporated into the Plant System Identification (PSI) software package, developed at the Midlands Research Station of British Gas plc (Porch, 1991) (see also Chapter 14). A multi-sine signal is also available in the Advantest TR98201 signal generator which accompanies the TR9403 digital spectrum analyzer. This consists of 400 consecutive equal-amplitude harmonics (1 to 400), so that $P_k = 1/400$, $(P_k)^{1/2} = 0.05$ and $(2P_k)^{1/2} = 0.0707$, $k = 1, 2, \ldots, 400$. The square root of power spectrum of this signal with its highest harmonic set at 2 kHz and the total power set as near as possible to 1 volt^2 is shown in Figure 2.16(a). It may be seen from this figure that the power spectrum is approximately constant, with a cursor placed (arbitrarily) at 1 kHz showing that $1.(P_{200})^{1/2} = 0.0492$ volts. Note that the spectrum appears continuous rather than a line spectrum because all harmonics are present and the spectral line spacing on the analyzer is automatically set to be the same as that of the multi-sine signal from the signal generator.

One period of the waveform itself is shown in Figure 2.16(b), from which it may be seen that the amplitude range $(x_{max} - x_{min})$ is 6.293 volts, giving a peak factor of $6.293/2\sqrt{2} = 2.225$. This is certainly much lower than that of an unphased signal (for example, a cosine unphased signal would have had x_{max}, at $t = 0$, equal to $400\,(2P_k)^{1/2} = 28.28$ volts), but it is not as low as that which could have been achieved from optimal phasing. A peak factor close to 1 could have been achieved and this would have corresponded to an amplitude range of approximately $2\sqrt{2}$ volts. This signal will be discussed further in Application Example 2.3 in Section 2.5.3.

2.4.3 Multi-frequency signal design using a MATLAB Toolset

A Frequency Domain System Identification Toolset has recently been developed for use with MATLAB by Kollar of the Technical University of Budapest, in collaboration with Schoukens and Pintelon of the Vrije Universiteit, Brussels. The Toolset has a section on excitation signal design, among which is the design of both binary

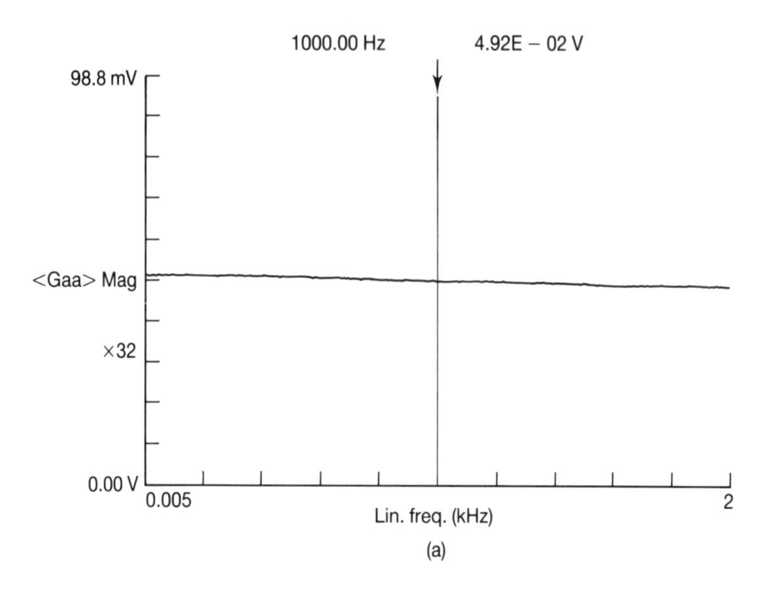

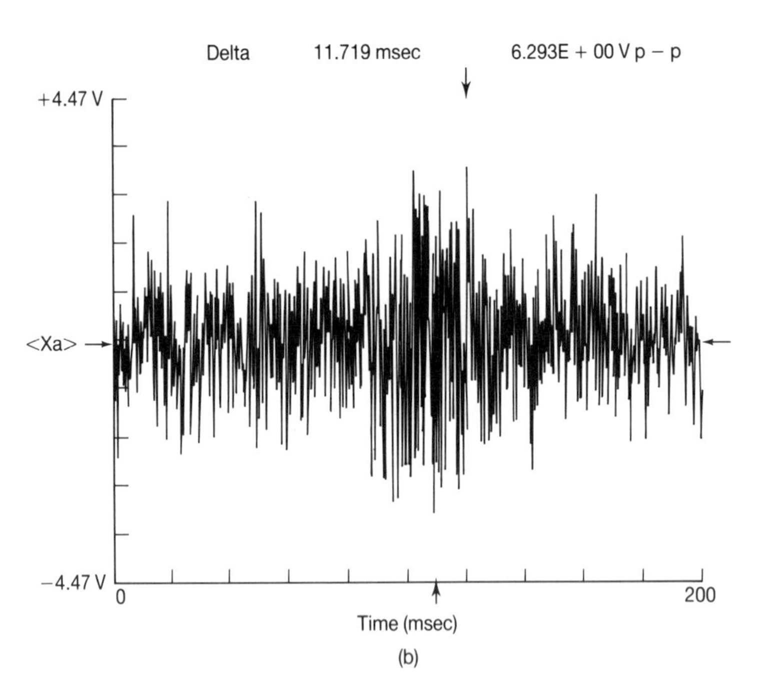

Figure 2.16 Multi-sine signal from the Advantest TR98201 signal generator with harmonic 400 set to 2 kHz and with total power set to 1 volt^2. (a) Square root of the power spectrum; (b) one period of the signal.

multi-frequency signals and sum of harmonics signals. The binary signal design uses the technique of van den Bos and Krol (1979), with possible improvement using a method described by Paehlike and Rake (1979), while the sum of harmonics design uses an algorithm based on swapping between the time domain and the frequency domain (van der Ouderaa *et al.*, 1988a).

2.5 Linear system identification in the frequency domain

2.5.1 Theory

In Section 1.1 we saw that, for a linear system of the form of Figure 1.1, the input signal $u(t)$ and the system response signal $y(t)$ were related by a convolution integral with the unit impulse response $h(t)$ of the system, giving the system relationship:

$$\begin{aligned} z(t) &= y(t) + n(t) \\ &= \int_0^{T_S} h(\lambda)u(t-\lambda)\,\mathrm{d}\lambda + n(t) \end{aligned} \tag{1.4}$$

In Section 2.2.4 we saw that multiplication of a data segment by a window function in the time domain became a convolution in the corresponding Fourier transforms in the frequency domain. The converse is also true, i.e. a convolution in the time domain becomes a multiplication in the frequency domain.

Thus, taking the Fourier transform of equation (1.2),

$$Z(\mathrm{j}\omega) = H(\mathrm{j}\omega)\,.\,U(\mathrm{j}\omega) + N(\mathrm{j}\omega) \tag{2.41}$$

where $Z(\mathrm{j}\omega)$, $U(\mathrm{j}\omega)$ and $N(\mathrm{j}\omega)$ are the Fourier transforms of the signals $z(t)$, $u(t)$ and $n(t)$, respectively, and $H(\mathrm{j}\omega)$, the Fourier transform of $h(t)$, is the *frequency response* of the system. This equation forms the basis of the most traditional method of measuring $H(\mathrm{j}\omega)$, by injecting (separately) sine waves of different frequencies and determining the gain and phase of the frequency response at those frequencies. This is a time-consuming procedure, and the noise term can often prove substantial in practice, so leading to poor estimation accuracy.

As we saw in Section 1.3.1, a feasible alternative is to use the convolution integral relationship involving correlation functions:

$$R_{uz}(\tau) = \int_0^{\infty} h(\lambda)\,.\,R_{uu}(\tau-\lambda)\,\mathrm{d}\lambda + R_{un}(\tau)$$

in which $R_{un}(\tau) = 0$ (over an infinite averaging time) if $u(t)$ and $n(t)$ are uncorrelated. For non-periodic signals, we obtain, by taking Fourier transforms,

$$S_{uz}(\mathrm{j}\omega) = H(\mathrm{j}\omega)\,.\,S_{uu}(\omega) + S_{un}(\mathrm{j}\omega) \tag{2.42}$$

where $S_{uu}(\omega)$ is the power-spectral density of $u(t)$, as given by equations (2.3) or (2.5), $S_{uz}(j\omega)$ is the *cross-power-spectral density* between $u(t)$ and $z(t)$, given by

$$S_{uz}(j\omega) = \int_{-\infty}^{\infty} R_{uz}(\tau)\, e^{-j\omega\tau}\, d\tau \tag{2.43a}$$

$$= \int_{-\infty}^{\infty} R_{uz}(\tau) \cos \omega\tau\, d\tau - j\int_{-\infty}^{\infty} R_{uz}(\tau) \sin \omega\tau\, d\tau \tag{2.43b}$$

and $S_{un}(j\omega)$ is the cross-power-spectral density between $u(t)$ and $n(t)$, similarly defined.

For periodic perturbation signals the above expressions have to be modified along the lines expected from Section 2.1.2, with consideration being restricted to cyclic frequencies $f = k/T$, where T is the period of the signal and k is an integer. The power-spectral densities become power spectra and the system relationship of equation (2.42) is in terms of power rather than power-spectral densities.

Similarly, when using the Discrete Fourier Transform (Section 2.2.1) the system relationship becomes:

$$S_{uz}(jk) = H(jk) . S_{uu}(k) + S_{un}(jk) \tag{2.44}$$

where, from equation (2.16a),

$$S_{uu}(k) = \frac{1}{N} \sum_{r=0}^{N-1} R_{uu}(r) \exp\left(-j\frac{2\pi kr}{N}\right) \tag{2.45a}$$

From equation (2.24), $S_{uu}(k)$ is also given by

$$S_{uu}(k) = U_k^* U_k \tag{2.45b}$$

where U_k is the DFT of the sampled data sequence u_r and the asterisk denotes complex conjugate. Similarly, the cross-power-spectral sequence is given by

$$S_{uz}(jk) = \frac{1}{N} \sum_{r=0}^{N-1} R_{uz}(r) \exp\left(-j\frac{2\pi kr}{N}\right) \tag{2.46a}$$

$$= U_k^* Z_k \tag{2.46b}$$

It is important to emphasize that the discrete relationships only apply to cyclic frequencies $f = k/T_0 = k/N . \Delta$, where Δ is the sampling interval, N is the number of samples within one segment of data and k is an integer. Since $n(t)$ is not, in general, periodic with period $N . \Delta/k$, the cross-power-spectrum term $S_{un}(jk)$ in equation (2.44) is variable over each data segment and this contributes to variability of the frequency-response estimates. The technique for reducing the variance of these estimates is to average results from several segments of data.

An important function in system identification in the frequency domain is the (squared) coherence function, defined by

$$\gamma_{uz}^2(k) = \frac{|S_{uz}(jk)|^2}{S_{uu}(k) \,.\, S_{zz}(k)} \tag{2.47}$$

γ^2 can take values between 0 and 1 and the closer γ^2 is to 1 at any frequency point $f = k/N\,.\,\Delta$, the greater the reliance that can be placed on the frequency-response estimate at that frequency. Three effects can cause γ^2 to be less than 1. These are input and output noise, non-linearities in the system, and inputs other than $u(t)$ which affect the system response signal $y(t)$.

Determining a value of γ^2 above which the frequency-response estimates can be regarded as reliable in a particular experimental situation is quite a complicated calculation (Wellstead, 1986) and it is more usual to work to a rule of thumb, which says that a value above, say, 0.8 indicates good reliability. The coherence function is very valuable for comparing results obtained with different perturbation signals and/or experimental conditions, as we will see in Section 2.5.3, and as will be demonstrated by Patton, Miles and Taylor in Chapter 10.

In the time domain, the impulse-response sequence, h_r, was a non-parametric model of a system, with the most familiar corresponding parametric model being the pulse transfer function $H(z)$, given by

$$H(z) = \frac{z^{-k}(b_1 z^{-1} + b_2 z^{-2} + \ldots + b_n z^{-n})}{1 + a_1 z^{-1} + a_2 z^{-2} + \ldots + a_n z^{-n}} \tag{2.48}$$

(See Application Example 1.4 and Section 1.10.) In a similar way, the frequency-response sequence $H(jk)$ is a non-parametric model in the frequency domain, with the most familiar corresponding parametric model being the (Laplace) transfer function $H(s)$, given by

$$H(s) = K_p \,.\, e^{-sT_D} \frac{(s - z_1)(s - z_2) \ldots (s - z_u)}{(s - p_1)(s - p_2) \ldots (s - p_v)} \tag{2.49}$$

where K_p is the steady-state gain of the system, T_D is the time delay, $z_1, \ldots, z_u$ are the zeros of the transfer function and $p_1, \ldots, p_v$ are the poles of the transfer function. To arrive at $H(s)$, a continuous-frequency parametric model is fitted to the frequency-response sequence and the $j\omega$ is replaced by s. In many texts, the (non-parametric) frequency response is referred to as a transfer function, but it is better to call the non-parametric model a frequency response, reserving ‘transfer function’ for the parametric model.

2.5.2 Applications of multi-frequency signals

There are many published applications of multi-frequency signals to system identification in the frequency domain. Binary multi-frequency signals have been applied to a small-scale distillation process (van den Bos, 1970), two nuclear reactors – the molten salt reactor experiment (Buckner and Kerlin, 1972) and the high-flux isotope reactor (Chen *et al.*, 1972) – the electric drive of a model position-control system

(Paehlike and Rake, 1979), a bench-scale stirred tank system and a liquid-level system (Harris and Mellichamp, 1980), a large water-heated crossflow heat exchanger (Franck and Rake, 1985), and an electric diffusion furnace (Sankowski, 1989c). In the two nuclear reactor applications, comparison is made between the use of a 128-digit binary multi-frequency signal, a 127-digit PRBS and a 126-digit binary inverse repeat signal. Bezanson and Harris (1986) have used a pair of uncorrelated binary multi-frequency signals (van den Bos, 1970) in the identification of a plastics extrusion process with two inputs (screw speed and valve position) perturbed simultaneously.

Early applications of low-peak-factor sum of harmonics signals were described by Flower and his co-workers, who used Schroeder-phased signals in dynamic measurements on hydrofoils (Flower *et al.*, 1976, 1978b) and a nuclear reactor (Flower *et al.*, 1978a). More recently, an application to a warm-air flow system has been reported (Mercer and Mailey, 1986), while Patton and his co-workers have made an assessment of the use of such signals in helicopter system identification (see Chapter 10).

2.5.3 Applications of system identification in the frequency domain

In this section three examples of system identification are presented, each making a different point about frequency-domain identification.

APPLICATION EXAMPLE 2.1 EFFECT OF SPECTRAL LEAKAGE ON FREQUENCY-DOMAIN IDENTIFICATION

In Section 2.2.4 we saw the effect of spectral leakage on the power spectrum of a sine wave and of a PRBS. In this Example the effect of leakage on the frequency-response gain of a system, measured using Fast Fourier Transformation, is illustrated.

The system was an eighth-order Butterworth low-pass filter with cut-off frequency 2 kHz, having a gain of 1 at low frequencies and a roll-off of -48 dB/octave at high frequencies. Measurements were made using an Advantest TR9403 digital spectrum analyzer, with maximum displayed frequency set to 5 kHz. With this setting, the discrete harmonic spacing is 12.5 Hz, data segments are of length 80 ms and the sampling interval Δ is 0.078125 ms.

Two signals with somewhat similar power spectra *when averaging over a long time* (i.e. over a large number of data segments) were used. The first, signal A, was a 2047-digit PRBS with voltage levels ± 1 volt and clock frequency 30 kHz ($\Delta t = 0.03$ ms) obtained from a Hewlett Packard HOI3722A noise generator. The period of the signal, 68.23 ms, was thus slightly less than the length of the data segments, while the harmonic spacing of 14.655594 Hz was slightly greater than the discrete harmonic spacing. The measured (power spectrum)$^{1/2}$ when averaging over 256 segments of data is shown in Figure 2.17(a); as in Examples 2G and 2H, averaging over more segments made no noticeable difference to the measured power spectrum.

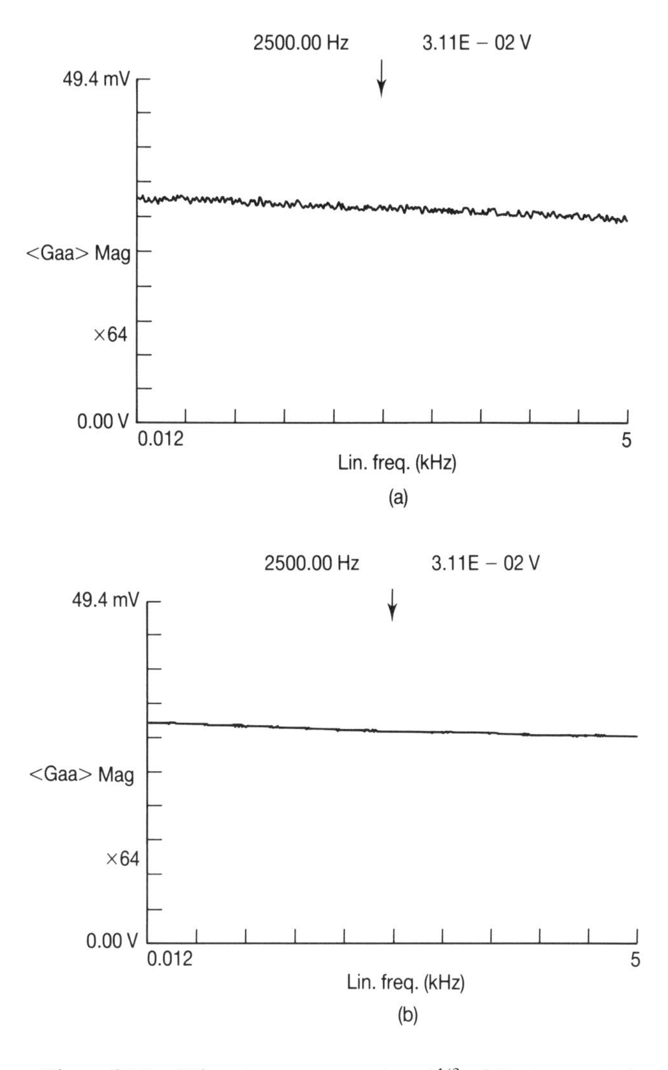

Figure 2.17 (Discrete power spectrum)$^{1/2}$ of the two pertubation signals used in Application Example 2.1. (a) Signal A: 2047-digit PRBS;. (b) signal B: multi-frequency signal.

The square root of the power in the first harmonic of signal A is

$$\frac{\sqrt{4096}\ \sin(\pi/2047)}{\pi} = 3.13 \times 10^{-2} \text{ volts}$$

The maximum displayed frequency of 5 kHz is closest to signal harmonic 341, the power in which is

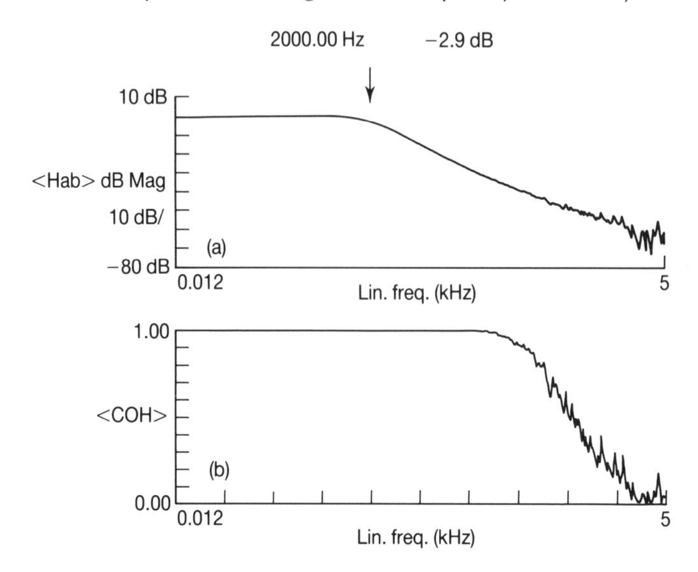

Figure 2.18 (a) Frequency-response gain and (b) coherence function of the system using signal A, averaging over 256 segments.

$$\frac{\sqrt{4096}\ \sin(341\pi/2047)}{341\pi} = 2.99 \times 10^{-2} \text{ volts}$$

The second signal (signal B) was the multi-sine waveform from an Advantest TR98201 signal generator (see Section 2.4.2), with the amplitude level adjusted to give approximately the same value of power spectrum at 2.5 kHz as for signal A. The measured power spectrum is shown in Figure 2.17(b), and it may be seen that the two power spectra are very similar over the frequency band up to 5 kHz.

The frequency response gain and coherence function using signal A are shown in Figure 2.18, with corresponding quantities obtained using signal B being shown in Figure 2.19. In both cases, averaging was over 256 segments of data, and a rectangular window was used.

The estimates of the frequency-response gain and the coherence functions using the two signals are very similar. The fall-off in coherence function at high frequencies in both cases is due to very low system gain coupled with a small amount of general laboratory electrical noise.

The picture is very different when averaging over a much shorter time, and corresponding results obtained when averaging over two segments of data (a total of 160 ms) are shown in Figure 2.20 (signal A) and Figure 2.21 (signal B).

When using signal B, there is no spectral leakage; averaging is over precisely two periods of the signal, and the results in Figure 2.21 show almost no variability from experiment to experiment. They are also very little different, up to 4 kHz, from the results obtained when averaging over 256 segments (Figure 2.19). There is more

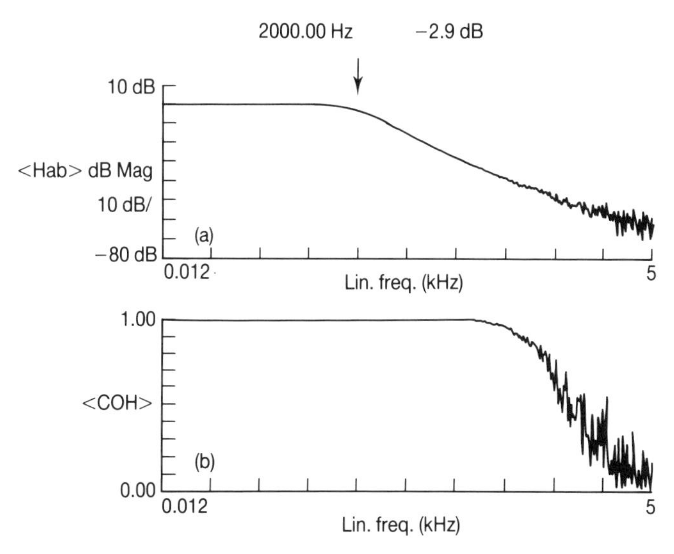

Figure 2.19 (a) Frequency-response gain and (b) coherence function of the system using signal B, averaging over 256 segments.

variability above 4 kHz due to the very low value of system gain (less than −48 dB) combined with general laboratory electrical noise. When using signal A, spectral leakage affects the results and both the gain and the coherence function vary considerably from experiment to experiment, depending on the point in the PRBS at which the averaging is commenced. The results shown in Figure 2.20 are typical and it can be seen that they are much less good than those obtained when averaging over a longer time (Figure 2.18).

The results using signal A are improved somewhat using a flat-pass window, and a typical result averaging over two segments of data is shown in Figure 2.22. Both the frequency response and the coherence function are smoother than in Figure 2.20, as expected using this type of window. Also, the coherence function is acceptably close to 1 up to about 3.8 kHz, compared with approximately 2.6 kHz in Figure 2.20(b).

The Example has demonstrated the desirability of avoiding spectral leakage if at all possible in frequency-domain system identification. In this Example this has been achieved using signal B, obtained from the Signal Generator associated with the Spectrum Analyzer. As noted at the end of Section 2.2.4, leakage could have been avoided when using signal A if a Discrete Fourier Transform of suitable length had been used instead of a radix-2 FFT. Such a DFT can be obtained using three FFT operations by means of the chirp z-transform (Rabiner *et al.*, 1969).

APPLICATION EXAMPLE 2.2 COMPARISON OF A PRBS AND A BINARY MULTI-FREQUENCY SIGNAL IN THE IDENTIFICATION OF A NOISY PROCESS

A binary multi-frequency signal with $N = 128$ (specified harmonics 1, 2, 4, 8, 16 and 32; Paehlike (1980) signal T12 – see Figure 2.14(a)) and a PRBS with $N = 127$ were

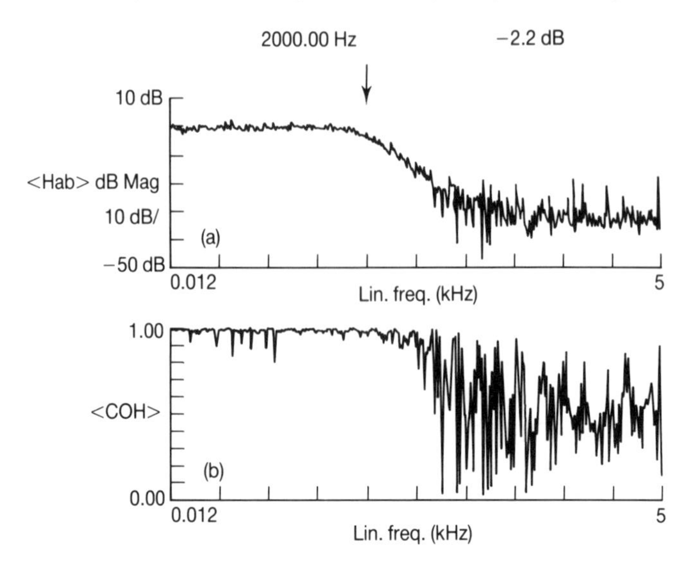

Figure 2.20 (a) Frequency-response gain and (b) coherence function of the system using signal A, averaging over 2 segments.

applied in turn to the heater input of a warm-air flow system (Feedback Process Trainer PT326), operating open-loop. The output signal was the temperature of air at the far end of the tube. With this configuration, the system has, for a 60° opening of the air-flow regulator, a pure time delay of approximately 0.16 s, a dominant time

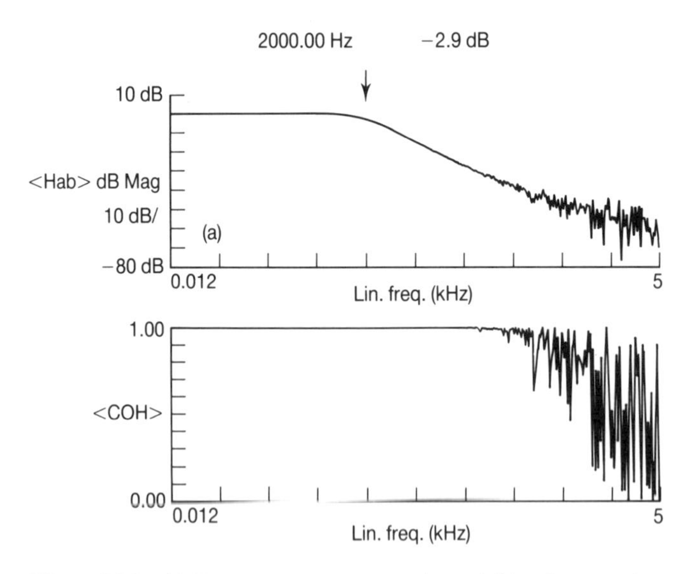

Figure 2.21 (a) Frequency-response gain and (b) coherence function of the system using signal B, averaging over 2 segments.

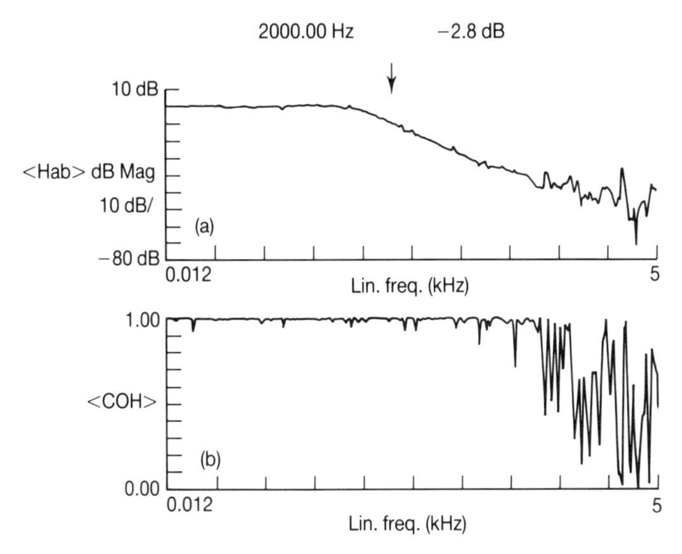

Figure 2.22 (a) Frequency-response gain and (b) coherence function of the system using signal A, averaging over 2 segments and using a flat-pass window.

constant of some 0.4 s and a steady-state gain of approximately 1 volt/volt. The period ($N \,.\, \Delta t$) of both signals was set to approximately 10 s and the amplitude to ± 1 volt.

Two separate experiments were performed using both input signals. In the first, no noise was added, while in the second, Gaussian noise with variance 1 volt^2 was added at the output. In each case, the magnitude of the cross-power spectrum was measured using a Hewlett Packard 5420B digital signal analyzer.

For the experiment with no noise added, averaging was over 10 segments of data, with each segment longer than the period of the perturbation signal, so that the effects of spectral leakage on the cross-power spectrum when using the PRBS were comparatively small. The magnitude of the cross-power spectrum is shown in Figure 2.23(a) for the binary multi-frequency signal input and in Figure 2.23(b) for the PRBS input. The low-pass characteristic is very much in evidence in both plots, but the cross-power spectrum is considerably smaller using the PRBS, as expected given the greater power in the specified harmonics of the multi-frequency signal.

The low-pass characteristic is still clear in Figure 2.24(a), but virtually nothing could be deduced from the cross-power spectrum when the PRBS is used (Figure 2.24(b)). It is possible that the latter result might have been improved had a longer averaging time been used, but to obtain the results in Figure 2.24, 20 segments of data were averaged, with each segment longer than the period (10 s) of the perturbation signal. It seems likely, therefore, that a very long averaging time would have been necessary to deduce much from the experiment with the PRBS input. It is also questionable whether much improvement would in fact have been obtained,

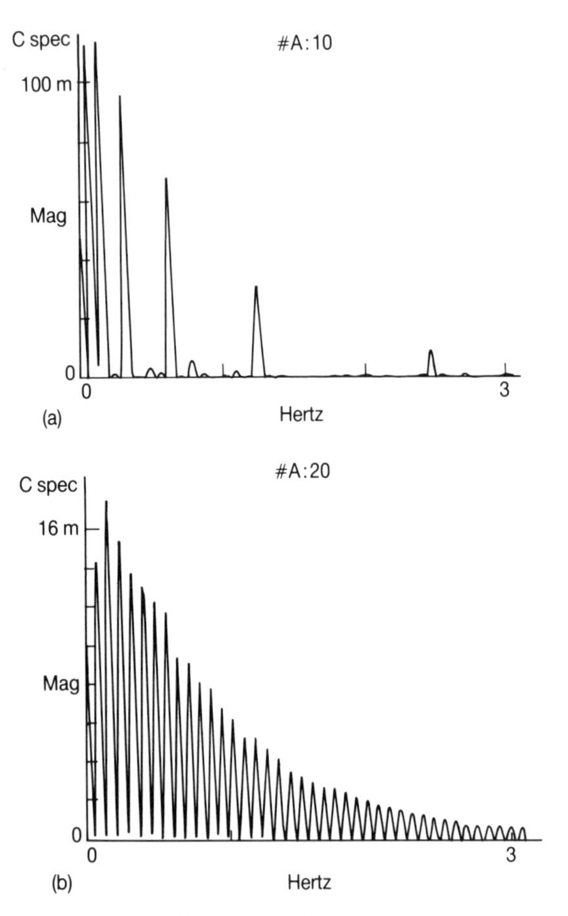

Figure 2.23 Magnitude of cross-power spectrum with no noise added. (a) Binary multi-frequency signal input ($N = 128$); (b) PRBS input ($N = 127$).

because this particular process does have a tendency to drift when operated open-loop, so violating the assumption of stationarity of the input and output signals.

APPLICATION EXAMPLE 2.3 USE OF A MULTI-SINE (SUM OF HARMONICS) SIGNAL

This example is designed to illustrate the problems which may arise when using a signal which does not have a particularly low peak factor. The multi-sine signal was obtained from an Advantest TR98201 signal generator and, as we have already seen in Section 2.4.2, this signal consists of the sum of 400 consecutive harmonics (1 to 400), with equal power in each, but its peak factor of 2.225 is not particularly low. A peak factor close to 1 could have been obtained using optimal phasing.

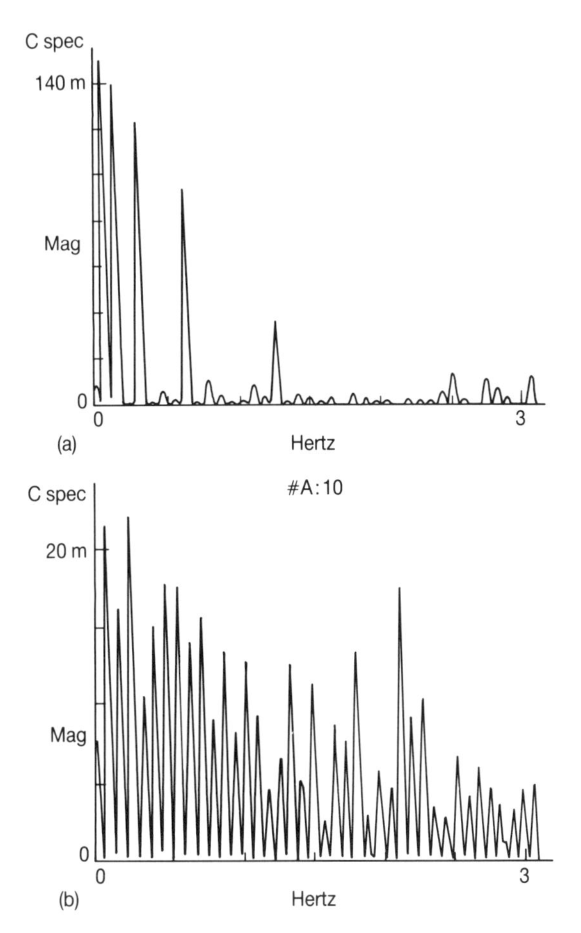

Figure 2.24 Magnitude of cross-power spectrum with noise added. (a) Binary multi-frequency signal input ($N = 128$); (b) PRBS input ($N = 127$).

The system in this example was third order with transfer function $H(s) = 1/(1 + sT_p)^3$ with $T_p = 10$ ms, which is a low-pass system with a cut-off frequency of 8.12 Hz. The maximum frequency was set at 200 Hz, so as to reasonably span the frequency range of the system. The harmonics are thus at intervals of 0.5 Hz from 0.5 Hz to 200 Hz and the period of the signal is 2 s. The total power in the signal was set to 1 volt^2, so that the square root of the power spectrum is as in Figure 2.16(a), with the frequency scale divided by 10, and one period of the signal itself is as in Figure 2.16(b), with the time scale multiplied by 10.

As noted above, a signal with peak factor close to 1 could have been obtained through optimal phasing; this peak factor would have corresponded to an amplitude range of $2\sqrt{2}$ volts. A second signal was therefore used in the system identification;

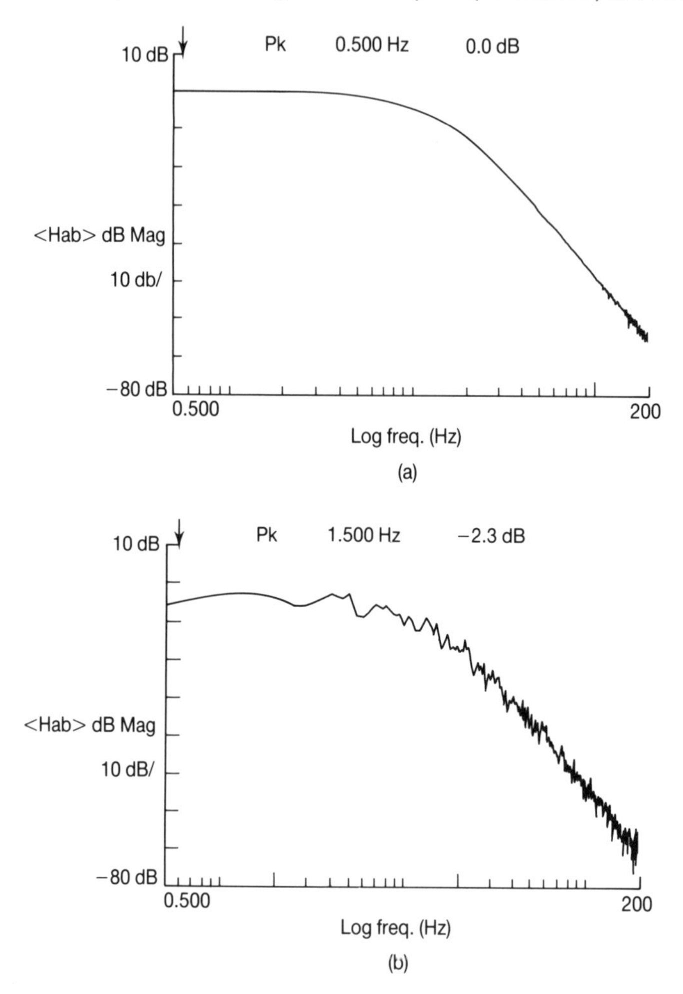

Figure 2.25 Bode gain diagrams in Application Example 2.3 using (a) the unclipped and (b) the clipped input signals.

this was the multi-sine signal obtained from the signal generator, clipped at $\pm\sqrt{2}$ volts. To emphasize the thinking behind this, it is expected that the system-identification results will be considerably affected by use of the clipped multi-sine signal, but another (hypothetical) signal with the same power spectrum as the unclipped signal could have been obtained by using computer optimization of the phases. This would have had its amplitude within the range $\pm\sqrt{2}$ volts, and so would have been unaffected by the clipping.

The gain of the frequency response of the system was measured, with averaging over 64 segments of data (each 2 s long). The Bode plot of the gain is shown in Figure 2.25(a) for the unclipped signal and in Figure 2.25(b) for the clipped signal,

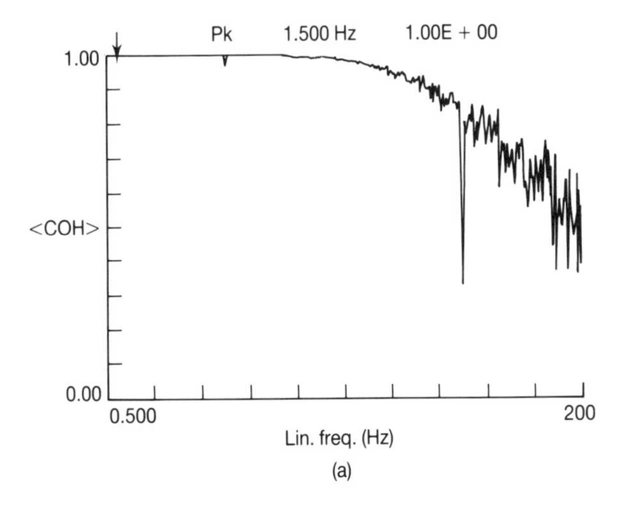

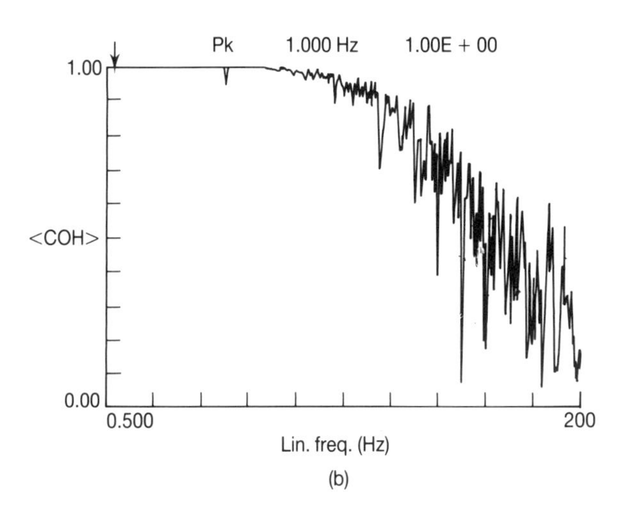

Figure 2.26 Coherence functions (γ^2) in Application Example 2.3, using (a) the unclipped and (b) the clipped input signals.

from which it may be seen that clipping introduced very noticeable distortion.

The corresponding coherence functions are shown in Figure 2.26. From Figure 2.26(a) it can be seen that when using the unclipped signal, γ^2 falls off as frequency increases; it is greater than 0.9 up to about 130 Hz and greater than 0.7 up to about 165 Hz (except for an isolated low value at 150 Hz). The fall-off in γ^2 is a result of the response signal becoming small as the frequency increases (as in Application Example 2.1) due to the system gain decreasing, while the isolated low value at 150 Hz is a combination of this together with a significant presence of third harmonic of mains frequency (50 Hz) from other equipment operating nearby. With the clipped signal, it may be seen from Figure 2.26(b) that the fall in γ^2 starts at lower frequencies, with 0.9 being reached near 100 Hz and 0.7 at approximately 130 Hz.

2.6 Non-linear system identification in the frequency domain

We saw in Section 1.7 that one of the most general mathematical representations of the relationship between the input $u(t)$ and output $y(t)$ of a non-linear system is the Volterra functional series expansion of equation (1.69). In the frequency domain, on Fourier transformation, the relationship becomes:

$$\begin{aligned} Y(\mathrm{j}\omega) = {} & H(\mathrm{j}\omega) \,.\, U(\mathrm{j}\omega) \\ & + H_2(\mathrm{j}\omega_1, \mathrm{j}\omega_2) U(\mathrm{j}\omega_1) U(\mathrm{j}\omega_2) \\ & + \dots \\ & + H_n(\mathrm{j}\omega_1, \mathrm{j}\omega_2, \dots, \mathrm{j}\omega_n) U(\mathrm{j}\omega_1) U(\mathrm{j}\omega_2) \dots U(\mathrm{j}\omega_n) \\ & + \dots \end{aligned} \tag{2.50}$$

where $H(\mathrm{j}\omega)$ is the linear system frequency response, as considered in Section 2.5, and

$$H_2(\mathrm{j}\omega_1, \mathrm{j}\omega_2) = \int_{-\infty}^{\infty} h_2(\lambda_1, \lambda_2) \exp(-\mathrm{j}\omega_1\lambda_1 - \mathrm{j}\omega_2\lambda_2)\, \mathrm{d}\lambda_1\, \mathrm{d}\lambda_2 \tag{2.51}$$

is the second-order Volterra kernel transformation, $h_2(\lambda_1, \lambda_2)$ being the second-order Volterra kernel, identification of which was considered in Section 1.7.2. Higher-order Volterra kernel transformations can be formed similarly.

2.6.1 Identification of the linear frequency response of a non-linear system

As noted in Section 1.7.1, in many experimental situations involving non-linear systems the desired result is as accurate a characterization as possible of the linear dynamics of the system. In the frequency domain this characterization is of the frequency response $H(\mathrm{j}\omega)$ or a corresponding parametric model. As in the time domain, the use of a signal with an inverse-repeat property is desirable, because the contributions from the even-order non-linearities then disappear. Buckner and Kerlin (1972) designed all their binary multi-frequency signals with an inverse-repeat property. These signals were used in frequency-response measurements of nuclear reactors which were considered to be significantly non-linear. Barker and Davy (1975) showed that two experiments with inverse-repeat signals of different amplitudes could be used to eliminate the effect of a third-order non-linearity on a frequency response. There is no obvious procedure for removing errors due to higher-order odd non-linearities, but the third-order would normally be expected to predominate.

For a multi-sine (sum of harmonics) signal it is desirable for the peak factor to be low to reduce possible distortion from saturation-type non-linearities; this was illustrated in Application Example 2.3. For most types of non-linearity a component with (cyclic) frequency f in the input signal results in components in the response signal at some of the harmonically related frequencies kf, where k is an integer ($\geqslant 1$). This could not be seen in Application Example 2.3 because the input signal contained power at all the displayed harmonics (1 to 400) but it is illustrated in the next Application Example.

APPLICATION EXAMPLE 2.4 EFFECT OF A SATURATION NON-LINEARITY ON AN OUTPUT POWER SPECTRUM

This example compares the use of two 16-harmonic, flat-power-spectrum signals with harmonic spacing of four (i.e. harmonics 1, 5, 9, ..., 61) in the identification of the fourth-order Butterworth low-pass filter of Application Example 2.1. Both signals have $P_k = 1/16$ and total power 1 volt^2, so that

$$u(t) = \sum_k 0.354 \cos\left(\frac{2\pi kt}{T} + \varphi_k\right) \text{ volts}$$

where $k = 1, 5, 9, \ldots, 61$.

Signal A (Figure 2.27(a)) had phases determined from Schroeder's formula (equation (2.38)) and a peak factor of 1.52. Note that this is higher than that (1.18) of the 20 consecutive harmonics signal of Figure 2.15(c), confirming that Schroeder's formula generally works less well as the harmonic spacing is increased. Signal B (Figure 2.27(b)) had all its phases set to $-90°$ (i.e. a sum of unphased sines) and a peak factor of 3.88.

The input signal to the filter was clipped at ± 3 volts to simulate input saturation. This clipping level does not affect signal A but it does affect signal B. The power spectrum of both signals is shown in Figure 2.28(a), and the power spectrum of the filter output signal is shown in Figure 2.28(b) for the signal A input and Figure 2.28(c) for the signal B input. Results were obtained using a Hewlett Packard 5420B digital signal analyzer, with the signal period (T) set at a sub-multiple of data-segment length to avoid spectral leakage.

Comparing Figures 2.28(b) and 2.28(c), it is seen that the amplitude of the output power spectrum at the harmonics present in the input signal is much reduced when using signal B. It is also seen that there are additional harmonics in the output power spectrum in Figure 2.28(c), including harmonics at frequencies above that of the highest harmonic in the input power spectrum.

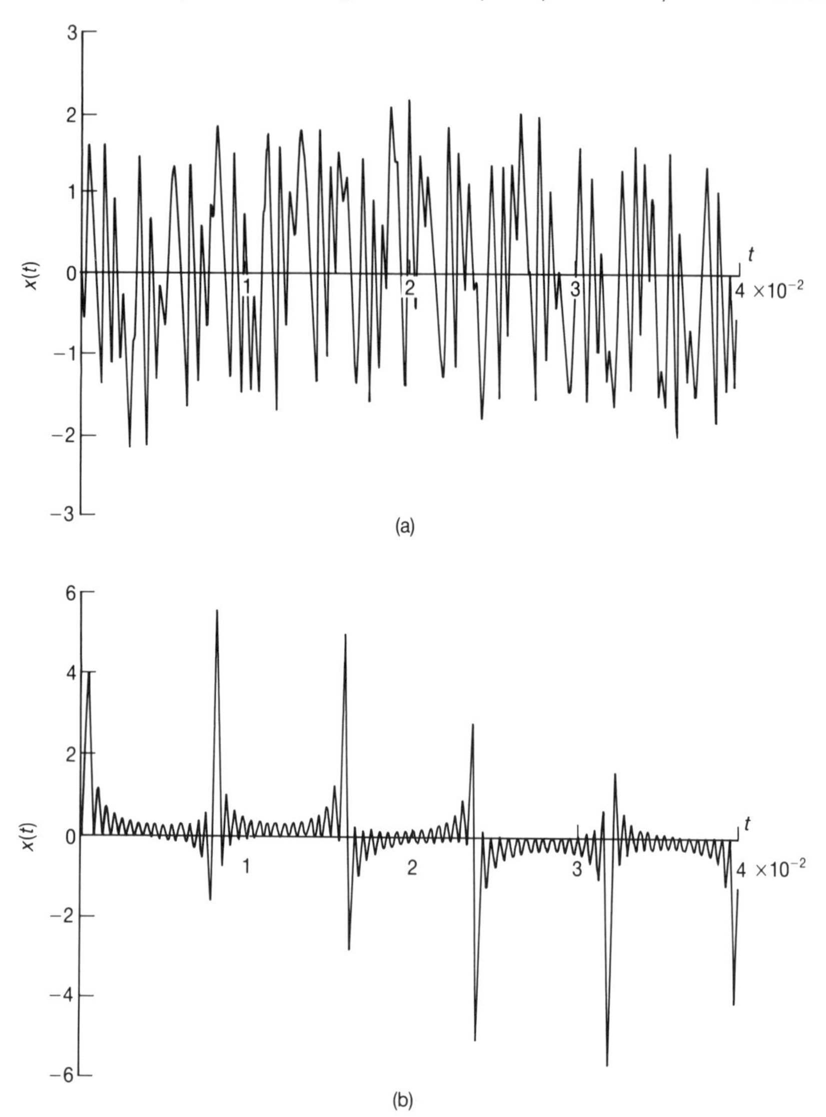

Figure 2.27 Multi-sine signals used in Application Example 2.4. (a) Signal A: Schroeder-phased waveform; (b) signal B: unphased sine waveform.

2.6.2 Identification of non-linear characteristics in the frequency domain

The second-order kernel transformation $H_2(j\omega_1, j\omega_2)$ of equation (2.49) could, in principle, be obtained by Fourier transformation of the second-order Volterra kernel

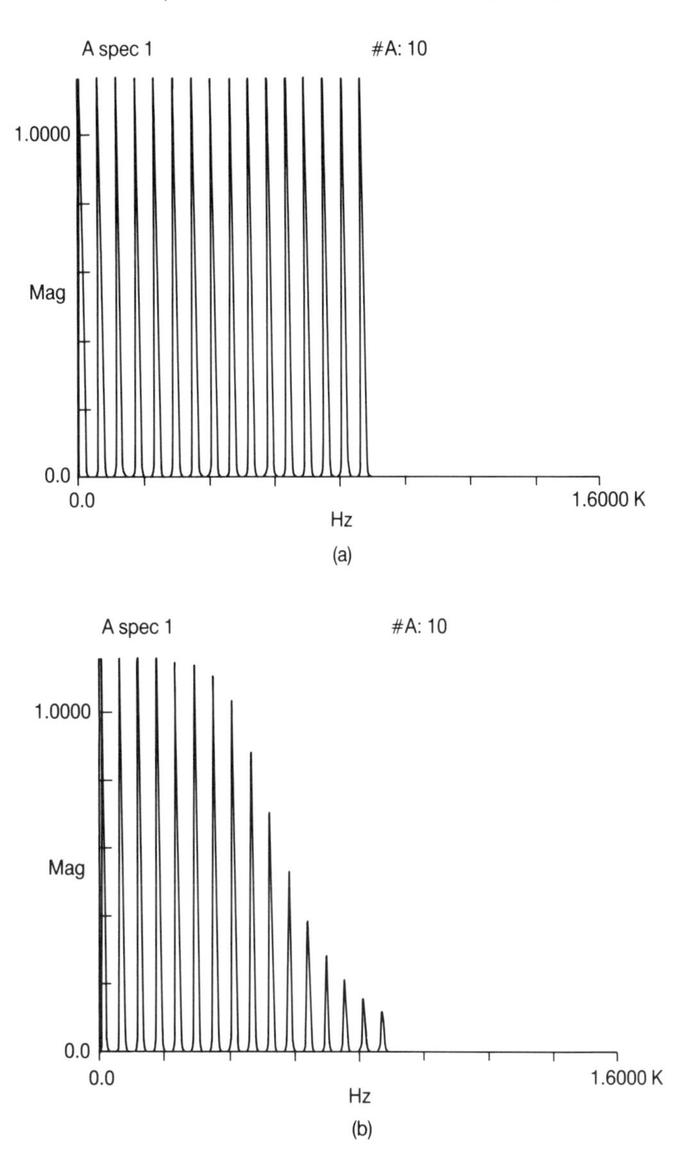

Figure 2.28 Power spectra of signals in Application Example 2.4. (a) Power spectrum of input signal (A, or B before clipping); (b) power spectrum of response to signal A.

$h_2(\lambda_1, \lambda_2)$, the identification of which was considered in Section 1.7.2. It was seen there that the requirements for a perturbation signal were that it should be of inverse-repeat form and that it should have at least three levels. Higher-order Volterra kernel transformations can be found similarly, but they are difficult to display and to interpret so that, in practice, attention is usually restricted to $H(j\omega)$ and

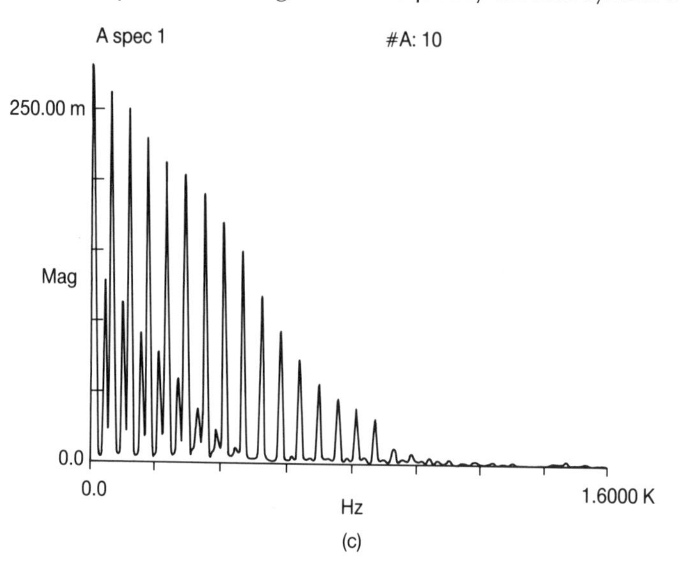

Figure 2.28 (cont.) (c) Power spectrum of response to signal B.

$H_2(j\omega_1, j\omega_2)$. The latter has two frequency axes (ω_1 and ω_2), but in most cases it is difficult to interpret the phase, so that attention is restricted to the gain. This can be displayed in a number of ways (for example, a contour plot or using three-dimensional graphics).

The contour plot of the gain H_2 is particularly suitable for examining for gains which are large for certain relationships between frequencies (for example, a small difference between frequencies). It is possible to envisage an application in which, on the basis of *linear* systems theory, two frequencies f_1 and f_2 occurring in an input signal are well above a particular resonance in a system, but because the system is *non-linear*, the difference $f_2 - f_1$ excites the resonance. This could be important in the design of oil or gas rigs operating at sea, where the input signal is sea-wave height. The potential of this type of analysis is only beginning to be realized.

Billings (1986) describes a different approach to computing $H_2(j\omega_1, j\omega_2)$, which is rather more tractable in most practical situations. This involves estimating a Non-linear Autoregressive Moving Average (NARMA) model relating the input sequence u_r and the response sequence y_r of a system:

$$y_r = F\{y_{r-1}, y_{r-2}, \ldots, y_{r-n_y}, u_{r-1}, u_{r-2}, \ldots, u_{r-n_u}\} \tag{2.52}$$

where $F\{.\}$ denotes some non-linear function of $\{y\}$ and $\{u\}$. The model is readily extended to systems with noise to provide a relationship which gives the measurable output sequence $z_r = y_r + n_r$. The model can involve non-linear terms in the input, past values of the output and, in the case of a noisy system, multiplicative terms between input and noise and between measurable output and noise.

Frequency responses can be computed readily from such a model (Billings and Tsang, 1989). To estimate $H(\mathrm{j}\omega)$, u_r is set to $\exp(\mathrm{j}\omega r\Delta)$, where Δ is the sampling interval, and coefficients of $\exp(\mathrm{j}\omega r\Delta)$ are equated. Similarly, to estimate $H_2(\mathrm{j}\omega_1, \mathrm{j}\omega_2)$, u_r is set to $\exp(\mathrm{j}\omega_1 r\Delta) + \exp(\mathrm{j}\omega_2 r\Delta)$ and coefficients of $\exp[(\mathrm{j}\omega_1 + \mathrm{j}\omega_2)\, r\Delta]$ in the model are equated. A good deal of data is required to obtain a reasonable model, and the perturbation signal must adequately span (in amplitude) any of the non-linearities it is required to model. Practical applications of this approach are described by Billings (1986) and Billings and Tsang (1989).

A further approach to estimating non-linear characteristics in the frequency domain has been described by Barker and Al-Hilal (1985). They used signals obtained from multi-level maximum length sequences, and were able to control the harmonic content to the extent that harmonics which are multiples of specified primes could be suppressed. This was achieved by appropriate mapping of the elements of the Galois field of the m-sequence into the levels of the corresponding pseudo-random signal, a technique considered further by Barker in Chapter 11. For example, for a system with a linear pathway together with second- and third-order non-linearities, the contribution of the second-order non-linearity to the system could be completely separated and the contributions of the linear and third-order non-linearities to the system output could be partially separated. Barker and Al-Hilal (1985) give an example in which a seven-level pseudo-random signal was used both to identify and to estimate the parameters of a model with a cubic non-linearity preceding linear dynamics.

3

Design of Broadband Excitation Signals

Johan Schoukens, Patrick Guillaume and Rik Pintelon

3.1 Introduction

In most system-analysis applications the dynamic behaviour of the system is derived from measurements of the input and output signals. Sometimes the input signal is imposed by the environment and it is impossible to excite the device under test with an arbitrarily chosen input (for example, in biological systems, where the choice of excitation is very limited). In other situations, only binary signals may be applicable. However, in a wide variety of cases, the only restriction on input signals is that of a limitation in the permitted amplitude range.

A very common method used in measuring transfer functions up to the end of the 1960s was that of the combination of a slowly swept sine with a tracking filter. Since the development of advanced digital signal processing algorithms, and especially since the efficient implementation of the Discrete Fourier Transform (DFT) with the Fast Fourier Transform (FFT), it has become possible to use more complex input signals. Instead of exciting the unknown system one frequency at a time, sophisticated waveforms with a broadband spectrum are generated, enabling collection of all the required spectral information from a single measurement using a digital-to-analog converter. This results in a considerable reduction in the measurement time, but also in an undesired loss of accuracy if no precautions are taken.

We will analyze the trade-off between accuracy and measurement time, but before starting, we must choose between a non-parametric or a parametric modelling approach. In the non-parametric representation the system is characterized by measurements of the frequency response at a large number of frequencies, while in a parametric model the system is described by a mathematical transfer function model with a limited number of parameters. It is precisely these parameters which have to be estimated in the parametric modelling approach. The optimum spectrum of the excitation in the parametric case will be different from that in the non-parametric case: this is principally because the parametric model combines the information available from all frequencies in only a few parameters. In a direct non-parametric frequency-response measurement there is no relation between the measurements at the various frequencies, and the excitation should be designed to achieve a predefined accuracy in the frequency bands of interest: for example, maximize the absolute or relative accuracy of the measurements. In a parametric approach the energy will be concentrated at those frequencies where it contributes most to the knowledge about the model parameters.

In this chapter we will first focus our attention on the design of excitation signals for non-parametric measurements. The parametric modelling approach will be studied in the second part.

3.2 Optimization of excitation signals for non-parametric frequency-response measurements

3.2.1 Introduction

The non-parametric measurement problem is studied in this part. In the first section we will consider some frequency-response function (FRF) measurement techniques and analyze their sensitivity to disturbing noise. In the second section these results will be used to determine some important design parameters of excitation signals which will finally result in the time factor (Tf) as a quantitative measurement of the quality of an excitation signal. This will allow the comparison of different excitation signals in the third section. Next, we will concentrate on the multi-sine as an excitation signal, and finally we will illustrate the ideas developed in this chapter by an experiment.

3.2.2 Measurement of FRF: study of the systematic and stochastic errors

From system theory it is known that a linear time-invariant system is completely characterized by its impulse response. The Fourier transform of the impulse-response function is called the FRF of the system. Since there exists a unique relation between both the time- and frequency-domain representations, the impulse response can be obtained from measurements in both domains. In practice, most instruments use a

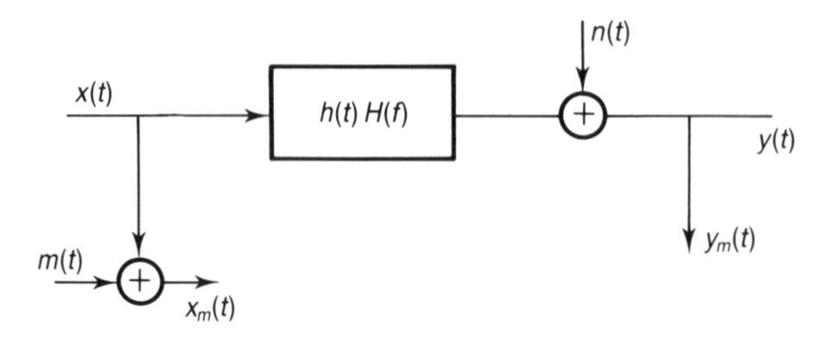

Figure 3.1 Signal flow diagram.

frequency-domain approach because this results in a higher signal-to-noise ratio (SNR). Initially, these measurements were made frequency by frequency but in the late 1970s FFT analyzers appeared on the market, making it possible to measure the FRF at a large number of frequencies at once.

The signal flow diagram of a typical measurement set-up is shown in Figure 3.1. The principle of the FFT analyzers is quite simple: the device under test (DUT) is stimulated using a broadband excitation signal. Both the input $x(t)$ and the output $y(t)$ are measured and transformed to the frequency domain. The noise on the input and output measurements is represented by $m(t)$ and $n(t)$ respectively. The FRF $H(\mathrm{j}\omega)$ is given by the ratio of the output spectrum $Y(\mathrm{j}\omega)$ to the input spectrum $X(\mathrm{j}\omega)$:

$$H(\mathrm{j}\omega) = \frac{Y(\mathrm{j}\omega)}{X(\mathrm{j}\omega)} \tag{3.1}$$

where $X(\mathrm{j}\omega)$, $Y(\mathrm{j}\omega)$ and $H(\mathrm{j}\omega)$ are the Fourier transforms of $x(t)$, $y(t)$ and $h(t)$, respectively. If the measurements are disturbed by noise, then random error terms due to the noise appear in the previous expression, and the measured FRF $H_m(\mathrm{j}\omega)$ becomes

$$H_m(\mathrm{j}\omega) = \frac{Y_m(\mathrm{j}\omega)}{X_m(\mathrm{j}\omega)} = \frac{Y(\mathrm{j}\omega) + N(\mathrm{j}\omega)}{X(\mathrm{j}\omega) + M(\mathrm{j}\omega)} = H(\mathrm{j}\omega)\,\frac{1 + [(N(\mathrm{j}\omega))/(Y(\mathrm{j}\omega))]}{1 + [(M(\mathrm{j}\omega))/(X(\mathrm{j}\omega))]} \tag{3.2}$$

where $M(\mathrm{j}\omega)$ and $N(\mathrm{j}\omega)$ are the noise on the spectra of $X(\mathrm{j}\omega)$ and $Y(\mathrm{j}\omega)$. If the SNR at the input is low, then the denominator $(1 + M(\mathrm{j}\omega)/X(\mathrm{j}\omega))$ can become very small or even zero, resulting in unreliable measurements. To avoid this problem, the auto- and cross-correlations $R_{xx}(\tau)$, $R_{yy}(\tau)$, $R_{xy}(\tau)$, $R_{yx}(\tau)$ were introduced together with their spectral representations $S_{xx}(\omega)$, $S_{yy}(\omega)$, $S_{xy}(\omega)$, $S_{yx}(\omega)$ (Bendat and Piersol, 1980). Using these quantities the transfer function can be calculated as

$$H_1(\mathrm{j}\omega) = \frac{S_{xy}(\mathrm{j}\omega)}{S_{xx}(\omega)} \quad \text{or} \quad H_2(\mathrm{j}\omega) = \frac{S_{yy}(\omega)}{S_{yx}(\omega)} \tag{3.3}$$

At this moment all the FFT analyzers use these relations to measure the FRF of a system. For periodic excitation signals equation (3.3) becomes

$$H_1(j\omega) = \frac{\frac{1}{N}\sum_{k=1}^{N} Y_{mk}(j\omega)X_{mk}^*(j\omega)}{\frac{1}{N}\sum_{k=1}^{N} X_{mk}(j\omega)X_{mk}^*(j\omega)} \qquad H_2(j\omega) = \frac{\frac{1}{N}\sum_{k=1}^{N} Y_{mk}(j\omega)Y_{mk}^*(j\omega)}{\frac{1}{N}\sum_{k=1}^{N} X_{mk}(j\omega)Y_{mk}^*(j\omega)} \tag{3.4}$$

The asterisk denotes the complex conjugate and $H_1(j\omega)$ is equal to $H_2(j\omega)$ if there is no noise on the measurements. Here, N is the number of averaged measurements. Under noisy measurement conditions the following results will be obtained:

$$H_1(j\omega) = H(j\omega)\frac{1}{1 + [(S_{MM}(\omega))/(S_{XX}(\omega))]} \qquad H_2(j\omega) = H(j\omega)\left(1 + \frac{S_{NN}(\omega)}{S_{YY}(\omega)}\right) \tag{3.5}$$

Due to the disturbing noise, systematic errors appear in both expressions. Only if the input is noise free will $H_1(j\omega)$ become equal to $H(j\omega)$, and $H_2(j\omega)$ equals $H(j\omega)$ if the noise on the output is zero. However, in many situations there will be noise on the input as well as on the output and then neither H_1 or H_2 can be used to measure the FRF without systematic errors.

Under these conditions a logarithmic averaging procedure gives much better results. It is shown that

$$\begin{aligned} H_{\log}(j\omega) &= \exp\left(\frac{1}{N}\sum_{k=1}^{N} \log H_{mk}(j\omega)\right) = \exp\left(\frac{1}{N}\sum_{k=1}^{N} \log \frac{Y_{mk}(j\omega)}{X_{mk}(j\omega)}\right) \\ &= \left(\prod_{k=1}^{N} \frac{Y_{mk}(j\omega)}{X_{mk}(j\omega)}\right)^{1/N} \end{aligned} \tag{3.6}$$

gives an estimate of $H(j\omega)$ with a very small bias (systematic error) if the SNR is not too small and the noise on the Fourier coefficients is normally distributed (Schoukens and Pintelon, 1990). For example, the relative bias is smaller than 10^{-3} (linear) if the SNR is larger than 7.1 dB and smaller than 4×10^{-6} (linear) if the SNR is larger than 10 dB (Guillaume, 1991). The assumption of normal distributed noise is very well approximated after an FFT with a sufficiently large number of time points (e.g. 512) for a wide class of the time-domain noise distributions (Brillinger, 1975; Schoukens and Renneboog, 1986). To avoid phase wrapping it is better to calculate it by the cross-spectrum that gives an unbiased estimate of the phase. In Figure 3.2 the amplitude behaviour of H_1, H_2 and $H_{\log}$ is compared for varying noise levels on the input (a) or the output (b), respectively.

To analyze the uncertainty of the measurements the complex variance $\sigma_H^2(\omega)$ on $H(j\omega)$ is defined as

$$\sigma_H^2(\omega) = E[(H(j\omega) - E[H(j\omega)])(H(j\omega) - E[H(j\omega)])^*] \tag{3.7}$$

which can also be written as

$$\sigma_H^2(\omega) = \sigma_{|H|}^2(\omega) + |H(j\omega)|^2\sigma_{\angle H}^2(j\omega) \tag{3.8}$$

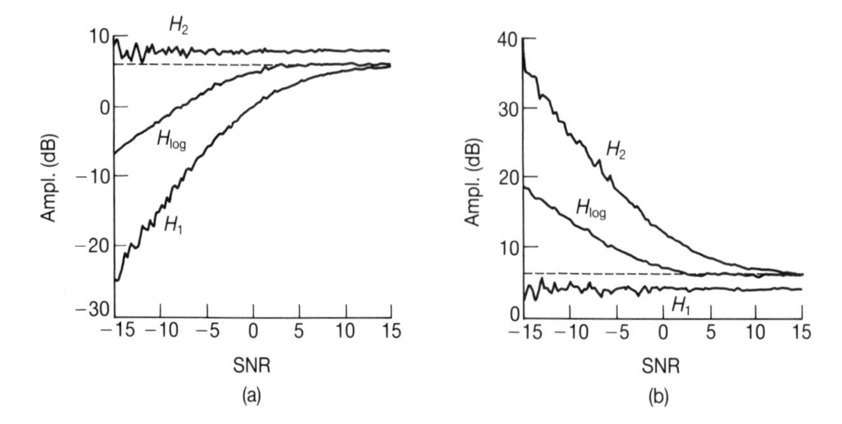

Figure 3.2 Study of the influence of the SNR on the bias of the FRF measurements. (a) SNR of the output is 6 dB, SNR of the input is varying; (b) SNR of the input is 6 dB, SNR of the output is varying. - - - True value (6 dB).

with $\sigma^2_{|H|}$ the variance of the amplitude of H and $\sigma^2_{\angle H}$ the variance of the phase. Making linear approximations the variance of the estimators H_1, H_2 and $H_{\log}$ is calculated for excitation signals whose Fourier transforms exist (including periodic signals) (Appendix 3.1). It was found that with this approximation they all have the same variance:

$$\sigma^2_H(\omega) = 2\,\frac{|H|^2}{N}\left(\frac{\sigma^2_X(\omega)}{|X(\mathrm{j}\omega)|^2} + \frac{\sigma^2_Y(\omega)}{|Y(\mathrm{j}\omega)|^2}\right) = \frac{2}{N\,|X(\mathrm{j}\omega)|^2}\,(|H(\mathrm{j}\omega)|^2\sigma^2_X(\omega) + \sigma^2_Y(\omega)) \tag{3.9}$$

where σ_X and σ_Y are the standard deviation of the noise on the real or imaginary part of the Fourier coefficients at the input or the output, respectively (and not the complex variances) (see Note 1, p. 159.) If random noise excitation is used, an additional term should be added to equation (3.9) because the exact power spectrum of noise signals can only be obtained if an infinite number of records is averaged in the frequency domain. If only a finite number of records is averaged the actual measured spectrum will vary around the final spectrum which generates an additional uncertainty. This will be analyzed in more detail in the next section. Note that FRF measurements with random noise excitation would still be wrong, even after an infinite number of averages due to leakage errors.

3.2.3 Optimization of excitation signals for non-parametric measurements

To design an optimized excitation signal it is necessary to specify the final objective. In this chapter we will look for signals that *maximize the minimum accuracy obtained in a fixed measurement time for a specified maximum peak value of the excitation*:

$$\min\left[\max_{\omega\in\mathbf{W}}(\sigma_H^2(\omega))\right] \qquad \max_t |x(t)| \leqslant x_{\max} \tag{3.10}$$

where **W** is the set of frequencies at which the frequency response is measured. From equation (3.9) it is seen that this means that $|X(j\omega)|$ should be as large as possible. This can be done by increasing the amplitude of the excitation signal, until the peak value constraint is reached. To increase $|X(j\omega)|$ further, it is necessary to compress the signal in order to inject more energy into the system for the same peak value of the excitation. On the other hand, it is also possible to decrease the uncertainty by making a proper selection of the amplitude spectrum of the excitation. At this moment, we do not know a global solution to the minimization problem in equation (3.10), but it is possible to obtain a sub-optimal solution in two steps, the first being the selection of an amplitude spectrum and the second consisting of the compression of the signal:

1. Choose the amplitude spectrum proportional to the standard deviation of the noise referred to the output:

$$|X(j\omega)| \div \sqrt{|H(j\omega)|^2\sigma_X^2(\omega) + \sigma_Y^2(\omega)} \tag{3.11}$$

2. Compress the signal obtained in 1.

Choosing the amplitude spectrum as specified in the first step will result in a constant absolute uncertainty (σ_H) on the measurements. If it is more desirable to keep the relative uncertainty constant (σ_H/H) the amplitude spectrum should be chosen as

$$|X(j\omega)| \div \sqrt{\sigma_X^2(\omega) + \frac{\sigma_Y^2(\omega)}{|H(\omega)|^2}} \tag{3.12}$$

It is obvious that it is also possible to specify an arbitrary uncertainty profile as a function of the frequency.

The compression of a signal is described by its crest factor Cr (see Note 2, p. 159), which is defined as

$$Cr = \frac{u_{\text{peak}}}{u_{\text{rmse}}} \qquad u_{\text{peak}} = \max_t(|u(t)|) \tag{3.13}$$

Here u_{rmse} is the effective value of the signal consisting only of this part of the signal that contributes to the measurement:

$$u_{\text{rmse}} = U_{\text{rms}}\sqrt{\frac{P_{\text{interest}}}{P_{\text{all}}}} \tag{3.14}$$

where P_{interest} is the power at the frequencies of interest and P_{all} the power of the complete signal. This means that the power at those frequencies where we are not interested in the measurements is not considered. The crest factor is not sufficient to compare different excitation signals because it does not incorporate the influence of the shape of the amplitude spectrum of the signal and the noise. Using equation (3.9) it is possible to derive a more general criterion. The number of averages, N, required to achieve an accuracy $\leqslant \sigma^2_{H\min}$ is given by

$$N = \max_{\omega \in W} \left\{ \frac{2}{\sigma^2_{H\min}} \left(\frac{|H(j\omega)|^2 \sigma^2_X(\omega) + \sigma^2_Y(\omega)}{|X(j\omega)|^2} \right) \right\} \tag{3.15}$$

The measurement time is proportional to this number of averages. Hence the time per measured frequency point will be proportional to N/F if F frequencies are measured in one experiment ($F = \#(W)$):

$$\frac{N}{F} = \max_{\omega \in W} \left\{ \frac{2}{\sigma^2_{H\min}} \left(\frac{|H(j\omega)|^2 \sigma^2_X(\omega) + \sigma^2_Y(\omega)}{|X(j\omega)|^2} \right) \frac{1}{F} \right\} \tag{3.16}$$

This expression can also be written as

$$\frac{N}{F} = \max_{\omega \in W} \left\{ Cr^2 \left(\frac{(|H(j\omega)|^2 \sigma^2_X(\omega) + \sigma^2_Y(\omega))/\sigma^2}{(|X(j\omega)|^2)/X^2_{\text{rms}}} \right) \left(\frac{4\sigma^2}{\sigma^2_{H\min} \cdot x^2_{\text{peak}}} \right) \right\} \tag{3.17}$$

where

$$\sigma^2 = \frac{1}{F} \sum_{k=1}^{F} (|H(j\omega_k)|^2 \sigma^2_X(\omega_k) + \sigma^2_Y(\omega_k)) \qquad X^2_{\text{rms}} = \frac{1}{F} \sum_{k=1}^{F} |X(j\omega_k)|^2 \tag{3.18}$$

and

$$Cr^2 = \frac{x^2_{\text{peak}}}{2FX^2_{\text{rms}}} \tag{3.19}$$

In equation (3.17) the last factor is dimensionless and is independent of the excitation signal if the peak value is fixed. To compare different excitation signals it can be set to an arbitrary number. The time factor Tf is defined as

$$Tf = \max_{\omega \in W} \left\{ 0.5Cr^2 \left(\frac{(|H(j\omega)|^2 \sigma^2_X(\omega) + \sigma^2_Y(\omega))/\sigma^2}{(|X(j\omega)|^2)/X^2_{\text{rms}}} \right) \right\} \tag{3.20}$$

The Tf allows a qualitative comparison of the different excitation signals. The measurement time needed to obtain a specified accuracy with two different excitation signals will be proportional to their respective time factors. It should be noted that the time factor does not account for the time required to obtain the steady-state response. From equation (3.20) it can be seen that the Tf will depend strongly upon the problem itself. This is because the optimum power spectrum depends upon the noise spectra and the transfer function of the studied system (if the input noise is not equal to zero). To simplify the comparison of different excitation signals we will assume that $|H(j\omega)|^2 \sigma^2_X(\omega) + \sigma^2_Y(\omega)$ is a constant independent of ω. Then the Tf reduces to

$$Tf = \min_{\omega \in W} \left\{ 0.5Cr^2 \frac{X^2_{\text{rms}}}{|X(j\omega)|^2} \right\} = \min_{\omega \in W} \left\{ 0.5 \frac{x^2_{\text{peak}}}{|X(j\omega)|^2} \frac{1}{2F} \right\} \tag{3.21}$$

In this simplified case the optimum signal has a flat power spectrum with a minimum crest factor. In the next section we will compare different excitation signals under this supposition.

3.2.4 Comparison of different excitation signals

In this section we will study and compare the properties of some excitation signals. From the previous section we know already that an optimum signal should have a low Tf which is, in general, only possible if the amplitude spectrum of the signal can be artibrarily chosen and the crest factor can be reduced. Besides these two conditions it is also important that no leakage appears during the analysis of the measurements. This is a problem typically related to FFT-based network analyzers: the energy of a spectral line seems to be smeared throughout the frequency domain due to the finite measurement time. This can be reduced by proper selection of a time window and by averaging of the results, but important errors may still remain (see Chapter 2, p. 77, p. 110.) A better solution is to restrict the excitation to periodic signals. If an integral number of periods is measured, there will be no leakage at all (Brigham, 1974). Leakage errors cannot be avoided if aperiodic signals are used, in which case it will be necessary to average over a large number of measurements even if a non-uniform time window is used. This will considerably increase the measurement time required for a specified accuracy. Burst or time-limited signals are exceptions to this rule: the continuous spectra of these signals are correctly sampled with the Discrete Fourier Transform (DFT) if the amplitude spectrum is sufficiently band-limited for the aliasing effect to be neglected (Brigham, 1974). (It should be remembered that it is impossible to have a time-limited signal with a band-limited spectrum.)

(STEPPED) SINE

Definition: a pure sine wave whose frequency is changed between measurements:

$$x(t) = 2A \sin 2\pi ft \tag{3.22}$$

Properties: crest factor $\sqrt{2}$, $Tf = 1$, no leakage.

With a sine wave we can make measurements at the maximum SNR, but it is necessary to wait until the steady-state response is obtained whenever the measurement frequency is changed. Leakage is avoided if an integral number of periods is measured. An important advantage is the possibility of analyzing the non-linear behaviour of the measured system, because the harmonic distortions are directly visible.

SWEPT SINE (ALSO CALLED PERIODIC CHIRP)

Definition: this is a sine sweep test, where the frequency is swept up and/or down in one measurement period, and this is repeated in such a way that a periodic signal is created (Brown *et al.*, 1977):

$$x(t) = 2A \sin\{[(at + b)t]\} \qquad 0 \leqslant t < T \tag{3.23}$$

where T = period, $a = (\pi(f_2 - f_1))/T$, $b = \pi f_1$ and f_1 and f_2 are the lowest and the highest frequency.

Properties: crest factor typically 1.45, Tf typically between 1.5 and 4, no leakage.

A swept sine has a low crest factor (comparable to the crest factor of a sine wave) but the amplitude spectrum is not really flat. This introduces frequency components with a lower SNR, resulting in a longer measurement time for a given accuracy. With a swept sine it is possible to create band spectra, but it is not possible to generate a signal with an arbitrary amplitude spectrum. A second drawback is that not only are the frequency lines of interest excited but also a number of other spectral lines appear. This is unimportant with linear systems, but it can be very disturbing in systems with a non-linear behaviour.

MULTI-SINE

Definition: a multi-sine is the sum of a number of harmonically related sinusoids with programmable amplitudes. The phases of the frequency components can be changed to reduce the crest factor of the signal (Schroeder, 1970; Van der Ouderaa *et al.*, 1988a,b; Guillaume *et al.*, 1991) or they can be chosen at random with a uniform distribution over the interval $[0, 2\pi]$ (Brown *et al.*, 1977; Van Brussel, 1975):

$$x(t) = \sum_{k=1}^{F} 2A_k \cos(2\pi f_k + \phi_k) \tag{3.24}$$

where f_k is a multiple of $1/T$ with T the period of the multi-sine.

Properties: crest factor typically between 1.45 and 2, Tf between 1 and 1.5, no leakage.

From the previous discussions it is clear that the crest factor of the excitation signal is a crucial parameter together with the amplitude spectrum of the signal. The multi-sine is the only periodic broadband excitation signal that allows an arbitrary choice of the amplitude spectrum. All the other signals discussed here can only approximate a desired amplitude spectrum. However, to take full profit from this advantage it is necessary to make the crest factor small. This can be realized by making a proper choice for the phases. In the literature many crest factor minimization methods are presented. A first group consists of explicit formulas which allows a direct calculation of the phases. A second group searches for an optimal set of phases using iterative algorithms. Both classes are discussed in Appendix 3.1.

Another major advantage of the multi-sine excitation is the possibility of controlling the crest factor of multiple multi-sines linked by linear systems. The most simple example is the minimization of the (scaled) peak value (or crest factor) of the excitation and response of a linear system. This is illustrated in Figure 3.3. A signal with a flat amplitude spectrum in the band 16–63 Hz (48 components) is integrated, and both the input and the output are compressed. If only the input signal is compressed the crest factors are 1.47/2.85; after combined compression the crest factors are 1.66/1.66. It is also possible to minimize the scaled peak values of both multi-sines as shown in Appendix 3.1. In Appendix 3.2 a list of phases is given for some examples of compressed multi-sines.

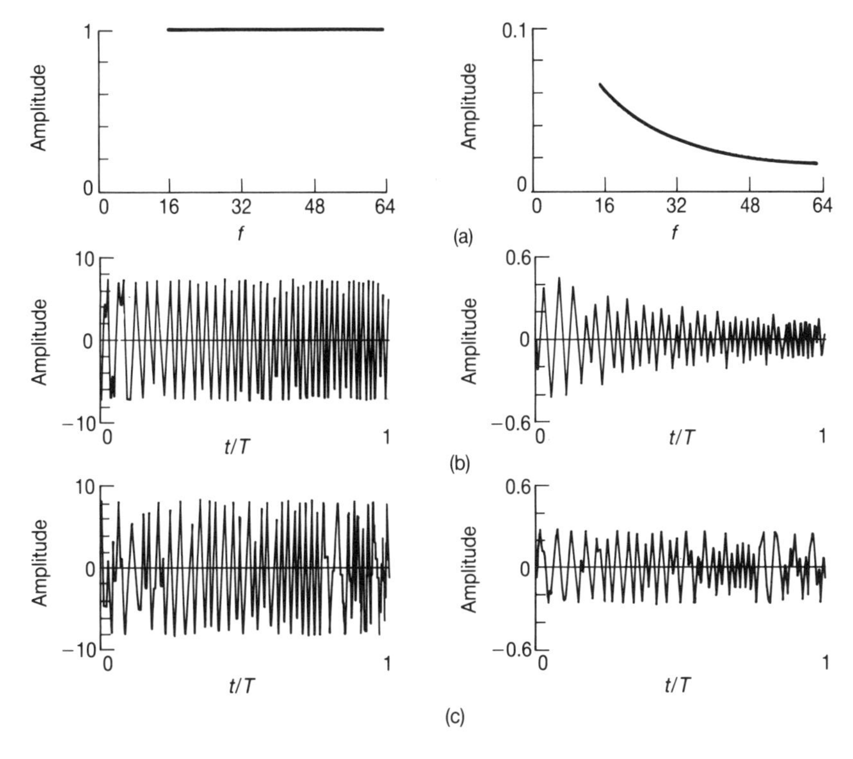

Figure 3.3 Optimization of two multi-sines related by an integrator. (a) Input/output spectrum; (b) crest factor minimization input (1.47/2.85); (c) crest factor minimization input/output (1.66/1.66).

The results of the optimization of different kinds of signals illustrate the flexibility of the multi-sine if it is combined with an adequate compression algorithm. In Figure 3.4 it is seen that for a wide range of signals the crest factor can be reduced to about 1.5. Good results are even obtained for signals with sparse spectra such as log-tone, coloured amplitude spectrum, etc. The l_∞ method was used to optimize the phase (see Appendix 3.1).

PERIODIC NOISE

Definition: a finite noise sequence is periodically repeated until the transients are damped out (Brown *et al.*, 1977), at which point a measurement is made. Next, a new random noise sequence is generated, and the procedure repeated until sufficient measurements are collected.

Properties: crest factor typically 3, Tf converges to 4.5 if a large number of measurements are averaged, no leakage.

A periodic noise signal behaves similarly to random noise, except that there is no leakage because of the periodicity of the noise. However, if the SNR is not good

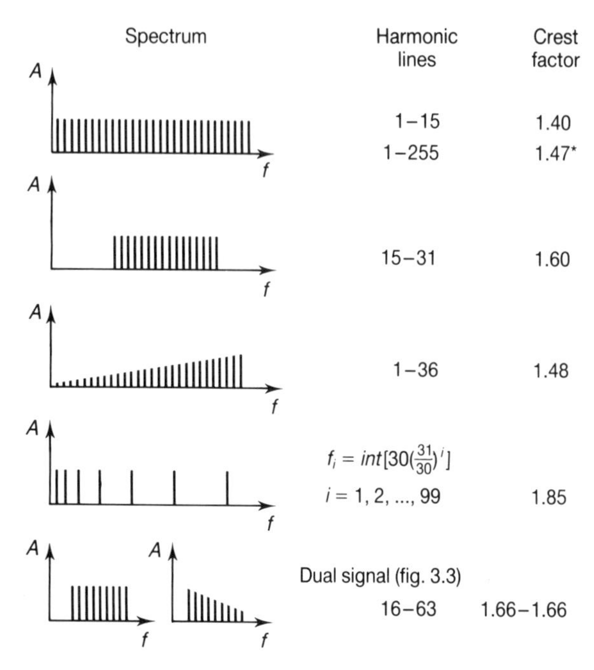

Figure 3.4 Some examples of optimized multi-sines. Crest factor minimization with l_∞ (see Appendix 3.2), except (*) * Crest factor minimization with algorithm 1 (see Appendix 3.2).

it is still necessary to average over a number of experiments because the amplitude spectrum of an individual experiment is non-flat due to the stochastic behaviour of the input signal. The (normalized) power spectrum is a random variable with a χ^2 distribution with $2N$ degrees of freedom if N measurements are averaged.

In Table 3.1 the 95% uncertainty regions of the amplitude spectrum are described by their upper and lower bounds. The ratio of the lower bound to the RMS value is also tabulated, to give an idea of the loss in SNR of the weakest components due to the stochastic behaviour.

In Figure 3.5 the loss in the SNR for random signals when compared to deterministic signals is shown as a function of the number of averaged measurements. The time factor will converge to its limit only if this number is sufficiently large.

MAXIMUM LENGTH BINARY SEQUENCE (MLBS) (see also Chapter 1)

Definition: this is a periodic binary sequence with an autocorrelation function which is an approximation of a Dirac pulse for a given clock frequency and register length (Godfrey, 1969a, 1980; Eykhoff, 1974; Holmes, 1982; Norton, 1986). From all possible binary sequences which can be generated with a fixed register length, the MLBS has the longest period and the shortest correlation length. This means that the spectrum

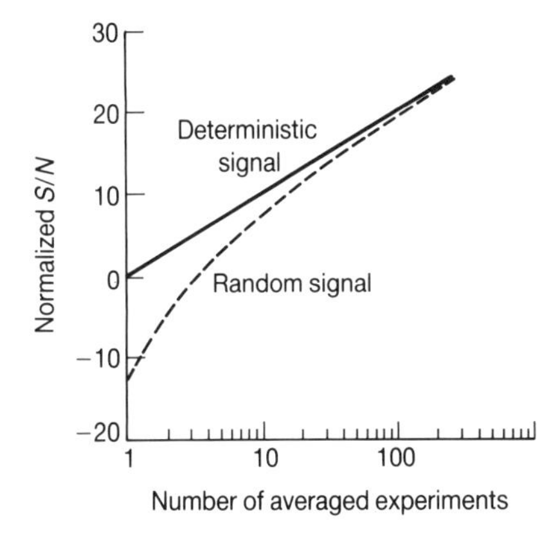

Figure 3.5 Loss in SNR due to the stochastic behaviour of the spectrum of a random signal.

Table 3.1 Study of the stochastic behaviour of the averaged spectrum of a random signal

N	Ratio 95% upper/95% lower bound (dB)	Ratio V_{rms}/(95% lower bound) (dB)
1	22	13
2	14	7.5
4	9	4.7
8	6.2	3.0
16	4.3	2.1
32	3.1	1.4
64	2.1	1.0
128	1.5	0.7
256	1.1	0.5

is as flat as possible. Such a sequence can be generated using a shift register, as illustrated in Figure 3.6. The feedback choice will determine whether a sequence with the maximum period

$$T_{max} = (2^R - 1)T_{clock} \tag{3.25}$$

is generated. Here R is the register length. A table of feedback connections can be found in Table 1.2 of Chapter 1 for register lengths of 2 to 100, and 127.
Properties: minimum crest factor = 1, minimum $Tf \approx 1$, no leakage.

The MLBS has a spectrum whose components decrease inversely proportionally to the frequency. The amplitude spectrum of a MLBS is given by

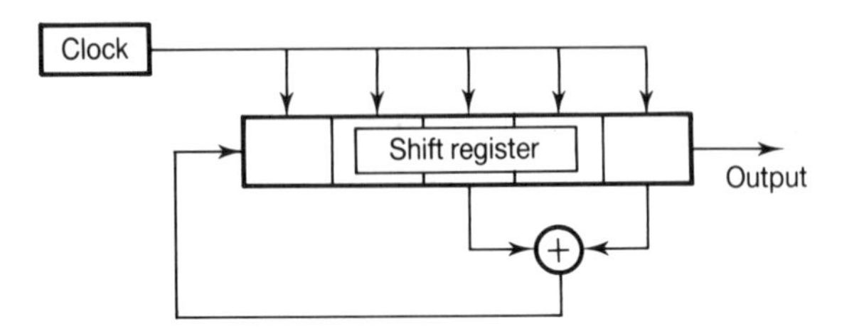

Figure 3.6 Generation of a maximum-length binary signal with a shift register.

$$A(0) = \frac{a}{N}$$

$$A(f_k) = a\sqrt{\frac{1 + 1/N}{N}}\ \frac{\sin[k(\pi/N)]}{k(\pi/N)} \qquad k = 1, 2, \ldots, N - 1 \tag{3.26}$$

where $N = 2^R - 1$, R is the register length, $2a$ the peak-to-peak amplitude of the sequence, and $f_k = k(f_{\text{clock}}/N)$

In Figure 3.7 details of the first lobe of the amplitude spectrum are given for a MLBS generated from register lengths of 4, 5 and 6. The amplitude of the individual components decreases with increase in the register length. The crest factor varies as a function of the spectral band ($0 < f \leqslant f_{\max}$) in use, decreasing to 1 as the bandwidth increases to infinity. However, the Tf has a different behaviour, as seen in Figure 3.7(b): it decreases for low frequencies but increases to infinity if $f_{\max}$ approaches f_{clock}, as the amplitudes decrease to zero for this frequency. The Tf is less than 1.5

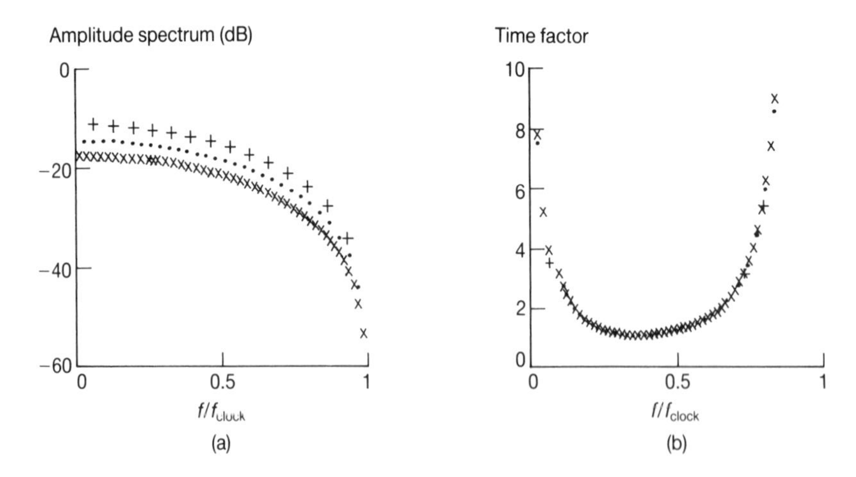

Figure 3.7 Part of the amplitude spectrum (a) and the time factor (b) of an MLBS as a function of the bandwidth used $0 \rightarrow f_{\max}$ (register lengths 4 (+), 5 (.) and 6 (X)).

if the upper limit of the frequency band is taken between 0.2 and 0.6 of the MLBS generator clock frequency. The optimal value of the upper frequency limit is around 0.4 f_{clock}, resulting in a Tf of 1.1, suggesting a choice of clock frequency f_{clock} equal to 2.5 times the maximum frequency of interest.

The period of the MLBS should be chosen long enough in order to obtain a sufficient resolution in the frequency domain. It is $N = 2^R - 1$ clock cycles, so the generator and the measurement clocks cannot be the same if an FFT analysis is made because the FFT requires 2^n samples. In order to avoid leakage the chirp-z transform can be used which permits efficient calculation of the DTF for an arbitrary number of points using three FFT operations (Rabiner and Gold, 1975; Oppenheim and Schafer, 1975).

MULTI-FREQUENCY BINARY SEQUENCE (MFBS)

Definition: this is a periodic binary sequence, where the sign can only change at an equidistant discrete set of points in time (Van den Bos, 1974; Paehlike and Rake, 1979; Van den Bos and Krol, 1979). The amplitude spectrum of the sequence can be optimized by choice of a good switching sequence so that the energy is concentrated within the frequency band of interest.

Properties: crest factor 1.1–1.2, Tf 2–4, no leakage.

The generation of a MFBS is based on an iterative algorithm proposed by Van den Bos and Krol (1979). The procedure is begun a number of times from different starting values, and the best signal is retained. With a MFBS it is possible to concentrate the energy in a discrete set of spectral lines. The crest factor is greater than one because a significant number of spectral lines other than those of interest are generated, but even so, most of the energy can be confined to the frequency band required, which is not possible with the MLBS. Paehlike and Rake (1979) have presented an iterative scheme for putting more of the energy into the weakest spectral lines, thus improving the SNR and decreasing the time factor.

PULSE (ALSO CALLED IMPACT TESTING)

Definition: the impulse response is measured directly in the time domain by exciting the DUT with a short pulse (Halvorsen and Brown, 1977). For example,

$$x(t) = \begin{cases} A & (0 \leqslant t < T_1) \\ 0 & (T_1 < t \leqslant T) \end{cases} \tag{3.27}$$

where T_1 is the pulse width and T the repetition time.

Properties: minimum crest factor $\sqrt{T/T_1}$, minimum $Tf(T/T_1)$, no leakage.

The autocorrelation of the impulse response is the same as that of the MLBS, so their amplitude spectra are identical. To obtain the same input energy, the amplitude must be increased by a factor of $\sqrt{T/T_1}$. The minimum time factor is for the same upper frequency limit as for the MLBS. More sophisticated impulse-generation techniques are given by Halvorsen and Brown (1977), but the general characteristics remain the same.

The main advantage compared to other excitations is the simple technical requirements for generation of an impulse in mechanical problems: no shakers or other expensive equipment are needed to create the input.

RANDOM BURST

Definition: a noise sequence is imposed on the system during part of the measurement sequence, and a zero input is applied for the rest of the measurement period (Herlufsen, 1984). To avoid leakage errors the burst and the measurement time, T, must be chosen so that the system response is sufficiently damped by the end of the measurement period:

$$x(t) = w(t)r(t) \tag{3.28}$$

where $r(t)$ is a random variable and $w(t)$ a window function:

$$w(t) = \begin{cases} 1 & (0 \leqslant t < T_1) \\ 0 & (T_1 \leqslant t < T) \end{cases} \tag{3.29}$$

Properties: crest factor typically $3\sqrt{T/T_1}$, minimum Tf is $Tf \geqslant 4.5\ T/T_1$, no leakage.

The crest factor of a random burst sequence is equal to that of the random sequence multiplied by $\sqrt{T/T_1}$. For systems with low damping factors, the relative width T_1/T of the burst must be very small, resulting in a high crest factor. The greatest advantage of using a random burst is that there are no leakage errors (a uniform window should be used to calculate the DFT). The power spectrum of a random burst is a random variable, as it is for a periodic noise sequence, and so the same restrictions are valid as those mentioned for periodic noise.

RANDOM NOISE

Definition: a noise sequence, where the power spectrum can be influenced by digital filters (Brown *et al*., 1977; Van Brussel, 1975).

Properties: crest factor typically 3 and Tf 4.5, leakage.

For good results using random noise a window is required to minimize the influence of leakage, but even then it is necessary to average over a large number of measurements to reduce leakage errors. Gade and Herlufsen (1987a,b) give a good introduction to the use of windows and their influence on DFT, and Harris (1978) provides a review of the properties of a large number of different windows; a discussion of different averaging techniques is given by Allemang (1985). The use of a window also introduces systematic errors. In most cases if the measured characteristics do not vary too much from one spectral line to another these errors will be small, but at sharp resonant peaks or at a notch the influence is greater and can introduce errors of 1 dB or more.

Many people prefer a random noise excitation to a periodic one if the measured linear system is disturbed by non-linear distortions. However, a detailed study shows that only under very restrictive conditions is it possible to eliminate a part of the non-linear contributions. But the peak value of the random excitation required to inject the same energy as a well-compressed signal is much larger. This will mostly result in a greater contribution of the non-linearity (Schoukens *et al.*, 1988b).

3.2.5 Discussion

The most important properties of the signals are compared in Table 3.2. Two classes of signals can be distinguished: deterministic and stochastic. The main advantage of deterministic signals is that their amplitude spectra are constant during a complete experiment, which is not so in the stochastic case. It is always necessary to average over a sufficient number of experiments in order to meet the limit properties for stochastic excitations. In the class of deterministic excitations the multi-sine is the only broadband signal which permits the creation of an arbitrary amplitude spectrum with very low time and crest factors.

Table 3.2 Properties of the signals studied

Signal	Crest factor	Time factor	Leakage	Averaging	Arbitrary amplitude spectrum
Stepped sine	$\sqrt{2}$	1	No	–	Yes
Swept sine	1.45	1.5–4	No	–	No
Multi-sine					
Schroeder	1.7	1.5	No	–	Yes
optimized	1.45	1	No	–	Yes
Periodic noise	3	4.5	No	Necessary	No
MLBS	1	1	No	–	No
MFBS	1.1	2–4	No	–	Yes
Impulse	$\sqrt{T/T_1}$	T/T_1	No	Advised	No
Random burst	$3\sqrt{T/T_1}$	$3.5\ (T/T_1)$	No	Advised	No
Random noise	3	4.5[a]	Yes	Necessary	No

[a] Time factor after a sufficient number of experiments has been averaged.

EXAMPLE

To illustrate the properties of the multi-sine excitation and the random burst, the transfer function of a low-pass linear system has been measured in its pass-band. The experimental set-up is given in Figure 3.8. At the output of the system a white-noise source (up to 25 kHz) with a RMS value of 95 mV is added to the output (SNR typically 20 dB or less in the time domain). The peak value of the measured input signals (before the anti-alias filter) is set to 1 volt for both excitations. The results are shown in Figure 3.9. From this figure it is noted that averaging is always necessary with random signals, even for random burst (which introduces no leakage), due to the dips in the spectrum after a single measurement. These dips do not introduce large errors if the noise level is sufficiently low, but otherwise averaging is necessary

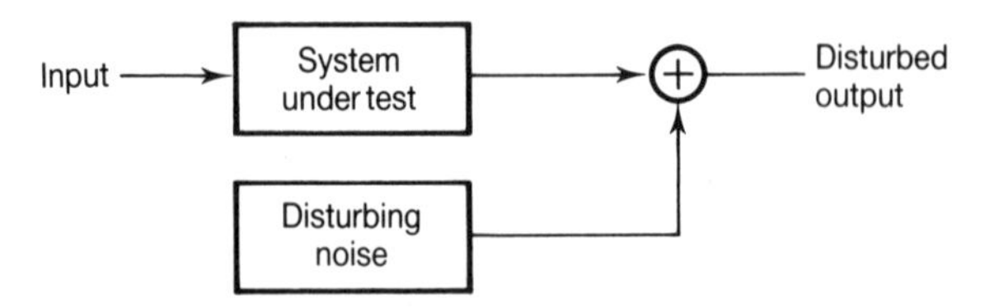

Figure 3.8 Experiment set-up to compare the noise sensitivity of the excitation signals.

to obtain a sufficiently flat amplitude spectrum. Even after a significant number of averages the multi-sine gives much better results, which is to be expected from its superior time factor. The errors for the random excitation after averaging over 16 experiments are still as large as those achieved with the multi-sine after a single experiment! It is necessary to carry out 64 experiments with random excitation to obtain results as good as those for only 16 averaged experiments using a multi-sine. This is in agreement with Figure 3.5, where it is shown that the SNR increases considerably faster for the random sequence than for deterministic signals over the first averages, and from Table 3.2 it is known that after a sufficient number of measurements the time factor for a random excitation is about 4.5, compared to a value of 1–1.5 for the multi-sine.

This conclusion can be generalized to all the deterministic signals we have considered in this chapter. It can be concluded that deterministic signals are more suitable for this kind of measurement than random signals because they have better time factors, which results in smaller errors. If an arbitrary amplitude spectrum is needed, the multi-sine becomes obviously superior because it is the only signal which gives complete freedom to the user to choose its amplitude spectrum.

3.3 Optimization of excitation signals for parametric measurements

3.3.1 Introduction

In this part the parametric measurement problem is studied. We will concentrate on the parameters $\mathbf{P}$ of the mathematical model $H(j\omega, \mathbf{P})$ which describes the measured transfer function $H_m(X_m, Y_m, j\omega)$. To fit the model H on the measurements H_m a cost function $K(X_m, Y_m, \mathbf{P})$, which is an index of the quality of the fit, is minimized. There are many possible choices for K and the statistical properties of the estimated parameters will strongly depend on the particular selection. A simple and very popular choice for K is the least squares method in which the squared differences between the model and the measurements are summed together. Another possibility is to

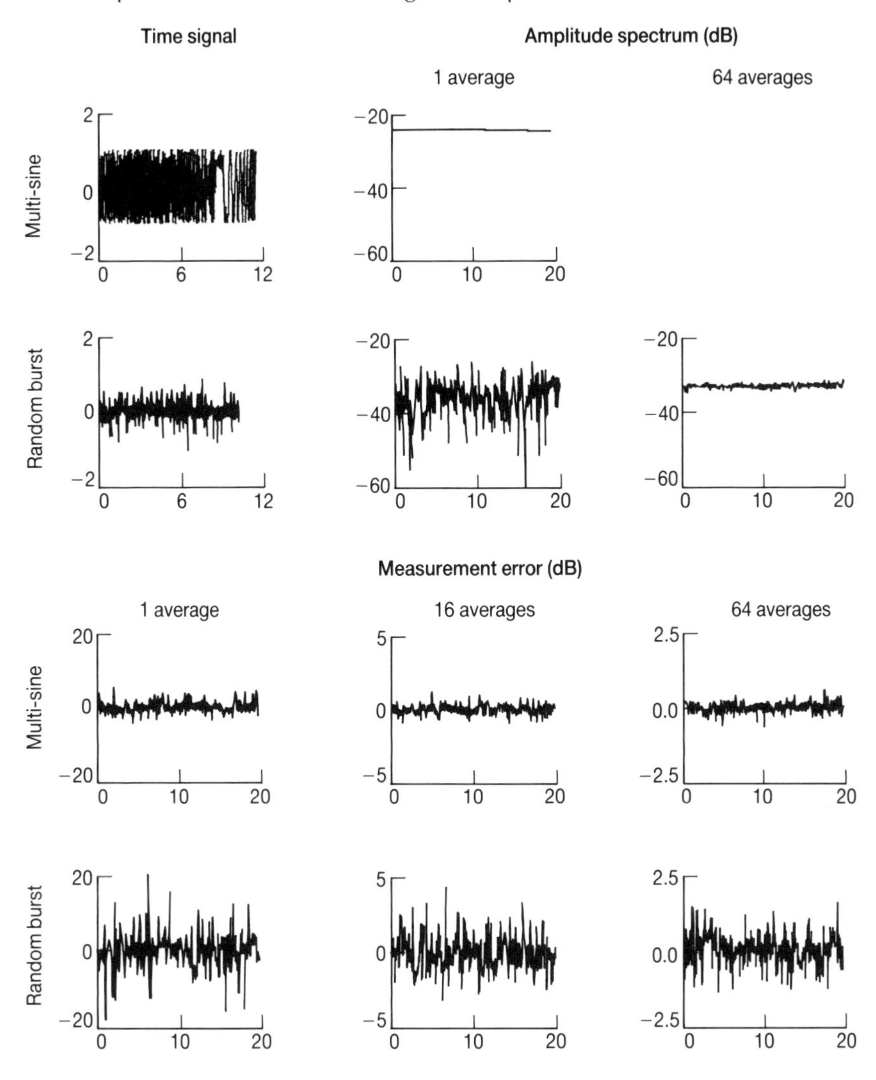

Figure 3.9 Experimental results of noise-sensitivity comparisons of the multi-sine and noise excitation (both signals with their amplitude spectrum in dB; measurement errors (measured – exact amplitude in dB)). The frequencies are given in kHz and the time in ms.

embed the choice of the cost function in a statistical framework. The maximum likelihood estimator (MLE) results from such an approach. The MLE and the (weighted) least squares estimator are equivalent if the disturbing noise is normally distributed (Eykhoff, 1974; Norton, 1986). In the literature a number of methods are developed to identify transfer function models of linear systems, using discrete time-domain models (mostly time-domain methods) or continuous time-domain models (mostly frequency-domain methods) (Schoukens and Pintelon, 1991). Only a few of them consider disturbing noise on both the input and the output (Van den Bos, 1974; Pintelon and Schoukens, 1990; Schoukens and Pintelon, 1991). The quality of an estimator strongly depends on the excitation signals applied during the experiment. As for the non-parametric case, the excitation signal will be optimized in two steps, the first being the selection of an optimized power spectrum followed by a crest factor minimization of the signals involved in the second step.

To optimize the input spectrum, a scalar criterion is needed which is sensitive to the accuracy of all the parameters of the system. The determinant of the covariance matrix, which is equal to the volume of the uncertainty ellipsoid, is such a criterion. It can be shown that minimization of the determinant will result in minimum dispersion of the frequency characteristic of the system calculated from the estimated parameters. The dispersion function $v(\omega)$ is a scaled value defined as the ratio of the variance of the calculated frequency characteristic to the variance of the measurement noise at the angular frequency ω. A number of important properties will be detailed later.

A range of criteria other than the determinant can be found in the literature, the optimization of the trace being the most popular. For the sake of brevity, we will limit ourselves in this chapter to examination of the minimization of the determinant of the covariance matrix. For more information on other criteria, the reader is referred to other publications (Fedorov, 1972; Goodwin and Payne, 1977; Zarrop, 1979).

For computational simplicity, the covariance matrix is approximated by the Cramér–Rao lower bound (inverse information matrix) because the latter can be calculated much more easily (Eykhoff, 1974; Kendall and Stuart, 1979; Norton 1986). General expressions of the information matrix can be obtained (without specifying an estimator) and the problem of minimizing the determinant of the covariance matrix is replaced by maximizing the determinant of the information matrix. This approximation is valid if the covariance matrix of the actual estimator approximates the Cramér–Rao lower bound sufficiently close for the experiments considered.

3.3.2 Optimization of the power spectrum of a signal

PRELIMINARY ASPECTS

The information matrix is the kernel of optimizing algorithms. It is a real symmetric and semi-positive definite ($n_p \times n_p$) matrix, where n_p is the number of unknown model parameters. Each optimal design in the frequency domain can be reduced to one consisting of a discrete set of $[n_p(n_p + 1)/2 + 1]$ frequencies (Fedorov, 1972; Goodwin and Payne, 1977), which corresponds to the number of free parameters in a symmetric ($n_p \times n_p$) matrix. In many applications it is even possible to create an optimum design requiring fewer frequencies than this, because of relations existing between the different elements of the matrix which reduce the number of free parameters even further. The minimum number of frequencies required in order to avoid a non-singular information matrix is $[n_p/2 + 1]$ (with $[x]$ the integer part of x). When using classical optimizing algorithms the computer time needed to search for an extreme value depends strongly on the number of frequencies. From a modelling point of view, however, the minimum number is undesirable, because if an estimate of n_p parameters is made using $(n_p/2 + 1)$ frequencies there is no possibility of detecting model errors. A second drawback of working with the minimum number of frequencies is that it is more difficult to compress the signals in the time domain.

Algorithms previously presented in the literature have always searched for optimal designs with the minimum number of frequencies. We will present a method for designing optimal power spectra based on a discrete frequency grid: this is not in itself a restriction because we look for periodic signals which have discrete spectra. The method can be applied in the Laplace domain (continuous-time systems) as well as in the z-domain (discrete-time systems). In order to stress this equivalence, we will use Ω as the frequency variable in the following interchangeable manner:

$$\Omega = j\omega \text{ (Laplace)} \qquad \text{or} \qquad \Omega = e^{j\omega T_s} \text{ (z-domain)} \tag{3.30}$$

The dispersion function The design of optimal input spectra is based on the dispersion function, which is a normalized variable describing the uncertainty on the fitted model.

Definition: The dispersion function for a given input power spectrum

$$\chi(\Omega) = (|X(\Omega_1)|^2 \ldots |X(\Omega_F)|^2) \qquad \text{with} \qquad \sum_{k=1}^{F} |X(\Omega_k)|^2 = \mathscr{P} \tag{3.31}$$

is

$$v(\chi, \Omega) = \text{trace}\{[\mathbf{Fi}(\chi)]^{-1}\mathbf{fi}(\Omega)\} \tag{3.32}$$

where $\mathbf{Fi}(\chi)$ is the information matrix resulting from the power spectrum $\chi(\Omega)$, $\mathbf{fi}(\Omega)$ the information matrix corresponding to a single frequency input with a normalized power spectrum $|X(\Omega)|^2 = \mathscr{P}$, and Ω the angular frequency. The dispersion can be interpreted as the ratio of the variance of the system frequency response, calculated with the estimated parameters, to the noise power of the measurements referred to the output of the system at the angular frequency Ω.

Properties of the dispersion function

- The dispersion function can be related to the input and output noise on the measurements (Schoukens and Pintelon, 1991) as

$$v(\chi, \Omega) = \frac{\sigma_H^2(\Omega)\mathscr{P}}{\sigma_X^2(\Omega)|H(\Omega)|^2 + \sigma_Y^2(\Omega)} \tag{3.33}$$

  In this expression σ_H is the complex standard deviation of H calculated from the estimated parameters, and σ_X and σ_Y the standard deviation of the noise on the real and imaginary part. The dispersion can here be interpreted as the variance of the calculated frequency-response function scaled by the variance of the measurement noise on the real and imaginary part of Fourier coefficients referred to the output.

- The dispersion function is a normalized quantity:

$$\sum_{k=1}^{F} v(\chi, \Omega_k) \frac{|X(\Omega_k)|^2}{\mathscr{P}} = n_p \tag{3.34}$$

where n_p is the number of unknown parameters (Goodwin and Payne, 1977).
- The maximum of the dispersion function $v(\chi, \Omega_k)$ over the frequency grid is larger than or equal to the number of parameters n_p (Goodwin and Payne, 1977).

These three properties will be used in the algorithm for designing an optimized excitation signal.

AN EFFICIENT ALGORITHM FOR MAXIMIZING THE INFORMATION MATRIX

Although the optimal input may be found analytically for simple situations, in general, no closed-form solution can be found so that an iterative design is required. Most algorithms carry out a search in the continuous frequency space to find the frequency with the maximum dispersion, and then add extra energy at this frequency. The resulting spectrum is normalized, and the procedure is repeated until the variations are negligible. More sophisticated algorithms combine this procedure with a mechanism which removes components from the spectrum (Fedorov, 1972; Zarrop, 1979). The search for a maximum is very time consuming, and the final spectrum is difficult to generate because the optimal frequencies are not harmonically related; arbitrary waveform generators can only generate periodic signals. For these reasons, it is better to reduce the frequency space to a discrete set of frequencies in the analysis. The implications of this restriction on the attainable accuracy is studied in more detail by Van den Eijnde and Schoukens (1991), and it is found that there is no significant loss in attainable accuracy if the discrete set of frequencies is sufficiently dense.

In general, any discrete set of frequencies can be used but if only periodic signals are considered it is obvious that the selected frequencies should be harmonically related. Some degree of prior knowledge concerning the optimal amplitude spectrum can also be incorporated but the simplest choice is that of equally spaced spectral lines within the frequency band of interest, with the total fixed input power uniformly distributed over the F frequencies in this set. The resulting spectrum constitutes the initial design χ_0. The response dispersion function $v(\chi, \Omega_k)$ is computed for every spectral line Ω_k in the set, and the available power is redistributed over all spectral lines proportionally to the corresponding values of the dispersion function. The optimal input is found by repeating this procedure; the iteration can be stopped when the variation of the determinant of the information matrix is small.

If we express this approach in mathematical terms, we end up with an algorithm with the following consecutive steps:

- Select a set W of F angular frequencies $\Omega_1, \ldots, \Omega_F$ within the frequency band of interest: $W = \{\Omega_1, \ldots, \Omega_F\}$. Distribute the input power equally over these F angular frequencies. This constitutes the initial design χ_0. Put $i = 0$.
- Compute the response dispersion function $v(\chi_i, \Omega_k)$ for $k = 1, \ldots, F$.

- If $\max(v(\chi_i, \Omega_k) - n_p) < \varepsilon$ with ε sufficiently small and $\Omega \in S$, the optimum design is found. If not, continue.
- Compose a new design in the following way:

$$\chi_{i+1}(\Omega_k) = \frac{v(\chi_i, \Omega_k)}{n_p} \chi_i(\Omega_k) \qquad k = 1, \ldots, F \tag{3.35}$$

- Set $i = i + 1$, and return to the second step.

Delbaen (1990) has shown that each run of this algorithm yields a superior input design, and that consecutive designs converge monotonously to one with the optimum dispersion function and hence a minimum determinant of the Cramér–Rao bound.

IMPORTANCE OF CREST FACTOR MINIMIZATION

In a second step, after the selection of the power spectrum the crest factor of the corresponding multi-sine(s) should be minimized. To compare different excitations it is necessary to scale the determinant $|\mathbf{C}|$ of the covariance matrix and the dispersion function with the optimized crest factor so that all signals are compared for the same peak value:

$$|\mathbf{C}|_{\text{scaled}} = |C| \, (\text{crest factor})^{2n_p} \tag{3.36}$$

PRACTICAL IMPLEMENTATION

It is obvious that the calculation of the optimum amplitude spectrum is only possible if a good knowledge of the system is available. In most situations a two-step procedure is required. In the first step the unknown parameters are estimated using a multi-sine with a flat-amplitude spectrum, and in the second step these estimated values are used to optimize the amplitude spectrum. The covariance matrix of the estimated unknown model parameters should be close enough to the Cramér–Rao lower bound. In Pintelon and Schoukens (1990) and Schoukens and Pintelon (1991) we described a robust Gaussian maximum likelihood estimator which allows the estimation of the parameters of the transfer function of a linear (discrete- or continuous-time) system starting from the measured input and output Fourier coefficients which may be disturbed by noise. Also, the explicit expressions to calculate the information matrix **Fi** are given. These are valid if the disturbing noise is independent and normally distributed with zero-mean value.

EXAMPLE: AN EXPERIMENTAL VERIFICATION

The power spectrum optimization for a parametric measurement is illustrated by the following example:

$$H(s) = \frac{\alpha_2 s^2 + \alpha_3 s^3 + \alpha_4 s^4}{\beta_0 + \beta_1 s + \ldots + \beta_6 s^6} \tag{3.37}$$

where

$$(\alpha_2, \alpha_3, \alpha_4) = (8.973\text{E-}10, 5.5155\text{E-}12, 3.2010\text{E-}17)$$

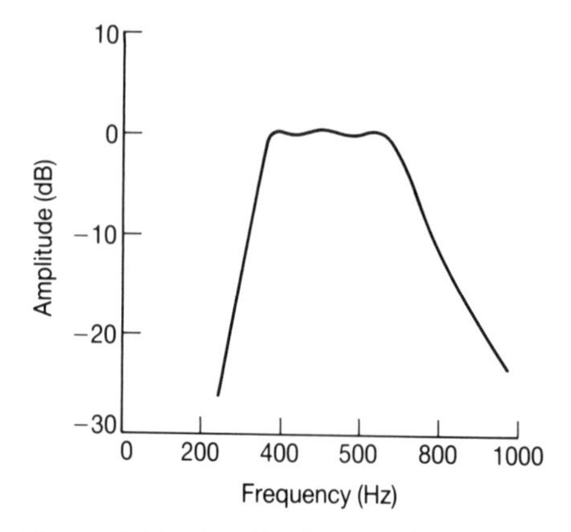

Figure 3.10 Amplitude transfer characteristic of the system studied.

$$(\beta_0, \beta_1, \ldots, \beta_6) = (1, 2.5017\text{E-}4, 3.5869\text{E-}7, 5.5550\text{E-}11, 3.36031\text{E-}14, 2.5351\text{E-}18, 1.0131\text{E-}21) \quad (3.38)$$

The corresponding amplitude characteristic is given in Figure 3.10. The system is excited with a multi-sine at the frequencies $f_k = kf_0$, with $k = 25, 26, \ldots, 100$ and $f_0 = 1/(2048 \times 50\text{E}{-6})$ Hz. The RMS value of the multi-sine is set equal to $1/\sqrt{2}$. Two multi-sines are considered, the first having a flat-amplitude spectrum and the second being optimized following the procedure described above. The evolution of the power spectrum-optimization process is given in Figure 3.11. The optimization is stopped before the final convergence is reached (after 3 iterations) to avoid signals with a sparse spectrum which are very difficult to compress. Therefore the crest factor would remain high and from equation (3.36) it is seen that this would jeopardize the accuracy gain due to spectrum optimization. The determinant of the corresponding Cramér–Rao lower bound was reduced by a factor of 43 after three iterations.

The crest factors or peak values (not the peak factor!) of the multi-sine at the input and output are minimized using the second iterative algorithm of Appendix 3.1 (l_∞) and the results are given in Table 3.3. Three situations are considered:

- Minimization of the crest factor of the input signal
- Simultaneous minimization of the crest factors of the input and output
- Simultaneous minimization of the peak values of the input and output

For our purpose the last possibility is the most important because it will determine the settings of the full scale of the measurement instruments. In Table 3.3 it is seen that the peak values of the multi-sine with the optimized power spectrum are equal to those of the multi-sine with flat-power spectrum (optimization (c)). Therefore the settings of the measurement instruments can remain the same for both excitations and consequently the noise on both measurements will be equal. However, the

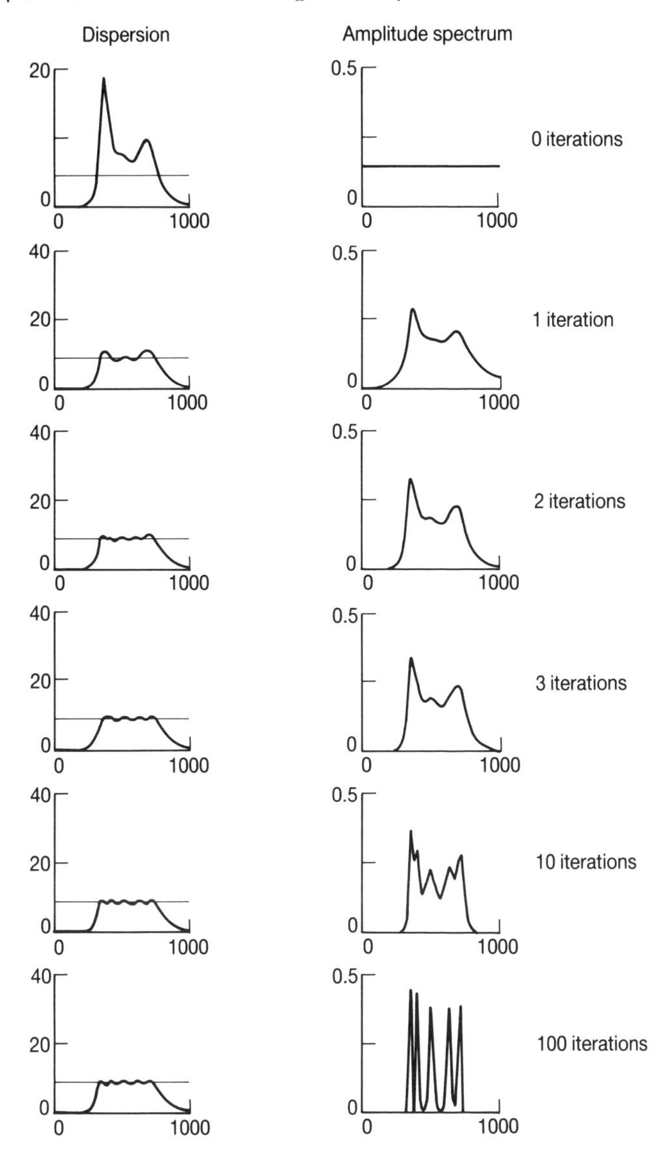

Figure 3.11 Evolution of the power spectrum optimization process.

uncertainty on the estimated parameters will be smaller in the second case because the determinant $|\mathbf{C}|$ is much smaller than in the first case, resulting in a smaller uncertainty on the calculated transfer characteristics.

Figure 3.12 shows the input and output multi-sine with optimized power spectrum and minimized peak values. The RMS value of the input multi-sine is put equal to $1/\sqrt{2}$ in all these examples.

Table 3.3 Minimization of the crest factor(s) or peak values of two multi-sines, related by the linear system (3.37). (a) Minimization of the crest factor of the input; (b) simultaneous minimization of the crest factors of the input and output; (c) simultaneous minimization of the peak values of the input and output

	Input		Output	
	Crest factor	Peak value	Crest factor	Peak value
Flat input power spectrum				
(a)	**1.459**	1.031	2.749	1.418
(b)	**1.667**	1.170	**1.667**	0.862
(c)	1.509	**1.067**	2.065	**1.067**
Optimized input power spectrum				
(a)	**1.459**	1.031	1.860	1.200
(b)	**1.582**	1.118	**1.582**	1.026
(c)	1.508	**1.066**	1.643	**1.066**

From experimental tests it was found that these signals can be easily generated in practice; small disturbances at the amplitudes or the phases do not result in an excessive growth of the crest factor. In Figure 3.12(b) measurements of the calculated multi-sines are given. They were generated with a 12-bit arbitrary waveform generator with 2048 points in one period (sampling frequency 20 kHz). The generator was followed by a reconstruction filter (a Cauer filter with a cut-off frequency of 2 kHz) (see note 3, p. 160). No phase or amplitude compensation was made for the distortion introduced by this reconstruction filter. If this amplitude/phase distortion becomes disturbing, it is always possible to give a precompensation to the amplitudes/phases of the multi-sine. The measurements were made with an 8-bit digitizer (full scale ± 1 volt) at 512 points with a sampling frequency of 5 kHz.

Figure 3.13(a) compares the $\sigma_H(\omega)$ when a multi-sine with a flat and an optimized amplitude spectrum is used. These results were experimentally verified using the set-up described previously. Sixty measurements were made to determine the standard deviation of the FRF measurement and the results are shown in Figure 3.13(b). It is obvious that this result is only relevant if the model errors of the parametric model in the identification step are smaller than the identification uncertainty due to the noise.

COMPARISON OF THE PARAMETRIC AND NON-PARAMETRIC APPROACH

For the non-parametric measurement of H the standard deviation $\sigma_{H\text{non-par}}(\omega)$ is given by equation (3.9), which becomes

$$\sigma^2_{H\text{non-par}}(j\omega) = \frac{2}{|X(j\omega)|^2}\left(|H(j\omega)|^2\sigma^2_X(\omega) + \sigma^2_Y(\omega)\right) \tag{3.39}$$

if no averages are made ($N = 1$), while for the parametric case it is found from equation (3.33) that

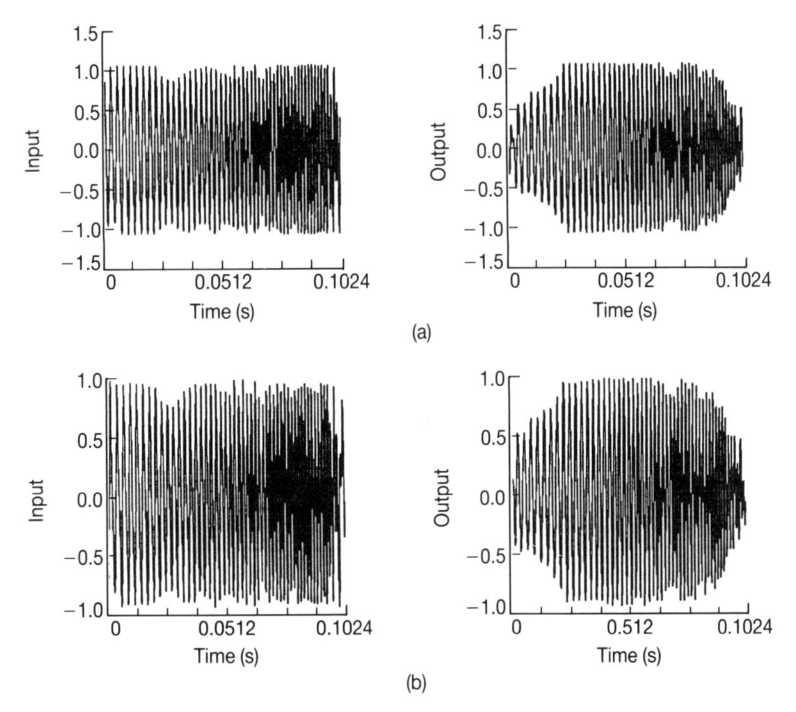

Figure 3.12 Input and output multi-sine with optimized power spectrum and minimized peak values (simultaneous at the input and the output). (a) Theoretical signals; (b) measured signals.

$$\sigma^2_{H\text{par}}(\omega) = \frac{v(\chi, \text{j}\omega)}{\mathscr{P}} \left(| H(\text{j}\omega) |^2 \sigma^2_X(\omega) + \sigma^2_Y(\omega) \right) \tag{3.40}$$

The ratio between both uncertainties is

$$\frac{\sigma^2_{H\text{non-par}}(\omega)}{\sigma^2_{H\text{par}}(\omega)} = \frac{2\mathscr{P}}{v(\chi, \omega) \,| X(\text{j}\omega) |^2} \tag{3.41}$$

If an excitation signal with a flat-amplitude spectrum is used for the non-parametric measurement ($| X(\text{j}\omega) |^2 = \mathscr{P}/F$) and an optimum spectrum for the parametric measurement is selected ($v(\chi, \omega) \leqslant n_p$), expression (3.41) becomes

$$\frac{\sigma^2_{H\text{non-par}}(\omega)}{\sigma^2_{H\text{par}}(\omega)} = \frac{2F}{n_p} \tag{3.42}$$

This shows that, due to a parametric approach, the uncertainty on the transfer function can be reduced. The improvement is proportional to the number of measurement points, F, in the non-parametric measurement and inversely proportional to the required model complexity (n_p). Note that this expression does not account for possible

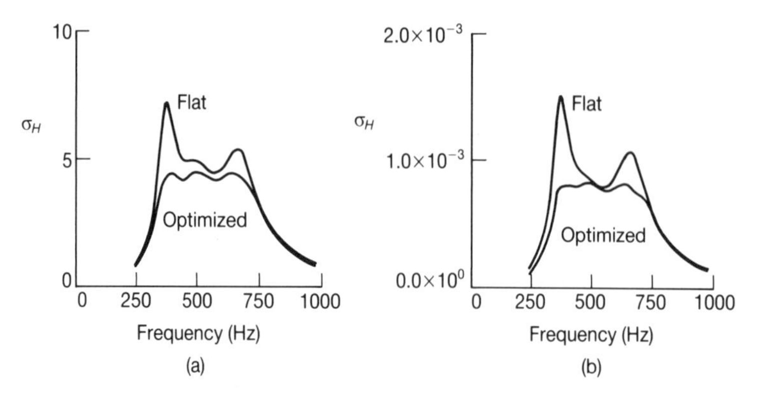

Figure 3.13 Comparison of the model uncertainty with the flat and the optimized power spectra. (a) Theoretical (scaled) results; (b) experimental results.

model errors which will introduce systematic errors into the parametric measurement. These results were verified using the measurements of the previously described experiment ($F = 76$, $n_p = 9$, a flat-amplitude spectrum for the non-parametric measurement, the optimized amplitude spectrum for the parametric measurement). The results are shown in Figure 3.14.

3.4 Conclusion

It is shown that it is possible to design optimized excitation signals for transfer function measurements. It was found that the multi-sine is a very flexible signal with good properties. It is the only signal which gives complete control of the amplitude spectrum. The proposed excitation design method consists of two steps:

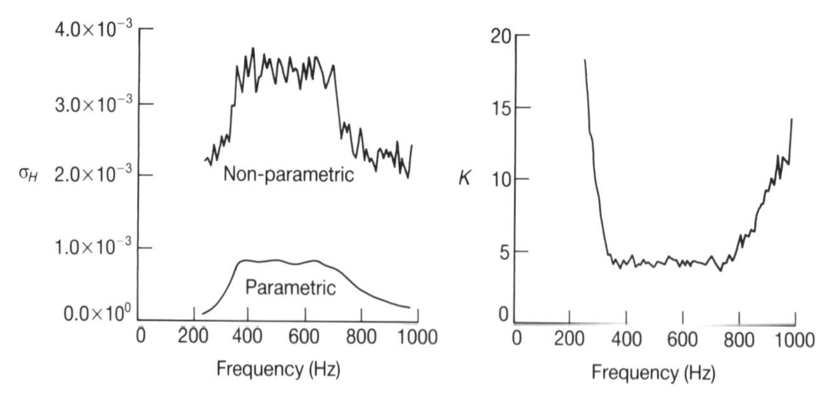

Figure 3.14 Comparison of the parametric and the non-parametric measurement approach ($K = \sigma_{\text{non-param}}/\sigma_{\text{par}}$).

- Selection of an optimum power spectrum
- Reduction of the crest factor (or peak values)

The second step gives considerable improvement in the measurement results and is easiest to perform because no prior information is required and the crest factor reduction algorithms are computationally efficient. The first step gives, in general, a less significant improvement of the measurement uncertainty while it requires more prior information – a noise analysis (for the non-parametric and for the parametric measurement) – while for the parametric measurement a good transfer function model is also required. Only if a large amount of almost identical systems has to be measured it is worth performing this optimization. Otherwise it is more reasonable to propose a power spectrum based on the user's experience.

Appendix 3.1 Uncertainty on frequency response measurements

In this appendix the variance of the frequency response measurements will be analyzed using linear approximations. In the first part it will be shown that all measurement techniques have the same linear approximation. This means that for sufficiently large SNR they will also have the same variance. This result may not be generalized for the bias because here at least the second-order contributions should be considered. In the second part the variance is calculated.

Linear approximations

H_1

$$H_1(j\omega) = \frac{\frac{1}{N}\sum_{k=1}^{N} Y_{mk}(j\omega)X^*_{mk}(j\omega)}{\frac{1}{N}\sum_{k=1}^{N} X_{mk}(j\omega)X^*_{mk}(j\omega)} \tag{3.A1}$$

Define $X_{mk} = X + M_k$ and $M'_k = M_k/X$ and $Y_{mk} = Y + N_k$ and $N'_k = N_k/Y$. With these definitions H_1 becomes

$$H_1 = \frac{\frac{1}{N}\sum_{k=1}^{N} YX^*(1 + N'_k)(1 + M'^*_k)}{\frac{1}{N}\sum_{k=1}^{N} |X|^2(1 + M'_k)(1 + M'^*_k)} \tag{3.A2}$$

In this expression the dependence of ω is not explicitly mentioned to simplify the notation. Defining H as the true value of the frequency response, this expression can be approximated by

$$H_1 \approx H \frac{\sum_{k=1}^{N} (1 + N'_k + M'^*_k)}{\sum_{k=1}^{N} (1 + M'_k + M'^*_k)} \tag{3.A3}$$

Making a linear approximation, this expression becomes

$$H_1 \approx H\left(1 + \frac{1}{N} \sum_{k=1}^{N} (N'_k - M'_k)\right) \tag{3.A4}$$

H_2

$$H_2(j\omega) + \frac{\frac{1}{N} \sum_{k=1}^{N} Y_{mk}(j\omega) Y^*_{mk}(j\omega)}{\frac{1}{N} \sum_{k=1}^{N} X_{mk}(j\omega) Y^*_{mk}(j\omega)} \tag{3.A5}$$

$$H_2 = \frac{\frac{1}{N} \sum_{k=1}^{N} |Y|^2 (1 + N'_k)(1 + N'^*_k)}{\frac{1}{N} \sum_{k=1}^{N} XY^*(1 + M'_k)(1 + N'^*_k)} \tag{3.A6}$$

which can again be approximated by

$$H_2 \approx H\left(1 + \frac{1}{N} \sum_{k=1}^{N} (N'_k - M'_k)\right) \tag{3.A7}$$

$H_{\log}$

$$H_{\log}(j\omega) = \exp\left(\frac{1}{N} \sum_{k=1}^{N} \log H_{mk}(j\omega)\right) = \left(\prod_{k=1}^{N} \frac{Y_{mk}(j\omega)}{X_{mk}(j\omega)}\right)^{1/N} \tag{3.A8}$$

With the previous definitions this expression becomes

$$H_{\log} = \left(\prod_{k=1}^{N} H \frac{1 + N'_k}{1 + M'_k}\right)^{1/N} \tag{3.A9}$$

and can be approximated by

$$H_{\log} \approx H\left(\prod_{k=1}^{N} (1 + N'_k)(1 - M'_k)\right)^{1/N} \approx H \prod_{k=1}^{N} \left(1 + \frac{N'_k}{N}\right)\left(1 - \frac{M'_k}{N}\right) \tag{3.A10}$$

Approximating the product by its linear terms gives

$$H_{\log} = H\left(1 + \sum_{k=1}^{N} \left(\frac{N'_k}{N} - \frac{M'_k}{N}\right)\right) \tag{3.A11}$$

which finally results in the same result as before.

This shows that all three methods can be approximated by the same expression.

Calculation of the variance

The variance is given by

$$\sigma^2_{H_m} = E\{(H_m - E\{H_m\})(H_m - E\{H_m\})^*\} \tag{3.A12}$$

If the input and output noise is uncorrelated and neglecting all higher-order (>2) noise contributions, this expression becomes

$$\sigma^2_{H_m} = \frac{|H|^2}{N^2} \sum_{k=1}^{N} E\{(N'_k - M'_k)(N'_k - M'_k)^*\} \tag{3.A13}$$

or

$$\sigma^2_{H_m} = 2\frac{|H|^2}{N}\left(\frac{\sigma^2_Y}{|Y|^2} + \frac{\sigma^2_X}{|X|^2}\right) \tag{3.A14}$$

Appendix 3.2 Crest factor minimization algorithms

Many crest factor minimization methods have been presented in the literature. A first group consists of explicit formulas which allows a direct calculation of the phases. A second group searches for an optimal set of phases using iterative algorithms. Both classes will be discussed briefly.

Explicit formulas

Schroeder (1970) proposed the following phase selection:

$$\phi_k = \phi_1 - 2\pi \sum_{n=1}^{k-1} (k-n)P_n \tag{3.A15}$$

where P_n is the relative power of the nth component, which reduces to

$$\phi_k = -\frac{\pi(k-1)k}{F} \tag{3.A16}$$

for a multi-sine with a flat-amplitude spectrum. From simulations, it was found that equation (3.A16) gives good results (crest factor 1.6–1.8) if the amplitude spectrum is dense and flat. For multi-sines with a sparse spectrum, where the frequency lines are few and far apart, or for multi-sines with an amplitude spectrum that is not flat, equations (3.A15) or (3.A16) give no better results than those obtained with a random phase selection, uniformly distributed in $[0, 2\pi]$. In these situations more sophisticated methods are needed and no explicit formulas are available.

Iterative procedures

CLIPPING ALGORITHM

Van der Ouderaa *et al.* (1988a,b) developed an iterative method to optimize the phases which is very closely related to an algorithm presented by Van den Bos (1987). The basic concept in this method is a clipping procedure, which is illustrated in Figure 3.15. For a given amplitude spectrum a time signal has to be found with a minimum peak value. The iteration procedure is started from the specified amplitude spectrum, and arbitrary phases are taken as starting values. Using the inverse Fourier transform the signal is calculated at a set of discrete equidistant times. A new time signal is then generated by clipping off all the values larger than a given maximum, and the new modified spectrum and phases are calculated using the FFT. These new phases are retained as a first approximation to the solution, but the modified amplitude spectrum is rejected in favour of the original one. This procedure is repeated until no further significant reduction in the crest factor is obtained. In general, the algorithm needs a few hundred iterations to obtain useful signals (for example, a flat multi-sine with a crest factor of 1.5), but in order to obtain near-optimal crest factors (of 1.4) a few hundred thousand iteration steps are more likely to be required. This algorithm will be called the clipping algorithm.

l_{2p} ALGORITHM

Guillaume *et al.* (1991) developed an algorithm based on the minimization of the l_{2p} norm:

$$l_{2p} = \|u\|_{2p} = \sqrt[2p]{\frac{1}{T}\int_0^T u^{2p}(t)\,\mathrm{d}t} \qquad (3.A17)$$

where T is the period of the multi-sine and $p \in \mathrm{IN}$. It is shown that the l_{2p} norm is equal to

$$l_{2p} = \sqrt[2p]{\frac{1}{N}\sum_{n=0}^{N-1} u^{2p}(t_n)} \qquad (3.A18)$$

if

$$t_n = n\frac{T}{N} \qquad N \geqslant 2p\frac{\omega_{\max}}{2\pi}T + 1 \qquad (3.A19)$$

where $\omega_{\max}$ is the maximum angular frequency occurring in the multi-sine (3.24) and N the number of samples in one period. Equation (3.A19) expresses that no alias contribution may appear on the d.c. component.

The l_{2p} norm is minimized with respect to the phases using a Marquardt algorithm for values of p which are gradually increased during the iteration process (e.g. $p = 2$,

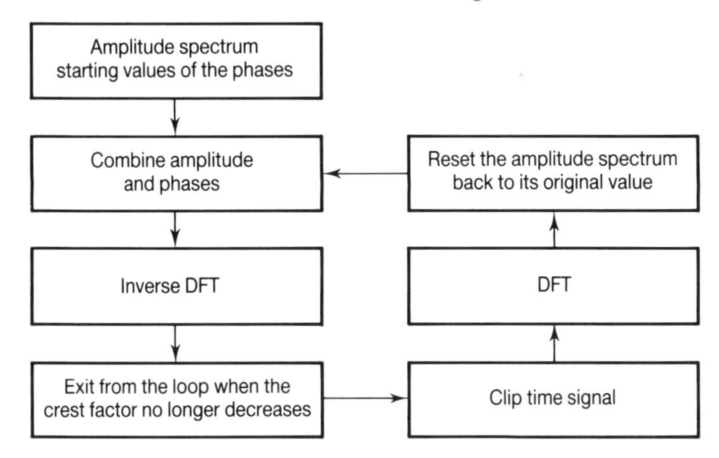

Figure 3.15 Minimization of the crest factor of a multi-sine: clipping algorithm.

4, 8, 16, 32, ...). This defines a descent algorithm that converges to a local minimum. From our experience it was found that the results of this algorithm were better than those obtained with the previous method. In practice, conditions (3.A19) may be violated as long as a sufficiently large number of points are considered (e.g. $N \geqslant 16\,\omega_{max}T/(2\pi) + 1$), which allows a significant reduction in the calculation time.

However, the major advantage of the l_{2p} algorithm is its suitability to optimize multiple multi-sines linked by linear systems. The most simple example is the optimization of the input and the output of a system $Y(\omega) = H(\omega)X(\omega)$. At that moment the loss function is generalized to

$$K = \left\| \left(\frac{x(\phi)}{x_{rms}}, \frac{y(\phi)}{y_{rms}} \right) \right\|_{2p} \tag{3.A20}$$

where ϕ is the phases of the multi-sine x. In Guillaume *et al.* (1991) it is shown that the minimum of K with respect to ϕ for p growing to infinity will result in two multi-sines with equal and minimum crest factors. Sometimes, it is more advantageous to minimize the scaled peak values of both multi-sines, which is done by minimizing

$$K = \left\| \left(x(\phi), \frac{y(\phi)}{S} \right) \right\|_{2p} \tag{3.A21}$$

where S is a scaling factor. This allows an optimal use of the full scale of the measurement equipment. When S is chosen as the ratio of the RMS values, signals with equal crest factors are obtained. Note that when S is chosen too large (or too small), the problem reduces to the minimization of $\|x(\phi)\|_{2p}$ or $\|y(\phi)\|_{2p}$.

With the clipping algorithm it is also possible to minimize the crest factor of multiple multi-sines related by a linear multiple-input, multiple-output system if the pseudo-inverse of the transfer function matrix exists (this is not required by the l_{2p}

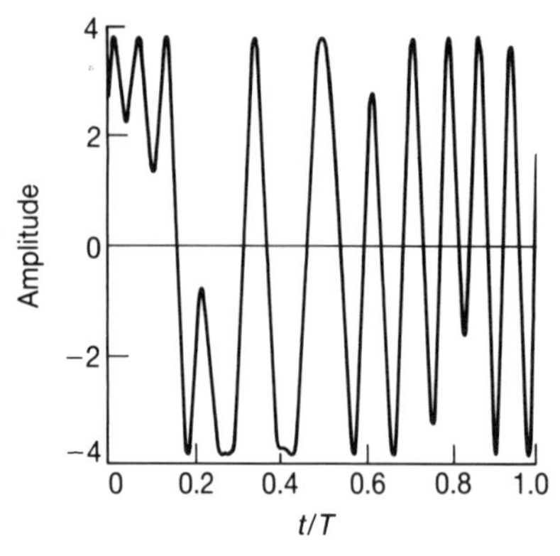

Figure 3.16 Compressed multi-sine with flat amplitude spectrum.

algorithm). However, the minimization is not as straightforward as it is with the l_{2p} algorithm.

Appendix 3.3 Some examples of phases

In this appendix the phases of two compressed multi-sines are tabulated to allow readers to compare these results with their own experience. The signals were compressed with the l_∞ algorithm.

EXAMPLE 1

A multi-sine with a flat amplitude spectrum, $f_k = k$, $k = 1, 2, \ldots, 15$ (Figure 3.16). The crest factor of the compressed signal is $c_r = 1.402$

Harmonic number	*phase* (°)
1	3.6066E + 01
2	−2.9698E + 01
3	−1.0449E + 02
4	−9.3379E + 01
5	1.6674E + 02
6	−4.1798E + 00
7	−1.4671E + 02
8	6.6054E + 01
9	−1.1615E + 02
10	−7.4311E + 01
11	1.3901E + 02
12	−1.3170E + 02
13	−5.8560E + 01
14	−6.4019E + 00
15	−1.6918E + 01

EXAMPLE 2:

Input/output optimization of two multi-sines related by an integer (see Figure 3.3(c)): phases of the input multi-sine:

Harmonic number	Phase (°)	Harmonic number	Phase (°)
16	−3.7824E + 01	17	2.0601E + 01
18	1.4225E + 01	19	2.4552E + 01
20	−1.0076E + 02	21	−8.9025E + 01
22	−1.1560E + 02	23	6.8384E + 01
24	1.5958E + 02	25	8.4457E + 01
26	−5.0065E + 01	27	−1.4354E + 02
28	8.6808E + 01	29	5.7490E + 01
30	−9.8009E + 01	31	−1.7369E + 02
32	1.0555E + 02	33	−6.2377E + 01
34	1.6763E + 02	35	−1.8621E + 01
36	−1.4928E + 02	37	4.9366E + 01
38	−1.0912E + 02	39	1.4428E + 02
40	−2.8594E + 01	41	3.6124E + 01
42	−1.1967E + 02	43	2.5180E + 01
44	1.5991E + 02	45	1.5954E + 01
46	9.3937E + 01	47	−1.6275E + 02
48	−3.2984E + 01	49	9.6499E + 01
50	1.2770E + 02	51	−1.2556E + 02
52	5.1556E + 01	53	1.5034E + 02
54	−1.1423E + 02	55	−1.0483E + 02
56	−3.5427E + 01	57	1.1040E + 02
58	9.9409E + 01	59	8.2832E + 01
60	1.1264E + 02	61	−1.7693E + 02
62	1.2849E + 02	63	1.3891E + 02

Notes

1. It can be shown that the standard deviation on the real and imaginary parts of the Fourier coefficients after a DFT are asymptotically (number of time points $\to \infty$) equal to each other (Schoukens and Renneboog, 1986; Schoukens and Pintelon, 1991).
2. The crest factor is closely related to the peak factor, which is defined as $(u_{max} - u_{min})/(2\sqrt{2}u_{rms})$. The crest factor is a better measure for the compactness of a signal. An improvement of the peak factor definition would be to replace u_{rms} by u_{rmse}. The peak factor also does not fully account for asymmetric peaks. The peak factor can increase by more than 20% if $(u_{max} - u_{min})/2$ is replaced by $\max(|u(t)|)$. If only the dynamic range of the signal has to be described, $(u_{max} - u_{min})/2$ is preferable over $\max(|u(t)|)$. To describe

the absolute behaviour of the signal (to minimize, for example, the influence of non-linearities $\max(|u(t)|)$ is to be preferred.

3. When a digital-to-analog converter (DAC) is used as a generator, it is necessary to eliminate the higher-order harmonics which are created by the zero-order-hold reconstruction, which keeps the signal constant between two successive sample points generated by the DAC. This should be done with an analog filter, called a reconstruction filter.

4

Periodic Test Signals – Properties and Use

Adriaan van den Bos

4.1 Introduction

One of the purposes of this chapter is to discuss a number of *properties* of periodic test signals relevant to the identification of linear dynamic systems. In this discussion the emphasis will be on *multi-frequency signals*. These are periodic, multi-harmonic signals. The spectrum of multi-frequency signals may be freely chosen. In this respect, they are different from standard periodic test signals such as maximum-length binary sequences. For a particular sequence length these have a fixed spectrum. The spectrum of multi-frequency signals, on the other hand, may be chosen on the basis of *a priori* knowledge about the system and the spectrum of the noise in observations of the response.

A further purpose is to describe aspects of the *use* of periodic test signals for the identification of linear dynamic systems. This concerns both the estimation of the Fourier coefficients of the steady-state response to a multi-frequency signal and the use of these Fourier coefficient estimates for system identification.

The outline of this chapter is as follows. In Section 4.2 relevant properties of multi-frequency signals are described. In particular, this concerns digitally generated, discrete-interval multi-frequency signals. The subject of Section 4.3 is the synthesis of discrete-interval multi-frequency signals. Methods for synthesis of low-peak-factor and, closely related, binary multi-frequency signals are also included. Section 4.4 is a review of methods for estimating Fourier coefficients from noise-disturbed responses to multi-frequency signals. In Section 4.5 two examples are given of methods for estimating parameters of linear discrete-time systems from the resulting estimated Fourier coefficients. Conclusions are drawn in Section 4.6.

4.2 Discrete-interval multi-frequency signals

Jensen (1959) and Levin (1959) were probably the first to propose multi-frequency signals for estimating properties of dynamic linear systems. They used analog experimental set-ups, the only possibility available at that time. Today analog equipment has been replaced by digital alternatives. This solves the problem of synchronization of analog harmonics and, at the same time, provides accurate sampling facilities essential to harmonic analysis. On the other hand, the test signals now become *discrete-interval* as a result of the presence of a digital-to-analog converter. Therefore in this section attention will be paid to the properties of discrete-interval multi-frequency signals.

Let $s(t)$ be a discrete-interval signal periodic with T. Define Δt and N as the clock-pulse interval and the integer number of clock-pulse intervals in a period, respectively. Therefore $T = N\,\Delta t$. Assume that in a period the value of $s(t)$ can only change at points $(n + 1/2)\Delta t$, $n = 0, \ldots, N - 1$. Define the Fourier coefficient of the kth harmonic of $s(t)$ as:

$$\gamma_{ks} = \frac{1}{T}\int_0^T s(t)\exp(-\mathrm{j}2\pi kt/T)\,\mathrm{d}t \tag{4.1}$$

where $\mathrm{j} = \sqrt{-1}$. Straightforward calculations then show that

$$\gamma_{ks} = S_k\,\mathrm{sinc}(\pi k/N) \tag{4.2}$$

where $\mathrm{sinc}(x) = \sin(x)/x$ and S_k, $k = 1, \ldots, N - 1$ is the Discrete Fourier Transform:

$$S_k = \frac{1}{N}\sum_n s_n \exp(-\mathrm{j}2\pi kn/N) \tag{4.3}$$

of the sequence $s_n = s(n\,\Delta t)$, $n = 0, \ldots, N - 1$ (see Van den Bos and Krol, 1979). Furthermore, since this sequence is real,

$$S_k = S^*_{N-k} \tag{4.4}$$

where the superscript asterisk denotes complex conjugation. Then the Fourier coefficients γ_{ks}, $k = 0, \pm 1, \pm 2, \ldots$ and all S_k, $k = 0, \ldots, N - 1$ are defined by

$$\gamma_{ks},\ k = 0, \ldots, [N/2] \tag{4.5}$$

or equivalently by

$$S_k,\ k = 0, \ldots, [N/2] \tag{4.6}$$

where $[N/2]$ is the largest integer smaller than or equal to $N/2$. Since $s(t)$ and s_n are real, equations (4.5) and (4.6) are defined by N real and imaginary parts, representing the N degrees of freedom of $s(t)$ or s_n in the frequency domain.

Since equations (4.5) or (4.6) are the only Fourier coefficients that can be freely chosen, they completely determine the higher harmonic content of the Fourier coefficient spectrum. For example, let $0 \leqslant k' \leqslant [N/2]$. Then, by equations (4.2) and

(4.4), all Fourier coefficients of the harmonics $k' \pm lN$, $l = 1, 2, \ldots$, and $-k' \pm lN$, $l = 1, 2, \ldots$, are determined by $\gamma_{k's}$. The Fourier coefficients of these *associated harmonics* are equal to zero if $\gamma_{k's}$ is. This is important for the synthesis of *orthogonal* multi-frequency signals. Two multi-frequency signals of the same period will be called orthogonal if a non-zero complex Fourier coefficient of a particular harmonic of the one signal implies a zero-valued Fourier complex coefficient for the same harmonic of the other. Hence, if two discrete-interval multi-frequency signals are orthogonal with respect to the harmonics on $[0, [N/2]]$, they are orthogonal with respect to all harmonics. If two or more orthogonal multi-frequency signals are applied to a linear multi-input multi-output system, the orthogonality thus defined also extends to the steady-state responses to them.

4.3 Synthesis of discrete-interval multi-frequency signals

4.3.1 Discrete-interval signals having a specified harmonic content

In this section the properties of discrete-interval multi-frequency signals described in Section 4.2 will be used for the synthesis of signals having a specified harmonic content. Suppose that the specified Fourier coefficient spectrum is

$$\gamma_{k_1 s}, \ldots, \gamma_{k_L s} \tag{4.7}$$

where $k_l < k_m$ if $l < m$. Now select N so that $k_L \leqslant [N/2]$ and construct a Discrete Fourier Transform sequence S_k, $k = 0, \ldots, N-1$ such that

$$\begin{aligned} S_k &= \gamma_{ks}/\text{sinc}(\pi k/N) && \text{for } k = k_l \\ S_{N-k} &= \gamma_{ks}^*/\text{sinc}(\pi k/N) && \text{for } k = k_l \\ S_k &= 0 && \text{for } k \neq k_l \\ S_{N-k} &= 0 && \text{for } k \neq k_l \end{aligned} \tag{4.8}$$

with $l = 1, \ldots, L$. Next, compute the Inverse Discrete Fourier Transform sequence s_n, $n = 0, \ldots, N-1$:

$$s_n = \sum_k S_k \exp(j2\pi nk/N) \tag{4.9}$$

where $k = 1, \ldots, N-1$. Then these s_n are the midpoint values on the intervals of the desired discrete-interval signal and define this signal completely.

It is observed that the total power of a selected Fourier coefficient $\gamma_{k_l s}$ and all harmonics associated with it is equal to

$$|S_{k_l}|^2 \sum_m \text{sinc}^2\{\pi(k_l + mN)/N\} = |S_{k_l}|^2 \tag{4.10}$$

where $m = 0, \pm 1, \pm 2, \ldots$. Hence, the ratio of the power of the undesirable associated harmonics to the power of the desired harmonic is

$$\text{sinc}^{-2}(\pi k_l/N) - 1 \tag{4.11}$$

For $k_l/N = 1/4$, $1/8$ and $1/16$ this ratio is equal to 0.23, 0.053 and 0.013, respectively. These results emphasize the importance of keeping k_L/N as small as possible for effective signal construction. Note, however, that the specified Fourier coefficients $\gamma_{k_l s}$ of multi-frequency signals obtained along the lines of this section are exactly realized.

4.3.2 Binary and low-peak-factor multi-frequency signals

The peak factor of a signal $s(t)$ will be defined as

$$\frac{s_{\max} - s_{\min}}{2\sqrt{2}s_{\text{rms}}} \tag{4.12}$$

where $s_{\max}$, $s_{\min}$ and s_{rms} are the maximum, the minimum and the root-mean-square values of the signal, respectively. It is clear that a low peak factor may be desirable. The maximum allowable signal value at the input of the system is always subject to physical limitations such as input transducer non-linearities. Therefore the signal with the lowest peak factor best exploits the allowable amplitude range to introduce maximum power to the system. This improves the signal-to-noise ratio in the measured response.

If $s_{\max} = -s_{\min}$, the peak factor is equal to 0.707 and minimal for binary signals with levels $s_{\max}$ and $s_{\min}$. A further advantage of binary signals is the ease of introduction into the system. Binary multi-frequency signals were probably first constructed by Jensen (1959) by clipping multi-sinusoids. Systematic procedures for their construction were developed later by Van den Bos (1967) and Van den Bos and Krol (1979).

The statement that binary multi-frequency signals have the lowest peak factor implies that in the computation of the root-mean-square value all harmonics have been included. However, *all* harmonics are seldom used in practice. Therefore computation of the peak factor relative to the root-mean-square value of all harmonics may lead to an unfair comparison of different signals. Preferably, in the peak factor formula s_{rms} should be taken as the root-mean-square value of the *specified harmonics* or that of the *harmonics used.* For discrete-interval multi-frequency signals a peak factor thus computed will be larger than the peak factor based on the power of all harmonics. The reason is the power concentrated in the harmonics associated with the specified ones. Binary discrete-interval multi-frequency signals developed thus far contain, in addition to the power in the unavoidable associated harmonics, power in undesirable spurious harmonics between those specified.

If the phase angles of the specified harmonics may be freely chosen, they may be manipulated to minimize the peak factor. This is, by definition, a minimax problem, and methods and software are available for its local solution (Overton, 1982; Sharda *et al.*, 1986). However, the very large amount of local minima of the peak factor and the relatively long computation times make this approach less attractive, therefore other procedures have been developed. One of the simplest of these, described in Van den Bos and Krol (1979) and Van den Bos (1987), will now be outlined.

The basic idea is that the synthesis of a binary multi-frequency signal approximating a specified Fourier amplitude spectrum $|\gamma_{k_l s}|$, $l = 1, \ldots, L$ and the synthesis of a low-peak-factor non-binary multi-frequency signal having the same spectrum are closely related. In the latter case, intuitively one tries to construct a signal that has the specified harmonic content and is as similar as possible to a binary signal because of the latter's low peak factor. This is achieved as follows. First, for arbitrary initial phase angles $\phi_{k_l s}$ the Fourier coefficients $\gamma_{k_l s} = |\gamma_{k_l s}| \exp(-j\phi_{k_l s})$, $l = 1, \ldots, L$ are computed. Next, from these Fourier coefficients the Discrete Fourier Transform sequence S_k, $k = 0, \ldots, N-1$, is formed such as described by equation (4.8). This sequence is subsequently inversely discrete Fourier transformed into the sequence s_n, $n = 0, \ldots, N-1$. From this sequence a new sequence u_n, $n = 0, \ldots, N-1$, is computed as follows. If s_n is positive u_n is equal to $+1$. If s_n is negative u_n is equal to -1. If s_n is equal to zero u_n may arbitrarily be taken equal to $+1$ or -1. The sequence u_n, $n = 0, \ldots, N-1$, is subsequently discrete Fourier transformed into the sequence U_k, $k = 0, \ldots, N-1$. Next, the phase angles $\phi_{k_l s}$ are taken as the corresponding $\phi_{k_l u}$ and all described steps are repeated. This is continued until the phase angles cease to change. This procedure converges with certainty in a finite number of steps. It produces simultaneously a binary discrete-interval multi-frequency signal locally best approximating the specified amplitude spectrum in the least-squares sense and a non-binary discrete-interval multi-frequency signal best approximating the binary signal in the least-squares sense. The procedure requires two Discrete Fourier Transforms in every iteration. Therefore it can be carried out very efficiently using a Fast Fourier Transform. It is, however, local and has to be repeated from random initial conditions until a satisfactory solution has been obtained. Other approaches are reported by Schroeder (1970, 1984), Van der Ouderaa (1988), Van der Ouderaa *et al.* (1988a) and Lenstra (1987).

The rule for phase-angle adjustment described by Schroeder (1970, 1984) is explicitly heuristic. It yields closed-form results for the phase angles, is extremely simple and has been designed for dense spectra consisting of a large number of lines. Its results could be tried as initial conditions for the other methods which are all iterative.

The method of Van der Ouderaa (1988) and Van der Ouderaa *et al.* (1988a) bears a close resemblance to that described above. The main difference is that the signal s_n is not clipped but is two-sidedly hard-limited at a certain level. This level is then gradually reduced during the iteration process.

Lenstra (1987) applies the *simulated annealing method* to peak-factor minimization. His results are summarized in the next section since they are not easily accessible to most readers.

4.3.3 Peak-factor minimization using simulated annealing

The simulated annealing method was originally designed to solve a minimization problem in statistical mechanics. Later, Kirkpatrick *et al.* (1983) showed that it can also be used for many multivariable minimization problems outside this field. For a brief description of simulated annealing and a demonstration program see Press *et al.* (1986).

For the problem at hand, the most important property of the simulated annealing method is that it is far less likely to remain fixed in a relative minimum than most conventional minimization methods. The reason is that in every step of the procedure it may, with a certain probability, go uphill. This probability is slowly diminished as the absolute minimum is approached. A disadvantage of the method is that it is very slow, and a single peak-factor minimization takes many hours of computing time.

Three illustrative numerical examples in Lenstra (1987) are now summarized which concern discrete-interval signals with 512 steps in a period. The peak factors are computed with respect to the root-mean-square value of the selected harmonics. Computations using equation (4.11) show that with 512 steps in a period the contribution of the associated harmonics to the RMS value is of the order of tenths of a per cent or less in the three examples chosen.

EXAMPLE 1

This concerns a spectrum used by Schroeder (1970):

$$|\gamma_{ks}|^2 = \sin^2(\pi(2k-1)/32) \qquad k = 1, \ldots, 16 \tag{4.13}$$

Therefore this is a sixteen harmonic bandpass spectrum. Simulated annealing produces a peak factor equal to 1.02. The minimum peak factors found by Schroeder (1970) and Van den Bos (1987) are equal to 1.17 and 1.07, respectively.

EXAMPLE 2

The spectrum is again one used by Schroeder (1970):

$$|\gamma_{ks}|^2 = 1 \qquad k = 1, \ldots, 31 \tag{4.14}$$

with phase angles restricted to 0 and π. Therefore the signals are symmetric with respect to the origin. The minimum peak factor found with simulated annealing is now 1.17. Using his rule, Schroeder (1970) finds a peak factor equal to 1.39. Schroeder (1984), using a number-theoretic method restricted to flat spectra and phase angles 0 and π, attains a peak factor equal to 1.21. Van den Bos (1987) reports a minimum peak factor equal to 1.25.

EXAMPLE 3

A third example in Lenstra (1987) concerns the spectrum:

$$|\gamma_{k_l s}|^2 = 1 \qquad k_l = 2^{l-1} \qquad l = 1, \ldots, 6 \tag{4.15}$$

The peak factors found with Schroeder's rule, the Van den Bos method and simulated annealing are now 1.61, 1.48 and 1.41, respectively.

From these numerical examples the following conclusions may be drawn. In all cases the peak factors attained by simulated annealing are lowest. Moreover, Lenstra (1987) reports that repeated application of simulated annealing to the same spectrum yields minimum peak factors that differ by 1%, at most. It is also found that the corresponding time-domain signals are almost the same.

The minimum peak factors attained by Van den Bos (1987) are substantially lower than those produced by Schroeder's rule, but higher than those achieved by simulated annealing. However, they have been selected from 100, 200 and 100 repetitions in Examples 1, 2 and 3, respectively. With these numbers of repetitions the computation time is only a fraction of that used for simulated annealing. The occurrence of peak factors lower than those mentioned above would have been highly probable had the computation times been equal for both methods.

4.4 Estimation of Fourier coefficients

In practice, observed responses to test signals will always be subject to errors. Suppose that the observations of the steady-state response of a linear system to a periodic test signal are described by

$$w_n = y_n + h_n \qquad n = 0, \ldots, J-1 \tag{4.16}$$

where y_n is the steady-state response, h_n is a stochastic error and J is an integral multiple of the supposedly integer number of samples in a period of the test signal. Since the system is assumed to be linear, the steady-state response y_n has the same period as the test signal. Then, as a rule, the estimator of the Fourier coefficients of y_n is taken as

$$W_k = \frac{1}{J} \sum_n w_n \exp(-\mathrm{j}2\pi kn/N) \tag{4.17}$$

with $n = 0, \ldots, J-1$ being the finite Discrete Fourier Transform of the w_n. The properties of this estimator are well known. In the first place, it is unbiased for γ_{ky} if no aliasing occurs. Furthermore, if it is assumed that the width of the autocovariance function of the h_n is small as compared with J, W_k has a variance (Brillinger, 1981)

$$\mathrm{var}(W_k) \approx S_{hh}(k)/J \tag{4.18}$$

where $S_{hh}(k)$ is the power-density spectrum of the h_n defined as the infinite discrete Fourier transform of the covariance sequence $R_{hh}(m)$, $m = 0, \pm 1, \pm 2, \ldots$, of the h_n.

To most users the choice of the DFT (4.17) as an estimator of the Fourier coefficients may appear almost self-evident. However, with this choice it is not certain that the available *a priori* knowledge about the errors is fully used. More specifically, the question arises if and (if so) how the available *a priori* knowledge can be used to construct estimators having a variance smaller than that described by equation (4.18). This question is relevant, since most estimators of the system parameters from observed periodic responses consist of two steps. In the first the Fourier coefficients are estimated. Then, in the second step, the system parameters are estimated from the Fourier coefficient estimates. A loss of precision in the first step cannot be compensated for in the second step, no matter which procedure is used. Therefore in Section 4.4.1 the statistical properties of the estimator (4.17) and related estimators are analyzed for various types of statistical errors. Then in Section 4.4.2 other approaches to estimating the Fourier coefficients are discussed.

4.4.1 Least squares estimation of Fourier coefficients

Equation (4.16) shows that the observations w_n are linear in the Fourier coefficients γ_{ky} since the y_n are. As the γ_{ky} are the parameters to be estimated, the observations w_n are the sum of a known linear combination of these unknown parameters and the statistical errors h_n. This is easily recognized as the classical *linear regression model* (see Goldberger, 1964). The usual estimator for the parameters in this model is the ordinary (uniformly weighted) least squares estimator. It is not difficult to show that in the problem at hand the ordinary least squares estimator is identical with the DFT estimator (4.17) (see Van den Bos, 1989). From regression theory, the ordinary least squares estimator is known to be the maximum likelihood estimator if the errors are i.i.d. (independent and identically distributed) and normal. It can also be shown that, under these conditions, this maximum likelihood estimator attains the Cramér–Rao lower bound *for any number of observations*. The Cramér–Rao lower bound is a lower bound on the variance of any unbiased estimator (see Stuart and Ord, 1991). It is often used as a measure of the relative precision of a particular estimator.

In conclusion, the DFT (4.17) is the most precise estimator if the errors h_n in equation (4.16) are i.i.d. and normal.

The question now arises as to what are the properties of the estimator (4.17) if these conditions are not met. First, let the errors be identically normally distributed but, different from previously, be covariant. In that case, the ordinary least squares estimator (4.17) can be shown to be different from the maximum likelihood estimator and not to attain the Cramér–Rao lower bound. Detailed computations show that in the presence of covariant errors the maximum likelihood estimator is the *weighted* least squares estimator with the inverse of the covariance matrix of the errors as weighting matrix (see Eykhoff, 1974). This estimator also attains the Cramér–Rao lower bound for any number of observations. In the special case considered here this weighting matrix is the inverse of the $J \times J$ covariance matrix with (m, n)th element $R_{hh}(m - n)$, where $R_{hh}(\cdot)$ is the covariance sequence of the stochastic process h_n.

The weighting makes the computation of the weighted least squares estimates much more complicated and time consuming than that of estimator (4.17). In particular, the computation of the inverse of the $J \times J$ covariance matrix is demanding since J may be large. Fortunately, there are sound reasons to assume that the difference of the variance of the ordinary least squares estimator (4.17) and the variance of the weighted least squares estimator vanishes asymptotically, that is, if the number of observed periods becomes large (see, for example, Grenander and Szegö, 1958). For what follows it is important that in their proof these authors make no assumptions with respect to the probability density function of the errors. Therefore, their result also applies to non-normal errors.

In conclusion, for i.i.d. and normal errors expression (4.17) is the maximum likelihood estimator and attains the Cramér–Rao lower bound for any number of periods. For covariant normal errors (4.17) is not the maximum likelihood estimator but it approaches the Cramér–Rao lower bound asymptotically.

In the regression literature, the weighted least squares estimator described above is called *best linear unbiased* (see Goldberger, 1964). The reason is that it can be shown that this estimator has the smallest variance among all estimators that are both unbiased and linear in the observations. This result is valid for any distribution of the errors. Therefore, if both the i.i.d. and the normality assumption with respect to the h_n in equation (4.16) are dropped the weighted least squares estimator has the smallest variance within the class of all unbiased estimators linear in the observations. Furthermore, as has been mentioned above, the variance of the weighted least squares estimator approaches that of the ordinary least squares estimator if the number of observed periods increases. Hence, if the ordinary least squares estimator (4.17) is used, the results will asymptotically have the smallest variance within the class of linear unbiased estimators for any distribution of the errors. However, if the errors are not normally distributed, this asymptotically best linear unbiased estimator will, as a rule, not be the maximum likelihood estimator and will not attain the Cramér–Rao lower bound. Therefore in Section 4.4.2 a brief review is presented how available statistical *a priori* knowledge about non-normal errors may be used to select an estimator for the Fourier coefficients.

4.4.2 Other approaches

Generally, maximum likelihood estimators attain the Cramér–Rao lower bound, at least asymptotically. In this sense they are most precise and may be preferable to least squares estimators. On the other hand, they require a considerable amount of *a priori* knowledge: the probability density function of the errors. The criterion of goodness of fit to be used in a maximum likelihood method is constructed from this probability density function. For an extensive description of the theory of maximum likelihood estimation, see Stuart and Ord (1991) and Norden (1972, 1973).

For non-normal errors the maximum likelihood criterion is usually different from the least squares criterion. For example, if the errors are i.i.d. with Laplacian probability density function $\lambda \exp(-2\lambda |h_n|)$ with $\lambda > 0$, the maximum likelihood criterion is the least moduli criterion.

Maximum likelihood estimators require and use *complete* knowledge of the probability density function of the errors. They have not been designed to be used when this knowledge is partial. It has also been found that relatively small deviations of the assumed probability density function of the errors from the one actually present may result in a variance substantially larger than the Cramér–Rao lower bound. It is said that the maximum likelihood estimator is not *robust* with respect to the probability density function of the errors. A well-known example is the use of the least squares estimator if the errors are largely normally distributed but outliers occasionally occur. Then, although the least squares estimator is identical with the maximum likelihood estimator if the errors are normal, the variance of the resulting estimates may considerably exceed the Cramér–Rao lower bound.

In the absence of reliable, complete knowledge of the probability density function of the errors, it may be advisable to use so-called *robust* estimators. These have been specially designed to incorporate partial statistical *a priori* knowledge of the errors. With respect to their variance, they have optimal properties for the whole *class* of error probability density functions compatible with the assumed *a priori* knowledge (see Poljak and Tsypkin, 1980; Zypkin, 1987). Robust estimators have been constructed that can handle outliers. A further example is a robust estimator for the case that it is only known that the errors do not exceed a particular amplitude level. It has also been found that the least moduli estimator is the robust estimator in the absence of any *a priori* knowledge.

The least moduli estimator is illustrative for a disadvantage connected with many maximum likelihood and robust estimators: they are relatively difficult to handle numerically. The least moduli criterion is not everywhere differentiable. As a result, simple closed-form solutions such as estimator (4.17) cannot be established and standard minimization procedures based on derivatives cannot be used. Zanakis and Rustagi (1982) review numerical methods for least moduli estimation. The methods discussed in their review are all more demanding and time consuming than estimator (4.17). On the other hand, for *linear* least moduli problems, such as the estimation of the Fourier coefficients considered here, reliable software is available (see NAG, 1990).

In conclusion, least squares estimation of Fourier coefficients using estimator (4.17) is simple and fast but may be imprecise. If the precision of the Fourier coefficient estimates is important and computation time is less so, the estimators described in this section may be preferable.

4.5 Use of estimated Fourier coefficients

The purpose of this section is to give an example of the *use* of estimated Fourier coefficients for discrete-time dynamic system identification. The procedure for continuous-time systems is analogous and is extensively described in Van den Bos (1974, 1991). A related approach is described by Schoukens *et al.* (1988a).

Define the system difference equation as

$$a_0 y_n + a_1 y_{n-1} + \ldots + a_P y_{n-P} = u_n + b_1 u_{n-1} + \ldots + b_Q u_{n-Q} \tag{4.19}$$

where y_n is the response to u_n. Then the exact Fourier coefficients of a periodic input and those of the corresponding steady-state response satisfy

$$A(r_k) Y_k - B(r_k) U_k = 0 \tag{4.20}$$

where Y_k and U_k are the Fourier coefficients of the kth harmonic of y_n and u_n, respectively, $A(r) = a_0 + a_1 r + \ldots + a_P r^P$ is the denominator polynomial, $B(r) = 1 + b_1 r + \ldots + b_Q r^Q$ the numerator polynomial, P and Q are the orders concerned, and r_k is defined as $\exp(-j2\pi k/N)$. It will be assumed throughout that $a_0, \ldots, a_P, b_1, \ldots, b_Q$ are the parameters to be estimated. Note that equation (4.19) can alternatively be written as $y_n = H(z^{-1}) u_n$, where $H(z^{-1}) = B(z^{-1})/A(z^{-1})$ is the transfer function of the system and z^{-1} is the backward shift operator.

Now assume that the observations of the periodic input and those of the steady-state response are described by

$$w_n = y_n + h_n \quad \text{and} \quad v_n = u_n + g_n \quad n = 0, \ldots, J-1 \tag{4.21}$$

respectively, where h_n and g_n are stationary stochastic processes defined as

$$h_n = h_{n1} + h_{n2} + h_{n3} \tag{4.22}$$

and

$$g_n = g_{n1} + g_{n2} \tag{4.23}$$

where h_{n2} and g_{n2} are measurement errors which are neither mutually covariant nor covariant with any of the other components of h_n and g_n. The process h_{n1} is the response to the additional input g_{n1} which may be the normal operating input. The process h_{n3} is assumed to be covariant with g_{n1} and, therefore, with h_{n1}. However, it is assumed not to be causally related with g_{n1}. A process such as h_{n3} may occur in closed-loop.

Next, two different methods are presented for estimating the system parameters from observations thus defined. The first, described in Section 4.5.1, is a frequency-domain instrumental variable method. The second, described in Section 4.5.2, is a frequency-domain least squares method. The relation of both methods is also explained and their advantages are discussed in Section 4.5.3.

4.5.1 A frequency-domain instrumental variable method

Assume that $P + Q + 1$ different signals $f_{n1}, \ldots, f_{n,P+Q+1}$ are constructed, periodic with N and having Fourier coefficients F_{km} with $k = 0, \ldots, N - 1$ and $m = 1, \ldots, P + Q + 1$. A frequency-domain estimator using these signals as instrumental sequences may be constructed as follows. Cross-correlating both sides of equation (4.19) with each of the $P + Q + 1$ instrumental sequences yields

$$\frac{1}{N} \sum_n (a_0 y_n + a_1 y_{n-1} + \ldots + a_P y_{n-P} - u_n - b_1 u_{n-1} + \ldots - b_Q u_{n-Q}) f_{nm} = 0 \qquad (4.24)$$

where $n = 0, \ldots, N - 1$ and $m = 1, \ldots, P + Q + 1$. These are $P + Q + 1$ linear equations in the $P + Q + 1$ exact system parameters $a_0, a_1, \ldots, b_1, \ldots, b_Q$. The coefficients and the right-hand sides of these equations are all values of the cross-correlation functions of the instrumental sequences and y_n or u_n for various shifts:

$$\frac{1}{N} \sum_n y_{n-p} f_{nm} \quad \text{and} \quad \frac{1}{N} \sum_n u_{n-q} f_{nm} \qquad n = 0, \ldots, N - 1 \qquad (4.25)$$

where $p = 0, \ldots, P$, $q = 0, \ldots, Q$ and $m = 1, \ldots, P + Q + 1$. By the discrete Parseval theorem the inner products (4.25) can easily be shown to be equal to, respectively:

$$\sum_k (r_k^*)^p Y_k^* F_{km} \quad \text{and} \quad \sum_k (r_k^*)^q U_k^* F_{km} \qquad k = 0, \ldots, N - 1 \qquad (4.26)$$

where Y_k and U_k are, as before, the Fourier coefficients of the kth harmonic of y_n and u_n, respectively, and F_{km} is the Fourier coefficient of the kth harmonic of the mth instrumental sequence. Assume that expressions (4.26) are substituted for the corresponding expressions (4.25) in (4.24). Then expression (4.24) is transformed into:

$$\sum_k D^*(Y_k, U_k) F_{km} = 0 \qquad (4.27)$$

where $k = 0, \ldots, N - 1$, $m = 1, \ldots, P + Q + 1$ and

$$D(Y_k, U_k) = A(r_k) Y_k - B(r_k) U_k \qquad (4.28)$$

Now assume that W_k and V_k are estimates of the Fourier coefficients Y_k and U_k from observations (w_n, v_n) $n = 0, \ldots, J - 1$. For example, W_k and V_k may be the Discrete Fourier Transforms (4.17) of these observations. Then the *frequency-domain instrumental variable (IV) estimator* of the a_p and b_q is obtained by substituting W_k for Y_k and V_k for U_k in equation (4.27):

$$\sum_k D^*(W_k, V_k) F_{km} = 0 \qquad (4.29)$$

The $D(W_k, V_k)$ are easily recognized as the frequency-domain residuals of equation (4.20). Note that estimator (4.29) is linear and closed form. The conclusion is that with periodic inputs both the IV estimator and the construction of the required instrumental sequences are extremely simple. First, $P + Q + 1$ linearly independent, but otherwise arbitrary, vectors of Fourier coefficients $(F_0 F_1 \ldots F_{N-1})_m$, $m = 1, \ldots, P + Q + 1$ are constructed with $F_{N-k,m} = F^*_{km}$. All that is left to be done is to compute the estimates of the Fourier coefficients W_k and V_k and solve equation (4.29).

4.5.2 A frequency-domain least squares method

An alternative estimator is, for continuous-time systems, presented in Van den Bos (1974, 1991). Modified for discrete-time systems, it may be described as follows. The least squares (LS) criterion

$$\sum_k |D(W_k, V_k)|^2 \tag{4.30}$$

is minimized with respect to the a_p and b_q. The result is the *frequency-domain least squares estimator*:

$$\sum_k D^*(W_k, V_k) r_k^p W_k = 0 \qquad \text{and} \qquad \sum_k D^*(W_k, V_k) r_k^q V_k = 0 \tag{4.31}$$

where $p = 0, \ldots, P$ and $q = 1, \ldots, Q$. Note that the LS estimator (4.31) is linear and closed form. Furthermore, from a comparison of equations (4.29) and (4.31) it is concluded that the LS estimator is equivalent to an IV estimator using the shifted versions of the estimated harmonic components of $u(t)$ and $y(t)$ as instrumental sequences.

Asymptotic properties of the frequency-domain residuals for continuous-time systems are derived in Van den Bos (1974). Modifying these properties for the discrete-time systems studied here shows that, asymptotically, the residuals $D(W_k, V_k)$ are non-covariant and have a variance equal to

$$\frac{1}{J}\,[\,|A(r_k)|^2\{S_{h_2 h_2}(k) + S_{h_3 h_3}(k)\} + |B(r_k)|^2 S_{g_2 g_2}(k)] \tag{4.32}$$

Note that $g_1(t)$ and $h_1(t)$ do not influence these variances. This is a consequence of the causal relation of these processes.

Subsequently, Van den Bos (1974) proposes an additional *weighted* LS step with the reciprocals of equation (4.32) as weights for the terms in criterion (4.30). The purpose is to decrease the variance of the estimates of the a_p and b_q or to attain the Cramér–Rao lower bound on the variance for normally distributed $g(t)$ and $h(t)$. Expression (4.32) shows that this requires knowledge of both the a_p and b_q and of the power spectra concerned. The a_p and b_q may be taken as their estimates obtained in the first step. The spectra may be known or may have been measured before applying the test signal. Alternatively, the variances (4.32) of the residuals may have been directly estimated in the first step. The latter approach is chosen by Van den Bos (1974). The additional step, of course, increases computational effort. However, approximate *a priori* knowledge about the spectra may be sufficient. For example, assume that $u(t)$ is the test signal itself. Then $g_2(t) = 0$. In addition, assume that the spectrum of $h(t)$ is known to be more or less flat over the bandwidth of $h(t)$. Then the weights may be taken as $|A(r_k)|^{-2}$. All this illustrates how *a priori* knowledge of system and noise may be used to improve the precision of the estimates of the system coefficients.

4.5.3 Discussion of the proposed frequency-domain estimators

Van den Bos (1991) extensively discusses the favourable properties of the continuous-time counterparts of the discrete-time frequency-domain estimators (4.29) and (4.31) of this chapter. Since most of these properties are similar in both cases, only a summary is given here.

First, the estimation of the system coefficients described by estimators (4.29) and (4.31) is preceded by estimation of the Fourier coefficients. This reduces the available time-domain observations to a relatively small number of frequency-domain ones – the Fourier coefficient estimates – which makes the observations much easier to handle. Furthermore, the Fourier coefficient estimates are very suitable for checking the usefulness of the observations available. For example, this may be done by plotting Bode or Nyquist diagrams, which improves control of the experiment. If needed, one or more faulty periods may be omitted.

Furthermore, as opposed to most other time-domain estimators of the system coefficients, estimators (4.29) and (4.31) are linear and closed form. They are, therefore, unique and require little computation time and only standard software.

Also, useful extensions are practically feasible, since the computation time for estimators (4.29) and (4.31) is so modest. For example, (4.29) and (4.31) may be applied repetitively for a number of values of a pure time delay to find the optimum solution for the system coefficients and the delay.

Finally, the least squares estimator (4.31) is *formally* similar to the classical linear regression estimators. This greatly facilitates the derivation of expressions for the covariance matrix of the system coefficient estimates, as is shown by Van den Bos (1974, 1991) for the continuous-time frequency-domain least squares estimator. These continuous-time results are easily transformed into those for discrete-time by replacing the differentiation operator $-j2\pi k/T$ by the shift operator $\exp(-j2\pi k/N)$.

4.6 Conclusions

Use of periodic, multi-harmonic test signals provides the experimenter with a substantial freedom with respect to the choice of test signal, choice of estimator and use of a discrete- or continuous-time model. Moreover, before estimating the system parameters the time-domain observations are transformed into Fourier coefficient estimates which are much easier to interpret and appreciate. In addition, the estimation of both the Fourier coefficients and the system parameters may be formulated as a linear estimation problem. It is concluded that with periodic test signals the parameters of linear systems can be estimated in a simple, flexible and, at every stage of the identification process, controlled way.

5

Periodic and Non-periodic, Binary and Multi-level Pseudo-random Signals

Michael Darnell

5.1 Introduction

In Chapter 1 the basic concepts associated with both periodic and non-periodic pseudo-random signals are introduced, as are examples of binary and multi-level (>2) pseudo-random signals and their applications. It is the objective of this chapter to examine in greater depth the various types of pseudo-random signals available for system-identification purposes. These signals are discussed under the three headings of

1. Periodic binary and multi-level signals;
2. Non-periodic binary signals;
3. Non-periodic multi-level signals.

An assumption is made that the primary requirement for all the pseudo-random signal types considered here is that of an impulsive, or near-impulsive, autocorrelation function (ACF). However, other requirements, such as energy–time distribution and the ability to synthesize uncorrelated signal sets for multi-input system-identification applications, will also be incorporated as appropriate. The latter requires that the signals in the set should have low (ideally, zero) cross-correlation function (CCF) values.

Implementational complexity has, until comparatively recently, been an important factor in determining the choice of system-identification stimulus. Consequently, binary pseudo-random signals, particularly maximum-length signals (otherwise known as m-signals), have been applied widely for many years because of their simplicity of generation (Hoffmann de Visme, 1971). Now, with the advent of cheap and powerful digital signal processing (DSP) devices, the capability exists for the real-time generation and processing of a much greater variety of pseudo-random signals (e.g. with multiple levels and complex elements) employing a wide range of clock rates. It is also possible to synthesize different types of non-periodic binary and multi-level signals with pseudo-random properties. In some cases the use of a non-periodic test input may well form the basis of an identification procedure which is more efficient than if a periodic input were to be used.

The following section will discuss the definitions of periodic and non-periodic correlation functions, and how the two can be related for a given signal. Sections 5.3–5.5 will describe the available types of pseudo-random signals in the categories (1)–(3) above. Application aspects will then be considered in Section 5.6, where non-periodic identification considerations are stressed.

5.2 Periodic and non-periodic correlation function evaluation

It is now important to distinguish formally between 'periodic' and 'non-periodic' correlation function evaluation. This can be accomplished by means of a simple example involving the 3-digit binary signal $+1\ +1\ -1$, which will be represented by the polarity symbols $+\,+\,-$ only.

5.2.1 Periodic correlation

The evaluation of the periodic ACF of the signal $+\,+\,-$ is shown schematically below as a function of a discrete shift r, variable in units of one digit position:

```
. . .  + + − +|+ − +|+ − . . .      Periodic reference signal
       + + −  |     |               ACF = +3 (r = 0)
         + + −|     |               ACF = −1 (r = 1)
           + +|−    |               ACF = −1 (r = 2)
             +|+ −  |               ACF = +3 (r = 3)
              |+ + −|               ACF = −1 (r = 4)
  r  = 0 1 2 3|4 5 6|7                  etc.
          ——→ |     |←———————————   1 signal period
```

In the above process, the number of digits in the signal period, N, is 3. Periodic ACF computation involves digit-by-digit multiplication of the reference signal by a time-shifted version of itself, followed by summation (integration) of the products over one period of N digits. Note that here N reference digits are always correlated against N shifted digits. It is evident that the periodic ACF is itself a periodic function of the shift variable, r. Evaluation of the periodic CCF between two distinct signals having the same period would be carried out in the same manner

5.2.2 Non-periodic correlation

The evaluation of the non-periodic ACF of the same binary signal + + − is illustrated below. Now, a single period of the signal embedded in an 'all-zero' digit stream is correlated against a time-shifted version of itself:

```
. . . 0 0 0 | + + − | 0 0 0 . . .      Non-periodic reference signal
      + + − |       |                  ACF = 0 (r = 0)
        + + | −     |                  ACF = −1 (r = 1)
          + | + −   |                  ACF = 0 (r = 2)
            | + + − |                  ACF = +3 (r = 3)
            |   + + | −                ACF = 0 (r = 4)
            |     + | + −              ACF = −1 (r = 5)
            |       | + + −            ACF = 0 (r = 6)
  r = 0 1 2 | 3 4 5 | 6 7 8
    ──────→ |       | ←──────          1 signal interval
```

Here, the effect is that multiplication and product summation of the non-zero digits only takes place in the 'overlap' range of the reference and shifted signals. The non-periodic ACF is itself a non-periodic function of the discrete shift variable r. Non-periodic CCF evaluation is performed in a similar manner. (Similar evaluation of the non-periodic ACF of a binary signal with $N = 7$ is given in Chapter 1, Section 1.8.)

5.2.3 Relationship between non-periodic and periodic correlation functions

The form of the non-periodic ACF of an arbitrary, N-digit signal completely determines the form of the periodic ACF of that same signal. These ACFs can be simply related, as is demonstrated in the discussion which follows.

In Figures 5.1(a) and 5.1(b) two distinct non-periodic ACF shifts are shown, i.e. $r - i$ digits and $r = (i + N)$ digits, respectively. Let the corresponding non-periodic ACF values be $R_{NN(N)}(i)$ and $R_{NN(N)}(i + N)$. Figure 5.1(c) illustrates the periodic ACF condition for the same N-digit signal at a shift of $r = i$, for which the corresponding periodic ACF value is $R_{NN(P)}(i)$.

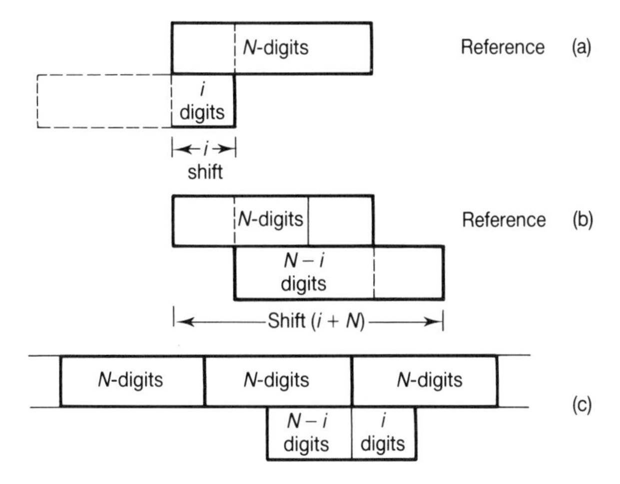

Figure 5.1 Non-periodic and periodic correlation functions.

Clearly, the periodic ACF value corresponding to Figure 5.1(c) can be obtained from the summation of the aperiodic ACF values of Figures 5.1(a) and 5.1(b), i.e.

$$R_{NN(P)}(i) = R_{NN(N)}(i) + R_{NN(N)}(i + N) \tag{5.1}$$

where $N \leqslant (i + N) \leqslant 2N$. When $i = 0$ at the 'in-phase' position for the periodic ACF

$$R_{NN(P)}(0) = R_{NN(N)}(N) \tag{5.2}$$

since $R_{NN(N)}(0) = 0$ at the 'no overlap' position for the non-periodic ACF. Hence the peak values of the periodic and non-periodic ACFs are identical. In general,

$$R_{NN(P)}(r) = R_{NN(N)}(r) + R_{NN(N)}(r + N) \tag{5.3}$$

If the normalized peak-to-maximum side-lobe modulus ratio (PSR) of the non-periodic ACF is $N{:}a$, where

$$a \leqslant N \tag{5.4}$$

simple reasoning dictates that the minimum bound on the corresponding periodic ACF normalized PSR will be $N{:}2a$. A corollary of this result is that a signal having a non-periodic ACF with zero side-lobe level will also have a periodic ACF with zero side lobes. The converse is not true, as can be seen from a consideration of the 4-digit binary signal $+ + + -$ having a non-periodic ACF given by

$$\begin{aligned} R_{44(N)}(r) &= 0 \quad -1 \quad 0 \quad +1 \quad +4 \quad +1 \quad 0 \quad -1 \quad 0 \\ r &= 0 \quad 1 \quad 2 \quad 3 \quad 4 \quad 5 \quad 6 \quad 7 \quad 8 \end{aligned} \tag{5.5}$$

The corresponding periodic ACF is

$$R_{44(P)}(r) = +4 \quad 0 \quad 0 \quad 0 \quad +4 \quad 0 \quad 0 \quad 0 \quad +4 \quad \ldots$$
$$r = 0 \quad 1 \quad 2 \quad 3 \quad 4 \quad 5 \quad 6 \quad 7 \quad 8 \qquad (5.6)$$

By examining equation (5.3) in more detail it is seen that for N odd, when r is even then $(r + N)$ must be odd, and vice versa. Similarly, when N is even, if r is odd it follows that $(r + N)$ must also be odd; when N is even and r is even, $(r + N)$ must also be even. Thus it is necessary to examine appropriate non-periodic ACF values at odd and even shifts in order to predict the form of the corresponding periodic ACF.

For the signal whose non-periodic ACF is given by equation (5.5), Table 5.1 indicates how the individual periodic ACF values can be calculated ($N = 4$). As a second example, consider the 7-digit binary signal $+ + + - - + -$ with non-periodic ACF

$$R_{77(N)}(r) = 0 \quad -1 \quad 0 \quad -1 \quad 0 \quad -1 \quad 0 \quad +7 \quad 0 \quad -1 \quad 0 \quad -1 \quad 0 \quad -1 \quad 0$$
$$r = 0 \quad 1 \quad 2 \quad 3 \quad 4 \quad 5 \quad 6 \quad 7 \quad 8 \quad 9 \quad 10 \quad 11 \quad 12 \quad 13 \quad 14 \qquad (5.7)$$

Table 5.1 Calculation of periodic ACF from the non-periodic ACF for a signal with $N = 4$

r	$(r + 4)$	$R_{44(N)}(r)$	$R_{44(N)}(r + 4)$	$R_{44(P)}(r)$
0	4	0	+4	+4
1	5	−1	+1	0
2	6	0	0	0
3	7	+1	−1	0
4	8	+4	0	+4

A similar tabulation (Table 5.2) allows the corresponding periodic ACF, $R_{77(P)}(r)$, to be computed. Thus prediction of a periodic ACF from a corresponding non-periodic ACF involves the systematic summation of pairs of non-periodic ACF values, separated by N-digit shifts. The same approach can be employed to relate non-periodic and periodic CCFs.

Table 5.2 Calculation of periodic ACF from the non-periodic ACF for a signal with $N = 7$

r	$(r + 7)$	$R_{77(N)}(r)$	$R_{77(N)}(r + 7)$	$R_{77(P)}(r)$
0	7	0	+7	+7
1	8	−1	0	−1
2	9	0	−1	−1
3	10	−1	0	−1
4	11	0	−1	−1
5	12	−1	0	−1
6	13	0	−1	−1
7	14	+7	0	+7

The reverse process, i.e. computation of the non-periodic correlation function from the periodic correlation function, is less straightforward in that the same off-peak periodic correlation values may be produced from different pairs of non-periodic values. To illustrate, the periodic ACF calculated in Table 5.2, $R_{77(P)}(r)$, corresponds to the non-periodic ACF of expression (5.7). However, a non-periodic ACF of the form

$$\begin{array}{lccccccccccccccc} R_{77(N)}(r) = & 0 & +1 & +2 & -1 & 0 & -3 & -2 & +7 & -2 & -3 & 0 & -1 & +2 & +1 & 0 \\ r = & 0 & 1 & 2 & 3 & 4 & 5 & 6 & 7 & 8 & 9 & 10 & 11 & 12 & 13 & 14 \end{array} \qquad (5.8)$$

will also give rise to the same periodic ACF. Hence, additional constraints will be required to eliminate this ambiguity, assuming that expression (5.8) represents a realizable ACF.

5.3 Periodic binary and multi-level pseudo-random signals

A comprehensive discussion of periodic binary, maximum-length, pseudo-random signals is given in Chapter 1; therefore only a brief resumé of their properties is given here within the more general context of multi-level, maximum-length signals. The design of multi-level pseudo-random signals is considered further in Chapter 11.

The general mathematical structure of q-level maximum-length (or m-) signals is presented in Zierler (1959) and Everett (1966). Of particular interest for system-identification applications are m-signals for which q is a prime ($\geqslant 2$). The widely used binary ($q = 2$) m-signals are specific (and, in some respects, untypical) examples of this general structure. For ease of reference, the important properties of q-level m-signals are summarized below (Darnell, 1991a):

1. Basic signals comprise the integer element $0, 1, 2, \ldots, (q-1)$.
2. Each cycle has $(q^n - 1)$ digits, where n is an integer (>1) corresponding to the number of stages in the equivalent q-level feedback shift-register (FSR) generator.
3. The number of zeros in each cycle is $(q^{n-1} - 1)$.
4. The number of each of the non-zero elements in a cycle is q^{n-1}.
5. Each cycle comprises $(q-1)$ sequential 'blocks' of digits of length $(q^n - 1)/(q-1)$ digits, and hence $(q-1)$ is always a factor of $(q^n - 1)$ for all q and n.
6. From any reference point in the cycle, the block comprising the subsequent $(q^n - 1)/(q-1)$ digits can be derived by multiplying (modulo q) all digits in the preceding block of length $(q^n - 1)/(q-1)$ by a primitive element g of the Galois field of q elements GF(q). Hence, it can be shown that

$$g^{q-1} = 1 \pmod q \qquad (5.9)$$

and

$$g^{(q-1)/2} = (q-1) \pmod q \qquad (5.10)$$

FSR feedback connections giving rise to binary m-signals for values of n in the range $2 \leqslant n \leqslant 127$ are given in Table 1.2 of Chapter 1. Also, some connections for q-level FSR generators with $3 \leqslant q \leqslant 31$ are shown in Tables 1.6 and 1.8.

As an example, the m-signal produced by the generator shown in Figure 5.2 will now be considered; here, $q = 5$, $n = 2$. This 24-digit periodic m-signal has the form:

$$\underset{\text{(i)}}{1\,0\,2\,4\,2\,2}\quad \underset{\text{(ii)}}{3\,0\,1\,2\,1\,1}\quad \underset{\text{(iii)}}{4\,0\,3\,1\,3\,3}\quad \underset{\text{(iv)}}{2\,0\,4\,3\,4\,4} \tag{5.11}$$

The partitioning of equation (5.11) into $(q - 1)$ (in this case 4) blocks is arbitrary. The value of the block multiplier, g, reading from left to right, is 3; in the reverse sense, the value of the block multiplier g is 2. Note that both 2 and 3 are primitive elements of GF(5). Thus, the signal can be specified completely by any block of $(5^2 - 1)/(5 - 1)$ contiguous digits, together with the appropriate primitive element of GF(5). Alternatively, it is specified completely by any block plus the first digit of the succeeding block. It should also be noted that blocks (iii) and (iv) are the modulo 5 complements of blocks (i) and (ii), respectively; this arises from equation (5.10).

For completeness, the procedure for calculating the primitive elements of GF(5) is now outlined. Table 5.3 is a table of reduced residues (mod 5) for powers 1 to 4 of the numbers 1 to 4. The primitive elements of GF(5) are those numbers, R, for which the reduced residue 1 occurs for the first time under R^4. In general, the primitive elements of GF(q) are obtained by calculating the set of reduced residues (modulo q) for the numbers $R = 1, 2, \ldots, (q - 1)$ raised to the powers $1, 2, \ldots, (q - 1)$; the numbers for which the reduced residue 1 occurs for the first time under R^{q-1} are the primitive elements.

Table 5.3 Reduced residues (mod 5) for powers 1 to 4 of the numbers 1 to 4

R	R^1	R^2	R^3	R^4
1	1	1	1	1
2	2	4	3	1
3	3	4	2	1
4	4	1	4	1

Clearly, the signal of equation (5.11) consists of all-positive integers, as will all q-level m-signals; in this form, they do not possess useful ACF properties. For many applications, bipolar pseudo-random signals with impulsive (or near-impulsive) ACFs are required; to synthesize such signals, it is necessary to employ a 'level transformation'. Two examples of appropriate level transformations are now presented.

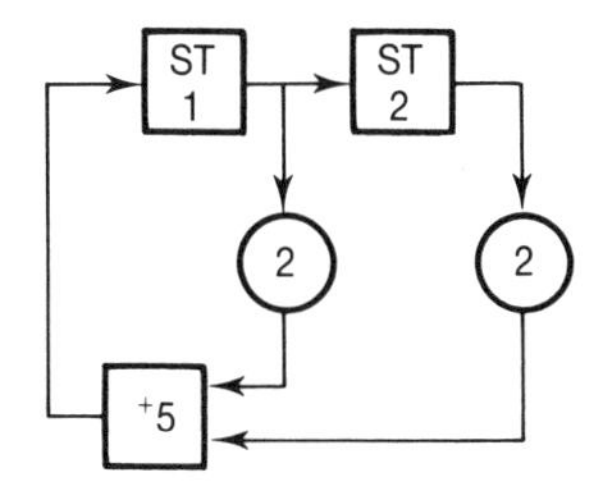

Figure 5.2 Schematic diagram of five-level m-signal generator.

5.3.1 Integer level transformation

For $q > 2$, bipolar signals can be produced from the basic unipolar m-signal by an integer level transformation applied to any integer element, k, of that basic signal, where $0 \leqslant k \leqslant (q-1)$. Three distinct ranges for k must be considered:

1. When $0 < k \leqslant (q-1)/2$, apply the level transformation

$$k \rightarrow [(q-1)/2 - (k-1)] \tag{5.12}$$

2. When $(q-1)/2 < k \leqslant (q-1)$, apply the level transformation

$$k \rightarrow [(q-1)/2 - k] \tag{5.13}$$

3. When $k = 0$

$$k \rightarrow 0 \tag{5.14}$$

Therefore a seven-level m-signal would transform as follows:

$$\begin{array}{cccccccc} k & 0 & 1 & 2 & 3 & 4 & 5 & 6 \\ \rightarrow & 0 & +3 & +2 & +1 & -1 & -2 & -3 \end{array} \tag{5.15}$$

Because the second half of any basic (all-positive integer) q-level m-signal is the modulo q complement of the first half, the integer level transformation of expressions (5.12)–(5.14) will always create a bipolar m-signal in which the second half is the 'inverse-repeat' (IR) of the first half; e.g. the basic signal of expression (5.11) becomes

$$\begin{array}{cccccccccccc} +2 & 0 & +1 & -2 & +1 & +1 & -1 & 0 & +2 & +1 & +2 & +2 \\ -2 & 0 & -1 & +2 & -1 & -1 & +1 & 0 & -2 & -1 & -2 & -2 \end{array} \tag{5.16}$$

when transformed. The periodic ACF for any transformed IR m-signal will itself have an IR format; for $q \leqslant 5$, the ACF will be quasi-impulsive in that it has only one major peak at zero delay, and an equal magnitude negative peak at a delay of half the signal period. For $q \geqslant 7$, significant subsidiary peaks occur in the periodic ACF, thus limiting the usefulness of such transformed m-signals in system-identification applications. In Darnell (1968) a list of typical normalized ACF profiles is given for integer transformed m-signals for $q \leqslant 31$ and these are reproduced in Table 5.4. Note that for each of these periodic ACF profiles, a single primitive element, g, of $GF(q)$ was chosen; other values of g for a given q will give rise to different subsidiary peak patterns.

Table 5.4 Normalized ACF peak amplitudes for typical signals derived from q-level m-sequences using the integer level transformation

Discrete shift	Values of q and g for typical sequences										
(Block Intervals)	$q = 2$ $g = 1$	3 2	5 3	7 5	11 2	13 2	17 3	19 2	23 5	29 2	31 3
0	+1.000	+1.000	+1.000	+1.000	+1.000	+1.000	+1.000	+1.000	+1.000	+1.000	+1.000
1		−1.000	0	−0.071	+0.255	+0.319	+0.250	+0.369	+0.091	+0.421	+0.246
2			−1.000	+0.071	+0.200	−0.121	+0.363	+0.207	+0.394	+0.098	+0.095
3			0	−1.000	−0.200	0	−0.069	−0.063	−0.006	+0.061	−0.213
4				+0.071	−0.255	+0.121	0	−0.190	+0.200	+0.384	+0.003
5				−0.071	−1.000	−0.319	+0.069	+0.190	−0.273	+0.286	−0.043
6					−0.255	−1.000	−0.363	+0.063	+0.273	+0.203	+0.422
7					−0.200	−0.319	−0.250	−0.207	−0.200	0	+0.102
8					+0.200	+0.121	−1.000	−0.369	+0.006	−0.203	−0.102
9					+0.255	0	−0.250	−1.000	−0.394	−0.286	−0.422
10						−0.121	−0.363	−0.369	−0.091	−0.384	+0.043
11						+0.319	+0.069	−0.207	−1.000	−0.061	−0.003
12							0	+0.063	−0.091	−0.098	+0.213
13							−0.069	+0.190	−0.394	−0.421	−0.095
14							+0.363	−0.190	+0.006	−1.000	−0.246
15							+0.250	−0.063	−0.200	−0.421	−1.000
16								+0.207	+0.273	−0.098	−0.246
17								+0.369	−0.273	−0.061	−0.095
18									+0.200	−0.384	+0.213
19									−0.006	−0.286	−0.003
20									+0.394	−0.203	+0.043
21									+0.091	0	−0.422
22										+0.203	−0.102
23										+0.286	+0.102
24										+0.384	+0.422
25										+0.061	−0.043
26										+0.098	+0.003
27										+0.421	−0.213
28											+0.095
29											+0.246

5.3.2 *Sinusoidal level transformation*

To overcome the problem of subsidiary ACF peaks mentioned in the previous section, a sinusoidal (non-integer) transformation can be applied to the basic unipolar m-signal (Everett, 1966); this takes the form

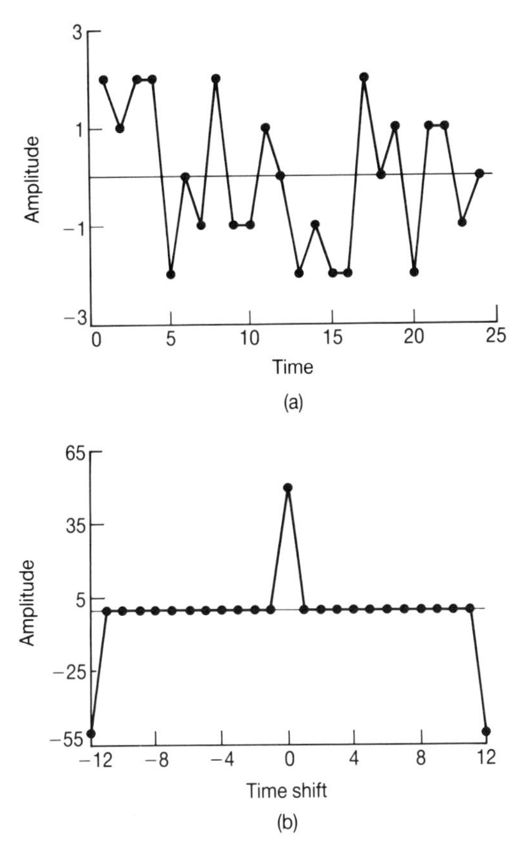

Figure 5.3 (a) Waveform and (b) periodic ACF of integer-transformed five-level m-signal.

$$k \to \sin[(2\pi k)/q] \tag{5.17}$$

where k is again specified by the values $0 \leqslant k \leqslant k(q-1)$. In general, with this level transformation, non-integer bipolar transformed m-signal levels result. For all $q > 2$, the transformed signal will have an IR format; it also has a periodic ACF which is IR, with a single positive peak at zero delay and an equal-magnitude negative peak at a delay equal to half the signal period and zero elsewhere.

To illustrate these level transformations, Figures 5.3(a) and 5.3(b), respectively, show waveform and periodic ACF for a 24-digit, five-level m-signal ($q = 5$, $n = 2$, $g = 3$) with the integer transformation applied; Figures 5.4(a) and 5.4(b) show the corresponding waveform and periodic ACF when the sinusoidal transformation is applied to the same unipolar m-signal.

The schematic diagram of the basic five-level m-signal generator is as shown in Figure 5.2. Figures 5.5(a) and 5.5(b), respectively, illustrate the waveform and periodic

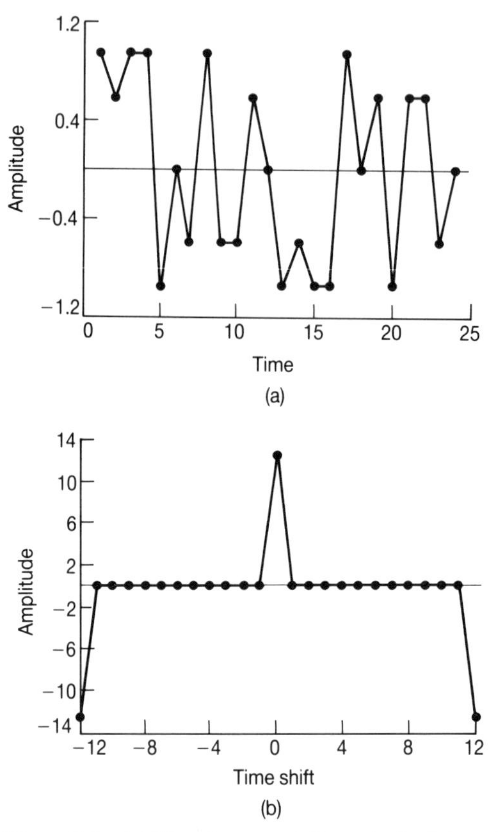

Figure 5.4 (a) Waveform and (b) periodic ACF of sinusoidally transformed five-level m-signal.

ACF for a 960-digit, 31-level m-signal ($q = 31$, $n = 2$, $g = 12$) with the integer transformation applied. Figures 5.6(a) and 5.6(b) show the corresponding waveform and periodic ACF when the sinusoidal transformation is used. The corresponding basic FSR 31-level m-signal generator is illustrated in Figure 5.7. It is seen that, in the 31-level case, the subsidiary ACF peaks arising in the ACF of the integer level transformed signal are eliminated through the use of the sinusoidal transformation. This subsidiary peak removal occurs for all $q \geqslant 7$.

Practically, the implementation of a non-integer transformation over a wide range of clock rates presents little difficulty, particularly if modern DSP devices are employed.

Although their consideration is beyond the scope of this chapter, it should be noted that transformed bipolar m-signals can also be used as the basis of methods for generating sets of completely uncorrelated pseudo-random signals, which find application in the identification of multi-input linear systems (Darnell, 1989). It is also possible to generate m-signals with elements taken from GF(q^m), where m is an

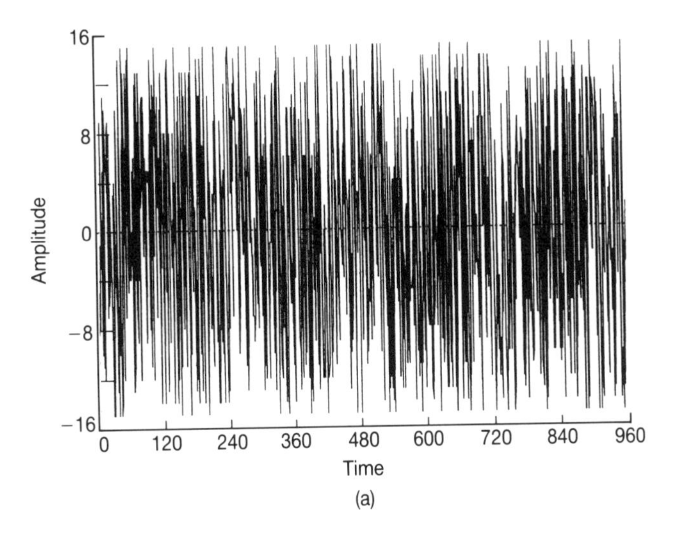

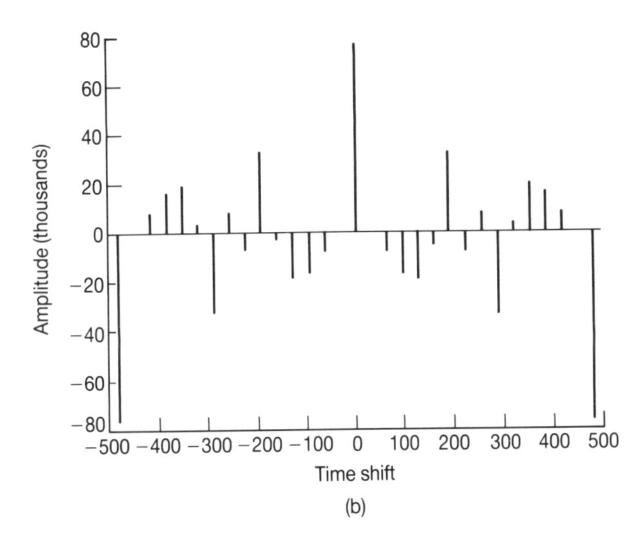

Figure 5.5 (a) Waveform and (b) periodic ACF of integer-transformed 31-level m-signal.

integer (Komo and Liu, 1990). These signals have properties analogous to those of the more familiar m-signals with elements taken from GF(q), as described above.

5.4 Non-periodic binary pseudo-random signals

This section describes options for non-periodic binary test signals with appropriate

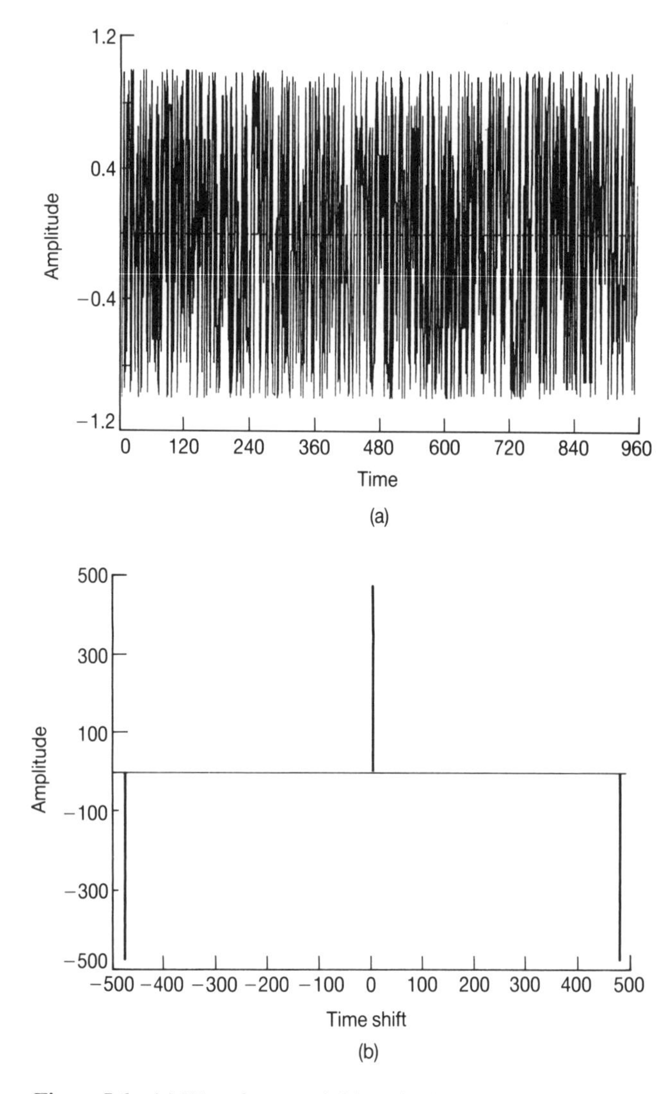

Figure 5.6 (a) Waveform and (b) periodic ACF of sinusoidally transformed 31-level m-signal.

non-periodic ACF and CCF properties. Three basic signal classifications will be considered, i.e. binary Barker signals, non-periodic binary signals with lengths greater than the longest binary Barker signal, and sets of binary complementary signals.

Two important fundamental differences between periodic and non-periodic pseudo-random signals should be stressed:

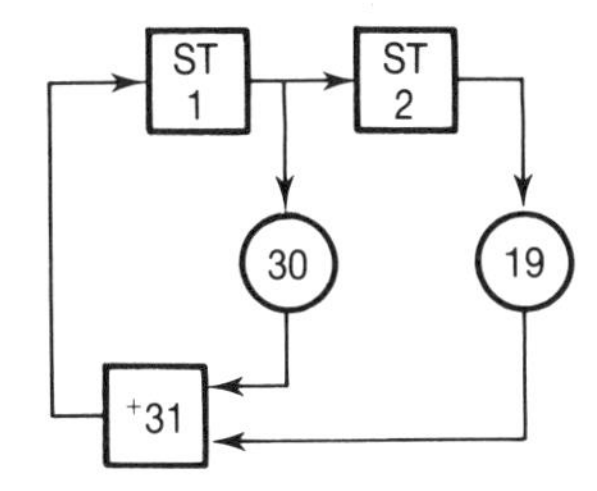

Figure 5.7 Schematic diagram of 31-level m-signal generator.

1. Periodic signals have discrete, or 'line', spectra whereas non-periodic signals give rise to continuous spectra.
2. Identification using non-periodic pseudo-random signals requires only a single application of the test input, in contrast to the use of periodic signals where the application transients must be allowed to decay before the input–output CCF becomes meaningful.

It is possible, therefore, that non-periodic test inputs may be more suitable if identification time is to be minimized, e.g. for systems with slowly time-varying parameters, or for systems in which a prolonged application of a test input may cause a shift in operating point. These considerations will be examined further in Section 5.6.

5.4.1 Binary Barker signals

A binary Barker signal of length N digits has the essential property that the PSR of its non-periodic ACF has a minimum value of $N:1$. No systematic method of generating Barker signals has been found; they have been derived by search methods, with the longest known Barker signal having $N = 13$. Table 5.5 lists all the binary Barker signals (Barker, 1953).

Table 5.5 Binary Barker signals and their peak to maximum side-lobe modulus ratio (PSR)

N	Barker signal format	PSR
2	$+-$ or $++$	2:1
3	$++-$	3:1
4	$++-+$ or $+++-$	4:1
5	$+++-+$	5:1
7	$+++--+-$	7:1
11	$+++---+--+-$	11:1
13	$+++++--++-+-+$	13:1

5.4.2 Binary signals with lengths >13

It is also possible to synthesize binary signals with lengths $N > 13$ having near-impulsive non-periodic ACFs. In Turyn (1968) signals with such ACF properties found by exhaustive computer search are tabulated; an example is the following 28-bit binary signal:

$$+ + - + + - + - - + - - - + - - - + - - - + + + + - - - \tag{5.18}$$

This has a normalized PSR of 14:1, which is better than that of the longest Barker signal. Other useful binary signals and their non-periodic ACFs (for $N \leqslant 40$) are given by Lidner (1975).

Longer binary non-periodic signals can also be formed from a concatenation of two Barker-type signals, i.e. replacing each digit of one component signal by the complete second signal; the polarity of the second signal inserted in each digit position is specified by the polarity of the digit it replaces. For example, concatenation of 7- and 11-bit Barker signals yields a composite 77-bit signal with the non-periodic ACF shown in Figure 5.8 (Chesmore, 1987). Note that the PSR of the composite signal ACF will not be better than the smallest ACF PSR of the two component signals, i.e. 7:1 in this case.

In general, the chief advantage of extending the length of the test input in this manner is an enhanced noise immunity. The process of concatenation can be extended further to encompass three or more binary signals, combined two at a time; e.g. 3-, 4- and 7-bit Barker signals can be concatenated to give an 84-bit composite signal by first concatenating the 3- and 4-bit signals, and then concatenating the resulting 12-bit composite signal with the 7-bit signal.

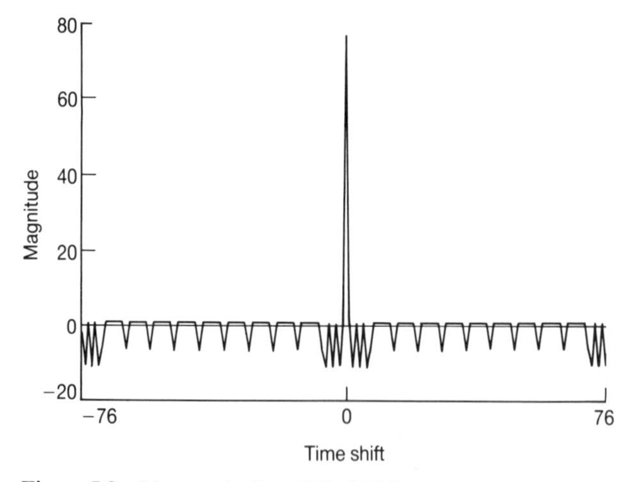

Figure 5.8 Non-periodic ACF of 77-bit concatenated (7, 11) Barker signal.

5.4.3 Binary complementary signal sets

Attention is now turned to a class of non-periodic binary signals known as 'complementary signals' which, to date, have found little application in the system-identification arena. However, they have considerable potential in this context and can, in principle, provide test signals with optimum non-periodic ACFs. In their basic form they can be synthesized with lengths $N = 2^n$, where n is any integer $\geqslant 1$.

The definition of a pair of binary complementary signals (CSs) (Golay, 1961) states that they are a pair of equal-length binary signals in which the number of pairs of *like* elements with a given separation in one signal is exactly equal to the number of pairs of *unlike* elements with the same separation in the other signal. By this definition, the following two signals form a complementary pair:

$$\begin{array}{lcccc} \text{Signal 1:} & +1 & +1 & -1 & +1 \\ \text{Signal 2:} & +1 & -1 & -1 & -1 \end{array} \tag{5.19}$$

This structure leads directly to the characteristic form of the summed non-periodic ACF for CSs, i.e. that the sum of the two individual non-periodic ACFs at corresponding delays is everywhere zero, except at the zero delay (in-phase) position where it takes the value $2N$, N digits being the length of each of the signals.

Assuming that the signals are regularly clocked, the non-periodic ACFs for the signals of expression (5.19), at delay increments of one digit interval, are:

$$\begin{array}{lccccccc} \text{Signal 1 ACF:} & +1 & 0 & -1 & +4 & -1 & 0 & +1 \\ \text{Signal 2 ACF:} & -1 & 0 & +1 & +4 & +1 & 0 & -1 \\ \hline \text{Summed ACF:} & 0 & 0 & 0 & +8 & 0 & 0 & 0 \end{array} \tag{5.20}$$

Note that the non-periodic CCF between the two individual signals in the set is generally non-zero; for the signals of expression (5.19) the non-periodic CCF is

$$\text{CCF 1 and 2:} \quad -1 \quad -2 \quad -1 \quad 0 \quad +1 \quad -2 \quad +1 \tag{5.21}$$

Thus if the impulsive summed ACF properties of CSs are to be exploited in system-identification applications it is essential that the signals are applied and processed in such a way that any potential cross-correlation effects are eliminated. The manner in which this can be achieved is discussed in Section 5.6.

It is also possible to synthesize another distinct pair of CSs which is completely uncorrelated, in a non-periodic complementary sense, with the original pair. To illustrate this, consider the following two pairs of binary CSs:

$$\begin{array}{cccccccccc} +1 & +1 & -1 & +1 & & +1 & +1 & +1 & -1 \\ & & & & \text{and} & & & & \\ +1 & -1 & -1 & -1 & & +1 & -1 & +1 & +1 \\ & \text{Pair (a)} & & & & & \text{Pair (b)} & & \end{array} \tag{5.22}$$

The overall non-periodic complementary CCF between pairs (a) and (b) is calculated simply by summing the non-periodic CCFs between corresponding signals in each pair at equivalent delays, i.e.

$$\begin{array}{lccccccc} \text{Upper CCF:} & +1 & 0 & +1 & 0 & +3 & 0 & -1 \\ \text{Lower CCF:} & -1 & 0 & -1 & 0 & -3 & 0 & +1 \\ \hline \text{Summed CCF:} & 0 & 0 & 0 & 0 & 0 & 0 & 0 \end{array} \qquad (5.23)$$

Since the summed CCF is everywhere zero, the two CS pairs are said to be uncorrelated in a complementary sense.

5.4.4 Recursive binary CS set synthesis

Methods of synthesizing binary CS sets of length $N = 2^n$ (n integer $\geqslant 1$) are now reviewed (Darnell, 1991b). Use will be made of a basic 'block structure' for CS sets, i.e.

$$\begin{array}{lcc} \text{Upper:} & \overrightarrow{+\text{A}} & \overrightarrow{+\text{B}} \\ \text{Lower:} & \overleftarrow{+\text{B}} & \overleftarrow{-\text{A}} \end{array} \qquad (5.24)$$

where A and B are blocks of length 2^{n-1} digits (half the signal length); the arrows indicate the sense in which all digits in a block should be read and the sign denotes the relative polarity of all digits in a block. In expression (5.24) therefore the lower signal is a reversed version of the upper, with the first half inverted.

Consider now a synthesis example which makes use of the arbitrary initial elements

$$+\text{A} = +1 \qquad \text{and} \qquad +\text{B} = +1 \qquad (5.25)$$

A first recursion places these values in the block structure of expression (5.24) as follows:

$$n = 1 \qquad \begin{array}{lcc} \text{Upper:} & +1 & +1 \\ \text{Lower:} & +1 & -1 \end{array} \qquad (5.26)$$

The two 2-bit signals now form a complementary pair. For the second recursion, the upper 2-bit signal is redefined as +A and the lower signal as +B, i.e.

$$\overrightarrow{+\text{A}} = +1 \quad +1 \qquad \text{and} \qquad \overleftarrow{+\text{B}} = +1 \quad -1 \qquad (5.27)$$

These are again substituted in the block structure of expression (5.24) to give two 4-bit CSs:

$$n = 2 \qquad \begin{array}{ll} \text{Upper:} & +1 \quad +1 \quad -1 \quad +1 \\ \\ \text{Lower:} & +1 \quad -1 \quad -1 \quad -1 \end{array} \tag{5.28}$$

Further redefinition of the blocks yields

$$n = 3 \qquad \begin{array}{ll} \text{Upper:} & +1 \quad +1 \quad -1 \quad +1 \quad -1 \quad -1 \quad -1 \quad +1 \\ \\ \text{Lower:} & +1 \quad -1 \quad -1 \quad -1 \quad -1 \quad +1 \quad -1 \quad -1 \end{array} \tag{5.29}$$

etc. for any value of n.

The definition of CSs can be further generalized to include sets with more than two signals (Tseng and Liu, 1972). To illustrate this, the pair of 2-bit binary CSs of expression (5.26) is now redefined as

$$\begin{array}{ll} +1 & +1 \\ +1 & -1 \end{array} = +\text{A} \quad \text{and} \quad +\text{B} \tag{5.30}$$

Substituting for A and B in the structure of expression (5.24) now results in four 4-bit signals:

$$\begin{array}{lcccc} \text{Signal 1:} & +1 & +1 & +1 & +1 \\ \text{Signal 2:} & +1 & -1 & +1 & -1 \\ \text{Signal 3:} & +1 & +1 & -1 & -1 \\ \text{Signal 4:} & -1 & +1 & +1 & -1 \end{array} \tag{5.31}$$

with the individual and summed non-periodic ACFs:

$$\begin{array}{lccccccc} \text{Signal 1 ACF:} & +1 & +2 & +3 & +4 & +3 & +2 & +1 \\ \text{Signal 2 ACF:} & -1 & +2 & -3 & +4 & -3 & +2 & -1 \\ \text{Signal 3 ACF:} & -1 & -2 & +1 & +4 & +1 & -2 & -1 \\ \text{Signal 4 ACF:} & +1 & -2 & -1 & +4 & -1 & -2 & +1 \\ \hline \text{Summed ACF:} & 0 & 0 & 0 & +16 & 0 & 0 & 0 \end{array} \tag{5.32}$$

It is seen that this set of four 4-bit (4×4) signals is complementary by virtue of the cancellation of the individual ACF side lobes leading to an impulsive summed ACF. The redefinition of A and B can again be employed recursively to produce (8×8), (16×16), etc. sets of CSs. A more comprehensive review of CS set synthesis is contained in Darnell and Kemp (1991).

5.4.5 Uncorrelated binary CS set synthesis

A number of techniques are available for the synthesis of an unlimited number of uncorrelated, multi-signal (>2) CS sets (Tseng and Liu, 1972); the simplest of these will now be outlined. The method requires an initial matrix $\mathbf{D}$ comprising a pair of uncorrelated CSs which, for example, can be obtained from the first and second halves of the signals of expression (5.28), i.e.

$$\mathbf{D} = \begin{pmatrix} +1 & +1 & & -1 & +1 \\ +1 & -1 & & -1 & -1 \end{pmatrix} \tag{5.33}$$

Let a matrix $\mathbf{D}'$ be defined as

$$\mathbf{D}' = \begin{pmatrix} +\mathbf{D}/+\mathbf{D} & -\mathbf{D}/+\mathbf{D} \\ -\mathbf{D}/+\mathbf{D} & +\mathbf{D}/+\mathbf{D} \end{pmatrix} \tag{5.34}$$

where $-\mathbf{D}$ indicates that all the elements of $\mathbf{D}$ are negated, and the symbol / denotes the interleaving of columns taken alternately from the two matrices specified. In Tseng and Liu (1972) it is demonstrated that the columns of $\mathbf{D}'$ are uncorrelated sets of CSs. Hence, substituting in equation (5.34) from (5.33) gives

$$\begin{pmatrix} +1 & +1 & +1 & +1 & & -1 & -1 & +1 & +1 & & -1 & +1 & -1 & +1 & & +1 & -1 & -1 & +1 \\ +1 & +1 & -1 & -1 & & -1 & -1 & -1 & -1 & & -1 & +1 & +1 & -1 & & +1 & -1 & +1 & -1 \\ -1 & +1 & -1 & +1 & & +1 & -1 & -1 & +1 & & +1 & +1 & +1 & +1 & & -1 & -1 & +1 & +1 \\ -1 & +1 & +1 & -1 & & +1 & -1 & +1 & -1 & & +1 & +1 & -1 & -1 & & -1 & -1 & -1 & -1 \\ & & \text{(a)} & & & & & \text{(b)} & & & & & \text{(c)} & & & & & \text{(d)} & \end{pmatrix} \tag{5.3}$$

Each of the four (4×4) set of signals 5.35(a)–(d) above is both complementary and uncorrelated with the other three sets in a complementary sense. If expression (5.35) is now redefined as $\mathbf{D}$, matrix (5.34) can be employed recursively to synthesize a (64×8) matrix which can then be partitioned into eight (8×8) uncorrelated CS sets, etc.

5.5 Non-periodic multi-level pseudo-random signals

Attention is now turned to various classes of non-periodic multi-level (>2) signals with useful ACF and CCF properties. The following types of signals will be considered: generalized Barker signals, polyphase signals, Huffman signals, multi-level complementary signal sets and trajectory-derived signals.

5.5.1 Generalized Barker signals

In the previous section, binary Barker signals of length N bits ($N \leqslant 13$) were introduced. The binary states of the signal can be interpreted as two real numbers, say 0 and 1, or $+1$ and -1; alternatively, they can be viewed as two phase values, say 0° and 180°.

A generalized Barker signal (Golomb and Scholtz, 1965) comprises a non-periodic sequence of complex numbers x_i, where i is an integer in the range $1 \leqslant i \leqslant N$ (N, the signal length, can now take any integer value); each x_i has a modulus of unity. Hence, there has in effect been an increase in the number of allowable phase states beyond the two of the basic binary Barker signal. In addition to the unity modulus property, the non-periodic ACF of the generalized Barker signal, $R_{xx(N)}(r)$, has a zero-delay peak of $+N$ and side lobes whose modulus does not exceed unity. This ACF is defined as

$$R_{xx(N)}(r) = \sum_{i=0}^{N-r-1} x_i x_{i+r}^* \tag{5.36}$$

for $r = 0, 1, 2, \ldots, (N-1)$, where the asterisk denotes a complex conjugate.

As an example, consider the 11-digit quaternary Barker signal

$$x_i = +1 \quad +\mathrm{j} \quad -1 \quad +\mathrm{j} \quad -1 \quad -\mathrm{j} \quad -1 \quad +\mathrm{j} \quad -1 \quad +\mathrm{j} \quad +1 \tag{5.37}$$

where $\mathrm{j} = (-1)^{1/2}$. Using the complex conjugate of this signal, i.e.

$$x_i^* = +1 \quad -\mathrm{j} \quad -1 \quad -\mathrm{j} \quad -1 \quad +\mathrm{j} \quad -1 \quad -\mathrm{j} \quad -1 \quad -\mathrm{j} \quad +1 \tag{5.38}$$

in expression (5.36) allows the ACF properties of the generalized Barker signal to be verified. Other signals with more phase states are described in Golomb and Scholtz (1965).

5.5.2 Polyphase signals

The polyphase signals described by Frank (1963) are, in some respects, similar to the generalized Barker signals introduced above. They have both periodic and non-periodic ACF properties which are near-impulsive; here, they will be considered initially as periodic signals.

Polyphase signals have a length $N = K^2$, where K is an integer. The modulus of all digits in the signal is again unity, with the possible phase states being selected from integer multiples of $(2\pi k)/K$, where k is an integer which is relatively prime to K. The polyphase signal can, in general, be specified by the following matrix:

$$\begin{pmatrix} 0 & 0 & 0 & 0 & \dots & 0 \\ 0 & 1 & 2 & 3 & \dots & (K-1) \\ 0 & 2 & 4 & 6 & \dots & 2(K-1) \\ 0 & 3 & 6 & 9 & \dots & 3(K-1) \\ \cdot & \cdot & \cdot & \cdot & & \cdot \\ \cdot & \cdot & \cdot & \cdot & & \cdot \\ 0 & (K-1) & 2(K-1) & 3(K-1) & \dots & (K-1)^2 \end{pmatrix} \tag{5.39}$$

Taking a specific example where $K = 5$ and $k = 1$: applying the structure of matrix (5.39), and writing all terms sequentially, yields the 25-digit signal

$$0\ 0\ 0\ 0\ 0\ 0\ 1\ 2\ 3\ 4\ 0\ 2\ 4\ 6\ 8\ 0\ 3\ 6\ 9\ 12\ 0\ 4\ 8\ 12\ 16 \tag{5.40}$$

This integer signal is now reduced modulo K, mod 5 in this case, to give

$$0\ 0\ 0\ 0\ 0\ 0\ 1\ 2\ 3\ 4\ 0\ 2\ 4\ 1\ 3\ 0\ 3\ 1\ 4\ 2\ 0\ 4\ 3\ 2\ 1 \tag{5.41}$$

The mod 5 states of expression (5.41) are then employed directly as the integer term, i, in

$$\exp(\mathrm{j}2\pi i)/5 \tag{5.42}$$

to give a complex signal with five phase states and unit modulus.

The periodic ACF of this signal has an in-phase peak value of $+25$ (generally $+K^2$) and an off-peak value of zero. Note, in the context of this section, that the same complex signal also has a useful non-periodic ACF with a PSR of approximately 15.6:1 (Frank, 1963).

5.5.3 *Huffman signals*

Huffman signals (Huffman, 1962; Ackroyd, 1970) are non-periodic signals comprising real or complex elements; in contrast to generalized Barker and polyphase signals discussed previously, the element magnitudes can take any values. The non-periodic ACF of a Huffman signal is a good approximation to a single impulse, i.e. zero everywhere, except at the in-phase position where there is a major peak, and at shifts of $\pm$(1 signal interval) with respect to the peak where the ACF has a magnitude of unity. A simplified resumé of the synthesis procedure for Huffman signals will now be presented.

Consider a series of $(N + 1)$ complex elements to be represented by a polynomial Q in the indeterminant D:

$$Q = C_0 + C_1D^1 + C_2D^2 + \dots + C_ND^N \tag{5.43}$$

The z-transform equivalent of equation (5.43) can be expressed as

$$Q(z) = C_0 + C_1z^{-1} + C_2z^{-2} + \dots + C_Nz^{-N} \tag{5.44}$$

which can be factored as

$$Q(z) = C_0 \prod_{i=1}^{N} (1 - r_iz^{-1}) \tag{5.45}$$

where r_i are the zeros of $Q(z)$. In Huffman (1962) it is shown that, in order for the signal to have the desired non-periodic ACF properties, the zeros of the polynomial must

1. Lie at equal angular separations in the complex plane;
2. Lie on one of two complex plane circles, centred on the origin, with radii X or $1/X$.

Hence, the values of the zeros are given by

$$r_i = \left.\begin{array}{l} X \exp[(\mathrm{j}2\pi i)/N];\ \text{if on circle with radius } X \\ 1/X \exp[(\mathrm{j}2\pi i)/N];\ \text{if on circle with radius } 1/X \end{array}\right\} \qquad (5.46)$$

Furthermore, for a set of zeros involving real values and complex conjugate pairs, the corresponding Huffman signal will have purely real elements; other zero configurations will give rise to complex signal elements.

A signal having any value of N, and fulfilling conditions (1) and (2) above, will have the required form of non-periodic ACF. The design problem therefore reduces to selecting N, selecting X, and then specifying on which of the two circles the zeros should be located; these choices dictate the exact nature of the time-domain signal. In addition to the non-periodic ACF property, system identification applications may require a reasonably uniform signal energy distribution throughout its duration. This again can be controlled, to some extent, by specification of X and the zero positions (Ackroyd, 1970). Figures 5.9(a) and (b) show, respectively the waveform of a purely real 27-digit Huffman signal, together with its non-periodic ACF. It is also possible to synthesize much longer Huffman signals directly, or via the concatenation of two or more relatively short Huffman signals in the same manner as that described for Barker signals in Section 5.4.2.

5.5.4 Multi-level complementary signal sets

In Section 5.4 the topic of binary complementary signal (CS) set synthesis was introduced. These sets have ideal non-periodic ACF properties, and multiple uncorrelated sets can also be generated. The same general synthesis techniques can be applied when the signal elements are multi-level (>2). Therefore the basic theory and definitions are not reiterated here, but rather extensions to the basic concepts and further examples are presented. A comprehensive treatment of multi-level CS sets and their applications is given in Darnell and Kemp (1991).

In Darnell and Kemp (1991) it is shown that the block structures

$$\begin{array}{cc} +\overrightarrow{\mathrm{A}} & +\overrightarrow{\mathrm{B}} \\ +\overrightarrow{\mathrm{B}} & -\overrightarrow{\mathrm{A}} \\ \multicolumn{2}{c}{\text{(a)}} \end{array} \quad \text{and} \quad \begin{array}{cc} +\overrightarrow{\mathrm{A}} & -\overrightarrow{\mathrm{B}} \\ +\overrightarrow{\mathrm{B}} & +\overrightarrow{\mathrm{A}} \\ \multicolumn{2}{c}{\text{(b)}} \end{array} \qquad (5.47)$$

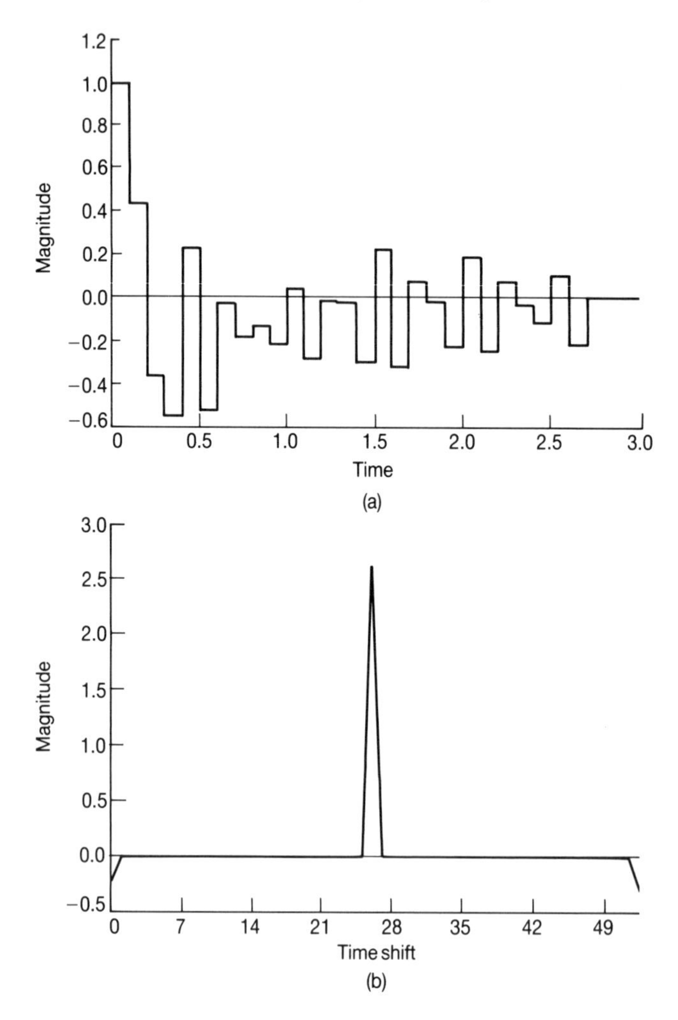

Figure 5.9 (a) Waveform and (b) non-periodic ACF for a 27-digit Huffman signal.

can both be used as the basis of a recursive method of CS set synthesis, in addition to the block structure already given in expression (5.24). If, arbitrarily, values of $+A = +1$ and $+B = +2$ are selected and then substituted in expression (5.47)(a), the following (2×2) non-binary CS set results:

$$\begin{array}{lrr} \text{Upper} & +1 & +2 \\ \text{Lower} & +2 & -1 \end{array} \tag{5.48}$$

If $\overrightarrow{+A}$ and $\overrightarrow{+B}$ are now both defined by expression (5.48) then, using the structure of (5.47)(a), a larger (4×4) multi-level CS set can be synthesized, i.e.

$$\begin{array}{cccc} +1 & +2 & +1 & +2 \\ +2 & -1 & +2 & -1 \\ +1 & +2 & -1 & -2 \\ +2 & -1 & -2 & +1 \end{array} \qquad (5.49)$$

Similar recursions can be used to provide sets of greater dimensions.

Methods given in Section 5.4.5 (Darnell and Kemp, 1991; Kemp and Darnell, 1989) allow the systematic synthesis of uncorrelated CS sets; for example, by employing the two structures of expression (5.47) with $+A = +3$ and $+B = +4$, two sets of uncorrelated (2×2) CSs can be created. The matrix $\mathbf{D}$, as defined for expression (5.33), then becomes

$$\mathbf{D} = \begin{pmatrix} +3 & +4 & +3 & -4 \\ +4 & -3 & +4 & +3 \end{pmatrix} \qquad (5.50)$$

Expression (5.34) can now be used for recursive synthesis of larger CS sets, e.g.

$$\mathbf{D}' = \begin{pmatrix} +3+3+4+4 & +3+3-4-4 & -3+3-4+4 & -3+3+4-4 \\ +4+4-3-3 & +4+4+3+3 & -4+4+3-3 & -4+4-3+3 \\ -3+3-4+4 & -3+3+4-4 & +3+3+4+4 & +3+3-4-4 \\ -4+4+3-3 & -4+4-3+3 & +4+4-3-3 & +4+4+3+3 \end{pmatrix} \qquad (5.51)$$

$$\text{(a)} \qquad \text{(b)} \qquad \text{(c)} \qquad \text{(d)}$$

Each of the four (4×4) CS sets of expression (5.51) is uncorrelated, in a complementary sense, with the other three sets. A (64×8) matrix will be generated by the next recursion; this can then be partitioned into eight (8×8) uncorrelated CS sets, etc.

Thus far, the CS sets described have been of even length, $N = 2^n$. It is possible to create odd-length and other even-length non-binary sets by means of a process which will be termed 'signal compression' (Darnell and Kemp, 1991). To illustrate this, consider the binary CS pair of expression (5.29):

$$\begin{array}{cccccccc} +1 & +1 & -1 & +1 & -1 & -1 & -1 & +1 \\ +1 & -1 & -1 & -1 & -1 & +1 & -1 & -1 \end{array} \qquad (5.52)$$

Pairs of 7-digit, 6-digit, etc. non-binary CSs can be produced by overlapping the two halves of the signals of expression (5.52) to a defined extent, and then summing in the overlap range; this is demonstrated below for a single-digit overlap:

$$\begin{array}{ccccccccccccccc} +1 & +1 & -1 & +1 & & & & & & & & & & & \\ & & & -1 & -1 & -1 & +1 & = & +1 & +1 & -1 & 0 & -1 & -1 & +1 \\ +1 & -1 & -1 & -1 & & & & & & & & & & & \\ & & & -1 & +1 & -1 & -1 & = & +1 & -1 & -1 & -2 & +1 & -1 & -1 \end{array} \qquad (5.53)$$

resulting in (7×2) CS pair.

It can be easily demonstrated that it is impossible to synthesize a binary CS set in which the number of signals in the set is odd. However, other basic block structures are available when non-binary elements are used to enable the direct synthesis of odd-number signal sets; these may be of odd or even length. As an example, the structure

$$\begin{array}{cc} +A & -2A \\ -2A & +A \\ -A & -4A \end{array} \tag{5.54}$$

when $+A = +1$, provides the (2×3) CS set

$$\begin{array}{cc} +1 & -2 \\ -2 & +1 \\ -1 & -4 \end{array} \tag{5.55}$$

If set (5.55) is now taken as $+A$, the structure of set (5.54) can be used recursively to yield a (4×9) CS set.

All the sets of CSs discussed in this chapter have purely real signal elements. Note that it is also possible to synthesize certain types of CS sets with complex signal elements (Sivaswamy, 1978).

5.5.5 *Trajectory-derived signals*

A new class of pseudo-random signals based upon rectilinear trajectories within a perfectly reflecting enclosure has been described in Al-Dabbagh and Darnell (1991). These signals, termed 'trajectory-derived (TD)' signals, are completely deterministic, have useful periodic and non-periodic ACF and CCF properties, can take any length, may have real or complex elements, and exhibit a continuum of element values. The last characteristic implies that, as with Huffman signals, the signals generated are effectively clocked 'analog' signals, where the possible states are not restricted to a limited discrete set.

Figure 5.10 shows a typical enclosure comprising an outer square boundary and an inner circular boundary, both centred on the origin. At some arbitrary starting point within the enclosure the trajectory is initiated at a defined angle with respect to an appropriate reference axis. The successive points of perfect reflection on the boundaries then define the states of the signal. In Figure 5.10 an example of an initial reflection, where the angle of incidence and angle of reflection are equal, is illustrated.

TD signals can be obtained by

1. Taking the horizontal or vertical co-ordinate at each reflection point to represent an element of a real-valued bipolar signal;
2. Using the horizontal and vertical co-ordinates at each reflection point to represent the real and imaginary parts of a complex signal element.

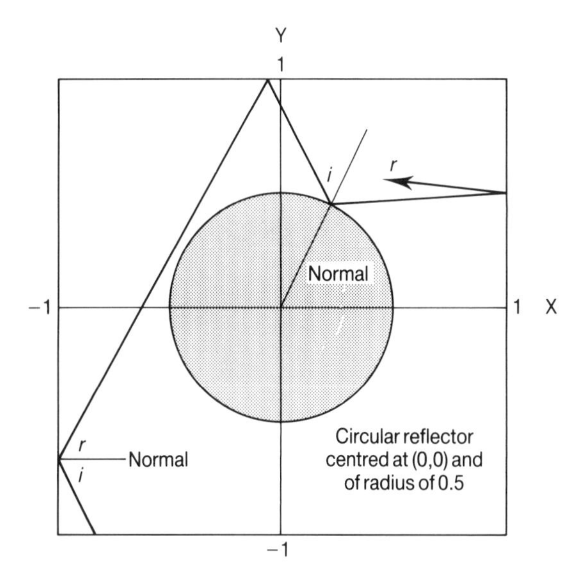

Figure 5.10 Typical enclosure configuration.

Figures 5.11(a), (b) and (c) show, respectively, an example of the waveform comprising the real components of a 1024-element (1024 reflection) complex TD signal obtained from the enclosure of Figure 5.10 rotated by 45°, its periodic ACF and its non-periodic ACF. It is seen that the ACF PSRs may be suitable for identification purposes. TD signals with low CCF values can be obtained by using different initial conditions within the enclosure.

The following points should also be noted:

1. The amplitude probability density function of a TD signal can be 'tailored' by altering the form of the enclosure and the number and shape of reflecting boundaries.
2. The format of a TD signal is completely defined by the geometry of the enclosure, the initial conditions (i.e. position within the enclosure and take-off angle) and the arithmetic precision of the computation of the reflection points.

Given the information in (2) above, the signal-generation process is completely deterministic and exactly repeatable.

5.6 System identification: application considerations

The input–output CCF method of evaluating the impulse response of an unknown linear system is well established (e.g. Lee, 1960), and will not be reiterated here.

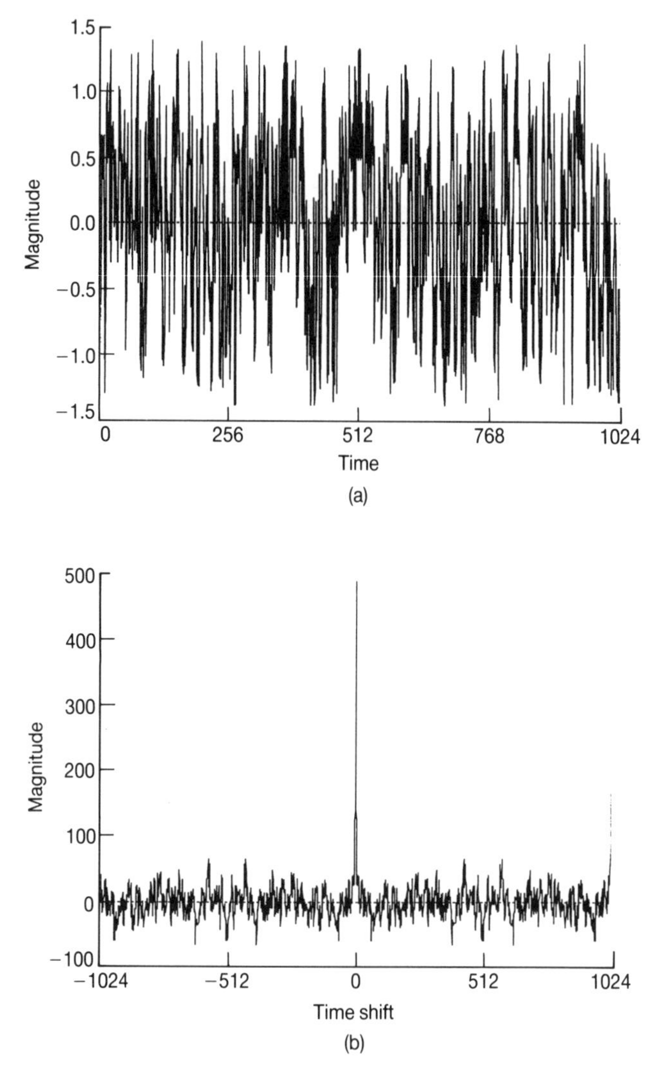

Figure 5.11 (a) Waveform, (b) periodic ACF for TD signal.

Essentially, any periodic pseudo-random signal, binary or multi-level, with a near-impulsive ACF can potentially be employed as a test input. Extensions to the basic single-input/single-output identification scenario involving multivariable linear systems are also well documented. The requirement in this case is for a set of uncorrelated periodic pseudo-random signals, all with near-impulsive ACFs and zero, or very low, CCFs (Darnell, 1989).

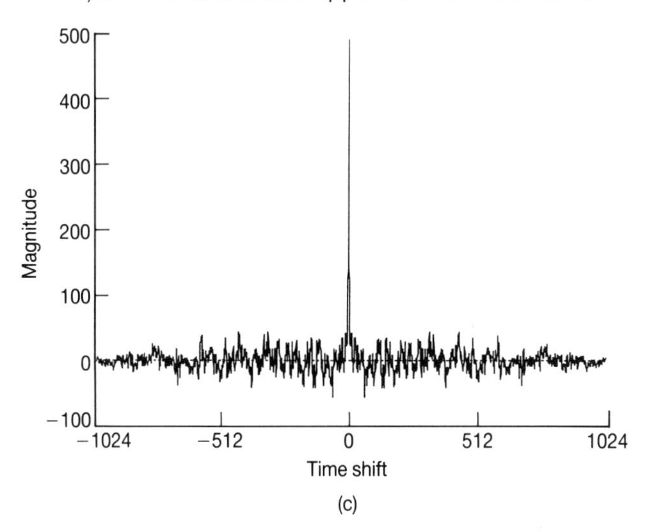

Figure 5.11 (cont.) (c) Non-periodic ACF for TD signal.

In this section attention will be largely confined to application considerations associated with the use of the various types of non-periodic test inputs. These have not yet been applied extensively to the identification problem. However, many types are available, as demonstrated in Sections 5.4 and 5.5 and they appear to offer some practical advantage when compared with periodic test signals.

5.6.1 *Efficiency of periodic and non-periodic system-identification procedures*

Figures 5.12 (a)–(e) show schematically a comparison of the efficiency of the linear system-identification process using periodic and non-periodic test inputs. Figure 5.12(a) shows the duration of the response of the system; in the case of periodic signal identification, the test signal period must exceed this response time, otherwise there will be 'spillover' of energy from one response to the next; a repetitive test signal of appropriate duration is shown schematically in Figure 5.12(b). After this signal is applied to the system under test, the steady state, when all effects of the application transient have decayed to negligible proportions, will be reached after about one test signal period, as indicated in Figure 5.12(c); at this point, periodic CCF identification can take place. When a non-periodic test input of similar duration (Figure 5.12(d)) is applied to the system, the discontinuity at the application point, together with the discontinuity at the end of the signal, are an integral part of the non-periodic signal format and contribute to its correlation properties; the corresponding system response is indicated in Figure 5.12(e).

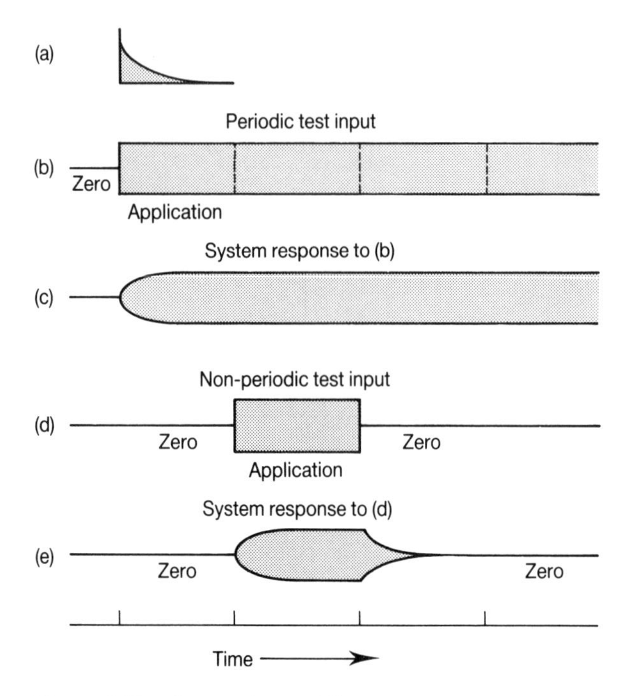

Figure 5.12 Comparison of linear system identification using periodic and non-periodic test inputs.

Thus it is seen that about two periods of the repetitive test signal are required here for a valid periodic CCF identification to be carried out, whereas a non-periodic signal with a duration equivalent to approximately one period of the repetitive signal will suffice. Under these circumstances, the non-periodic test signal enables the test procedure to be made significantly more efficient. It should also be noted that prolonged application of a periodic test input may possibly cause a drift in system parameter values; this should not occur with a non-periodic test stimulus.

5.6.2 Use of complementary signal sets in non-periodic system identification

The manner in which non-periodic test inputs such as Barker signals, concatenated Barker signals and Huffman signals can be used in system identification is straightforward. An appropriate length of system-output record must be stored and then cross-correlated non-periodically with the corresponding input test signal to yield an estimate of the system impulse response.

It may be desirable to use a binary or multi-level non-periodic CS set as a test input because of the optimum nature of its non-periodic ACF, and the fact that

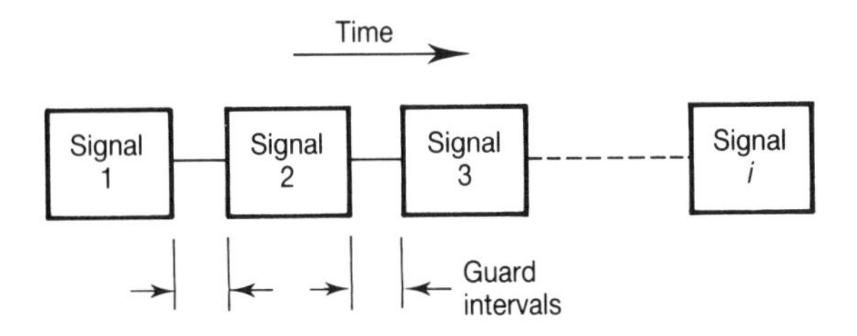

Figure 5.13 Application of a CS set to a system under test.

completely uncorrelated CS sets can be synthesized for multivariable applications. In this case, the input/output CCF identification procedure must be modified to take account of the specific properties of CS sets. Two important considerations apply: first, a set comprises two or more separate signals and, second, the cross-correlations between the signals comprising the set are generally non-zero, as demonstrated in Section 5.4.3. Therefore the signals must be applied to the system under test in such a way that any potential cross-correlations between component signals are eliminated. Figure 5.13 shows how this can be achieved.

The two or more component signals are applied serially to the system under test with guard intervals, equal to at least the maximum significant response time of the system, between them. The individual input/output CCFs for each of the component signals and its corresponding system output record are then computed, aligned employing a knowledge of the guard interval magnitude, and then summed to obtain an overall impulse response function estimate. Hence, for an L-signal CS set, the individual input/output CCFs, expressed as convolutions, are

$$\begin{aligned} \mathrm{CCF}_1 &= \int_{-\infty}^{\infty} h(u) R_{11(N)}(\tau - u)\, \mathrm{d}u \\ &\vdots \qquad\qquad \vdots \\ \mathrm{CCF}_L &= \int_{-\infty}^{\infty} h(u) R_{LL(N)}(\tau - u)\, \mathrm{d}u \end{aligned} \tag{5.56}$$

where u is a dummy time variable, τ is a continuous delay variable, $h(t)$ is the unknown unit impulse response function, and $R_{11(N)}(\tau)$ to $R_{LL(N)}(\tau)$ are the non-periodic ACFs for the individual signals of the CS set. Summing the left- and right-hand sides of equation (5.56) gives

$$\text{Sum CCF} = \int_{-\infty}^{\infty} h(u)[R_{11(N)}(\tau - u) + \ldots + R_{LL(N)}(\tau - u)]\, \mathrm{d}u \tag{5.57}$$

From the basic definition of a CS set, the summed ACF term in square brackets on the right-hand side of equation (5.57) is impulsive, and therefore the summed CCF will be proportional to the required system impulse response function.

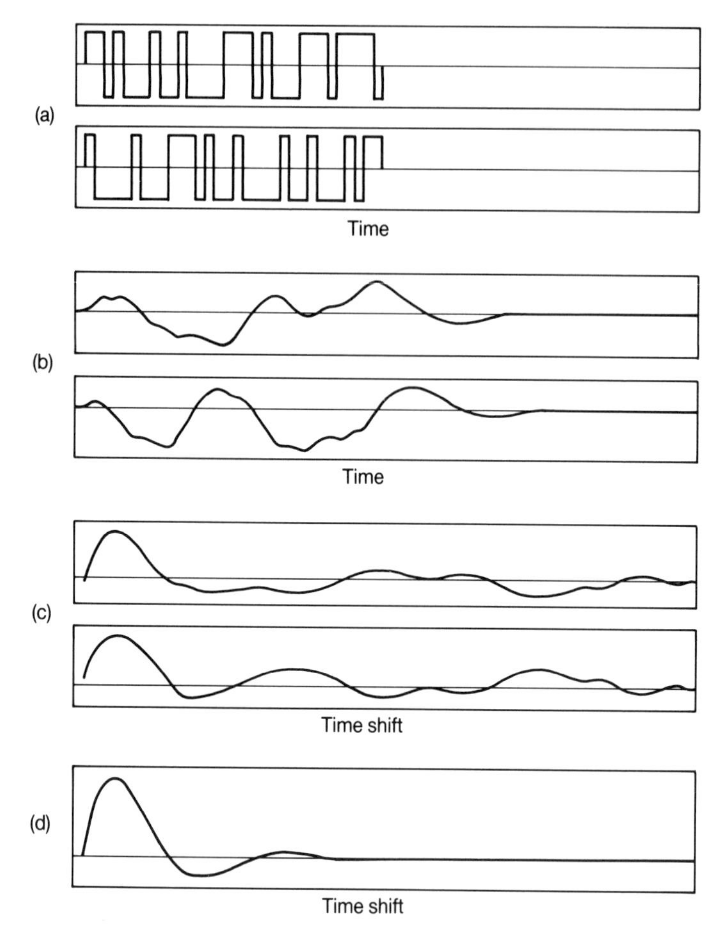

Figure 5.14 Example of system identification using binary CS pair.

Figures 5.14(a)–(d) show how a pair of 32-bit binary CSs can be used for the identification of a second-order linear system. Figure 5.14(a) shows the waveforms of the two 32-bit CSs and Figure 5.14(b) the corresponding second-order system output waveforms in response to the application of the two signals. The input/output CCFs

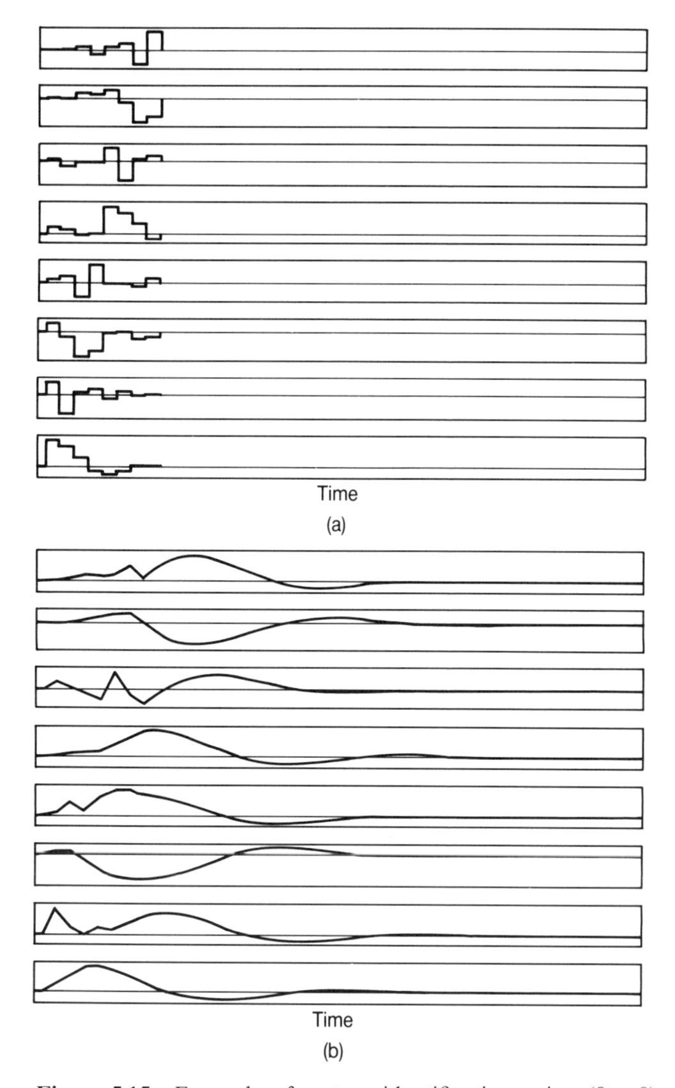

Figure 5.15 Example of system identification using (8 × 8) multi-level CS set.

computed for each of the signals separately are given in Figure 5.14(c), while the summed CCF is shown in Figure 5.14(d). Figures 5.15(a)–(d) show similar waveforms for an (8 × 8) multi-level CS set test input.

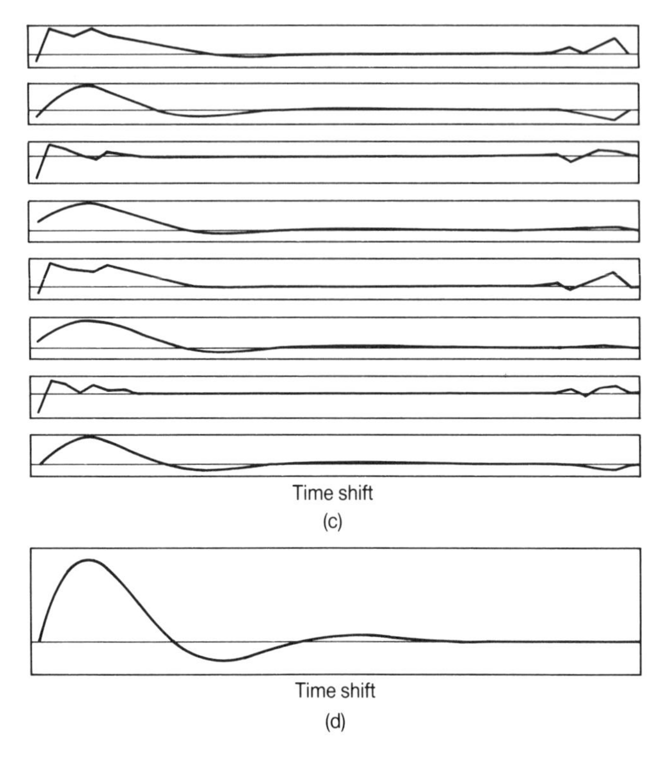

Figure 5.15 (cont.)

5.7 Conclusions

This chapter has presented a brief survey of the various classes of pseudo-random test signals with ACF and CCF properties making them potentially suitable for system-identification applications. Periodic and non-periodic, binary and multi-level signals have been described. In general, attention has been concentrated on signals with real elements, although complex element signals have also been introduced.

Whereas in the early days of pseudo-random signal testing, inputs were usually of a periodic binary form due to implementational considerations, the advent of cheap DSP and digital memory devices now makes multi-level and non-periodic test inputs a viable alternative.

6

Generation and Applications of Binary Multi-frequency Signals

Sandra L. Harris

6.1 Introduction

Careful planning is essential in executing identification and control experiments, or indeed, any experiments in which the dynamics (rather than the steady state alone) of the process play a major role. Persistent excitation is necessary when identifying a process, whether the objective is a system model or implementation of an adaptive or other control algorithm. Few processes can be identified using normal operating data. Thus experiments strictly for identification usually require an input perturbation. While some information may be obtained using simple signals such as a step, additional and more accurate information may be obtained if some care is taken in selecting the input.

Considerations in signal selection include range of frequency content and amplitude. Usually simple preliminary tests (such as step tests) are done to determine the time scale and thus the frequency range of interest for the process. The amplitude of the perturbation must be large enough to be detected over process noise but sufficiently small enough for the system to remain in a linear region (if a linear model is desired) and so that the output does not deviate more than necessary from the desired value. In many cases it is not necessary to have power at more than a few properly selected frequencies to obtain an accurate identification or parameter estimation. An advantage to using a signal with most of its power concentrated at a few frequencies is that a lower-amplitude signal may be used than that needed for a signal with power distributed at many frequencies. Alternatively, if the same amplitude signal is used, a higher signal-to-noise ratio is obtained, which would yield more accurate identification results.

Binary signals are often employed as an external perturbation because they have a favourable ratio of mean square value to amplitude and they are easy to use. Perhaps the most common binary signals are pseudo-random binary sequences (PRBS). Such a signal has its power fairly evenly distributed over all the frequencies in its bandwidth.

A binary multi-frequency signal (BMFS) has most of its power concentrated and evenly distributed over a limited number (3 to 10 or so) of frequencies. With a little *a priori* knowledge (system bandwidth, maximum acceptable perturbation amplitude) a low-level BMFS can be used to ensure persistent excitation of the process. Additionally, Van den Bos (1973) has shown that the accuracy of identification using a BMFS with properly chosen bandwidth is close to the accuracy achieved using an optimal signal.

6.2 Signals utilized

6.2.1 Early signals

Any published binary multi-frequency signal can be used for identification by selecting the amplitude and switching interval. Some of the early signals presented by Van den Bos (1967) were used in the work summarized below. These generally had power concentrated at frequencies spaced by a factor of 2 (e.g. at 2, 4, 8, 16, etc. times the fundamental frequency). However, often a more specialized power distribution will give more accurate results. For example, when the process is higher order or in a closed-loop, more power clustered in a mid-frequency (with respect to the system bandwidth) range is useful, since this region is more sensitive than other regions to process or controller variations. Thus several signals were also generated, the first two using the early method of Van den Bos (1967). In this method, a binary signal is generated using a random number generator and then one point changed at a time. If an improvement in power distribution (relative to the desired distribution) is obtained, the change is kept, otherwise the point is returned to its original value. When no improvements resulted from any change, the signal is retained and another starting sequence used. The best signal after trying several starting sequences is used. This method reliably produces signals having a high efficiency (percentage of total power concentrated at the selected harmonics), but it does take substantial computer time. Both signals generated using this method have additional frequency content in the mid-range, and these are shown in Table 6.1. In this table, as well as in other tables giving sequences, '+' indicates $+1$, '$-$' indicates -1, and the numeric value indicates the number of time intervals at the high or low level. The power at each of the selected harmonics is shown. As a measure of even power content at these harmonics, the ratio of the highest power at one of these frequencies to the lowest power at one of these frequencies is given. Ideally this ratio would be unity. The efficiency of each signal is also given.

Table 6.1 Sequences having power at additional mid-range frequencies: first half-period (512 intervals per period)

Frequencies	Signal	Power	P_h/P_l	E
1, 2, 4, 8, 12, 16, 24, 32 (extended)	44+, 12−, 16+, 14−, 7+, 4−, 14+, 10−, 2+, 23−, 13+, 6−, 4+, 46−, 6+, 5−, 14+, 8−, 8+	0.0925, 0.102, 0.0575, 0.081, 0.0597, 0.0691, 0.0807, 0.0642	1.77	60.7%
1, 2, 3, 4, 6, 8, 12, 16 (compact)	10+, 20−, 16+, 4−, 16+, 58−, 14+, 22−, 2+, 8−, 62+, 16−, 8+	0.0526, 0.0847, 0.0556, 0.0774, 0.0958, 0.0879, 0.1118, 0.1008	2.12	66.7%

The efficiencies of the two sequences are not particularly high. However, note that a substantial amount of the 'wasted power' (power not at the selected harmonics) is at least within the bandwidth defined by the selected harmonics. For example, the extended sequence has an efficiency of 60.7%, but 70% of the total power is concentrated in frequencies between (and including) the fundamental frequency and 32 times the fundamental frequency.

6.2.2 New signals

Additional signals were generated by first specifying the frequency spectrum of the desired sequence and then employing an inverse Fast Fourier Transform program. Since the BMFS is real, the frequency spectrum will be even. If the BMFS is also specified to be even, then the frequency spectrum will be real. Thus the power at the mid-frequency, $N/2$ times the fundamental frequency, where N is the number of points in the sequence, will be 0.0 and the second half of the spectrum will be constructed so as to be symmetrical about that mid-frequency. The total power of a binary sequence is 1.0. If there are N_F frequencies at which power is desired, then the required power at each of these frequencies is $1/N_F$ and zero power at each other frequency. Since the power contained in the kth harmonic of a periodic signal $x(t)$ is

$$P_k = 2|X_k|^2 \tag{6.1}$$

where X_k is the Fourier transform of $x(t)$ evaluated at the kth harmonic, the frequency spectrum will be

$$X_k = s(P_k/2)^{0.5} \tag{6.2}$$

At each of the N_F selected harmonics this will be

$$X_k = s(1/(2N_F))^{0.5} \tag{6.3}$$

and for all other frequencies X_k will have the value 0.0. The s in equations (6.2) and

(6.3) denotes either + or −, the only unknown in the analysis. In the cases presented here and by Harris (1987), alternative signs or previously published patterns were employed. A method of sign determination analogous to Van den Bos' (1987) phase angle assignment technique was attempted but with little success (starting values were always selected as the optimum).

Once the desired frequency spectrum has been constructed, an inverse FFT is used to generate a sequence in time. This sequence is not binary; thus, all points having a value greater than the average value of the signal are assigned as +1 and all points having a value below the average are assigned as −1.

Taking the FFT of this new signal shows that while the efficiency is around 70%, the power is not evenly distributed among the selected harmonics. To improve the distribution, the frequency spectrum used to generate the signal can be weighted to reflect the imbalance (frequencies with more than average power weighted less heavily and vice versa) and a new signal generated. Various empirical weighting factors were tried, and the one which gave the best results was

$$w_i = (P_{av}/P_i)^{0.25} \tag{6.4}$$

where P_{av} is the average value of the power at the selected harmonics and P_i is the power at the ith selected harmonic. This adjustment can be made sequentially several times. The resulting sequence for the case $N_F = 6$ (1, 2, 4, 8, 16 and 32 times the fundamental) is shown in Table 6.2(a), with no adjustment to the 'desired power spectrum' and after using the above weighting ten times. These results are similar to signals generated by Van den Bos (1967) and by a clipping method used by Jensen (1959). The signals are shown in Table 6.2(b) for comparison.

Table 6.2 Sequences having power at 1, 2, 4, 8, 16, 32: first half-period (512 intervals per period)

	Signal	Power	P_h/P_1	E (%)
(a) New sequences using FFT				
First pass	24+, 5−, 8+, 3−, 46+, 24−, 5+, 28−, 2+, 28−, 10+, 4−, 12+, 1−, 13+, 24−, 8+, 11−	0.114, 0.124, 0.161, 0.120, 0.101, 0.098	1.63	71.8
Tenth pass	24+, 5−, 7+, 4−, 46+, 24−, 5+, 27−, 4+, 27−, 10+, 4−, 11+, 3−, 12+, 24−, 8+, 10−, 1+	0.109, 0.110, 0.123, 0.124, 0.115, 0.130	1.19	71.2
(b) Previous signals				
Van den Bos	24+, 4−, 8+, 4−, 45+, 24−, 6+, 27−, 3+, 27−, 11+, 4−, 11+, 3−, 12+, 24−, 8+, 10−, 1+	0.116, 0.114 0.126, 0.114, 0.116, 0.122	1.105	70.7
Jensen	23+, 5−, 9+, 3−, 45+, 25−, 5+, 27−, 3+, 27−, 11+, 3−, 27+, 24−, 7+, 12−	0.107, 0.108, 0.185, 0.127, 0.101, 0.094	1.97	72.1

Two additional sequences were generated using this method, and these are shown in Table 6.3. The first signal has an efficiency of only 58.4%; however, 75% of the total power is concentrated at frequencies included within the bandwidth of the signal. The second signal has the same power distribution as the first sequence shown in Table 6.1, but has a higher efficiency. Use of the FFT method also produced a more even power distribution, as can be seen by comparison of the P_h/P_1 ratios of the two signals.

Table 6.3 Additional sequences using FFT: first half-period (512 intervals per period)

Frequencies	Signal	Power	P_h/P_1	E (%)
1, 2, 4, 8, 12, 16, 24, 32, 48, 64	1+, 14−, 2+, 10−, 15+, 3−, 8+, 1−, 6+, 2−, 5+, 2−, 15+, 3−, 17+, 7−, 4+, 5−, 4+, 2−, 5+, 3−, 3+, 6−, 2+, 9−, 16+, 4−, 6+, 4−, 2+, 20−, 5+, 6−, 11+, 27−, 1+	0.05, 0.0471, 0.0663, 0.0669, 0.0628, 0.0432, 0.0602, 0.0611, 0.0545, 0.0722	1.67	58.4
1, 2, 4, 8, 12, 16, 24, 32	23+, 11−, 37+, 2−, 16+, 21−, 4+, 12−, 5+, 38−, 13+, 8−, 6+, 7−, 14+, 20−, 6+, 9−, 4+	0.0942, 0.0919, 0.0855, 0.0869, 0.0708, 0.082, 0.0786, 0.0782	1.33	67

6.2.3 *Signal modifications*

Two modifications in the sequences have been investigated. Gillenwater (1988) used BMF sequences to identify an extruder, as discussed in a later section, and found that the switching caused undue strain on the motor. Thus another sequence was generated, having a zero at each switching point, to crudely ramp between levels. This seemed to alleviate strain on the motor. A typical signal, for the frequency distribution (1, 2, 4, 8, 16 and 32), is shown in Table 6.4. The total power at the selected harmonics is still near 70%, and it is quite evenly distributed. The ramping could be made even more gradual to reduce further strain on equipment.

A second modification is to use the technique to generate binary sequences having all power concentrated at selected frequencies according to a certain uneven power pattern. An example is shown in Table 6.4. The desired power for this sequence was one quarter at two times the fundamental, one half at four times the fundamental, and one quarter at eight times the fundamental. The resulting signal has 68% of the total power at the three frequencies, with that distributed as 24.2%, 49.6%, 26.2%. Some optimal signals for identification have similar power spectra.

Table 6.4 Modified sequences using FFT: first half-period (512 intervals per period)

Frequencies	Signal	Power	P_h/P_l	E (%)
1, 2, 4, 8, 16, 32	23+, 0, 5−, 0, 7+, 0, 1−, 0, 45+, 0, 23−, 0, 5+, 0, 25−, 0, 3+, 0, 27−, 0, 9+, 0, 3−, 0, 9+, 0, 3−, 0, 11+, 0, 23−, 0, 7+, 0, 9−, 0	0.123, 0.107, 0.116, 0.112, 0.119, 0.123	1.16	70
2, 4, 8	13+, 20−, 52+, 87−, 52+, 21−, 11+	0.165, 0.337, 0.178	1.08	68

The first signal has a zero each time the signal switches from +1 to −1 and vice versa.

The power distribution desired for the second signal was 1/4 at two and eight times the fundamental and 1/2 at four times the fundamental.

For the second signal, the power at the second frequency was divided by two before making the P_h/P_l comparison.

6.3 Applications to bench- and pilot-scale equipment

6.3.1 Identification

Identification of process dynamics may be defined as 'the determination of a model that adequately characterizes the dynamic performance of a process'. The meaning of 'adequate' depends on the purpose of the identification. All applications presented here are for the purpose of controller design or for tracking of process changes in adaptive control. Thus the results need not be as accurate as would be necessary if the purpose was to gain new physical insight into a process. In all cases linear models were found, although some of the processes were non-linear. However, linear approximations were sufficient for the control applications considered.

Important considerations are sampling rates and level of disturbance to the process. Preliminary step tests were performed on each process studied to determine the extent of the linear region of operation, any time delays, and the apparent system time constants. This information was used to select the frequency ranges and amplitudes of the binary multi-frequency perturbations.

In all cases, periodic binary multi-frequency signals were employed. Identification data were not taken until at least the second period.

LIQUID-LEVEL SYSTEM AND CONSTANT-VOLUME STIRRED TANKS

This work is described in more detail by Harris and Mellichamp (1980). These two bench-scale processes both consisted of two tanks connected in series. Control of level in the second tank of the liquid-level system was accomplished by manipulating the flowrate to the first tank, through a non-linear automatic valve. Each tank also had a manually adjusted flow line (kept constant during any run). Control of

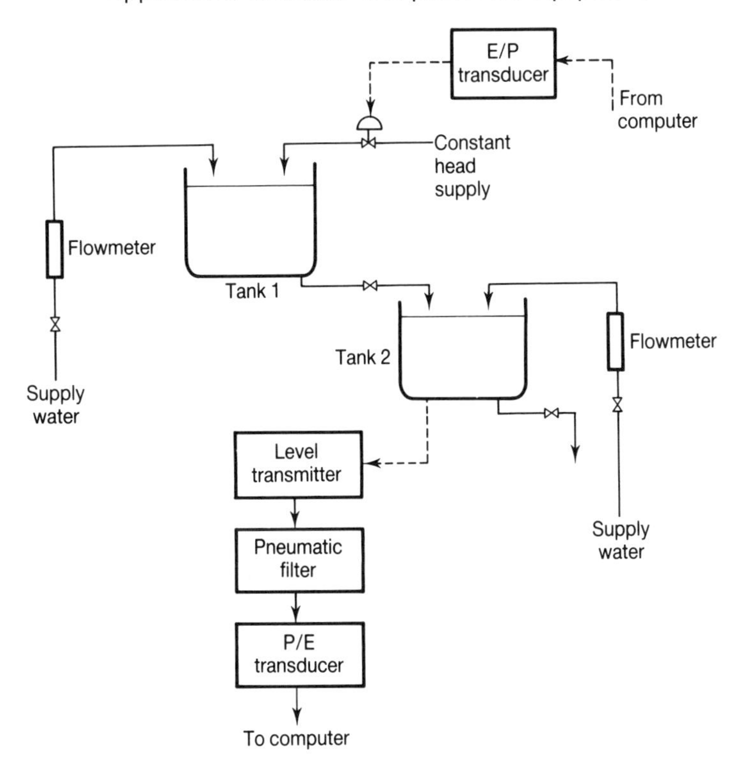

Figure 6.1 Bench-scale liquid-level system.

temperature in the second tank of the stirred tank system was accomplished by utilizing a silicon-controlled rectifier-driven heating element in the first tank. Each tank also had a manually adjusted heating adjustment, kept constant during any run. A long transfer line connected the two tanks. These two systems are shown in Figures 6.1 and 6.2.

The liquid-level system was perturbed using the compact signal (Table 6.1) and the stirred tank was perturbed using the extended signal (Table 6.1). Resulting output values were $\pm 3.5\%$ of the measurable range (± 0.2 inch for the liquid-level system and ± 0.8°F for the stirred tank) (standard deviations for steady-state operation were 0.02 inch and 0.1°F, respectively). Output and input signals were Fourier transformed to yield the frequency response at the frequencies perturbed. This was then fitted to a second-order with dead time parametric model using a direct search method. The results are shown in Table 6.5, which includes for comparison theoretical parameters (neglecting valve and measurement dynamics) and parameters found by Johnson and Mellichamp (1972) for the stirred tank process. Identification of the closed loop was also performed for both processes.

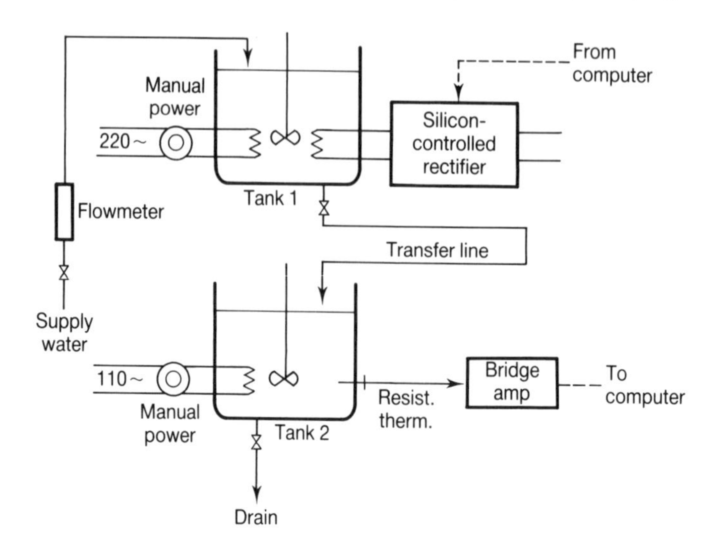

Figure 6.2 Bench-scale stirred tank system.

Table 6.5 Identification of bench-scale processes

Description	K_p	τ_d	τ_1	τ_2	τ	ζ
(a) Liquid-level process						
Theoretical	3.85	0	4.62	1.8	2.88	1.11
Measured	3.51	0.211	3.87	2.01	2.79	1.05
(b) Stirred tank process						
Theoretical	0.51	0.414	1.82	1.05	1.38	1.04
Johnson	0.665	0.733	2.37	1.33	1.78	1.04
Measured	0.735	0.824	2.22	1.28	1.68	1.04

Gains in volt/volt, time constants in minutes.
Liquid level $\omega_1 = 0.123$ rad/min.
Stirred tank $\omega_1 = 0.092$ rad/min.

BENCH-SCALE EXTRUDER

This work is described in more detail by Bezanson and Harris (1986). The extruder system used in the experiments consisted of a Haake single-screw extruder with an L/D (barrel length to diameter) ratio of 25:1 and a barrel diameter of 19.05 mm, completely interfaced to an IMSAI microcomputer. Temperature and pressure at the die could be read by the computer and screw speed and valve position (die restriction) could be manipulated from the computer. Barrel temperature profile was maintained by four proportionally controlled electrical heaters. A schematic is shown in Figure

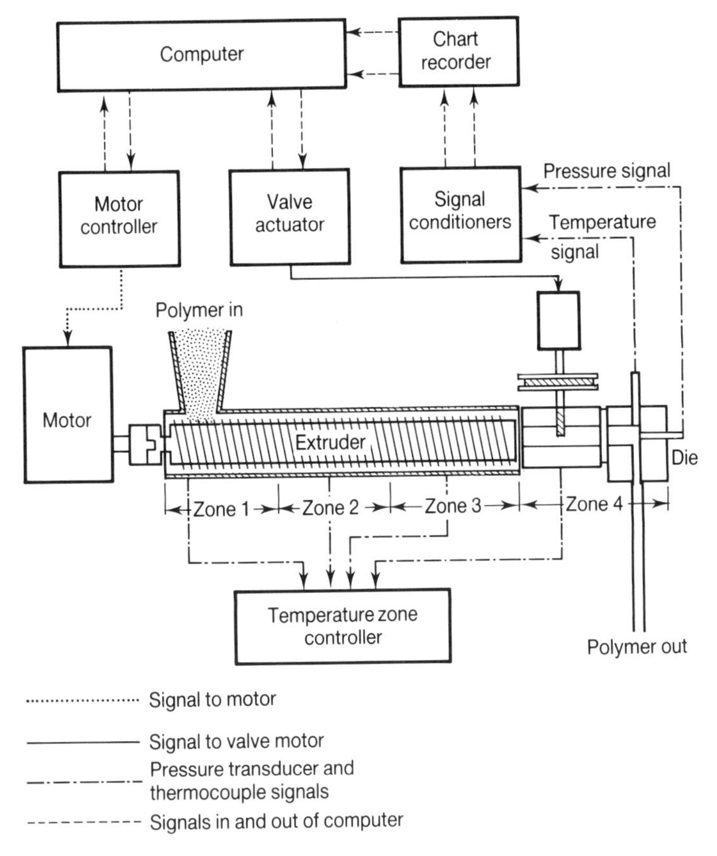

Figure 6.3 Bench-scale extruder system.

6.3. A polyethylene–polyethylene, ethyl acrylate copolymer was extruded. To obtain data for the auto-regressive moving average model determination the inputs were simultaneously perturbed using orthogonal binary multi-frequency signals given in Van den Bos (1970). The screw speed was varied ± 5 rpm (about a mean of 40 rpm) and the valve position was perturbed $\pm 22.5°$ about a mean of 45°. These perturbations resulted in a ± 5°C variation about a mean of 216°C and a ± 1.5 MPa variation about a mean of 29 MPa (standard deviations for steady-state operation were 2.19°C and 0.338 MPa, respectively).

The open-loop data were analyzed using recursive least squares and extended least squares methods. The model order was determined by computing the F (variance ratio) statistic (Söderström, 1977) for first-, second- and third-order models, and corresponding residual errors were tested for independence by computing the autocorrelation. The variance achieved with each model was also examined.

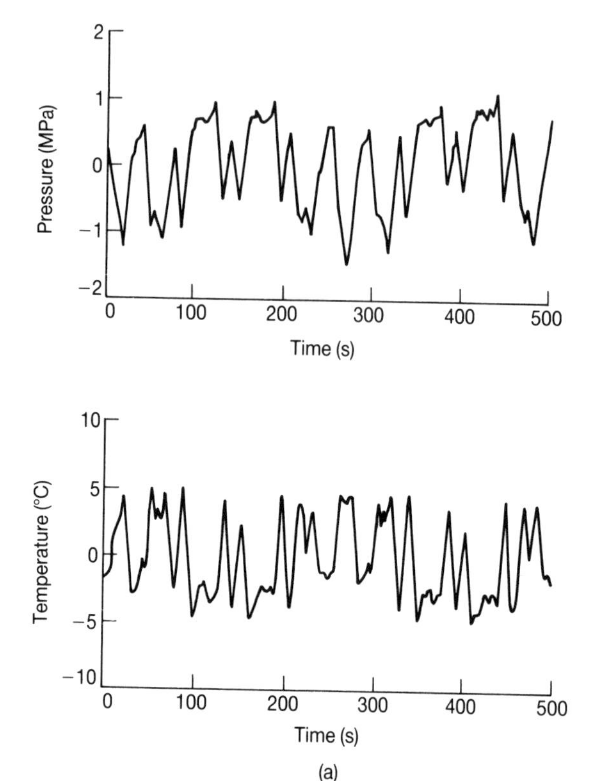

Figure 6.4 (a) Actual extruder response to BMFS inputs.

A second-order discrete multivariable process model, exhibiting non-minimum phase characteristics, resulted. Figures 6.4(a) and 6.4(b) show the actual response to BMF inputs and the simulated response, respectively, using the second-order model.

PILOT-SCALE EXTRUDER

A Killion single-screw extruder with an L/D ratio of 24:1 and a barrel diameter of 38 mm was interfaced to a Zenith Z-100 microcomputer. Melt temperature and pressure were read by the computer and screw speed was manipulated by the computer. Polyethylene was extruded. The screw speed was perturbed using a binary multi-frequency signal, shown in Table 6.3, having power at 1, 2, 4, 8, 12, 16, 24, 32, 48 and 64 times the fundamental frequency. Amplitude was ± 5 rpm about a mean value of 20 rpm. Output pressure fell within the range 8.1–11.2 MPa and output temperature fell within the range 139–139.5°C. Using verification techniques similar to those employed with the bench-scale extruder, two models were selected for use in later control studies: one second-order in output and one third-order. Details are given by Gillenwater (1988).

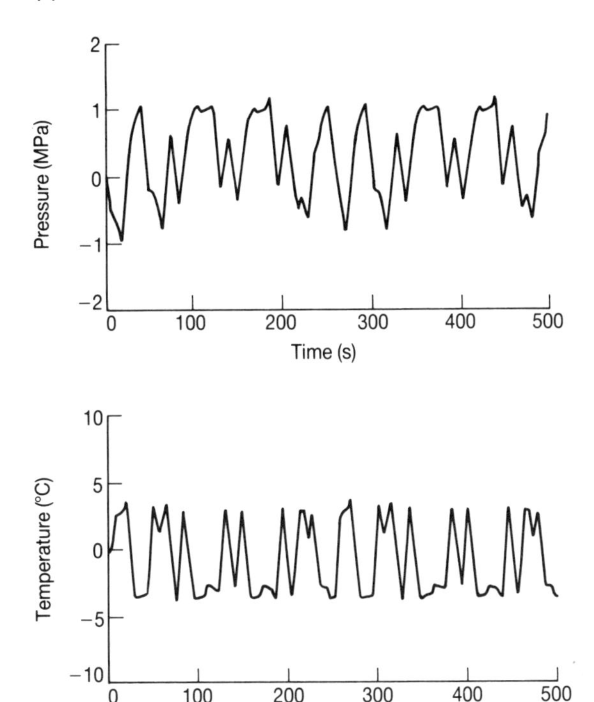

Figure 6.4 (cont.) (b) Simulated extruder response to BMFS inputs.

6.3.2 *Adaptive control*

An adaptive controller seeks to maintain a defined set of system characteristics through self-modification over a wider range of external conditions than a standard controller could satisfactorily handle. Some kind of an identification procedure is a part of any adaptive algorithm.

LIQUID-LEVEL SYSTEM AND STIRRED TANKS

A frequency-domain adaptive controller was implemented on the bench-scale processes described above. The closed-loop frequency response at a limited number of frequencies (six to eight) was used as the index of performance. The closed-loop was perturbed using the same binary multi-frequency signals as used in the open-loop identification experiments described above. New values of PID controller parameters were then found based on the index of performance. The method is described in detail by Harris and Mellichamp (1981). Figure 6.5 shows the results of an adaptation to a doubling of flowrate to the stirred tank system (this would have the effect of reducing

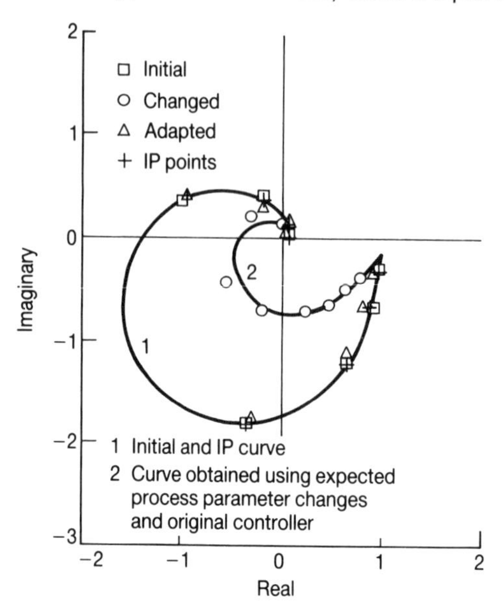

Figure 6.5 Adaptation to doubling of flowrate, stirred tank system.

gain, time constants and dead time by almost half). Identification and application of the index of performance at a limited number of carefully selected frequencies did provide a successful adaptation for cases of changes in the process gain, time constants and dead-time changes coupled with time constant changes, as well as small dead-time changes alone.

EXTRUDERS

The models found as described above were used to give the model form and, in some cases, initial parameter guesses in the application of several adaptive algorithms to the two extruders. In addition, a preliminary identification option, using a binary multi-frequency input signal, was part of two of the algorithms applied to the bench-scale extruder. Little or no improvement over use of initial values equal to zero resulted, probably because the model differed slightly when identified in the closed-loop from that found in the open-loop.

6.4 Applications using simulations

Each of the identification and control schemes above which were applied to physical processes were also tested successfully on simulated processes. Less common applications of binary multi-frequency signals were also investigated using simulations, and two of these are discussed here.

6.4.1 Prevention of burst and turn-off

It is possible, when using certain types of adaptive controllers, for the control to be so good that the quality of the estimation suffers because the system is not adequately excited. This in turn causes a deterioration in the control, and can lead to the complete 'turn-off' of the control signal. Another undesirable condition occurs when the output increases without constraint – known as 'burst' or 'escape'. One solution to these problems is to supply an external perturbation signal. This may be provided constantly (passive addition) or at intervals tied to the current quality of the estimation (active addition). Czekai (1984) considered the coupling of a control law presented by Wieslander and Wittenmark (1971) with addition of a perturbation signal when the elements of the estimation covariance matrix fell out of a predetermined range. The cases of no perturbation, squarewave, PRBS (active and passive) and BMFS (active and passive), were compared using simulations with a simple first-order process. The BMFS was very similar to those shown in Table 6.2(a), and the PRBS was generated by Czekai (1984). The signals covered similar frequency ranges and had the same amplitudes in time.

The control input and process output were examined for each case. With no perturbation, control signal turn-off occurred several times. Burst was also evident. Addition of a uniform squarewave reduced turn-off but did not eliminate burst. Passive addition of a PRBS also reduced turn-off but did not eliminate burst. Active addition of a PRBS further reduced turn-off but did not eliminate burst. Passive addition of a BMFS reduced turn-off and eliminated burst, and active addition of the BMFS further reduced turn-off (and eliminated burst). The accumulated loss was examined after 2500 samples. These results are shown in Table 6.6, and correspond to the previous discussion. This study should be considered preliminary; additional cases should be examined.

Table 6.6 Accumulated loss at $t = 2500$ for various perturbation signals

Perturbation signal	Accumulated loss
None	23 657
Squarewave	24 673
PRBS (passive)	21 681
PRBS (active)	18 178
BMFS (passive)	5 926
BMFS (active)	5 656

6.4.2 Periodic processing

Usually the aim of a control system is operation about some optimal or near-optimal steady state. However, in some cases unsteady operation results in an improvement; for example, in some reacting systems, higher yields may result using periodic operation. Three frequency ranges have been designated as of interest by Bailey (1973): frequencies much below the natural frequency (the inverse of the dominant system time constant), frequencies on the order of the natural frequency, and frequencies much above the natural frequency. Since the mid-range is that which is of the most interest from an identification and controls point of view, that range was investigated by McGreal (1989). He examined the use of different types of periodic signals on the production rates of three different non-linear kinetic systems. Most commonly, simple squarewaves or sinusoids are utilized in periodic processing. McGreal also considered the use of periodic binary multi-frequency signals. He used signals presented by McGhee *et al.* (1987) as well as signals generated using the FFT method described above. In two of the systems studied, ethanol dehydration and the reaction between tolylene diisocyanate and butanol, squarewaves and binary multi-frequency signals both gave an improvement of 10–20% in production rate over steady-state operation. Neither type of signal could be judged better than the other.

6.5 Conclusions

Binary multi-frequency signals can be generated having virtually any desired power spectrum. These signals, long overshadowed by pseudo-random binary sequences, have advantages in identification experiments, as a perturbation in some adaptive algorithms, and whenever perturbation within a certain frequency range would be advantageous.

6.6 Nomenclature

E	signal efficiency (percentage total power concentrated at selected harmonics)
K_p	process gain
N	number of points in the sequence
N_F	number of frequencies at which power is desired
P_{av}	power averaged over the selected harmonics
P_h	highest power at a selected harmonic
P_i	power at the ith selected harmonic
P_l	lowest power at a selected harmonic
P_k	power at the kth harmonic

s	sign (+1 or −1)
w_i	weighting factor for use at the ith selected harmonic
$x(t)$	signal in time
X_k	Fourier transform of $x(t)$ at the kth harmonic
τ	time constant
τ_1, τ_2	first-order time constants
τ_d	dead time
ω	frequency, radians/time
ω_k	kth radian frequency (k times the fundamental frequency)
ζ	damping coefficient

7

Multi-frequency Binary Sequence Identification of Simulated Control Systems

Ian A. Henderson and Joseph McGhee

7.1 Introduction

The microcomputer has become a main component in modern measuring systems. This means that information technology must exert a large impact on the design and practice of instrumentation. Information contrivances have a long history (Sydenham, 1979) and they form the main sub-systems in information technology (Finkelstein, 1985). It is generally accepted that the primary information operations of instruments are handled by four information machines (Finkelstein, 1977; McGhee *et al.*, 1986b; Henderson and McGhee, 1990e). A close symbiosis, which exists between the data machines for calculation, communication, measurement and control, is apparent due to their many similarities. Hence, a cross-fertilization of basic theory and techniques can assist perception in both teaching and research.

An examination of the existing pool of knowledge shows that the information technology ideas have evolved for the data-communication machine but not for the data-measurement machine. There is a dire need to advance information machine signals and data-measurement techniques. This may be achieved by developing coding theorems, information theory, modulation and eye patterns as applied to data communications but modified for use by a measurement-information machine. In order to more easily evaluate these new measurement signals and techniques, a simulation software package, which may be based on existing control system simulation and display techniques, is required. This chapter describes such a package and the resulting novel approach to system identification.

7.2 Identification coding theory and multi-frequency binary sequences

7.2.1 Baseband MBS

Non-compact Multi-frequency Binary Sequence (or MBS) microcomputer test signals have greater than/equal to 128 bits in their binary generating code. Table 7.1 gives the generating codes, waveforms and dominant harmonic information for some useful non-compact MBS. They have been successfully applied to reactors (Buckner and Kerlin, 1972), an electric resistance furnace (McGhee *et al.*, 1991b) and a process (Harris and Mellichamp, 1980). As the above signals include the fundamental as a dominant harmonic, they are baseband MBS. All of these MBS signals have the same number of 1's and 0's in their binary generating code. Therefore, they are MAXimum ENTropy (or maxent) MBS test signals.

Pseudo-Random Binary Sequences (or PRBS) which form approximate white-noise signals spread their binary energy as equally as possible among all the harmonic numbers. In contrast to PRBS, MBS signals concentrate the test signal energy in a known small number of dominant harmonics. This allows a higher energy per MBS dominant harmonic and a lower magnitude of binary interrogation of a process assuming the same noise conditions for the two binary signals. Henderson and McGhee (1990d) have shown that the improvement in signal-to-noise power ratio for an MBS test signal at dominant harmonic, n, and assuming the same binary amplitude and noise conditions, is given by

$$S_n \geqslant 10 \log \frac{P_{n\text{MBS}}}{P_{n\text{PRBS}}} \geqslant 20 \log \frac{|E_{n\text{MBS}}|}{|E_{n\text{PRBS}}|} \text{ dB} \tag{7.1}$$

where P_n and E_n are the power and magnitude associated with dominant harmonic n.

A systems engineering perspective provides a model for a system which has an identification boundary between the system and its environment. The demand input and the contaminating influences impact on this boundary and, from a measurement point of view, it forms an identification channel (Henderson *et al.*, 1987). Among the remarkable similarities between a data-communication channel and an identification channel are carrier generation, coding and decoding of spectral signatures and modulation (Henderson *et al.*, 1987; McGhee *et al.*, 1986a). However, there is a fundamental difference in the data communication and the identification premise (Henderson and McGee, 1990b). The identification of a control system channel requires the data to capture and reveal the spectral signature of the channel. This is in contrast to data communication, where reliability in data flow, with high fidelity

Table 7.1 Table of useful non-compact MBS

Name	Generating code, numberof bits, N		Harmonic amplitude, E_n (n = harmonic number)	Power (%)	Power per comp. (%)
Van den Bos octave	$1^6 0 1^2 0 1^{11} 0^6 1^2 0^{14} 1^3 0 1^6 0^6 1^2 0^3$ E	N = 128	$E_1 = 0.4821$, $E_2 = 0.4832$, $E_4 = -0.5586$ $E_8 = 0.4928$, $E_{16} = -0.4638$, $E_{32} = 0.3979$	69.7	11.6 ± 2.7
Lodz octave	$1^5 0^2 1^{14} 0^7 1 0^6 1 0^7 1^2 0 1^7 0^6 1^2 0^3$ E	N = 128	$E_1 = 0.4509$, $E_2 = 0.4705$, $E_4 = -0.6022$ $E_8 = 0.5170$, $E_{16} = -0.3706$, $E_{32} = 0.3979$	67.5	11.3 ± 3.7
Strathclyde extended	$1^{11} 0^3 1^4 0^3 1^2 0 1^3 0^2 1 0^6 1^3 0^2 1 0^{12}$ $1 0 1^4 0^2 1^2$E	N = 128	$E_1 = 0.4588$, $E_2 = 0.4519$, $E_4 = -0.3481$ $E_8 = -0.3687$, $E_{12} = -0.3383$, $E_{16} = 0.3183$ $E_{19} = -0.3664$, $E_{24} = 0.3428$, $E_{32} = -0.3183$	62.0	6.9 ± 1.9
Harris compact	$1^5 0^{10} 1^8 0^2 1^8 0^{29} 1^7 0^{11} 1 0^4 1^{31} 0^8 1^4$ E	N = 256	$E_1 = -0.3156$, $E_2 = 0.4170$, $E_3 = 0.3315$ $E_4 = -0.3962$, $E_6 = -0.4406$, $E_8 + 0.4269$ $E_{12} = 0.4509$, $E_{16} = 0.4412$	65.7	8.2 ± 1.9
Harris extended	$1^{44} 0^{12} 1^{16} 0^{14} 1^7 0^4 1^{14} 0^{10} 1^2 0^{23} 1^{13}$ $0^6 1^4 0^{46} 1^6 0^5 1^{14} 0^8 1^8$ E	N = 512	$E_1 = 0.4393$, $E_2 = 0.4467$, $E_4 = 0.3361$ $E_8 = -0.4058$, $E_{16} = -0.3488$, $E_{16} = 0.3664$ $E_{24} = 0.4095$, $E_{32} = -0.3321$	60.2	7.6 ± 1.7

Note: The full generating code for the Van den Bos octave signal is 6 ones, followed by 1 zero, 2 ones, 1 zero, 11 ones, 6 zeros, 2 ones, 14 zeros, 3 ones, 1 zero, 6 ones, 6 zeros, 2 ones, 6 zeros, 2 ones, 6 zeros, 2 ones, 6 zeros, 6 ones, 1 zero, 3 ones, 14 zeros, 2 ones, 6 zeros, 11 ones, 1 zero, 2 ones, 1 zero, 6 ones.

in reproduction of the spectral signature of the transmitted message, is required from the channel.

Multi-frequency binary testing of an identification channel (Henderson *et al.*, 1987; Henderson and McGhee, 1990b; McGhee *et al.*, 1987) requires the data to capture and reveal the spectral signature of the identification channel of the control system. Shannon and Weaver's source and channel coding theorems need to be modified (Henderson *et al.*, 1987; Henderson and McGhee, 1990b; McGhee *et al.*, 1986a) to account for the initial premise of an identification channel.

The *Identification Source Coding Theorem* states that

> The optimum test signal for the interrogation of a system identification channel must be source coded such that it is compact yet uniquely and instantaneously decodable, so that the evoked output response satisfies the aims required by the identification channel coding theorem.

The *Identification Channel Coding Theorem* states that

> The most effective digital test signal for identification will be source coded to arrange that the evoked response at the output of the system channel to be identified will have only those optimum characteristics which match the identification channel.

The concept of an identification channel and the application of the identification source coding theorem have been responsible for the discovery of compact and ultra-compact MBS test signals (Henderson and McGhee, 1990a,b, 1991c). Compact MBS have less than 128 bits while ultra-compact MBS are a subset of compact MBS with less than/equal to 16 bits in their binary generating code. Ultra-compact maxent MBS were given a UC number in the order in which they were discovered (Henderson and McGhee, 1991c) when the bit length was increased from 2 to 16. However, the MBS tables are given in sub-sets which are chosen and put in order according to their dominant harmonics. Table 7.2 gives compact MBS with octave/decade frequency ratios. Useful ultra-compact MBS with dominant even and odd harmonics are given in Tables 7.3 and Table 7.4, respectively. They include the fundamental or harmonic number 1 but Table 7.4 excludes the sub-set with no even harmonics which is given in Table 7.5. All of the MBS of this sub-set, which have only odd dominant harmonics, have a binary generating code where the second half is the complement of the first half. Table 7.5 includes a well-known and important member, UC1, which is the squarewave. The source coding theorem clearly shows that compactness of the generating code, with a unique, instantaneous decodability and maximization of the power per component of specified dominant harmonics, is required. This ensures fast, economical and efficient digital generation (Henderson *et al.*, 1987) and signal processing (McGhee *et al.*, 1987).

Table 7.2 Strathclyde compact MBS with octave/decade frequency ratios

Name	Generating code, half waveform	number of bits, N	Harmonic amplitude, E_n (n = harmonic number)	Power (%)	Power per comp. (%)
Odd short octave	$1^3 0$ (E)	$N = 8$	$E_1 = 0.9003$, $E_2 = 0.6366$, $E_4 = 0.6366$	81.1	27.0 ± 9.6
Even short octave	$1^3 0^3 1\ 0$ (E)	$N = 16$	$E_1 = 0.7633$, $E_2 = 0.6366$, $E_4 = -0.6366$	69.7	23.2 ± 4.2
Mid level-power octave	$1^3 0\ 1^7 0^4 1\ 0^7 1^4 0^3 1\ 0$ (E)	$N = 64$	$E_1 = 0.5038$, $E_2 = 0.5231$, $E_4 = -0.5091$, $E_8 = 0.6093$	57.9	14.5 ± 2.4
Mid high-power octave	$1^3 0\ 1^7 0^{11} 1^5 0^3 1\ 0$ (E)	$N = 64$	$E_1 = 0.4232$, $E_2 = 0.6111$, $E_4 = -0.7342$, $E_8 = 0.5434$	69.3	17.3 ± 6.5
Octave	$1^3 0\ 1^7 0^3 1\ 0^7 1\ 0\ 1^3 0^3 1\ 0$ (E)	$N = 64$	$E_1 = 0.5254$, $E_2 = 0.5039$, $E_4 = -0.5091$, $E_8 = 0.4775$, $E_{16} = -0.4775$	62.3	12.5 ± 0.9
Octave extended	$1^2 0\ 1^7 0^3 1\ 0^3 1\ 0^4 1\ 0\ 1^3 0^3 1\ 0$ (E)	$N = 64$	$E_1 = 0.4513$, $E_2 = 0.4391$, $E_4 = -0.4159$, $E_8 = 0.4115$, $E_{15} = 0.5046$, $E_{16} = -0.3183$, $E_{30} = -0.2774$	58.6	7.8 ± 3.0
Decade	$1^2 0^4 1\ 0$ (E)	$N = 16$	$E_1 = -0.6243$, $E_2 = 0.1865$, $E_3 = 0.7626$, $E_4 = 0.3183$, $E_5 = 0.2627$, $E_6 = 0.3623$, $E_7 = 0.4253$, $E_8 = 0.3183$, $E_9 = -0.3308$, $E_{10} = 0.2174$	87.3	8.7 ± 8.3

Table 7.3 Strathclyde ultra-compact MBS with fundamental and even dominant harmonics

Maxent MBS	Generating code number of bits, N		Component amplitude, E_n (n = harmonic number)	S_n	Power (%)	Power per comp. (%)
UC57	$1^2 01^3 0$ O	$N = 14$ odd	$E_1 = 0.6366$ $E_4 = 0.8744$	$S_1 \geqslant 1.60$ $S_4 \geqslant 5.34$	58.49	29.25 ± 8.98
UC25	$101^3 0$ O	$N = 12$ odd	$E_1 = 0.6366$ $E_6 = 0.8488$	$S_1 \geqslant 1.60$ $S_6 \geqslant 6.46$	56.29	28.14 ± 7.88
UC102	$10^2 10^4$ O	$N = 16$ odd	$E_1 = -0.6891$ $E_6 = 0.7245$	$S_1 \geqslant 2.29$ $S_6 \geqslant 5.08$	49.99	24.99 ± 1.25
UC3	10^3 O	$N = 8$ odd	$E_1 = -0.9003$ $E_2 = 0.6366$ $E_4 = 0.6366$	$S_1 \geqslant 4.61$ $S_2 \geqslant 1.79$ $S_4 \geqslant 2.58$	81.05	27.01 ± 9.55
UC18	$101^2 0^2$ E	$N = 12$ even	$E_1 = 0.6366$ $E_2 = -0.5513$ $E_4 = 0.8270$	$S_1 \geqslant 1.60$ $S_2 \geqslant 0.54$ $S_4 \geqslant 4.85$	69.66	23.22 ± 8.03
UC49	$1^1 0^5$ O	$N = 14$ odd	$E_1 = -0.7939$ $E_2 = 0.7783$ $E_4 = 0.6051$	$S_1 \geqslant 3.52$ $S_2 \geqslant 3.54$ $S_4 \geqslant 2.14$	80.11	26.70 ± 5.96
UC82	$101^3 0^3$ E	$N = 16$ even	$E_1 = 0.7633$ $E_2 = -0.6366$ $E_4 = 0.6366$	$S_1 \geqslant 3.18$ $S_2 \geqslant 1.79$ $S_4 \geqslant 2.58$	69.66	23.22 ± 4.18
UC96	1010^5 O	$N = 16$ odd	$E_1 = -0.7633$ $E_2 = 0.6366$ $E_8 = 0.6366$	$S_1 \geqslant 6.42$ $S_2 \geqslant 2.92$ $S_8 \geqslant 5.81$	69.66	23.22 ± 4.18
UC81	$1^2 01^2 0^3$ E	$N = 16$ even	$E_1 = 0.9003$ $E_4 = 0.6366$ $E_6 = -0.5123$	$S_1 \geqslant 4.61$ $S_4 \geqslant 2.58$ $S_6 \geqslant 2.07$	73.91	24.64 ± 11.61
UC103	$10^4 10^2$ O	$N = 16$ odd	$E_1 = -0.7633$ $E_4 = 0.6366$ $E_6 = 0.5123$	$S_1 \geqslant 3.18$ $S_4 \geqslant 2.58$ $S_6 \geqslant 2.07$	62.52	20.84 ± 6.55
UC112	$10^3 10^3$ O	$N = 16$ odd	$E_1 = -0.6891$ $E_4 = 0.6366$ $E_8 = 0.6366$	$S_1 \geqslant 5.54$ $S_4 \geqslant 5.07$ $S_8 \geqslant 5.81$	64.27	21.42 ± 1.64
UC52	$101^4 0$ O	$N = 14$ odd	$E_1 = 0.7939$ $E_6 = 0.7269$ $E_8 = 0.5452$	$S_1 \geqslant 6.77$ $S_6 \geqslant 6.53$ $S_8 \geqslant 4.46$	72.80	24.27 ± 6.97
UC97	$10^5 10$ O	$N = 16$ odd	$E_1 = -0.9003$ $E_6 = 0.5123$ $E_8 = 0.6366$	$S_6 \geqslant 7.86$ $S_6 \geqslant 3.49$ $S_8 \geqslant 5.81$	73.91	24.64 ± 11.61

Table 7.4 Strathclyde ultra-compact MBS with odd dominant harmonics

Maxent MBS	Generating code number of bits, N		Component amplitude, E_n (n = harmonic number)	S_n	Power (%)	Power per comp. (%)
UC23	$1^2 01^3$ O	$N = 12$ odd	$E_1 = 0.6366$ $E_3 = 0.8488$	$S_1 \geqslant -1.78$ $S_3 \geqslant 3.23$	56.29	28.14 ± 7.88
UC114	$10^2 10^3 1$ O	$N = 16$ odd	$E_1 = -0.5922$ $E_5 = 0.6848$	$S_1 \geqslant 0.97$ $S_5 \geqslant 3.82$	40.98	20.49 ± 2.96
UC50	$1^2 01^4$ O	$N = 14$ odd	$E_1 = 0.7627$ $E_3 = 0.5422$	$S_1 \geqslant 3.17$ $S_3 \geqslant 0.72$	58.40	19.47 ± 6.80
			$E_5 = 0.5407$	$S_5 \geqslant 1.77$		
UC56	$10^2 10^3$ O	$N = 14$ odd	$E_1 = -0.5805$ $E_3 = -0.6237$	$S_1 \geqslant 0.80$ $S_3 \geqslant 1.94$	55.37	18.46 ± 1.15
			$E_5 = 0.6176$	$S_5 \geqslant 2.92$		
UC98	$1^2 0^5 1$ O	$N = 16$ odd	$E_1 = -0.8034$ $E_3 = 0.5621$	$S_1 \geqslant 3.62$ $S_3 \geqslant 1.04$	62.23	20.74 ± 8.18
			$E_5 = 0.5322$	$S_5 \geqslant 1.63$		
UC105	$1010^4 1$ O	$N = 16$ odd	$E_1 = -0.6663$ $E_7 = 0.7162$	$S_1 \geqslant 5.24$ $S_7 \geqslant 6.60$	63.36	21.12 ± 4.21
			$E_9 = 0.5570$	$S_9 \geqslant 4.91$		

Table 7.5 Strathclyde ultra-compact odd MBS with no even harmonics

Maxent MBS	Generating code	number of bits, N	Component amplitude, E_n (n = harmonic number)	S_n	Power (%)	Power per comp. (%)
UC32	10^41 O	$N = 12$ dual	$E_1 = -0.9321\ E_5 = 0.6957$	$S_1 \geqslant 4.91\ S_5 \geqslant 3.96$	67.64	33.82 ± 9.62
UC1	1 O	$N = 2$ dual (squarewave)	$E_1 = 1.2730\ E_3 = 0.4244$ $E_5 = 0.2546$	$S_1 \geqslant 4.20\ S_3 \geqslant -2.79$ $S_5 \geqslant -0.87$	93.30	31.10 ± 35.40
UC5	10^21 O	$N = 8$ dual	$E_1 = -0.5274\ E_3 = 1.0246$ $E_5 = 0.6148$	$S_1 \geqslant -0.03\ S_3 \geqslant 6.25$ $S_5 \geqslant 2.88$	85.29	28.43 ± 17.13
UC13	10^31 O	$N = 10$ odd	$E_1 = -0.7869\ E_3 = 0.6867$ $E_5 = 0.7639$	$S_1 \geqslant 3.44\ S_3 \geqslant 2.77$ $S_5 \geqslant 4.77$	83.72	27.91 ± 3.15
UC76	1^301^3 O	$N = 14$ odd	$E_1 = 0.7066\ E_3 = 0.9536$ $E_7 = 0.5457$	$S_1 \geqslant 2.51\ S_3 \geqslant 5.63$ $S_7 \geqslant 3.57$	85.32	28.44 ± 12.72
UC77	101^301 O	$N = 14$ dual	$E_1 = 0.5666\ E_3 = -0.5292$ $E_7 = 0.9095$	$S_1 \geqslant 0.59\ S_3 \geqslant 0.51$ $S_7 \geqslant 8.01$	71.41	23.80 ± 12.44
UC79	10^210^21 O	$N = 14$ odd	$E_1 = -0.4544\ E_5 = 1.0310$ $E_9 = 0.5728$	$S_1 \geqslant 1.92\ S_5 \geqslant 9.40$ $S_9 \geqslant 5.15$	79.88	26.62 ± 18.92
UC150	$1^201^201^2$ O	$N = 16$ dual	$E_1 = 0.4471\ E_5 = 1.0853$ $E_{11} = 0.4933$	$S_1 \geqslant 1.78\ S_5 \geqslant 9.84$ $S_{11} \geqslant 4.50$	81.06	27.02 ± 22.56
UC78	$1^20^31^2$ O	$N = 14$ odd	$E_1 = -0.3145\ E_3 = 1.1892$ $E_5 = 0.3680\ E_{11} = 0.3243$	$S_1 \geqslant -1.28\ S_3 \geqslant 10.39$ $S_5 \geqslant 0.45\ S_{11} \geqslant 0.85$	87.69	21.92 ± 28.18
UC148	101^401 O	$N = 16$ dual	$E_1 = 0.7212\ E_3 = -0.5006$ $E_7 = 0.7752\ E_9 = 0.6029$	$S_1 \geqslant 5.93\ S_3 \geqslant 2.88$ $S_7 \geqslant 7.28\ S_9 \geqslant 5.60$	86.76	21.69 ± 6.80
UC75	10^51 O	$N = 14$ odd	$E_1 = -1.0211\ E_5 = 0.5722$ $E_7 = 0.5457\ E_9 = 0.3179$	$S_1 \geqslant 8.95\ S_5 \geqslant 4.28$ $S_7 \geqslant 4.24\ S_9 \geqslant 0.04$	88.45	22.11 ± 17.87

7.2.2 Shift-keyed modulation and eye patterns

Important coding decisions for designing new MBS are embodied in the identification channel coding theorem. By linking the design of MBS test signals to the channel statistics to be captured, it takes into account different channel characteristics. Digital shift-keyed modulation, which uses either a squarewave or compact MBS as the modulating signal of a squarewave carrier, allows the design of special MBS test signals for interrogation of a control system according to the identification channel coding theorem. These shift-keyed modulation techniques concentrate the binary test signal energy in useful spectral arrays with excellent signal-to-noise power ratios.

The design of phase-shift-keyed (or PSKMBS) signals was proposed by Henderson and McGhee (1988), and applied to control systems (Henderson and McGhee, 1989a,b, 1990a,d), where narrowband detail with a high spectral resolution is required. These use digital phase-shift-keyed modulation with a finite squarewave carrier to concentrate the test signal power in a narrow bandwidth about a specified central frequency. In an analogy to optical zooming, a factor, which is called the zoom factor, is used to determine the bandwidth of the PSKMBS signal.

When MBS are used, they ensure minimum interference (Henderson and McGhee, 1990d) in the normal operation of the process. As information machine signals, they are ideal computer-generated interrogation signals, where the calculated frequency estimates may be used to diagnose the satisfactory working of a control system. Identification may be regarded as a process of pattern as well as parameter measurement. It is concerned with acquisition, processing and utilization of data with a measurement emphasis. Eye patterns are used extensively for the practical examination of a data-communication channel. They have also been proposed (Henderson and McGhee, 1990b,c,d; Henderson *et al.*, 1991a) as time-domain aids for monitoring an identification channel as shown in Figure 7.1. Practical problems, when binary test signals are used to identify a control system, of minimizing the switch-on transient by prediction (Henderson and McGhee, 1990c) and trend elimination (Sankowski, 1989a) have been solved.

The compact and PSKMBS are the basis of simulation software, which is described in this chapter. It is used to investigate the identification channel of a

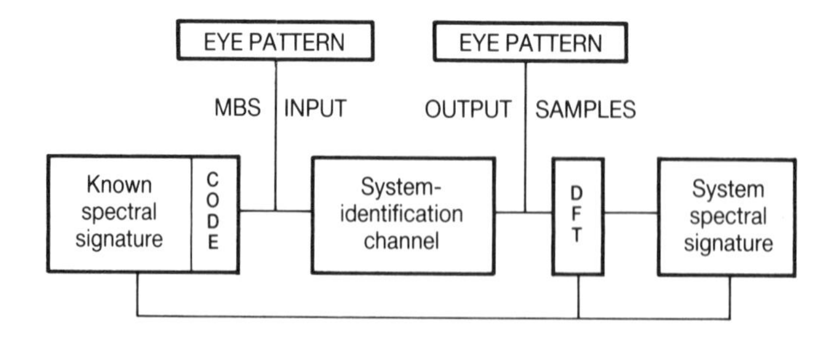

Figure 7.1 Block diagram of an identification channel.

simulated control system with the aid of eye patterns. The frequency information may be plotted on the computer screen and printed as points in a Nyquist, Inverse Nyquist, Bode or Nichols diagram.

7.3 Data measurement using MBS

7.3.1 Fourier analysis of MBS test signals

The amplitude and phase of each of the dominant harmonics, which are given in Tables 7.1 – 7.5, 7.7 and 7.8, were evaluated using a software package. A modified form of the Fourier transform is used. It applies to any typical finite binary signal such as that given in Figure 7.2. Here, the decimal code is given in terms of the position of the zero crossovers from the start of the signal at $s = 0$.

In Figure 7.2 the position of the zero crossovers in binary bits are 0, s_1, s_2, $s_3, \ldots, s_r$ with a total N bits in the finite binary signal. If the time of each bit is Δt, a continuous Fourier transform gives

$$f(s) = a_0 + \sum_{n=1}^{\infty} \left[a_n \cos\left(\frac{2n\pi s}{N}\right) + b_n \sin\left(\frac{2n\pi s}{N}\right) \right] \tag{7.2}$$

where

$$a_0 = [(1s_1) + (-1s_2) + (1s_3) + \ldots + (1s_{r-1}) + (-1s_r)]/N \tag{7.3}$$

and in terms of the Fourier decimal code

$$a_n = \frac{2}{N} \int_0^N f(s) \cos\left(\frac{2n\pi s}{N}\right) \mathrm{d}s \tag{7.4}$$

and

$$b_n = \frac{2}{N} \int_0^N f(s) \sin\left(\frac{2n\pi s}{N}\right) \mathrm{d}s \tag{7.5}$$

These integrals are simplified when the signal is binary where $f(s)$ takes the normalized values $+1$ and -1 alternatively. Hence, a_n and b_n become

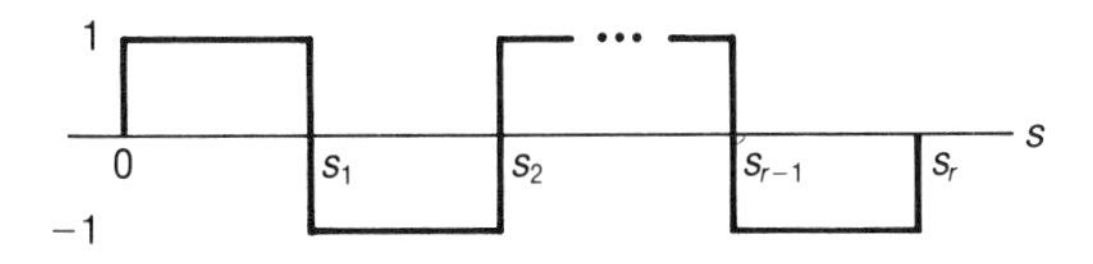

Figure 7.2 Decimal code for Fourier analysis.

$$a_n = \frac{1}{n\pi}\left\{\left[\sin\left(\frac{2n\pi s}{N}\right)\right]_0^{s_1} - \left[\sin\left(\frac{2n\pi s}{N}\right)\right]_{s_1}^{s_2} + \ldots - \left[\sin\left(\frac{2n\pi s}{N}\right)\right]_{s_{r-1}}^{s_r}\right\} \tag{7.6}$$

$$b_n = -\frac{1}{n\pi}\left\{\left[\cos\left(\frac{2n\pi s}{N}\right)\right]_0^{s_1} - \left[\cos\left(\frac{2n\pi s}{N}\right)\right]_{s_1}^{s_2} + \ldots - \left[\cos\left(\frac{2n\pi s}{N}\right)\right]_{s_{r-1}}^{s_r}\right\} \tag{7.7}$$

This is relatively simple to incorporate into a computer program. When the binary signal is symmetrical, the integral for the two halves is the same. Although equations (7.6) and (7.7) work for both asymmetrical and symmetrical binary signals, there is a time saving in having separate software for symmetrical signals. This is due to

$$a_n = 0, \quad \text{if } f(s) \text{ has odd symmetry} \tag{7.8}$$

and

$$b_n = 0, \quad \text{if } f(s) \text{ has even symmetry} \tag{7.9}$$

At each harmonic number, the frequency-response magnitude is

$$E_n = \sqrt{a_n^2 + b_n^2} \tag{7.10}$$

and the phase is

$$\phi_n = \tan^{-1}\left[-\frac{b_n}{a_n}\right] \tag{7.11}$$

An even symmetrical code produces a test signal which has an axis of symmetry where $f(t) = f(-t)$. The odd symmetrical code gives a signal with $f(t) = -f(-t)$. Each symmetrical code crossover from either logic '1' to '0' or logic '0' to '1' approximates to the zero crossover of an equivalent composite test signal with a sum of fundamental and defined harmonics of either sine or cosine terms (McGhee *et al.*, 1987). The associated phase shift is either 0° or 180°, which may easily be accounted for by a plus or minus sign, respectively. Naturally, if the software is designed to calculate the Fourier harmonics for an even-symmetry MBS, it will give +90° and −90° when the MBS signal has odd symmetry.

7.3.2 MBS generating codes with the same power-spectral density

Each even or odd code has two axes of symmetry and two complements (or reverse codes) to give four different symmetrical codes with the same power-spectral density as shown in Figure 7.3. Time shifting or delay may also reveal that some codes have dual symmetry. According to the Fourier time delay or shift (Mayham, 1984), a bit rotation will alter the phase but not the magnitude of each harmonic. Obviously, the complement will change the phase of all the harmonics by 180°. All these symmetrical and asymmetrical versions of the original binary code will have an unaltered power-

	Binary code	Decimal code
Even short octave (Table 7.2)		
	Axis of symmetry ↓	
$f(t) = f(-t)$	11100010 01000111	$^{+}3,3,1,2,1,3,3$
Complement	00011101 10111000	$^{-}3,3,1,2,1,3,3$
Other axis	10111000 00011101	$^{+}1,1,3,6,3,1,1$
Complement	01000111 11100010	$^{-}1,1,3,6,3,1,1$
Odd short octave (Table 7.2)		
$f(t) = -f(-t)$	1110 1000	$^{+}3,1,1,3$
Complement	0001 0111	$^{-}3,1,1,3$
Other axis	1000 1110	$^{+}1,3,3,1$
Complement	0111 0001	$^{-}1,3,3,1$
Dual squarewave (Table 7.5)		
Even axis, $f(t) = f(-t)$	10 01	$^{+}1,2,1$
Complement	01 10	$^{-}1,2,1$
Odd axis, $f(t) = -f(-t)$	11 00 (10)	$^{+}2,2$ or $^{+}1,1$
Complement	00 11 (01)	$^{-}2,2$ or $^{-}1,1$

Figure 7.3 Binary and decimal forms of symmetrical maxent codes with the same power-spectral density.

spectral density. Both the binary and decimal forms of these symmetrical versions of MBS codes are given in Figure 7.3.

7.3.3 Multi-frequency binary testing

In the design of a multi-frequency binary testing instrument, MBS interrogation signals require a fast signal-processing algorithm to obtain either on- or off-line frequency-response estimates. As it is only necessary to calculate the magnitude and phase of a small number of harmonics, their harmonic numbers are transferred to a Discrete Fourier Transform (or DFT) sub-routine. This ensures fast signal processing of the samples of the evoked response at the output of the identification channel of Figure 7.1. It also allows the software to use any MBS signal with a common generating and signal-processing format.

A normalized fundamental frequency of harmonic number 1 and a normalized amplitude of ± 1 V are used in the formation of the tables. This procedure means that maxent MBS signals have unlimited bandwidth before the burst time, Ts, of one MBS test signal is assigned. The fundamental frequency of harmonic number 1 is given by

$$f_1 = \frac{1}{T} \text{ Hz} \tag{7.12}$$

while the harmonic number, n, has a frequency of

$$f_n = \frac{n}{T}\,\text{Hz} \tag{7.13}$$

Each MBS signal has an equivalent composite signal which is composed of dominant harmonics. Here the penalty for simple binary generation and signal processing is the loss of power not contained in these harmonics. The percentage power in the dominant harmonics with amplitude E_n,

$$\eta = \frac{\Sigma\, E_n^2\, 100}{2}\,\% \tag{7.14}$$

To allow a comparison with other MBS the

$$\text{Percentage power per component} = \frac{\eta}{\text{Number of dominant harmonics}} \tag{7.15}$$

An indication of the evenness of the power distribution among the dominant harmonics is given by the standard deviation.

7.4 MBS using phase-shift-keyed modulation

7.4.1 Introduction

There is a need for an information machine signal that will give a precise measurement of a portion of the total spectrum of a control system. Such a signal will obtain frequency information in a way similar to that of an optical zoom lens. Modulation by phase-shift keying (Henderson and McGhee, 1990a), which normally uses a sinusoidal carrier, may also be applied to a squarewave carrier. In order to obtain the spectral signature of the phase shift-keyed modulated signal, only the fundamental of the squarewave, which contains 81% of the total energy, needs to be considered. In the same way as for communication signals, the spectral signature consists of upper and lower side bands about a suppressed carrier frequency. If the dominant harmonics of the modulating MBS are f_n and the suppressed carrier is f_c, the upper and lower frequency arrays consist of $f_c + f_n$ and $f_c - f_n$, respectively. Phase-shift-keyed modulation concentrates the binary energy into two frequency arrays. They contain twice the original number of dominant harmonics which are available with the chosen modulating signal. A bandwidth comparison for different PSKMBS signals may be made using the normalized bandwidth,

$$B_n = \pm\frac{100 B_z}{2 f_c}\,\% \tag{7.16}$$

where the bandwidth of the PSKMBS is B_z.

A zoom factor is used to detail a portion of the frequency-response record. This is achieved by making the carrier have a zoom factor, Z, which is an even number of bits per bit of the binary code of the modulating signal. By multiplying the binary code of the modulating signal with N bits by Z and repeating the '10' carrier binary code $NZ/2$ times, the two signals must be suitably synchronized for PSK modulation.

7.4.2 Graphical illustration of PSK modulation

Figure 7.4 gives two graphical illustrations of PSK modulation using the full 16-bit binary codes of the even short octave (UC82) from Table 7.2 and the odd harmonic MBS (UC148) from Table 7.5. The short octave has dominant harmonic numbers 1, 2 and 4 while the UC148 has dominant harmonic numbers 1, 3, 7 and 9. Because of the method used to synchronize the finite squarewave carrier and the modulating signal, the phase of the squarewave is easily changed by 180° when the modulating signal changes its logic level. Alternate levels of the squarewave carrier will become two identical levels straddling each zero crossover of the modulating keying signal. In between changes in the modulating logic level there must also be repetition of either the '10' or '01' full cycle of the squarewave. There must be one or more cycles of the squarewave for each bit of the modulating signal in the synchronizing procedure. Hence, $Z = 2, 4, 6, 8$, etc. This is all clearly shown for the short octave with even symmetry and the UC148 with odd symmetry. The same PSKMBS generating binary codes may also be obtained using XOR logic between the complement of the squarewave carrier binary code and the synchronized code of the modulating signal. This is also shown in Figure 7.4 and the XOR method may be compared with the graphical method.

7.4.3 Dominant harmonics

The dominant harmonics for the binary signals are also given in Figure 7.4. A squarewave carrier, whose processing window contains a '10' binary code, gives the fundamental at harmonic number 1. If the number of '10' cycles within the window is increased to $NZ/2$, because of the zoom factor, the carrier will now be at $NZ/2$. When $Z = 2$ this becomes harmonic number 16 for the two examples in Figure 7.4. As the dominant harmonics for the short octave are 1, 2 and 4, the dominant harmonics in the PSKMBS signal with $Z = 2$ will be $16 \pm 1, 2, 4$ or 12, 14, 15 and 17, 18, 20 for the two frequency arrays. In the case of the UC148 modulating signal with dominant harmonics 1, 3, 7 and 9, the dominant harmonics in the PSKMBS signal with $Z = 2$ are $16 \pm 1, 3, 7, 9$ or 7, 9, 13, 15 and 17, 19, 23, 25. For any zoom factor, Z, this becomes

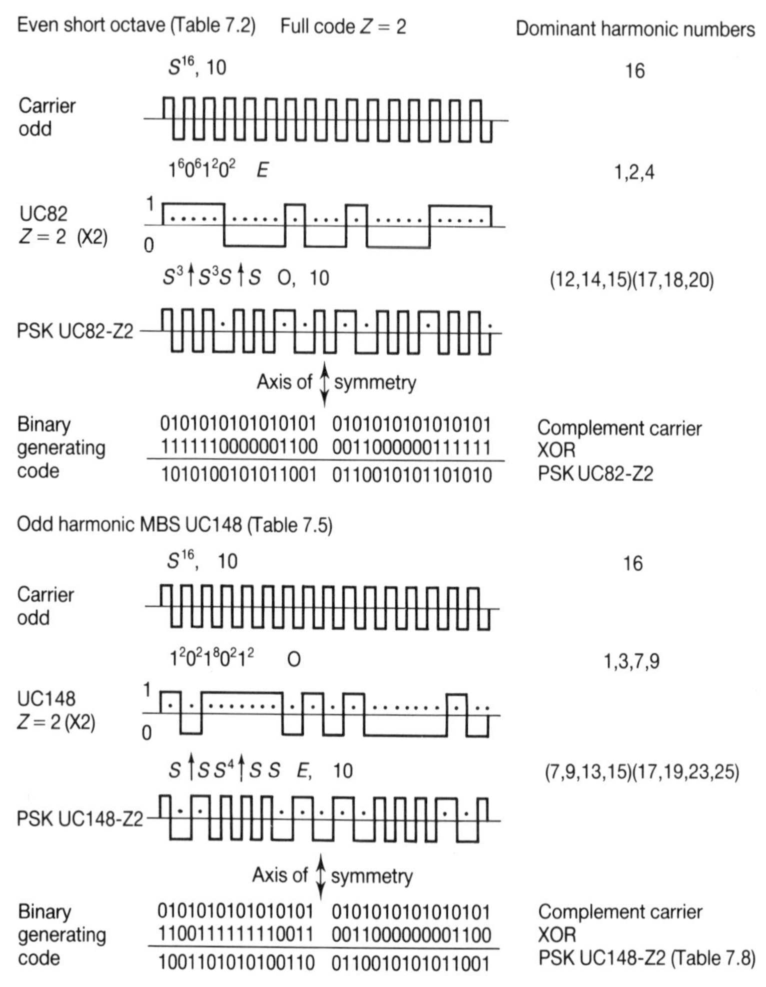

Figure 7.4 Formation of PSKMBS generating codes for symmetrical MBS. (The generating code notation is explained in Section 7.4.5.)

PSK dominant harmonic numbers

$$= \frac{NZ}{2} \pm \text{modulating signal dominant harmonic numbers} \qquad (7.17)$$

Any signal-processing software need only calculate the magnitude and phase at these dominant harmonic numbers.

7.4.4 Symmetrical relationships

Although asymmetrical MBS may also be used, symmetrical MBS are easier to generate by microcomputer and are simpler to use. They also allow a further simplification in the PSK modulation as only half the binary codes are required. However, it is necessary to examine what occurs at the axis of symmetry. A simple interpretation at the axis of symmetry is possible if even and odd are taken as no 180° change and 180° change, respectively. This means that two evens or two odds gives no change at the axis of symmetry and a PSKMBS binary code with even symmetry will be obtained. A mixture of odd and even will produce a PSKMBS binary code with odd symmetry. PSK modulation with an MBS, which has dual symmetry, gives a PSKMBS signal with dual symmetry. These combinations for PSK modulation with symmetrical binary codes are shown in Table 7.6. They will allow phase-shift-keyed modulation binary codes to be evaluated with half the binary codes and the aid of either an E or an O, where E and O indicate even and odd symmetry, respectively.

Table 7.6 Symmetrical relationships with PSK modulation

Carrier	Modulating signal	PSK signal
Odd	Odd	Even
Odd	Even	Odd
Odd	Dual	Dual
Even	Even	Even
Even	Odd	Odd
Even	Dual	Dual

7.4.5 PSKMBS using a descriptive language technique

All PSKMBS, which use a finite carrier, 10101010, etc., with odd symmetry, will have a simple decimal code consisting of 1's and 2's. The 1's are due to the original squarewave carrier while the 2's correspond to the 180° phase change in the modulating signal. Descriptive languages from information theory (Papentin, 1983; Kruger, 1983) use an exponent for repetition of a symbol or a sub-set of symbols. With this concept it is possible to improve on an earlier attempt (Henderson and McGhee, 1989a) to obtain a more concise mathematical approach to acquiring PSKMBS generating codes. The complement and reverse of a sub-set S are represented by $\uparrow S$ and $\leftarrow S$, respectively.

Consider the even short octave with the binary code 11100010 E. This becomes $1^3 0^3 1^1 0^1\ E$ where the exponents give the decimal form of the generating code. From Figure 7.4, when $Z = 2$ the PSKMBS binary code is (10)(10)(10)(01)(01)(01)(10)(01) O or $(10)^3(01)^3(10)(01)\ O$. Hence the PSKMBS code may be obtained from the modulating code by replacing each 1 by '10' and each 0 by '01'. If S is used for '10', $\uparrow S$ or $\leftarrow S$ may be used for '01'. This gives $S^3 \uparrow S^3 S \uparrow SO$, 10 where the sub-set is given after the comma. The description of the generating code for any Z is given by multiplying all the components by $Z/2$. Hence the

$$\text{PSKMBS generating code} = S^{3Z/2} \uparrow S^{3Z/2} S^{Z/2} \uparrow S^{Z/2} O, 10 \tag{7.18}$$

For the UC148 modulating signal the generating code is $101^4 01\, O$. With any Z the

$$\text{PSKMBS generating code} = S^{Z/2} \uparrow S^{Z/2} S^{2Z} \uparrow S^{Z/2} S^{Z/2} E, 10 \tag{7.19}$$

where the O and E refer to the symmetry of the full binary code.

For a given modulating signal and zoom factor, here is a simple method of obtaining the full binary generating code using software. Useful details for the PSKMBS test signal using the UC13 and UC148 MBS from Table 7.5 are given for $Z = 2$, 4, 6, 8 in Tables 7.7 and 7.8, respectively. Any ultra-compact MBS may be easily used to design PSKMBS test signals. Information on designs, which use the squarewave, multi-PSK squarewave MBS and the octave signals of Table 7.2, as modulating signals are published elsewhere (Henderson and McGhee 1989a, 1990a,b).

7.4.6 Relationship between uncertainty and zoom factor

If the maximum dominant harmonic number is m, the normalized bandwidth of equation (7.16), which is given in terms of frequencies, may be calculated in terms of harmonic numbers and the zoom factor:

$$B_n = \pm \frac{200m}{NZ} \% \tag{7.20}$$

This clearly illustrates the narrowing of the bandwidth as Z is increased. It is of equal importance to investigate the effect of Z on the uncertainty associated with experimental identification with PSKMBS. Van den Bos (1967), Rake (1980) and Ljung (1985) show that the variance of the estimators for data measurement by MBS is inversely proportional to the product of the observation time and the power concentrated at the frequency of the dominant harmonic. For a PSKMBS signal, the observation time

$$T_0 = \Delta t N Z \tag{7.21}$$

where Δt and N are the bit time and number of bits associated with one burst of the modulating MBS. If the uncertainty constant is k, the number of bursts, M, and the dominant harmonic amplitude, E_n, the uncertainty becomes

$$\sigma_{\text{PSK}} = \frac{k}{E_n \sqrt{\Delta t N Z M}} \tag{7.22}$$

Table 7.7 PSKMBS using Strathclyde ultra-compact odd harmonic MBS UC13

The PSK modulating code for UC13 (see Table 7.5) is $10^3 1$ O. For any zoom factor, Z, this gives the Strathclyde PSKMBS generating code as $S^{Z/2} \uparrow S^{3Z/2} S^{Z/2}$ E, 10.

Maxent MBS	Zoom factor	Harmonic amplitude, E_n	S_n	Power (%)	Power per comp. (%)	B_N (%)
UC13–Z2 $N = 20$	2	$E_5 = 0.7639$, $E_7 = 0.5776$ $E_9 = -0.5520$, $E_{11} = 0.4517$ $E_{13} = -0.3110$, $E_{15} = -0.2546$	$S_5 \geqslant 6.79$, $S_7 \geqslant 4.73$ $S_9 \geqslant 4.83$, $S_{11} \geqslant 3.73$ $S_{13} \geqslant 1.29$, $S_{15} \geqslant 0.52$	79.4	13.4 ± 8.7	± 50.00
UC13–Z4 $N = 40$	4	$E_{15} = 0.6148$, $E_{17} = 0.5048$ $E_{19} = -0.5262$, $E_{21} = 0.4761$ $E_{23} = -0.3731$, $E_{25} = -0.3689$	$S_{15} \geqslant 8.50$, $S_{17} \geqslant 7.03$ $S_{19} \geqslant 7.67$, $S_{21} \geqslant 7.11$ $S_{23} \geqslant 5.34$, $S_{25} \geqslant 5.62$	70.6	11.8 ± 4.2	± 25.00
UC13–Z6 $N = 60$	6	$E_{25} = 0.5702$, $E_{27} = 0.4817$ $E_{29} = -0.5178$, $E_{31} = 0.4844$ $E_{33} = -0.3942$, $E_{35} = -0.4073$	$S_{25} \geqslant 9.40$, $S_{27} \geqslant 8.36$ $S_{29} \geqslant 9.46$, $S_{31} \geqslant 9.39$ $S_{33} \geqslant 8.17$, $S_{35} \geqslant 9.07$	69.1	11.5 ± 2.9	± 16.67
UC13–Z8 $N = 80$	8	$E_{35} = 0.5487$, $E_{37} = 0.4704$ $E_{39} = -0.5135$, $E_{41} = 0.4885$ $E_{43} = -0.4048$, $E_{45} = -0.4267$	$S_{35} \geqslant 10.88$, $S_{37} \geqslant 9.68$ $S_{39} \geqslant 10.59$, $S_{41} \geqslant 10.30$ $S_{43} \geqslant 8.83$, $S_{45} \geqslant 9.46$	68.5	11.4 ± 2.3	± 12.50

Table 7.8 PSKMBS using Strathclyde ultra-compact odd harmonic MBS UC148

The PSK modulating code for UC148 (see Table 7.5) is $101^4 01$ O. For any zoom factor, Z, this gives the Strathclyde PSKMBS generating code as $S^{Z/2}\uparrow S^{Z/2}S^{2Z}\uparrow S^{Z/2}S^{Z/2}$ E, 10.

Maxent MBS	Zoom factor	Harmonic amplitude, E_n	S_n	Power (%)	Power per comp. (%)	B_N (%)
UC148–Z2 $N = 32$	2	$E_7 = 0.6362$, $E_9 = 0.7347$ $E_{13} = -0.3809$, $E_{15} = 0.4882$ $E_{17} = -0.4308$, $E_{19} = 0.2606$ $E_{23} = -0.2875$, $E_{25} = -0.1781$	$S_7 \geqslant 5.57$, $S_9 \geqslant 7.32$ $S_{13} \geqslant 3.05$, $S_{15} = 6.18$ $S_{17} \geqslant 6.27$, $S_{19} \geqslant 3.32$ $S_{23} \geqslant 8.07$, $S_{25} \geqslant 6.71$	84.8	10.6 ± 8.3	± 56.25
UC148–Z4 $N = 64$	4	$E_{23} = 0.4988$, $E_{25} = 0.6066$ $E_{29} = -0.3491$, $E_{31} = 0.4736$ $E_{33} = -0.4449$, $E_{35} = 0.2893$ $E_{39} = -0.3889$, $E_{41} = -0.2798$	$S_{23} \geqslant 7.86$, $S_{25} \geqslant 9.94$ $S_{29} \geqslant 6.03$, $S_{31} \geqslant 9.20$ $S_{33} \geqslant 9.22$, $S_{35} \geqslant 6.10$ $S_{39} \geqslant 10.10$, $S_{41} \geqslant 8.07$	73.7	9.2 ± 4.5	± 28.13
UC148–Z6 $N = 96$	6	$E_{39} = 0.4587$, $E_{41} = 0.5676$ $E_{45} = -0.3389$, $E_{47} = 0.4688$ $E_{49} = -0.4496$, $E_{51} = 0.2990$ $E_{55} = -0.4231$, $E_{57} = -0.3138$	$S_{39} \geqslant 9.51$, $S_{41} \geqslant 11.61$ $S_{45} \geqslant 7.46$, $S_{47} \geqslant 10.46$ $S_{49} \geqslant 10.28$, $S_{51} \geqslant 6.94$ $S_{55} \geqslant 10.38$, $S_{57} \geqslant 8.01$	71.8	8.98 ± 3.6	± 18.75
UC148–Z8 $N = 128$	8	$E_{55} = 0.4394$, $E_{57} = 0.5487$ $E_{61} = -0.3338$, $E_{63} = 0.4663$ $E_{65} = -0.4520$, $E_{67} = 0.3039$ $E_{71} = -0.4405$, $E_{73} = -0.3310$	$S_{55} \geqslant 10.71$, $S_{57} \geqslant 12.87$ $S_{61} \geqslant 9.05$, $S_{63} \geqslant 12.21$ $S_{65} \geqslant 12.22$, $S_{67} \geqslant 9.06$ $S_{71} \geqslant 12.91$, $S_{73} \geqslant 10.76$	71.2	8.9 ± 3.3	± 14.06

If only Z is altered, by increasing Z the uncertainty is reduced by $\sqrt{Z}$. It should be realized that an increase in Z is accompanied by an increase in the experiment duration. This explains the exceptional precision (Henderson *et al.*, 1991b) obtained from experimental frequency responses, when Z has been increased.

7.5 Simulation using SYDLAB

7.5.1 MBS time-domain analysis

Assume that a typical binary signal, which is shown in Figure 7.5, is in positive pulse form where a pulse starts at N_s bits and finishes at N_r bits. At the output, this pulse interval is an exponential rise from initial conditions and the following space is an exponential decay from initial conditions. It is easier to analyze the time response when all the pulses of the input signal start from zero. The time response of any other pulse waveform may be calculated by subtracting the vertical shift from the calculated values for the positive pulse version. In the case of a ± 1 V symmetrical meander, this is 1. For m positive pulses (Henderson and McGhee, 1990d) with amplitude 1 V and a sequence of N bits each of bit time λ s,

$$f_B(t) = \sum_{m=1}^{m} \{u(t - N_s\lambda) - u(t - N_r\lambda)\} \tag{7.23}$$

If the binary signal continues to infinity in the time domain, its Laplace transform is

$$F_B(s) = \sum_{m=1}^{m} \frac{\{e^{-N_s\lambda s} - e^{-N_r\lambda s}\}}{(1 - e^{-N\lambda s})} \tag{7.24}$$

A state-space representation of the simulated control system is required for the MBS identification software (Henderson and McGhee, 1991a). Consider a system with a steady-state coefficient of one, and three time constants of one second. The transfer function,

$$G(s) = \frac{y}{u} = \frac{1}{(s+1)^3} \tag{7.25}$$

and the initial conditions for each successive step change is the same as the number of system states. In the example, there are three states and three initial conditions.

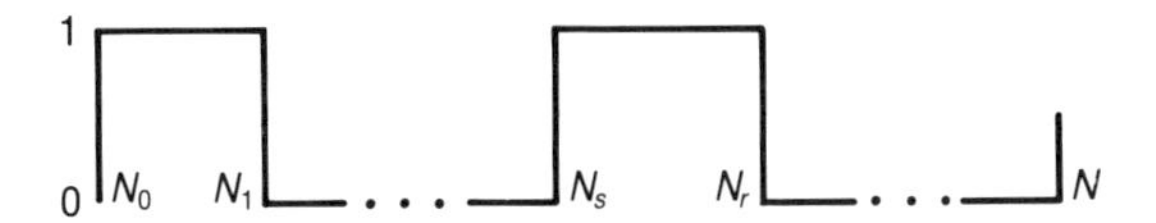

Figure 7.5 Pulse waveform for general binary signal.

The general state vector equations are

$$\dot{X}(t) = AX(t) + Bu(t) \tag{7.26}$$

and

$$y(t) = CX(t) + Du(t) \tag{7.27}$$

The Laplace transformation gives

$$X(s) = X_{\text{IC}} + X_{\text{d}}$$

$$X(s) = [sI - A]^{-1}X(0) + [sI - A]^{-1}Bs^{-1} \tag{7.28}$$

for a pulse and

$$X(s) = [sI - A]^{-1}X(0) \tag{7.29}$$

for a space. The initial condition, $X(0)$, must be taken into account at the beginning of each pulse and space.

7.5.2 The state-space quadruple of program CC

Program CC is used with the MBS simulation software. It uses the STATE command with a user-declared state-space quadruple for the simulated control system. If the three state variables of the above example each occur after a transfer function of $1/(s + 1)$, the simulated control system,

$$P(\text{number}) = \left[\begin{array}{c|c} \text{A} & \text{B} \\ \hline \text{C} & \text{D} \end{array}\right] = \left[\begin{array}{ccc|c} -1 & 1 & 0 & 0 \\ 0 & -1 & 1 & 0 \\ 0 & 0 & -1 & 1 \\ \hline 1 & 0 & 0 & 0 \end{array}\right] \tag{7.30}$$

The SIMULATION command produces a file, which is called $P(\text{number}).Y$, where P is for process and Y is the output variable. This contains the sample values at the output of the simulated identification channel. The total time of the MBS signal and the sample interval, Δt, are required by Program CC. A matrix exponential is used where the recursive digital simulation is updated every Δt s for the kth sample. Here

$$X(t_k + \Delta t) = \text{e}^{A\Delta t}X(t_k) + \int_0^{\Delta t} \text{e}^{AT}B\,\text{d}Tu(t_k) \tag{7.31}$$

and

$$y(t_k + \Delta t) = [1\;0\;0]X(t_k + \Delta t) \tag{7.32}$$

7.5.3 The system detection laboratory

A simulation environment, which allows a detailed examination of MBS identification, condition monitoring or diagnosis of an engineering system, is provided by the SYstem Detection LABoratory (SYDLAB) software package. An important feature of this simulation engine is the ability to conduct an investigation in a shorter time than that required by real-time experiments. Hence, there is no frequency limitation, which must be part of the specification of any MBS instrumentation.

In this chapter the simulation software package uses the *SIMULATION, PLOT* and frequency-response graphic commands of Program CC. SYDLAB may use other simulation and graphic software packages. Suitable simulation and graphic routines are being developed and will eventually be an integral part of the SYDLAB program. A simulated control system is entered using a linear state-space quadruple, *P*. Software modules, which correspond to the *MBS, DFT, PSK* and *EYE* commands are used (Henderson and McGhee, 1991a). The *MBS* command *CREATE* chooses an MBS signal and produces a time file *mbs.in*, which is used as an input to the CC *STATE SIMULATION* command. This produces a *P. Y* time file of the identification channel output, or *Y*, samples for the state quadruple, *P*. The *SIMULATION* command requires the time, T_{max}, and the sample interval, Δt, for the overall test signal. These are calculated by the *MBS* command but must be inserted using the keyboard. The samples are calculated using equations (7.31) and (7.32). PSKMBS generating binary codes may be designed using the *PSK* command before creating the input time file for the *SIMULATION* command.

AUTO within the *DFT* command automatically produces a frequency-response file of the identification channel. Naturally, it must be informed of the chosen MBS and corresponding *P.Y* files. There are other commands with *DFT* that will allow intermediate stages of the *AUTO* command to be evaluated. Any frequency files may be saved to or loaded from disk. A Program CC file may be produced from a frequency file or a number of previously saved files. In this way, the CC *NYQUIST, INVERSE NYQUIST, BODE* or *NICHOLS* commands may be used to produce a frequency response for the simulated control system. When the graphical display on the VDU screen has been perfected, a permanent record may be obtained by using the *PRINT SCREEN* command.

7.5.4 Eye patterns and the identification channel

In data communications information is carried in the form of symbols but it is not always recognized that the reverse is true. Symbols may be used to capture information for subsequent processing using a data-measurement machine. Identification is the process of pattern and parameter measurement, where pattern measurement is the assignment of symbols to pattern attributes and, in so doing, describes them. Parameter measurement is the assignment of symbol to the parameter attributes. Symbols in binary form are the main way data communication systems carry information and must similarly offer an enormous untapped potential for data measurement by an information machine. Data relating to identification patterns and characteristics of interest may be captured by using a suitable method of modulation. Towards this end, the identification-coding theorems highlight the possibilities of designing novel binary signals using shift-keyed modulation techniques.

Although the normal time-domain waveforms at the input and output of the system give a suitable identification pattern, the information, which the pattern contains, is diffuse. In a data-communication channel the effect of noise on the symbols is best examined by a total concentration of the information in the time allocated to one binary bit. The resulting pattern looks like a human eye. When the eye closes due to noise, information is lost in the data-communication channel. In an identification channel (Henderson and McGhee, 1990b–d; 1991a, b; Henderson *et al.*, 1991a) active condition monitoring and diagnosis may be accomplished by examining the pattern obtained with a suitable finite binary signal of N bits. If the time-domain information is concentrated in an $N/2$ eye pattern, a closed steady-state path is obtained. The resulting eyes in this pattern may more easily be used to monitor the identification channel. In this chapter it is shown that a further concentration with PSKMBS test signals to give a $N/4$ eye pattern (Henderson *et al.*, 1991a) is still meaningful.

Eye patterns are special types of Lissajous figures where the binary signal is applied to the vertical input and a suitable ramp or multi-ramp waveform is applied to the horizontal X input of an X–Y display. The horizontal input is not the normal sawtooth waveform with approximately zero flyback time. This means that one burst of the signal, which is applied to the vertical input, is plotted alternately from left to right then right to left 1, 2 and $N/2$ times for $N/2$, $N/4$ and 1-bit eye patterns, respectively. Figure 7.6(a) shows the generation of the 1- and $N/2$-bit eye patterns using the UC148 MBS from Table 7.5.

The 1-bit eye pattern of the squarewave, which has the binary code, '10', has the simplest fully open rectangular 1-bit eye pattern. It forms a closed path which starts and stops at the same point. As the UC5 from Table 7.5 with dominant harmonic numbers 1, 3, 5 has dual symmetry, it is used to demonstrate the perfect eye patterns obtained with either even or odd symmetry. All maxent MBS with either even or odd symmetry give the same perfectly rectangular 1-bit eye pattern which is obtained with the squarewave. This is the eye pattern which is mainly used for a practical investigation of a telecommunications channel. However, it is just as valuable when used with an identification channel.

The ideal input ADC waveform is determined by the 1-bit eye pattern, which is obtained at the input and output of the identification channel of Figure 7.1. On inspection of the output eye pattern, a completely closed eye and channel noise are obvious. They must influence the measurement ability of the ADC. Maximum accuracy is obtained when the mean level of the signal, which is applied to the ADC, is zero. In this ideal situation, the logic '1' and '0' voltage levels should be adjusted to match the limits of the ADC. At high frequencies, when the eye is almost closed, the specification of the ADC, oversampling (Tiefenthaler, 1980; Classen *et al.*, 1980) and

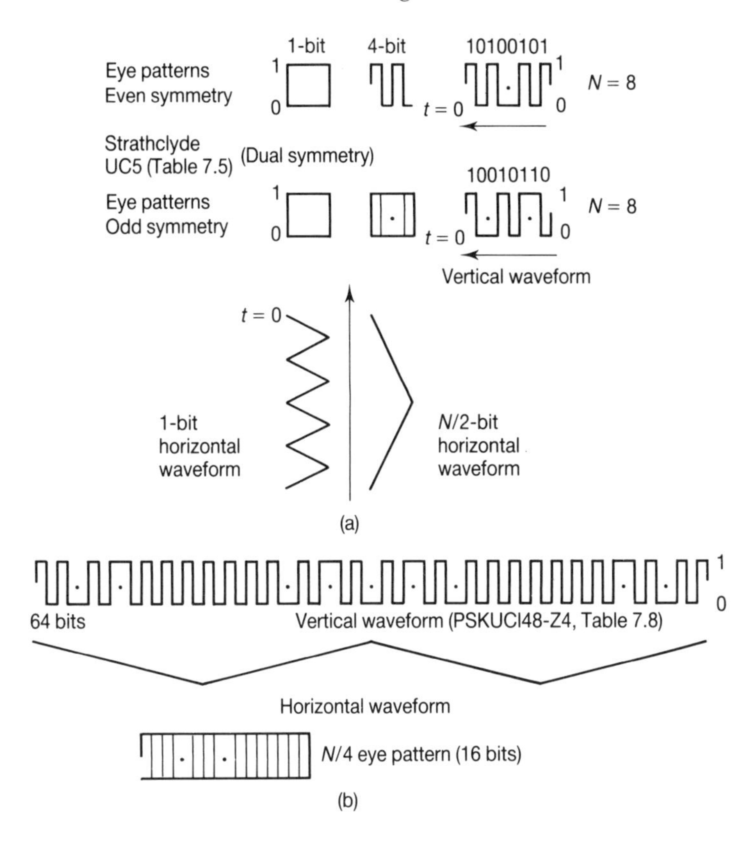

Figure 7.6 Eye patterns for symmetrical maxent MBS.

linear averaging (Rakshit *et al.*, 1985) will decide the precision of the measurement.

If the horizontal and vertical waveforms, which produce the Lissajous figure, are followed, a closed-loop path is obtained for all the eye patterns of Figure 7.6. For condition monitoring or diagnosis the maximum information, which may be obtained from the patterns of the measurement symbols, is required. This is obtained by revealing all the individual eyes which make up the 1-bit eye pattern. While all $N/2$ eyes are fully open from '1' to '0' binary levels with odd symmetry, even symmetry produces only level '1' or '0' half-eyes. The symbols of an asymmetrical MBS are not so useful, as they produce an input pattern which is an untidy mixture of both full and half-eyes.

A special eye pattern is obtained with dual-symmetry MBS test signals. If the dual signal used in Figure 7.6(a) is examined it becomes apparent that the information contained by the eyes is repeated. Because of the dual symmetry, it may be possible to use an $N/4$ eye pattern without any loss of information. This has been shown to be especially true when dual MBS are used to design PSKMBS, which also have dual symmetry. The full generating code of the Strathclyde PSKMBS compact UC148 with a zoom factor of 4, or PSK UC148-Z4, may be obtained using equation (7.19). This is $(10)^2(01)^2(10)^2(01)^2(10)^2(01)^2(10)^2(01)^4(10)^2(01)^2$ and the waveform is given in Figure 7.6(b). Both the horizontal ramp waveform and the corresponding 16-bit $N/4$ eye pattern are also given. This type of MBS test signal is later shown to be a sensitive active condition monitoring signal.

7.6 MBS identification monitoring and diagnosis

7.6.1 Identification

If the experimental frequency results of a control system are obtained in a frequency file which is similar to the simulation frequency file, the graphic software of Program CC may be used to plot the frequency response. It is then possible to obtain and plot frequency estimates by calculation, experimentation or simulation. In Figure 7.7, the Nyquist diagram of the control system, whose state quadruple is given in equation (7.30), shows negligible differences for the three methods over the same frequency range.

The *EYE* command takes the sample values of file *P.Y* and rearranges them to produce an eye pattern file. This may then be drawn using the *PLOT* command of Program CC. Time waveforms and $N/2$ eye patterns at the input and output of the simulated control system are given in Figure 7.8. Two bursts of the octave extended MBS are used to illustrate the switch-on transient and the steady-state closed path. The concentrated information, which is given by the eye pattern, clearly reveals the switch-on transient and the steady-state response.

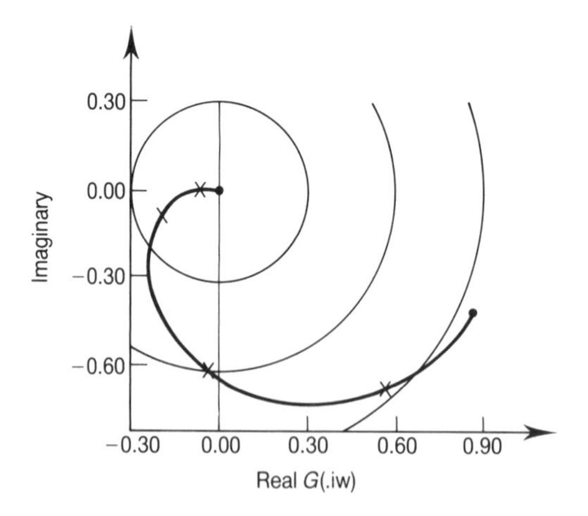

Figure 7.7 Calculated, simulated and experimental Nyquist diagrams for $G(s) = 1/[(s+1)^3]$ with no distinguishable differences.

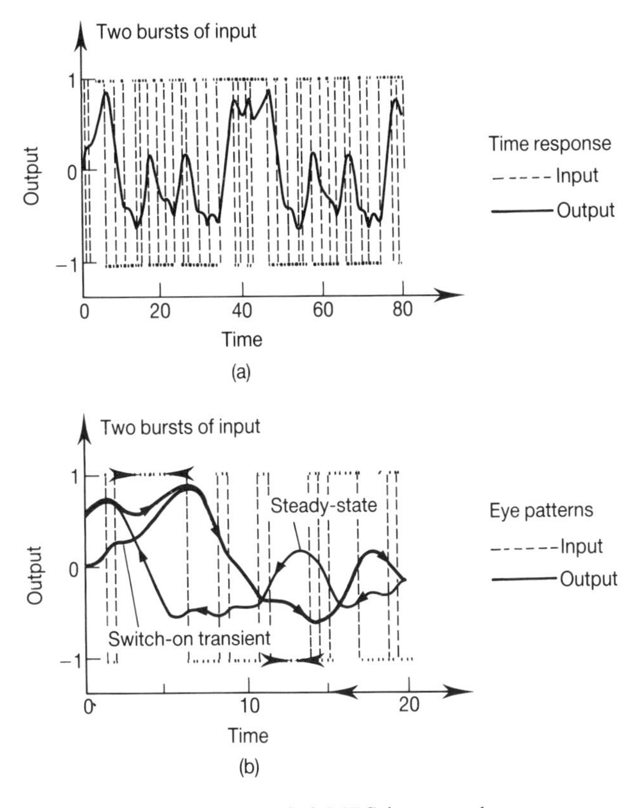

Figure 7.8 Octave extended MBS input and output patterns for $G(s) = 1/[(s + 1)^3]$.

7.6.2 Condition monitoring and diagnosis

Applying PSKMBS test signals to the same simulated control system illustrates the frequency zoom and the extraordinary precision which may be obtained by data measurement using these signals. They are ideal information signals for the condition monitoring or diagnosis of engineering systems. PSKMBS UC148-Z4 and UC148-Z8 from Table 7.8 were used to obtain the Nichols diagrams which are given in Figure 7.9. The zoom attribute and high precision are self-evident. In open-loop the most sensitive frequency range of a control system must include the region between the gain and phase crossovers. Here, there is a high rate of change of magnitude and phase. The central frequency was set at an approximate value for the phase crossover. This was estimated from the experimental results used to plot the Nyquist diagrams of Figure 7.7. The gain margin is 18.06 dB and the phase crossover occurs at a frequency of 0.276 Hz. In closed-loop, a suitably sensitive region for condition

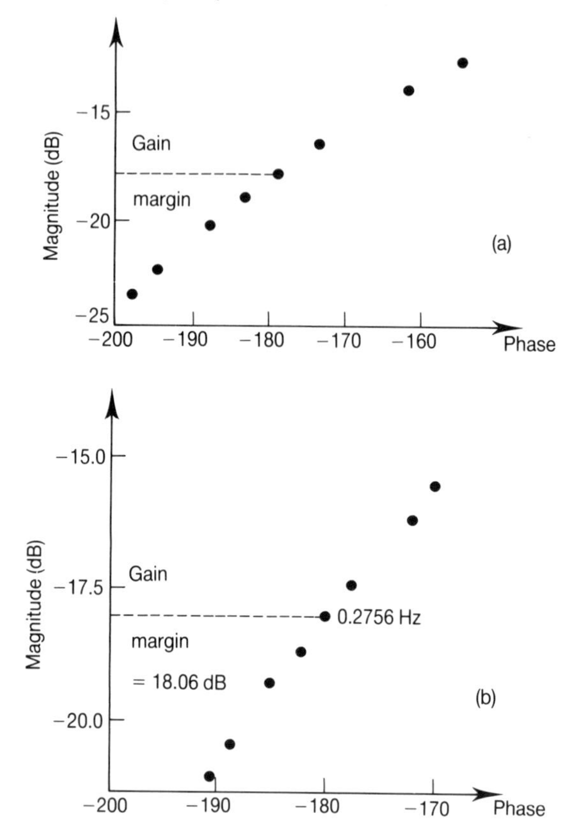

Figure 7.9 Nichols diagrams with PSKMBS test signals centred at phase crossover for $G(s) = 1/[(s+1)^3]$.
(a) UC148-Z4 PSKMBS plot with Bz = 0.1575 Hz
(b) UC148-Z8 PSKMBS plot with Bz = 0.0788 Hz

monitoring of a control system must contain the resonant frequency where the magnification is a maximum.

Eye patterns are a sensitive way of diagnosing parameter change (Henderson *et al.*, 1991a). Obviously, the maximum sensitivity to parameter change will occur where the rate of change of magnitude and phase is greatest. As the sensitive region is fairly narrow with most control systems, PSKMBS with their narrow bandwidth are ideal information machine signals to capture this information. It is worth plotting the eye patterns associated with the experimental results of Figure 7.9. In a recent paper (Henderson and McGhee, 1991a) it was noted that in these $N/2$ eye patterns the information could be further concentrated by using an $N/4$ eye pattern. This is especially true when PSKMBS test signals with a zoom factor of 4 or greater are used.

PSKMBS test signals with the suppressed carrier at the phase crossover were found to have eye patterns with sensitive regions or eyes where the PSK modulating

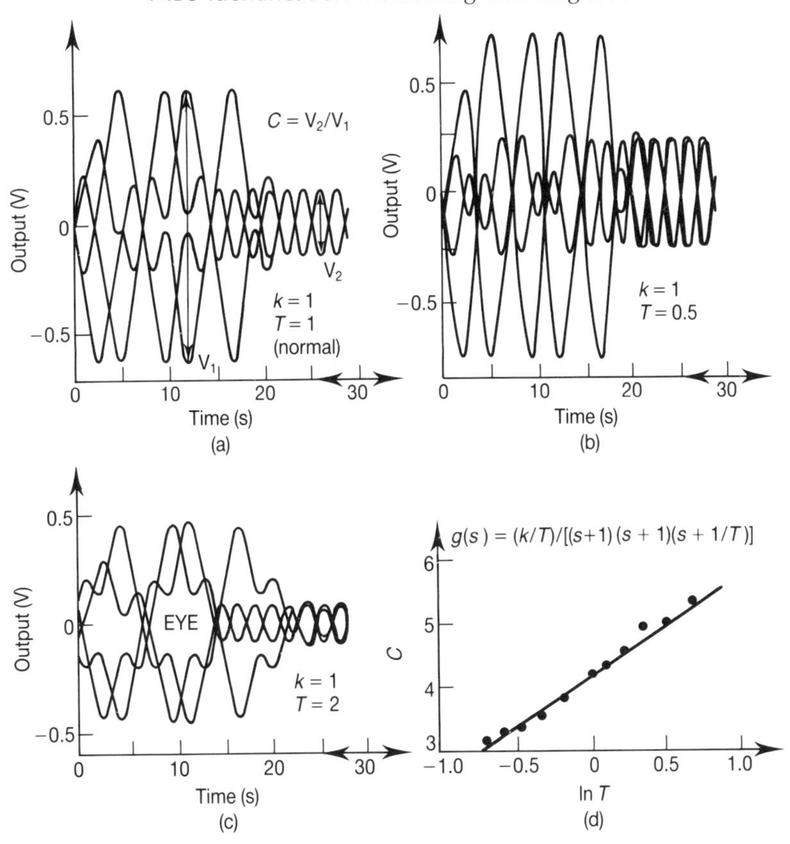

Figure 7.10 State eye and algorithm illustrating change of parameter T from normal for

$$G(s) = \frac{k}{T(s+1)^2\left(s+\frac{1}{T}\right)}.$$

signal changed the phase of the carrier. A reasonable approach to diagnose parameter change seemed to be an investigation of the shape of the eyes by changing the error constant and any time constant by octave variations. When the error constant was changed the shape of the eye pattern was returned to normal by rescaling the vertical axis. However, this was not the case when a time constant was altered. The $N/4$ eye pattern for the normal operation of the control system is shown in Figure 7.10(a). Three changing eyes and a phase shift appeared in all the eye patterns. The most obvious eye is marked in Figure 7.10(c).

Although the changing eyes are of value, there was also a crest factor which needed to be investigated. The crest factor is given by

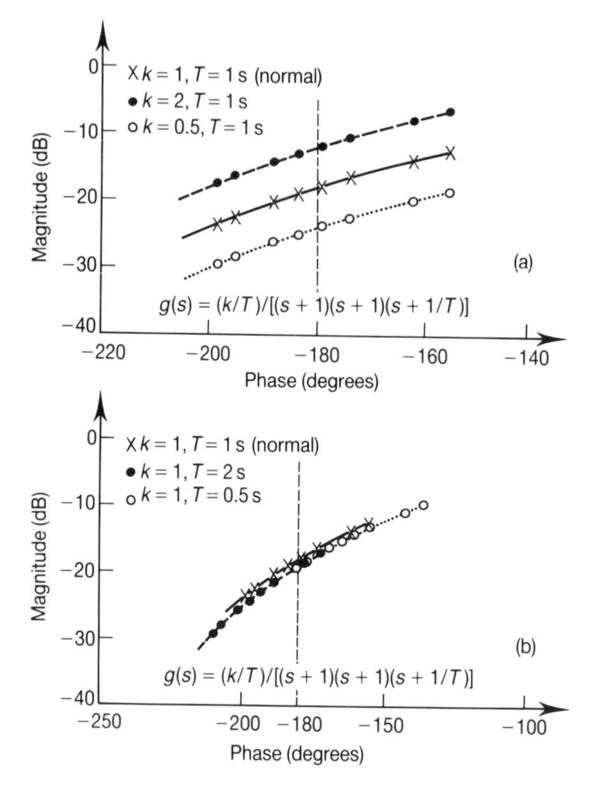

Figure 7.11 Nichols diagrams for octave changes from normal in parameters k and T for control system

$$G(s) = \frac{k}{T(s+1)^2\left(s + \frac{1}{T}\right)}.$$

$$C = \frac{V_2}{V_1} \tag{7.33}$$

where the position of V_1 and V_2 are shown in Figure 7.10(a). A simple algorithm

$$T = e^{(0.6462C - 2.67)} \text{ s} \tag{7.34}$$

was found to relate the crest factor, C, to the changing time constant, T. The linear graph of the crest factor when plotted against ln T is given in Figure 7.10(d). This simulation software package has clearly vindicated the use of PSKMBS test signals for diagnosing parameter change in a control system.

Nichols diagrams for the octave changes in error constant, k, and time constant, T, are given in Figures 7.11(a) and 7.11(b), respectively. The narrowband frequency information of Figure 7.11(a) shows the vertical change obtained as the error constant is changed while holding the time constant at its normal value. In the case of changing only the time constant, the phase is altered but the gain margin is unchanged. This simulation experiment clearly indicates that eye patterns using PSKMBS test signals with changes to a specified normal operation of a control system offer interesting possibilities in condition monitoring and diagnosis. Just as physical states are used to fully model a control system with state-vector differential equations, a state eye pattern may be used to monitor the parameter change associated with each state.

7.7 Conclusions

The software for the MBS simulation package, which is described in this chapter, is still being developed. It has been successfully used to simulate the identification of a d.c. position control system, a warm-air process and temperature sensors. Accurate linear models have been obtained and checked against the experimental frequency estimates using the same MBS signals. The MBS simulation package was the means of understanding the MBS switch-on transient and its minimization by rotating the MBS generating binary code. Eye patterns have been shown to produce a deeper understanding of the practical operation of an identification channel in terms of the data-acquisition system, switch-on transient and the design of new MBS signals.

Shift-keyed modulation may be used to design the generating symbols of a new kind of MBS. The binary energy is redistributed to suit the frequency information which is required by the identification channel (Henderson and McGhee, 1991b). PSKMBS test signals with a zoom attribute concentrate the binary power in the increasingly narrow bands of two frequency arrays. These are represented by an upper- and lower-frequency array of equal bandwidth about a central frequency which is also the suppressed carrier. When the central frequency is positioned at a sensitive part of the frequency response, PSKMBS signals have been shown to be excellent active condition-monitoring and diagnostic signals. The open-loop phase crossover or the closed-loop magnitude peak are suitable sensitive regions of a control system frequency response. A relationship has been established between the crest factor of the PSKMBS eye pattern and parameter variations. This offers interesting new possibilities for condition monitoring and diagnosis. By measuring the crest factor on-line, new algorithms for adaptive control need to be investigated.

Data-communication techniques have been successfully applied to the measurement-information machine using the MBS simulation software. This has resulted in new kinds of interrogation signals and new ways of monitoring the identification channel of a control system. Part of this new awareness is due to the high precision of the measurement available from these information machine techniques. This is under more detailed investigation in the environment available in a calibration laboratory (Henderson *et al.*, 1991b). The techniques, which are described in this chapter, are still being extended. New MBS signals using other shift-keyed modulation methods (Henderson and McGhee, 1991b) are being designed and investigated using the MBS simulation software. As ideal data-measurement machine signals, these novel multi-frequency binary test signals must offer new software and hardware design opportunities in measurement and control.

8

Application of Multi-frequency Binary Signals for Identification of Electric Resistance Furnaces

Dominik Sankowski, Joseph McGhee, Ian A. Henderson, Jacek Kucharski and Piotr Urbanek

8.1 Introduction

The application of electric resistance furnaces, which is well known in many branches of industry, requires a dynamic furnace model for the design of suitable controllers. Such a model can be determined using one of the many methods of process identification (Ljung, 1987). Identification of electric resistance furnaces consists of various contributory steps (Eykhoff, 1974). The first of these is preparation, which is concerned with choosing a model, and, in active identification, planning of the test signal. Active system identification uses external stimulating signals, while passive identification uses normal plant-operating records. Binary test signals, especially multi-frequency binary signals (or MBS) have worthwhile advantages over normal operating inputs (see Chapter 2, Section 2.4.1, and Chapter 7).

Multi-frequency binary signals constitute a versatile sub-group of the important class of binary test signals. These signals offer the advantages of excellent signal-to-noise power ratios for their dominant harmonics, zero offset and a wide variety of power-spectral distribution, as shown in Chapter 7. As they are based upon step functions, which are easily generated, they are ideally suited for a microcomputer environment, particularly for the digital measurement of the frequency response of thermal and other types of control systems. Among the many existing MBS test signals the Van den Bos (1967), the Strathclyde and the Lodz (Henderson *et al.*, 1987) forms have all given good results for the identification of electroheat systems. A cross-flow heat exchanger (Franck and Rake, 1985) and electric resistance furnaces (Sankowski, 1983; Lobodzinski *et al.*, 1983; Plaskowski and Sankowski, 1984; Sankowski *et al.*, 1991) such as the diffusion furnace used in the semiconductor industry (Sankowski, 1989c) have been identified. Several further applications of MBS are listed in Chapter 2 (Section 2.5.2) and the application of MBS to temperature sensor identification is considered in Chapter 9.

In the MBS method of identification, as in other methods, the experiment time should be as short as possible. This experiment time, equal to the number M of MBS sequences in the MBS method, can only be reduced to a minimum of one sequence duration. Because of switch-on effects the use of the results from the first MBS sequence is generally seriously limited by these switch-on transients. Moreover, in an electroheat system when the thermal steady state is usually not fully reached a thermal trend appears at its output, which causes serious bias in the frequency-response estimates. For an electroheat system such as an electric resistance furnace it was proved by Sankowski (1989a) that this trend can be described by a linear function of time. In this chapter two possibilities for minimizing the effect of trend are presented. The first method eliminates the trend from the output signal before any estimation while the second approach corrects the frequency-response estimates obtained with the trend still present.

The theory of switch-on transients given in this chapter complements previous work (Henderson and McGhee, 1990c; Sankowski, 1990) by showing that its effects can be minimized using MBS with the most appropriate starting point (rotated MBS). Application of MBS with the most appropriate switch-on point combined with the correction of the frequency-response estimates which are biased by the trend can minimize the experiment time to only one sequence duration. Their use in the identification of an electric resistance furnace, and the diffusion furnace, are described in this chapter.

8.2 System block diagram and dynamic model

Electric resistance chamber furnaces usually operate in the temperature range from 700°C to 1300°C. They are widely used in the thermal processing of metals and in the baking of pottery and ceramics. Previous work on the identification of these furnaces by Sankowski (1989b) has shown that they may be modelled as linear stationary systems for small deviations from the set-point temperature. Hence, Figure 8.1, which shows a single-input–single-output linear stationary process, is a valid representation for electric resistance furnace identification.

Also shown in Figure 8.1 is a generator of MBS test signals with its rotation operator. The MBS signals may be generated and the signal processing accomplished using software written in assembly language (McGhee *et al.*, 1987; Henderson *et al.*,

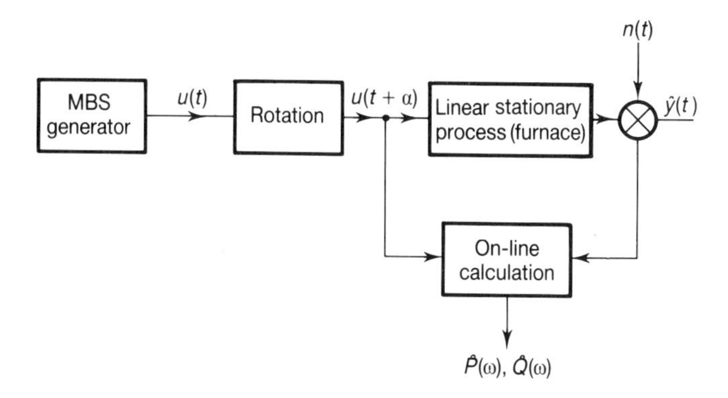

Figure 8.1 Single-input–Single–output linear stationary process.

1987). Such an approach has led to a versatile instrument for system detection called SYD (see Chapter 7, Section 7.5). The use of assembly language programming means that SYD has the widest possible bandwidth from a given hardware. In slower processes the generation and signal processing may be achieved using either interpreted BASIC routines or from compiled high-level language source code.

The dynamic behaviour of the electric resistance furnace considered in this chapter can be modelled by its transfer function, $G(s)$, with s as a Laplace variable. This transfer function, which may be written in the form of the most commonly applied model of electric resistance furnaces, consists of a first-order inertia system with a time constant τ, a time delay T_D and a static 'gain' K_s (Michalski *et al.*, 1985). Such a transfer function, $G(s)$, is given by

$$G(s) = K_s \exp(-sT_D)/(1 + s\tau) \tag{8.1}$$

8.3 Rotated MBS and switch-on transient

Let $u(t + \alpha)$ be a rotated MBS test signal, where α is the rotation in the time domain. For a better explanation, the rotation is presented in Figure 8.2. The MBS Strathclyde Short Sequence from Table 7.2 having base code $3^+, 1^-, 1^+, 3^-$ is rotated for the rotation α, and the direction of this rotation is shown in this figure. To allow the possibility of precisely controlling the switch-on time, the number of bits in this MBS sequence was increased to $N = 128$.

The observed output, $y(t)$, from any linear system, responding to some periodic excitation or forcing function, is composed of the steady-state response, $y_{ss}(t)$, and the transient response, $y_t(t)$, as follows:

$$y(t) = y_{ss}(t) + y_t(t) \tag{8.2}$$

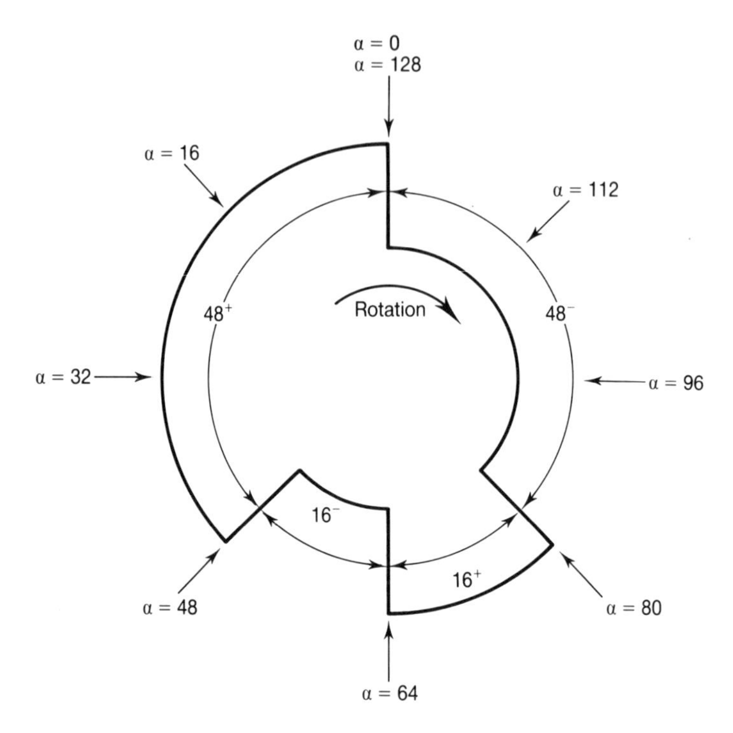

Figure 8.2 Rotation of the MBS signal.

As the transfer function, $G(s)$, has only one pole at $s = -1/\tau$ it is well known that the transient response, $y_t(t)$, in this case is given by

$$y_t(t) = \lim_{s \to -1/\tau} [G(s)u(s)(s + 1/\tau)\exp(st)] \tag{8.3}$$

where $-1/\tau$ is the pole of $G(s)$.

The MBS test signal $u(t + \alpha)$ has the Laplace transform, $u(s)$, given by

$$u(s) = V \sum_k \left[a_{ku} \frac{\cos(k\omega_0\alpha) - k\omega_0 \sin(k\omega_0\alpha)}{s^2 + (k\omega_0\tau)^2} - b_{ku} \frac{\sin(k\omega_0\alpha) - k\omega_0\tau \cos(k\omega_0\alpha)}{s^2 + (k\omega_0\tau)^2} \right] \tag{8.4}$$

where V is the amplitude of the MBS, a_{ku}, b_{ku} are the amplitudes of the kth harmonic cosine and sine Fourier coefficients of the unrotated MBS and ω_0 is the angular frequency of the fundamental with the value ($\omega_0 = 2\pi/N\Delta$).

Equation (8.3) can be written using equation (8.4) in a form, only valid for $t > T_D$, given by

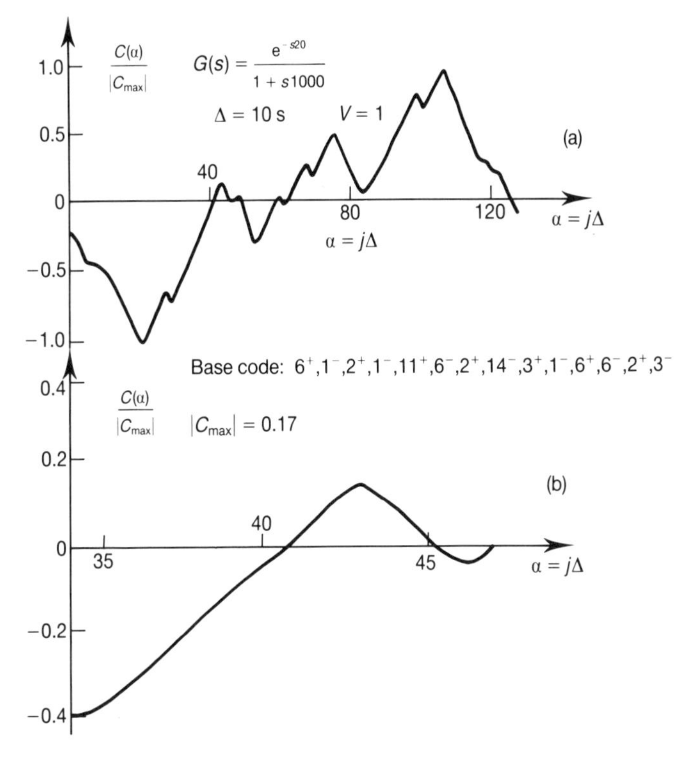

Figure 8.3 Initial value of the transient response for the MBS Van den Bos sequence.

$$y_t(t) = C(\alpha)\exp[-(t - T_D)/\tau] \tag{8.5}$$

where $C(\alpha)$ is the initial value of the transient response. This initial value may be expressed as

$$C(\alpha) = K_s V \sum_k \left[-a_{ku} \frac{\cos(k\omega_0\alpha) + k\omega_0\tau \sin(k\omega_0\alpha)}{1 + (k\omega_0\tau)^2} - b_{ku} \frac{\sin(k\omega_0\alpha) - k\omega_0\tau \cos(k\omega_0\alpha)}{1 + (k\omega_0\tau)^2} \right] \tag{8.6}$$

Obviously, equation (8.6) will be simplified when $u(t)$ is either an even or an odd function of time. When $u(t)$ is an even function the coefficients b_{ku} are equal to zero. When $u(t)$ is an odd function the coefficients a_{ku} are equal to zero. The coefficients a_{ku} and b_{ku} can be calculated before conducting an experiment.

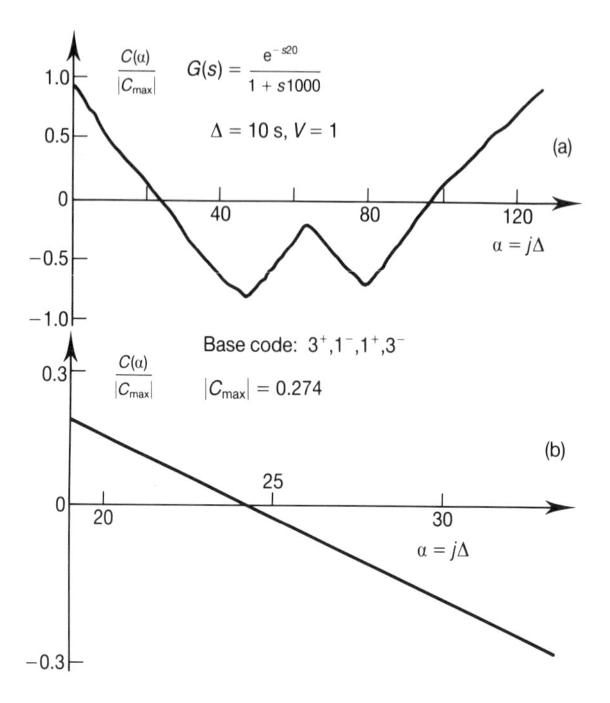

Figure 8.4 Initial value of the transient response for the Strathclyde Short Octave MBS sequence.

8.4 Removal or minimization of MBS switch-on transient

The transient response of equation (8.5) disappears when

$$C(\alpha) = 0 \tag{8.7}$$

Equation (8.7) was solved numerically. For example, let $K_s = 1$, $V = 1$, $\tau = 1000$ s, $T_D = 20$ s and apply the MBS of Van den Bos (1967) having base code 6^+, 1^-, 2, 1, 11, 6, 2, 14, 3, 1, 6, 6, 2, 3^-, as a test signal. This MBS signal, whose main power is concentrated at six angular frequencies ω_0, $2\omega_0$, $4\omega_0$, $8\omega_0$, $16\omega_0$, $32\omega_0$ has a number of bits $N = 128$, with the duration of one bit $\Delta = 10$ s. The solutions of equation (8.7) are nearby $\alpha_{opt} = 41\Delta$, 45Δ, 48Δ, 58Δ, 60Δ, 62Δ and 126Δ. Other 'nearly optimal solutions' occur nearby 46Δ, 47Δ, 59Δ, 61Δ, 86Δ. For this case Figure 8.3(a) shows the initial value of the transient response as a function of the shift $\alpha = j\Delta$, $j = 1, 2, \ldots, 128$, while Figure 8.3(b) is an enlargement of that part of Figure 8.3(a) where $\alpha = \alpha_{opt} = 41\Delta$.

For the Strathclyde Short Octave MBS Sequence from Table 7.2 having base code 3^+, 1^-, 1^+, 3^- the solutions of equation (8.7) are nearby $\alpha_{opt} = 24\Delta$, 96Δ. The main power of this signal is concentrated at three angular frequencies: ω_0, $2\omega_0$, $4\omega_0$.

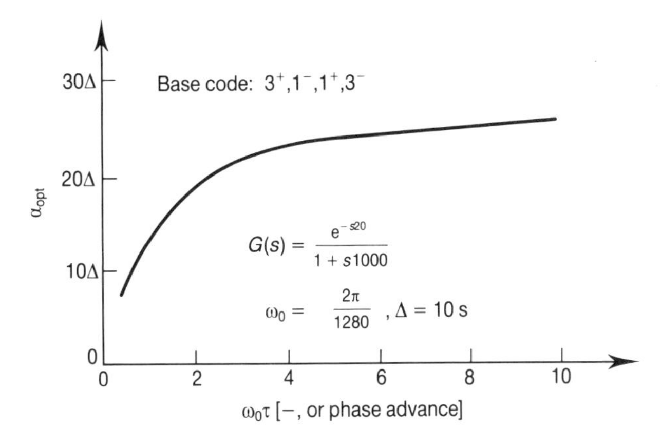

Figure 8.5 Optimal switch-on point as a function of time constant.

To allow the possibility of precisely controlling the switch-on time, the number of bits in this MBS sequence was increased to $N = 128$. For this case Figure 8.4(a) shows the initial value of the transient response as a function of the rotation $\alpha = j\Delta$, $j = 1, 2, \ldots, 128$ and Figure 8.4(b) is an enlargement of that part of Figure 8.4(a) where $\alpha = \alpha_{\text{opt}} = 24\Delta$.

Inspection of equations (8.6) and (8.7) indicates that *a priori* knowledge of the time constant, τ, is the only information about the system required to determine the optimal switching point, α_{opt}. This assumes that the time delay, T_{D}, is much smaller than the time constant τ. The accuracy of this knowledge of the time constant need not be so high. For example, in the case shown in Figure 8.4, taking $\tau = 2000$ s instead of 1000 s (100% error), the optimal value of the rotation is equal to 27Δ instead of 24Δ. Figure 8.5 shows the optimal switching point α_{opt} for different values of $\omega_0\tau$ with the same experiment time $N\Delta = 1280$ s for the model described by equation (8.1). The Strathclyde Short Octave MBS Sequence with base code $3^+, 1^-, 1^+, 3^-$ was used as a test signal.

Referring to Figures 8.3 and 8.4, with this signal the initial value of the transient response for $\alpha = 0$ is clearly indicated and much larger than with the Van den Bos MBS signal. Hence, correct rotation of any signal for minimization of the switch-on transient response effect is particularly important.

8.5 Frequency-response estimation by MBS test signal

The concept of a frequency-response function or a frequency-transfer function is fundamental in linear system theory. Frequency-response measurements with periodic test signals are generally regarded as the most powerful method available for determining non-parametric process models with high accuracy even from noisy measurements (Rake, 1980). Many methods for estimating transfer functions have been developed. These range from straightforward frequency-response analysis to spectral analysis and to various sophisticated time-domain, parametric identification methods (Ljung, 1987).

Figure 8.1 shows a resistance furnace represented as a single-input–single-output linear stationary process. Let $u(t)$ be a periodic MBS test signal, where the main power is concentrated in a few chosen frequencies. This group of test signals includes MBS test signals which are even or odd functions of time, as well as rotated MBS giving an optimal switch-on point. For estimation purposes, the furnace temperature, which is the observed output, $\hat{y}(t)$, is regarded as being composed of the true response $y(t)$ plus random disturbances $n(t)$. These disturbances, arising in the system itself or in the data-collection equipment, may be considered as random Gaussian variables with zero mean values. The dynamics of the system can be represented by its frequency response $G(j\omega)$ in the form:

$$G(j\omega) = \frac{Y(j\omega)}{U(j\omega)} = P(\omega) + jQ(\omega) \tag{8.8}$$

where $Y(j\omega)$, $U(j\omega)$ are the Fourier transforms of the output and input and $P(\omega)$, $Q(\omega)$ the real and imaginary parts of $G(j\omega)$.

The Fourier transform of the output signal is expressed as

$$Y(j\omega) \cong \int_0^{T_{\text{obs}}} y(t)\,e^{-j\omega t}\,dt \tag{8.9}$$

where $T_{\text{obs}} = MN\Delta$ (i.e. the total observation time) and M is the number of MBS equences applied to the input.

The signals $u(t)$ and $y(t)$ can be written in the form of the Fourier series:

$$u(t) = V \sum_k (a_{ku} \cos \omega_k t + b_{ku} \sin \omega_k t) \tag{8.10a}$$

$$y(t) = \sum_k (a_{ky} \cos \omega_k t + b_{ky} \sin \omega_k t) \tag{8.10b}$$

where

V	= the amplitude of MBS,
ω_k	= the angular frequency of kth harmonic ($\omega_k = k\omega_0 = 2\pi k/N\Delta$),
ω_0	= the basic angular frequency,
a_{ku}, b_{ku}	= the kth harmonic cosine and sine Fourier coefficients of the MBS, and
a_{ky}, b_{ky}	= the kth harmonic cosine and sine Fourier coefficients of the output.

It is easy to prove that the Fourier coefficients of the MBS a_{ku}, b_{ku}, calculated in advance of the experiment, have the values expressed by (Van den Bos, 1967):

$$a_{ku} = \frac{2}{\pi k} \sum_{n=1}^{N} u_n \sin \frac{\pi k}{N} \cos \frac{\pi k}{N}(2n+1) \tag{8.11a}$$

$$b_{ku} = \frac{2}{\pi k} \sum_{n=1}^{N} u_n \sin \frac{\pi k}{N} \sin \frac{\pi k}{N}(2n + 1) \tag{8.11b}$$

Similarly, the Fourier coefficients a_{ky}, b_{ky} are expressed as

$$a_{ky} = \frac{2}{T} \int_0^T y(t) \cos k\omega_0 t \, \mathrm{d}t \tag{8.12a}$$

$$b_{ky} = \frac{2}{T} \int_0^T y(t) \sin k\omega_0 t \, \mathrm{d}t \tag{8.12b}$$

Inserting relations (8.9), (8.10a) and (8.10b) into equation (8.8) and using the orthogonal properties of cosinusoids, relationship (8.8), after transformations, can be written

$$G(\mathrm{j}\omega_k) = \frac{a_{ky} - \mathrm{j}b_{ky}}{V(a_{ku} - \mathrm{j}b_{ku})} \qquad k = 1, 2, \ldots, N_{\mathrm{F}} \tag{8.13}$$

where N_{F} is the number of angular frequencies.

As measurements at the output are disturbed by Gaussian-type noise, $n(t)$, the coefficients a_{ky} and b_{ky} are, respectively, replaced by their estimators $\hat{\alpha}_{ky}$ and $\hat{\beta}_{ky}$, which are given by

$$\hat{\alpha}_{ky} = \frac{2}{T_{\mathrm{obs}}} \int_0^{T_{\mathrm{obs}}} [y(t) + n(t)] \cos k\omega_0 t \, \mathrm{d}t \tag{8.14a}$$

$$\hat{\beta}_{ky} = \frac{2}{T_{\mathrm{obs}}} \int_0^{T_{\mathrm{obs}}} [y(t) + n(t)] \sin k\omega_0 t \, \mathrm{d}t \tag{8.14b}$$

For computer processing the integral operations in equations (8.14a) and (8.14b) are replaced by numerical summation with values of $\hat{y}(t) = y(t) + n(t)$ which are samples at discrete time intervals $t = n\Delta$ ($n = 0, 1, \ldots, N - 1, \ldots, MN - 1$). These estimators are then given by

$$\hat{\alpha}_{ky} = \frac{2}{MN} \sum_{n=0}^{MN-1} [y(n\Delta) + n(\Delta)] \cos\left(\frac{2\pi}{N} kn\right) \tag{8.15a}$$

$$\hat{\beta}_{ky} = \frac{2}{MN} \sum_{n=0}^{MN-1} [y(n\Delta) + n(\Delta)] \sin\left(\frac{2\pi}{N} kn\right) \tag{8.15b}$$

Taking relationship (8.14) into account, where the coefficients a_{ky} and b_{ky} are replaced by their estimators $\hat{\alpha}_{ky}$ and $\hat{\beta}_{ky}$, the estimators $\hat{P}(\omega_k)$, $\hat{Q}(\omega_k)$ of the real and imaginary parts of $G(\mathrm{j}\omega)$ can be expressed, respectively, as

$$\hat{P}(\omega_k) = \frac{\hat{\alpha}_{ky} a_{ku} + \hat{\beta}_{ky} b_{ku}}{a_{ku}^2 + b_{ku}^2} \tag{8.16a}$$

$$\hat{Q}(\omega_k) = \frac{\hat{\alpha}_{ky} b_{ku} - \hat{\beta}_{ky} a_{ku}}{a_{ku}^2 + b_{ku}^2} \tag{8.16b}$$

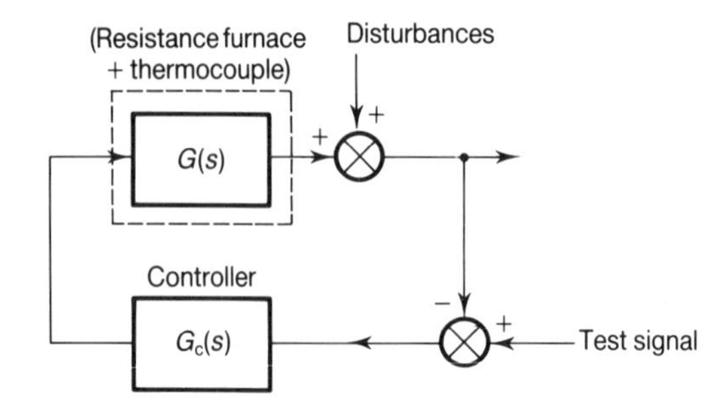

Figure 8.6 Closed-loop structure of the identified system.

This MBS estimation algorithm can be applied in an open-loop structure as well as in a closed-loop one and Figure 8.6 shows the closed-loop structure of the identified system. In this case the estimation of the open-loop frequency response is determined indirectly by a two-step algorithm. In the first step the closed-loop system is regarded as a whole, and its frequency response is identified using the MBS method. In the second the open-loop frequency response is calculated from the estimated closed-loop frequency response based upon knowledge of the controller parameters (Sankowski, 1989c).

In relationships (8.15a) and (8.15b) it can be seen that the MBS method corresponds to a kind of filtering. As the constituent operation is summation over n for every k after multiplying the output, $\hat{y}(t)$, by $\cos \omega_k t$, and $\sin \omega_k t$, respectively, the method is precisely digital correlation. Using equations (8.15), a simultaneous formation of numerical sums, for the number, N_F, of angular frequencies may be performed during the idle periods between sampling instants. By contrast, although the FFT (Fast Fourier Transform) algorithm is much quicker, a complete data record must be available before computation commences. In addition, as the MBS method requires only a limited number of computable frequency points there is no need to employ an algorithm as complex as the FFT. It has been shown that the $\hat{P}(\omega_k)$ and $\hat{Q}(\omega_k)$ estimates of the real and imaginary part of the frequency response, determined by MBS, are unbiased and consistent estimates, with variances which are inversely proportional to the product of the experiment time duration and the power concentrated at the kth frequency of the input signal (Sankowski, 1991).

8.6 Bias error of frequency-response estimates due to the trend

From the definition, bias errors (Bendat and Piersol, 1971), $\Delta\alpha_{ky}$, $\Delta\beta_{ky}$ of the $\hat{\alpha}_{ky}$, $\hat{\beta}_{ky}$ estimates of the parameters a_{ky}, b_{ky}, respectively are expressed as

$$\Delta\alpha_{ky} = E[\hat{\alpha}_{ky}] - a_{ky} \tag{8.17a}$$

$$\Delta\beta_{ky} = E[\hat{\beta}_{ky}] - b_{ky} \tag{8.17b}$$

where $E[\quad]$ is the mathematical expectation.

The observed temperature output, $\hat{y}(t)$, is regarded as being composed of the true response $y(t)$ plus random disturbances $n(t)$. Normally these disturbances are Gaussian in nature with zero mean value, so that $E[n(t)] = 0$.

Assuming that the observation time $T_{\text{obs}} = MN\Delta$, equations (8.14a) and (8.14b) can be written as

$$\begin{aligned}\hat{\alpha}_{ky} &= \frac{2}{MN\Delta}\int_0^{MN\Delta} [y(t) + n(t)] \cos \omega_k t \, dt \\ &= a_{ky} + \frac{2}{MN\Delta}\int_0^{MN\Delta} n(t) \cos \omega_k t \, dt\end{aligned} \tag{8.18a}$$

Similarly,

$$\hat{\beta}_{ky} = b_{ky} + \frac{2}{MN\Delta}\int_0^{MN\Delta} n(t) \cos \omega_k t \, dt \tag{8.18b}$$

The mathematical expectation of these estimates $\hat{\alpha}_{ky}$, $\hat{\beta}_{ky}$ can be expressed as (Sankowski, 1991)

$$E[\hat{\alpha}_{ky}] = E[a_{ky}] + \frac{2}{MN\Delta}\int_0^{MN\Delta} E[n(t)] \cos \omega_k t \, dt \tag{8.19a}$$

$$E[\hat{\beta}_{ky}] = E[b_{ky}] + \frac{2}{MN\Delta}\int_0^{MN\Delta} E[n(t)] \cos \omega_k t \, dt \tag{8.19b}$$

In the case where input disturbances are negligible, and output disturbances are Gaussian type with zero mean value, the estimates $\hat{\alpha}_{ky}$ and $\hat{\beta}_{ky}$ are unbiased.

If an electroheat system has not fully reached the thermal steady state, the trend appearing at its output causes a non-zero mean value of the disturbance corresponding to $E[n(t)] \neq 0$. For an electroheat system such as an electric resistance furnace, Sankowski (1989a) proved that this trend can be described by a linear function of time:

$$y_{\text{tr}} = E[n(t)] = b_1 t \tag{8.20}$$

where b_1 is the coefficient equal to the slope of the trend. Taking equations (8.17a) and (8.18a) into account, the bias error $\Delta\alpha_{ky}$ of the Fourier coefficient a_{ky} is expressed as

$$\Delta\alpha_{ky} = E[\hat{\alpha}_{ky}] - a_{ky} = \frac{2}{MN\Delta}\int_0^{MN\Delta} b_1 t \cos \omega_k t \, dt = 0 \tag{8.21a}$$

Similarly, for b_{ky}

$$\Delta\beta_{ky} = E[\hat{\beta}_{ky}] - b_{ky} = \frac{2}{MN\Delta}\int_0^{MN\Delta} b_1 t \sin \omega_k t \, dt = -\frac{2b_1}{\omega_k} \tag{8.21b}$$

Relationships (8.16a) and (8.16b) can be written as

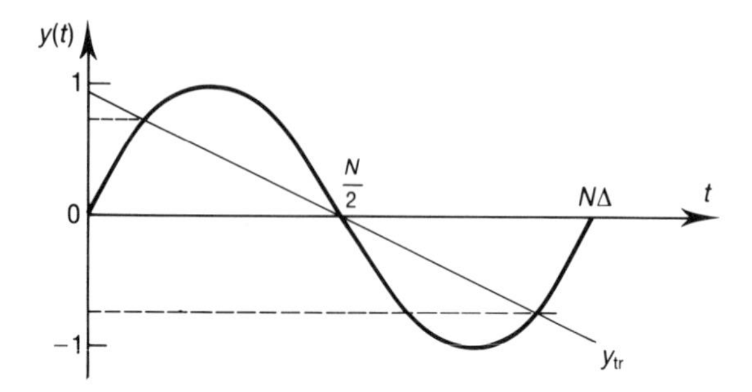

Figure 8.7 Least squares trend elimination.

$$\hat{P}(\omega_k) = \frac{(a_{ky} + \Delta\hat{\alpha}_{ky})a_{ku} + (b_{ky} + \Delta\hat{\beta}_{ky})b_{ku}}{a_{ku}^2 + b_{ku}^2} \tag{8.22a}$$

$$\hat{Q}(\omega_k) = \frac{(a_{ky} + \Delta\hat{\alpha}_{ky})b_{ku} - (b_{ky} + \Delta\hat{\beta}_{ky})a_{ku}}{a_{ku}^2 + b_{ku}^2} \tag{8.22b}$$

After transformations, these $\hat{P}(\omega_k)$ and $\hat{Q}(\omega_k)$ estimators are given by

$$\hat{P}(\omega_k) = P(\omega_k) + \Delta P(\omega_k) \tag{8.23a}$$

$$\hat{Q}(\omega_k) = Q(\omega_k) + \Delta Q(\omega_k) \tag{8.23b}$$

where the bias of each estimator has the form

$$\Delta P(\omega_k) = \frac{-2b_1}{V\omega_k} \frac{b_{ku}}{a_{ku}^2 + b_{ku}^2} \tag{8.24a}$$

$$\Delta Q(\omega_k) = \frac{2b_1}{V\omega_k} \frac{a_{ku}}{a_{ku}^2 + b_{ku}^2} \tag{8.24b}$$

For the even-function MBS signals like the Van den Bos sequence, the value of $b_{ku} = 0$ making $\Delta P(\omega_k) = 0$.

Relationships (8.24a) and (8.24b) show that the frequency-response estimates are seriously biased estimators due to the presence of output trend. As this bias is particularly present at the lowest frequencies, removal of the trend from the output or correction of the frequency response will improve the estimation. Error components, such as trend, cannot be removed by high-pass digital filtering. Hence a special trend-removal technique must be applied. The two techniques which are known (Bendat and Piersol, 1971) may be called least squares trend removal and average slope trend removal. In the former method, which is recommended by many authors (see, for example, Bendat and Piersol, 1971), incorrect results are found in those cases where test signals are periodic functions of time with zero mean value. Signals such as mono- or multi-frequent sinusoids, including MBS, belong to this group. The reason for this, explained in Sankowski (1989a), is shown in Figure 8.7.

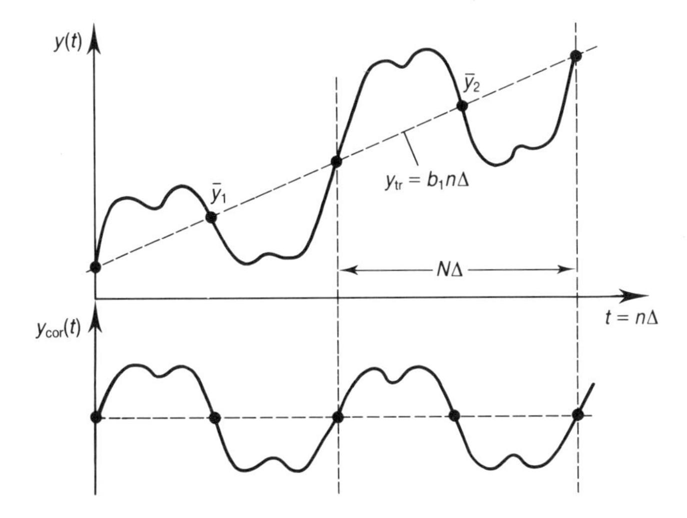

Figure 8.8 Linear trend and its elimination.

As a test signal a sinusoidal function was used. The b_1 value was computed by the least squares method, giving $b_1 \cong -0.3$. In theory, for the functions where the trend does not exist the value of b_1 should be zero. The average slope method is less complicated and gives good results. After determining the mean values $\bar{y}_1$ and $\bar{y}_2$ of the outputs for two sequence periods the value of b_1 is then calculated according to

$$b_1 = (\bar{y}_2 - \bar{y}_1)/N\Delta \tag{8.25}$$

Subsequently, corrected data are calculated using the relationship

$$y_{\text{cor}} = y(n\Delta) - b_1 n\Delta \tag{8.26}$$

The principles of this method are shown in Figure 8.8.

Correction of the frequency response is based upon knowledge of the values b_1 of the mean slope of the trend. In this chapter the mean slope b_1 is determined just after the first sequence, according to the relationship

$$b_1 = \left(\frac{1}{N} \sum_{n=1}^{N} y(n\Delta) - y(0) \right) \Big/ \frac{N}{2}\Delta \tag{8.27}$$

For the next sequence the mean slope b_1 is determined according to relationship (8.25). The values $\hat{P}(\omega_k)$ and $\hat{Q}(\omega_k)$ are determined after the first sequence and data are corrected according to equations (8.23) using (8.24). In this way the experiment time is much shorter. For further details, see Sankowski (1992).

8.7 Experimental work

Experiments were performed to identify in open-loop structure a laboratory resistance furnace of maximum working temperature 1000°C with an automatic data acquisition and processing system based upon an IBM-PC. Three different NiCr–NiAl thermocouples were used:

1. A bare thermocouple $\phi = 1.2$ mm
2. A mineral insulated thermocouple $\phi = 3.2$ mm
3. A mineral insulated thermocouple $\phi = 19$ mm.

Every combination of furnace and thermocouple can be regarded as a separate system. A block diagram of the furnace with an associated automatic data acquisition system is shown in Figure 8.9.

The furnace was allowed to reach a temperature of 800°C in almost ideal thermal steady state. For these conditions, in which the trend at the output can be neglected, the MBS Van den Bos test signal was applied for furnace identification. Figure 8.10 shows an example of this MBS input signal, $u(t)$, which corresponds to the power to the heating elements. The temperature output $y(t)$ for the NiCr–NiAl thermocouple $\phi = 3.2$ mm is also shown in this figure. It is clearly seen when the temperature starts to increase or decrease as a result of increasing or decreasing power input.

The frequency response of the furnace was determined according to equations (8.16a) and (8.16b) for different experiment times between one and ten sequences. Due to the effect of transient response, the measurements from the first two sequences

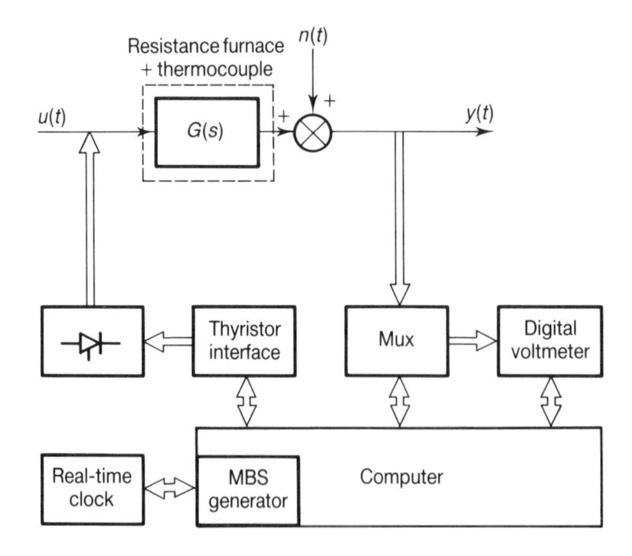

Figure 8.9 Identified furnace with automatic data-acquisition system.

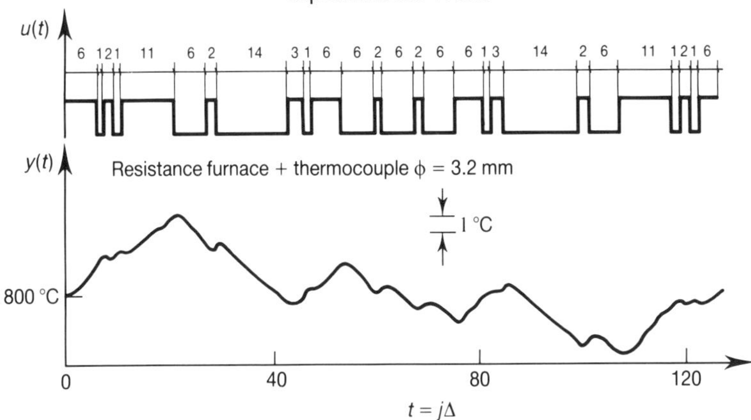

Figure 8.10 Van den Bos MBS power input test signal and output temperature of the furnace.

were not taken into account. The results for sequences 3–10 corresponding to a total of 8 sequences duration were used as reference results. Figure 8.11 shows the Bode diagram of the frequency-response estimates for this longest experiment time of 8 sequences duration using the three different thermocouples described earlier.

It is relevant to consider the level of noise in the signal. To do so, the results from different experiment times were compared with the reference results of Figure 8.9. A relative deviation $\delta(\omega)$ between reference results $G_r(j\omega)$ and these $G^{(i)}_{MBS}(j\omega)$ for different experiment times expressed in a number of sequences i is defined as (Ströbel, 1975)

$$\delta_G(\omega) = \frac{|G_r(j\omega) - G^{(i)}_{MBS}(j\omega)|}{|G_r(j\omega)|}\ 100\,\% \tag{8.28}$$

Figure 8.12 shows these relative deviations for $i = 3$, $i = 4$, $i = 3 + 4$, $i = 3 + 4 + 5$, $i = 3 + 4 + 5 + 6$ as functions of angular frequency ω.

The results for $i = 3 + 4 + 5 + 6$, which are almost the same as the reference result, give a relative deviation $\delta_G(\omega) < 3\%$ over a wide frequency range. For one sequence giving data during the period $i = 3$ the results are all good except one frequency-response point for $\omega = 32\omega_0$, where $\delta_a(\omega) > 15\%$. It is claimed that errors even as large as 15% are acceptable in electroheat systems (Sankowski, 1989b). Figure 8.13 shows the first sequence of the output signal with the transient response effect for the Van den Bos MBS and for the Strathclyde Short MBS Sequence, when a NiCr–NiAl thermocouple sensor of $\phi = 3.2$ mm was used.

From examination of Figures 8.3, 8.4 and 8.13, the transient response effect appears to be more pronounced in the case of the Strathclyde Short Octave MBS test signal. Therefore for this signal a proper choice of the optimal starting point, α_{opt}, for minimization of the transient response effect is particularly important.

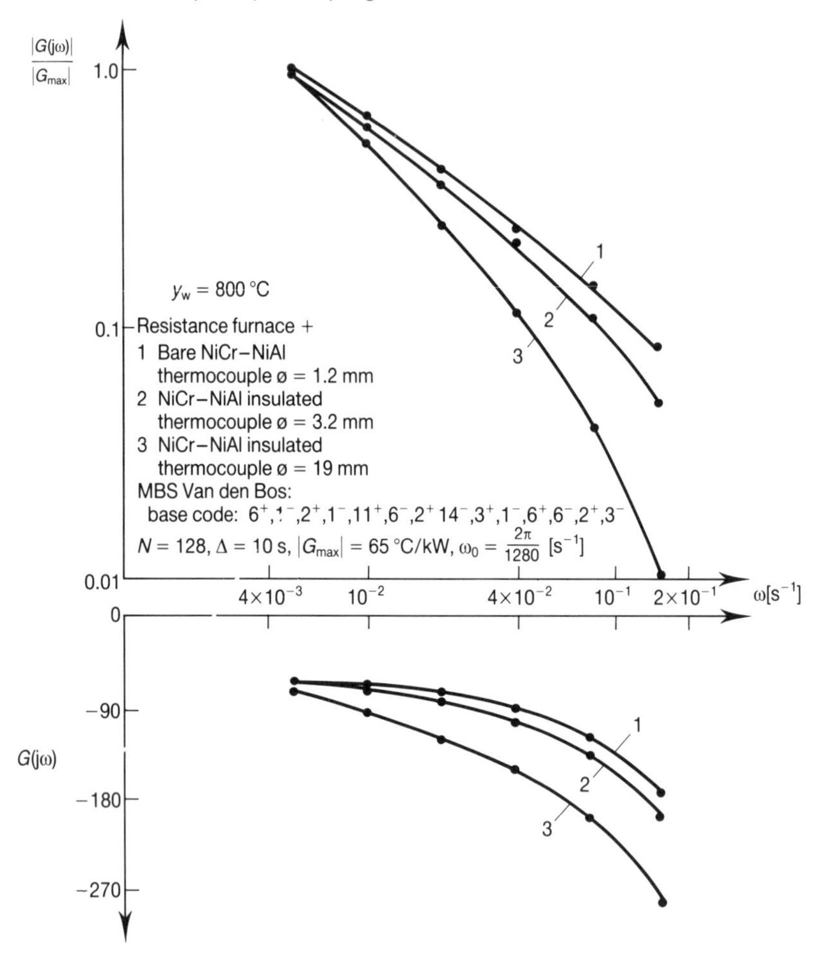

Figure 8.11 The Bode diagram of the frequency response of furnace.

To investigate the influence of trend on the accuracy of the frequency response, the furnace was studied while not fully in a thermal steady state. Under these conditions a trend appears at the furnace output. The Van den Bos MBS test signal was used to modulate the input power. After applying a correction of the frequency-response characteristics the bias error of the $\hat{P}(\omega_k)$, $\hat{Q}(\omega_k)$ estimators were calculated from the measurement. These errors, which are shown in Figure 8.14, give almost the same results as the theoretical calculations from equations (8.24a) and (8.24b).

The identification results in the form of a measured frequency response, determined from the first MBS sequence measurement with optimal switch-on time and from a third sequence, with the disappearing transient response, are compared with reference results. The trend effect was eliminated by correction of the frequency

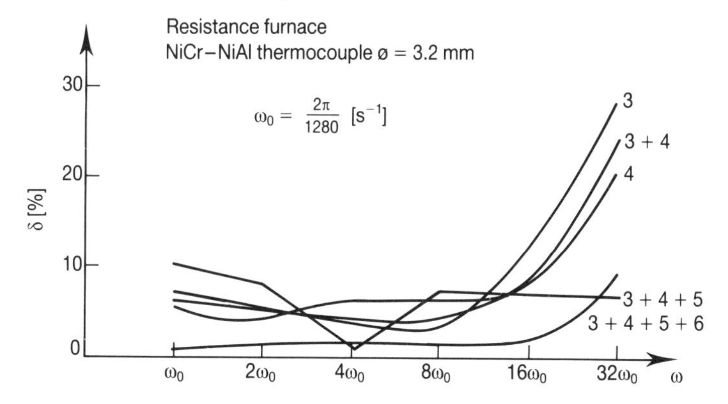

Figure 8.12 Relative deviation of the frequency-response result for different experiment times.

response in accordance with equations (8.24a) and (8.24b). As a measure of accuracy, the relative deviation expressed by equation (8.26) was used. The results, which are shown in Figure 8.15, give deviations of less than 15% over a wide frequency range with the trend removed for the first sequence and with optimal switch-on time.

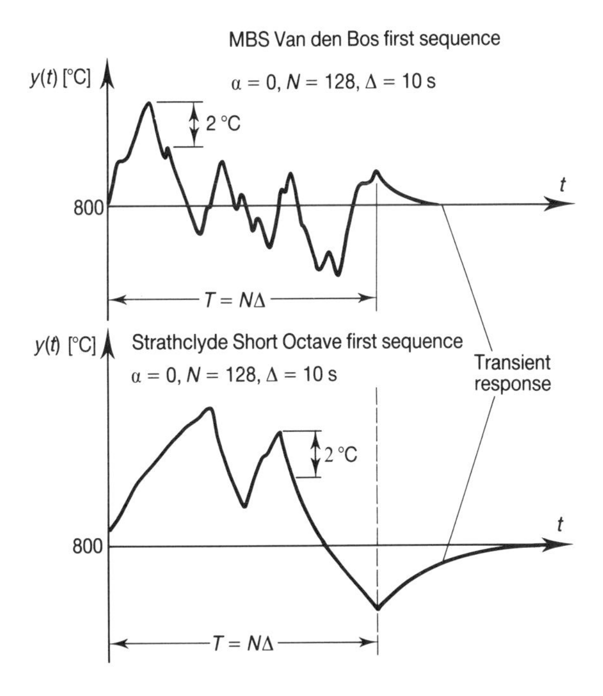

Figure 8.13 The temperature at the output with the transient response effect.

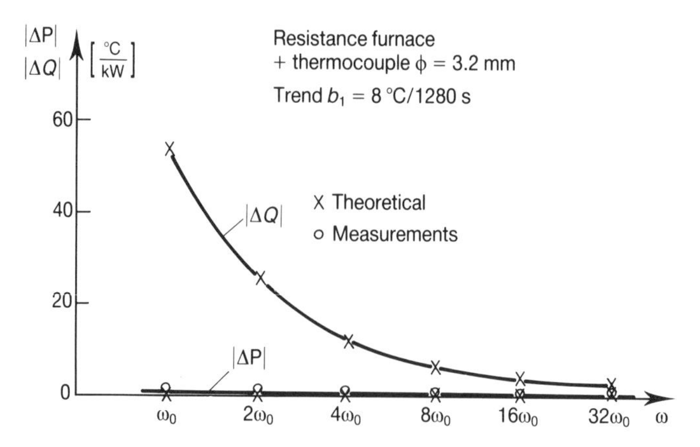

Figure 8.14 The bias error of the $\hat{P}(\omega_k)$ and $\hat{Q}(\omega_k)$ estimators.

Better results were obtained for the third sequence where the relative deviation was less than 5%. Both results are much better than in the case of the frequency response without trend elimination, where the relative deviation was more than 40% for the lowest frequency.

It will be recalled that deviations smaller than 15% are still acceptable in electroheat systems. These practical results confirmed the theoretical prediction that the experiment time can be minimized to only one sequence duration. The minimization uses a MBS test signal with an optimal switch-on point, combined with correction of the errors in the frequency-response estimate due to the presence of trend.

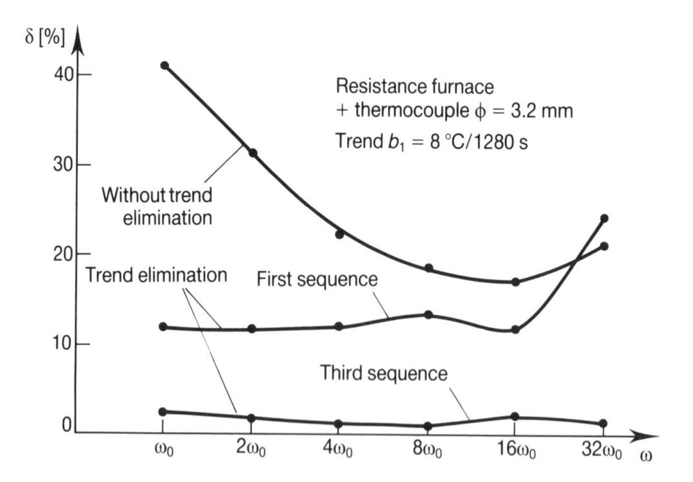

Figure 8.15 Relative deviation of frequency response with and without trend elimination; 0 = 800°C.

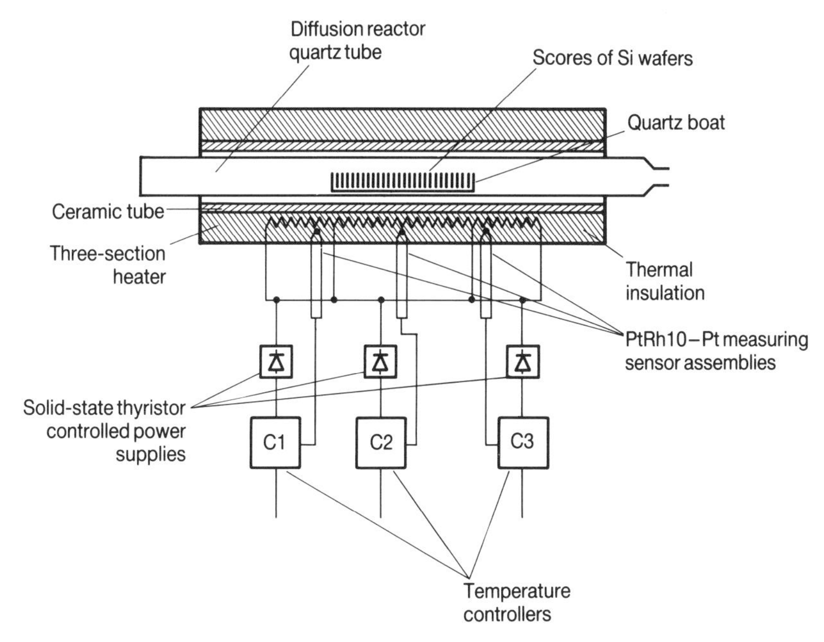

Figure 8.16 Cross-section of a three-zone diffusion furnace.

Another system of great practical importance is a diffusion furnace of the type used in the semiconductor industry. Figure 8.16 shows a typical structure for a diffusion furnace.

The input signals, u_1, and u_3, in Figure 8.17 represent the heating power input to the active thermal-insulating zones while u_2 represents that to the work zone. A set of outputs, y_1, y_2, y_3, represents the temperatures near the heater surface of each zone, respectively. *A priori* knowledge of this furnace justifies its representation as a linear stationary system for small deviations between a set-point temperature and the temperature to be measured (Sankowski, 1989a). Interactions between each of the zones and process regions of the furnace may be accounted for by representing the dynamics of this furnace by the multivariable transfer-function model given in Figure 8.17. The mathematical model is also indicated in this figure. Typical curves of MBS input signal, *R*, and output signal, *Y*, are illustrated in Figure 8.18.

Time-series representations of the MBS power input and the thermocouple output signal are obtained by a digital data-acquisition system. Subsequently, the transfer functions are computed from these data which are processed during the idling periods of the data acquisition using a DFT algorithm. A closed-loop frequency response of the work zone obtained in this manner is shown in Figure 8.19.

Open-loop frequency responses of the various forward- and cross-coupled elements of the model given in Figure 8.17 may be also computed easily. More details are given in Sankowski (1989c).

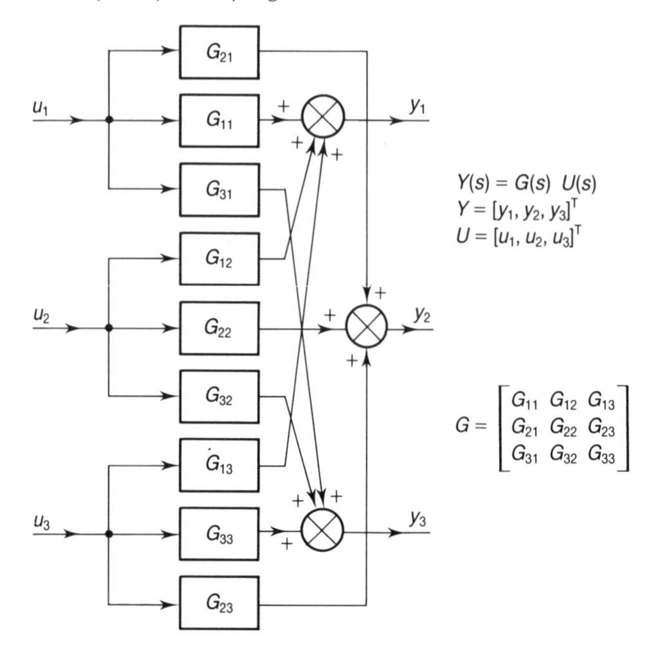

Figure 8.17 Block diagram and mathematical model of a three-zone diffusion furnace.

8.8 Conclusions

Multi-frequency binary signals constitute a versatile sub-group of the important class of binary test signals. They offer the advantages of excellent signal-to-noise power ratios for their dominant harmonics, zero offset and a wide variety of power-spectral distributions. As they are based upon step functions, which are easily generated, they are ideally suited for a microcomputer environment, particularly for the digital measurement of the frequency response of thermal and other types of control systems.

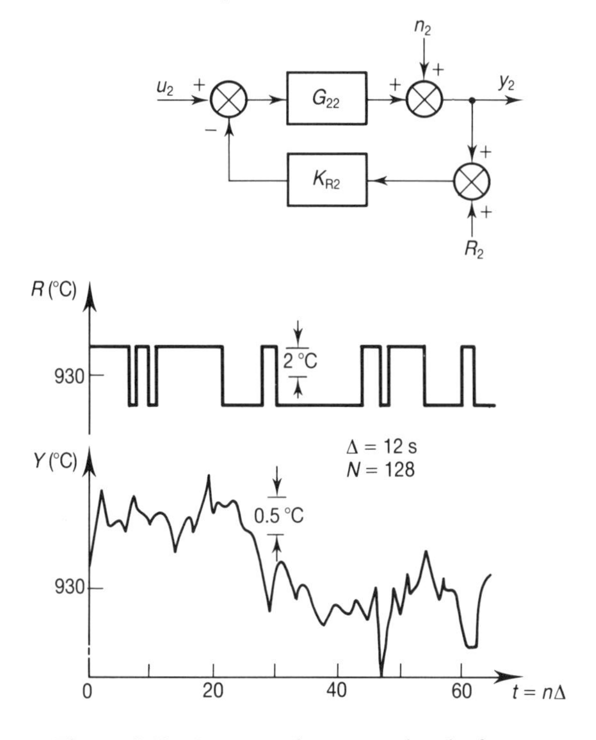

Figure 8.18 Input and output signals from a closed-loop experiment.

The MBS identification method corresponds to a kind of filtering. As the constituent operation is summation over n for every k after multiplying the output, $\hat{y}(t)$, by $\cos \omega_k t$, and $\sin \omega_k t$, respectively, the method is precisely digital correlation. Estimates of the real and imaginary parts of the frequency response determined by MBS are unbiased and consistent with variances which are inversely proportional to the product of the experiment duration and the power concentrated at the kth harmonic of the input signal.

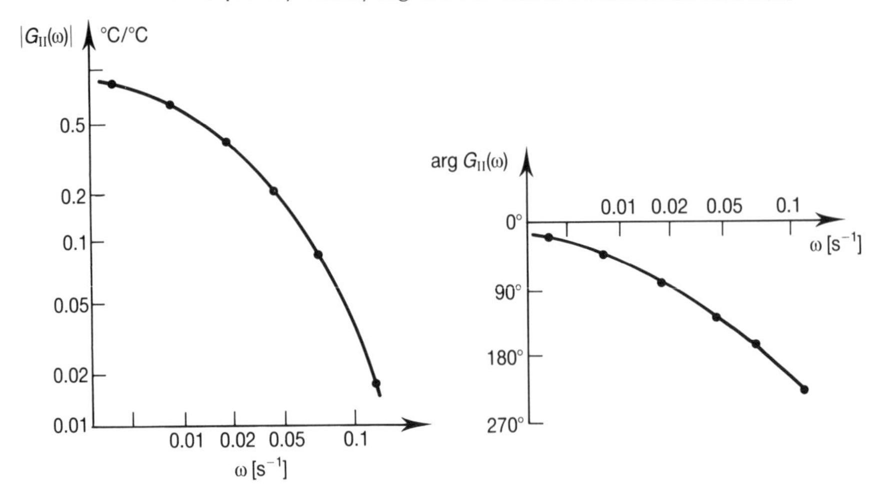

Figure 8.19 Closed-loop frequency response of the work zone.

Frequency-response estimates may be seriously biased in the presence of output trend. This effect can be minimized by the correction of frequency-response characteristics. The switch-on transient present in the output has been minimized by choosing the optimal switch-on point by rotation of the MBS input signal.

A priori knowledge of the time constant, τ, is the only information about the system required to determine the optimal switch-on point. The experiment time can be minimized to only one sequence duration using these rotated MBS, combined with correction of frequency-response characteristics against trend-induced errors.

9

Temperature Sensor Identification with Multi-frequency Binary Sequences

Joseph McGhee, Ian A. Henderson and Lidia M. Jackowska-Strumillo

9.1 Introduction

Although temperature sensors are of particular importance in temperature control (Michalski and Eckersdorf, 1987), Michalski *et al.* (1991) have asserted that they are also important in most branches of science and technology. Because of the relationship which exists between the dynamics of the temperature sensor used in measurement for temperature-control purposes, and the closed-loop performance of the temperature-control system, it is necessary to have a good knowledge of the dynamic behaviour of the temperature sensor (Michalski and Eckersdorf, 1987). Another very important aspect of temperature measurement is the accuracy of measurement. In addition to the error contributed by static inaccuracies, dynamic errors always occur due to the dynamic behaviour of the sensor employed. Hofmann (1976) has pointed out that this dynamic behaviour must be known in two main cases. The first occurs when the temperature to be measured is changing, as in temperature control, while the second is evident if a constant medium temperature is to be measured. In the second instance, knowledge of the dynamic behaviour is required so that any immersion time needed may be assessed before regarding the indicated temperature as the correct value. It is apparent that a knowledge of the sensor dynamics is essential in any temperature measuring or control application.

Design validation procedures are required during the manufacturing of temperature sensors. As dynamic behaviour is determined by the materials of construction and their manner of assembly, the dynamic behaviour of temperature sensors is both predictable in design and measurable in testing. By introducing a dynamic model for an idealized temperature sensor a lumped-parameter model for the tested sensor is given.

A theoretical basis for empirical estimation of sensor frequency response shows that the testing method is in fact frequency analysis by the correlation method. When the testing signals are deterministic this correlation estimate is shown to be identical to the periodogram estimate. Practical details of temperature sensor identification and testing are given in this chapter. A summary of the two existing groups of methods for sensor testing follows an account of empirical transfer function estimation. The testing method described in this chapter uses a test signal known as a multi-frequency binary sequence, sometimes called MBS or MFBS. Although any MBS may be used for testing purposes, the sequences used in the tests described later belong to a special group of MBS called Strathclyde MBS. These signals are described in Chapter 7. Strathclyde base-band MBS and PSKMBS signals are used. Base-band sequences have their energy concentrated in the base-band region with low-frequency components which may include the fundamental. Phase-shift-keyed (or PSK) operations as used in communications engineering may be applied using base-band MBS as the keying signal. Further tests use Strathclyde PSKMBS. Results for a typical sensor are compared with those for the same sensor enclosed by an additional plastic sleeve to model the effects of accumulated impurity layers. It is shown that Strathclyde base-band and phase-shift-keyed compact MBS are novel signals for the interrogation of temperature sensors. Future trends in sensor testing are also described.

9.2 Dynamic behaviour of contact temperature sensors

9.2.1 Frequency response of idealized temperature sensors

Temperature sensors extend the human faculties to sense hotness relations between bodies or entities in the real world. This functional extension of the human faculties is held in common with other instruments for measurement, calculation, communication and control (Henderson and McGee, 1993; McGhee, 1990; McGhee *et al.*, 1986b; McGhee and Henderson, 1989, 1991a,b). The sensor testing described in this chapter is exclusively concerned with the contacting group of sensors. Convective and conductive heat transfer characterize this group.

Although the testing methods to be described later may be applied to all kinds of electrical temperature sensors, the emphasis is upon the Resistance Thermometer Detector, or RTD.

All temperature sensors exhibit dynamic behaviour which depends upon both the types of materials and their method of construction. This behaviour is described using the laws of thermodynamics. The most general statement of these laws takes the form of non-linear partial differential equations, which express the balance of matter and the flow and storage of heat energy in time and space. Simplified versions of these equations, expressed as linear ordinary differential equations, will be used for the present purposes.

Consider what may be called an idealized temperature sensor. This is assumed to be a uniform homogeneous cylinder. It is made from material with an infinitely large thermal conductivity, k, has a mass, m, a specific heat, c, and a heat-transfer area, A. The sensor is further assumed to be totally immersed in the medium to be measured. This ensures that there is no heat exchange with any other medium at a different temperature from the measured medium. Setting up the differential equation uses the method of heat balance. At the time $t = 0^-$ (an infinitesimally small time before zero) assume that the sensor is in a steady thermal state with its temperature, ϑ_T, initially equal to the ambient temperature, ϑ_a. When $t = 0$ immerse the sensor in the medium at some temperature, $\vartheta_m > \vartheta_a$. The excess temperatures notated by Θ, henceforth simply referred to as temperature, have the values

$$\Theta_T = \vartheta_T - \vartheta_a \qquad \text{and} \qquad \Theta = \vartheta_m - \vartheta_a \tag{9.1}$$

Using Newton's law of cooling, when the sensor is immersed in the medium, it gives the heat, dQ_c, transferred to the sensor in the time dt as

$$dQ_c = \alpha A(\Theta - \Theta_T)\,dt \tag{9.2}$$

where α is the heat-transfer coefficient between the sensor and the medium and A is the heat-exchange area. The hcat stored in the sensor is

$$dQ_s = mc\,d\Theta_T \tag{9.3}$$

where m and c are the mass and specific heat of the sensor. Using this relation in equation (9.2) and rearranging leads to

$$\frac{mc}{\alpha A}\frac{d\Theta_T}{dt} + \Theta_T = N_T\frac{d\Theta_T}{dt} + \Theta_T = \Theta \tag{9.4}$$

where $N_T = mc/\alpha A$ is defined as the time constant of the sensor. Relation (9.4) is the differential equation for an idealized temperature sensor.

In most physical systems, dynamic behaviour is represented in the time domain by an impulse-response function, $h(t)$, in the s-domain by a Laplace transfer function, $G(s)$, and in the frequency domain by a frequency response, $G(j\omega)$. Because it is not possible to generate impulses in the thermal domain, an exclusively used time-domain representation of a temperature sensor is its step response. The frequency response, $F_T(j\omega) = \Theta_T(j\omega)\,/\,\Theta(j\omega)$, of the sensor's thermal dynamics, also known as the thermal conversion stage, is almost exclusively used in practice. This frequency response should not be confused with that of an electrical sensor given later as $G(j\omega) = v(j\omega)/\Theta(j\omega)$.

Hence, the importance of frequency-response analysis and estimation in temperature sensor testing emphasizes the relevance of MBS identification methods in this field. This thermal frequency response may be represented in the Cartesian and polar forms as

$$F_T(j\omega) = (1 + j\omega N_T)^{-1} \tag{9.5}$$

$$= Re[F_T(j\omega)] + jIm[F_T(j\omega)] = P(\omega) + jQ(\omega) \tag{9.6}$$

$$= |F_T(j\omega)| \exp[-j\phi(\omega)] \tag{9.7}$$

$$= [P^2(\omega) + Q^2(\omega)]^{1/2} \exp[-j \arctan(Q(\omega)/P(\omega))] \tag{9.8}$$

Sinusoidal sensor response, given by Michalski *et al.* (1991), is important. As the excitation used in the tests described later are periodic binary signals, they are of special interest. When an idealized temperature sensor is subjected to these input signals their typical responses are as shown in Figure 9.1. It can be seen that the fundamental component of the response to a squarewave has an amplitude and phase of

$$\Delta\vartheta_{T,1} = (\Delta\vartheta 4/\pi)[1 + (\omega_1 N_T)^2]^{-1/2} \quad \text{and} \quad \phi_1 = -\arctan(\omega_1 N_T) \tag{9.9}$$

Multi-frequency binary signals have their energy concentrated in only a small specific number of their harmonics. This type of excitation may be approximately written as

$$\Delta\vartheta(t) = \sum_{k=1}^{m} \Delta\vartheta_k \sin[k\omega_1 + \phi_k(\omega)] \tag{9.10}$$

Since the sensor is linear, superposition may be used to obtain the amplitude and phase of the response to each harmonic frequency, $\omega_k = k\omega_1$. Taking this into account, the composite response is

$$\Delta\vartheta_T(t) = \sum_{k=1}^{m} \Delta\vartheta_{T,k} \sin[k\omega_1 + \phi_{T,k}(\omega)] \tag{9.11}$$

where the amplitude and phase of each harmonic is

$$\Delta\vartheta_{T,k} = \Delta\vartheta_k[1 + (\omega_k N_T)^2]^{-1/2} \quad \text{and} \quad \phi_{T,k}(\omega) = \phi_k(\omega) - \arctan(\omega_k N_T) \tag{9.12}$$

9.2.2 Response and modelling of real temperature sensors

The idealized temperature sensor is a convenient means of describing the basic principles of temperature sensor dynamic behaviour. In contrast, real sensors exhibit dynamic behaviour which is different in a number of ways due to variations in structure and heat transfer conditions. For example, the assumption that real sensor material has infinite thermal conductivity means that the temperature throughout its whole mass is uniform, which rarely occurs in real sensors. The large majority of cylindrically structured, real sensors may belong to one of the three main groups

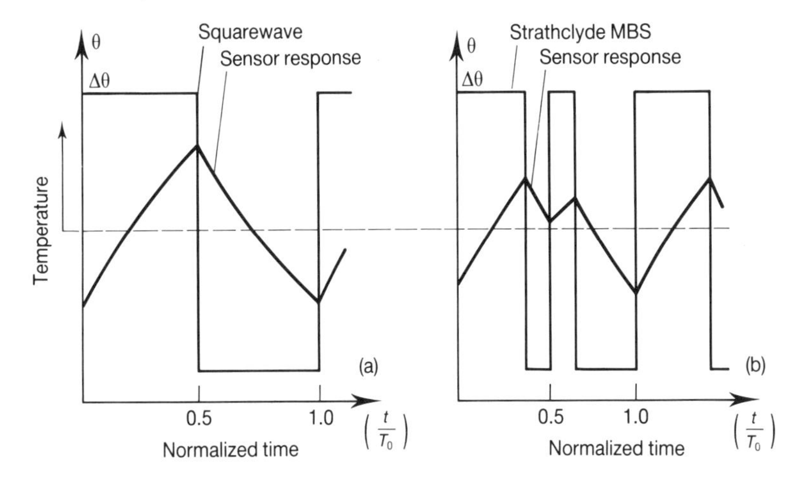

Figure 9.1 Response of an idealized temperature sensor to binary test signals. (a) Mono-frequent, (b) multi-frequent.

described by Lieneweg (1975). This classification is based upon the ratio $t_{0.9}/t_{0.5}$, where $t_{0.9}$ and $t_{0.5}$ are, respectively, the 90% and 50% rise times of the step transient response of the sensor. When deriving the differential equation (9.4) describing ideal sensor behaviour it was assumed that all effects were linear. This assumption cannot be justified for real sensors in any environment because the total heat-transfer coefficient contains components due to conductive, convective and radiative heat transfer (Hackforth, 1960; Michalski, 1966; Michalski *et al.*, 1991).

In electrical contact temperature sensors the physical thermal variable is transduced to an electrical signal. This allows an electrical temperature sensor to be modelled in terms of a cascaded combination of a thermal conversion stage and an electrical conversion stage. It is usually assumed that the dynamic behaviour is totally due to the transformation of the measured temperature, ϑ, into the sensor's temperature, ϑ_T, by the flow of heat in the thermal conversion stage. This sensor temperature is then converted into the electrical output signal, $v(t)$, in a purely static electrical conversion stage as shown in Figure 9.2 (McGhee *et al.*, 1992a,b; Michalski *et al.*, 1991).

Three main methods of sensor modelling have been developed. The first employs calculation, the second uses testing experimentation and the third combines calculation and experimentation.

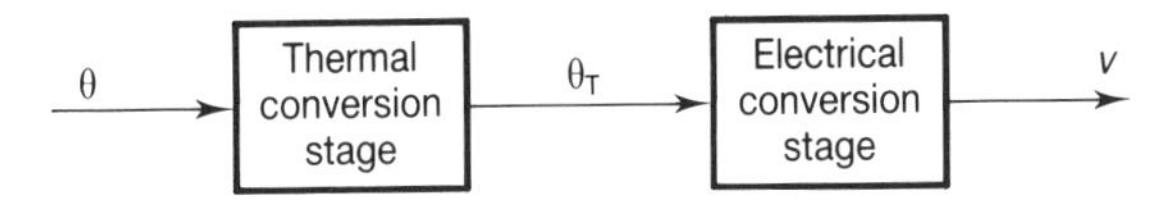

Figure 9.2 Block diagram of a real, linearized, contacting temperature sensor.

As the equations for the calculative model are the basis of the lumped model of the sensor which is tested in this chapter the briefest of outlines of only this method is necessary. In calculative modelling of temperature sensor heat flow dynamics the most important law used in the description of sensor dynamic behaviour is the Fourier diffusion equation with boundary conditions of the IIIrd kind. For the most important homogeneous or multi-layered sensor structures (Hofmann, 1976; Jakob, 1957, 1958; McGhee *et al.*, 1992a,b; Michalski *et al.*, 1991) this equation may be written in cylindrical co-ordinates as

$$\frac{\partial \Theta}{\partial t} = \frac{k}{\rho c}\left(\frac{\partial^2 \Theta}{\partial r^2} + \frac{1}{r}\cdot\frac{\partial \Theta}{\partial r}\right) + \frac{Q}{\rho c} \tag{9.13}$$

where Θ, t, r, k, ρ, c and Q are temperature, time, radial dimension, thermal conductivity, specific density, specific heat and internal heat source, respectively. The boundary conditions are

$$\frac{\partial \Theta(R,\ t)}{\partial t} = -\frac{\alpha}{k}\left[\Theta(R, t) - \Theta_e\right];\ \frac{\partial \Theta(0, t)}{\partial r} = 0 \tag{9.14}$$

and the initial condition is

$$\Theta(r,\ 0) = 0 \tag{9.15}$$

The above theory can also be applied to more complex multi-layer structures where boundary conditions between each layer are also required. The most commonly occurring boundary condition, taking account of the boundary condition between the sensor well and the fluid, is Newton's law of cooling, described by

$$-k\frac{\partial \Theta}{\partial t} = h[\Theta(r_0) - \vartheta] \tag{9.16}$$

where r_0 and θ are the external radius of the sheath and the fluid temperature, respectively, while h is the heat-transfer coefficient of the boundary film.

In the development of physical models of electrical temperature sensors a number of important assumptions must be made. First, it is assumed that dynamic behaviour is exclusively due to the flow and storage of heat in the thermal conversion stage. Also, as real sensors exhibit non-linear behaviour, it is necessary to make linearizing assumptions. For these purposes, it is assumed that materials which have unilateral thermal properties exhibit behaviour that is independent of the direction of heat flow. Mean values of material properties describe this assumed bilateral heat flow behaviour. In addition, the properties are assumed to be constant within the temperature range of interest. As heat flow is essentially diffusive in nature, distributed-parameter models should be regarded as most representative for temperature sensor modelling. To permit realistic and widespread use of models, these distributed-parameter models are approximated by lumped-parameter equivalents.

Lumped-parameter models with linearized parameters are based upon the translation of equations (9.13)–(9.16) using the electrothermal analogy. This allows

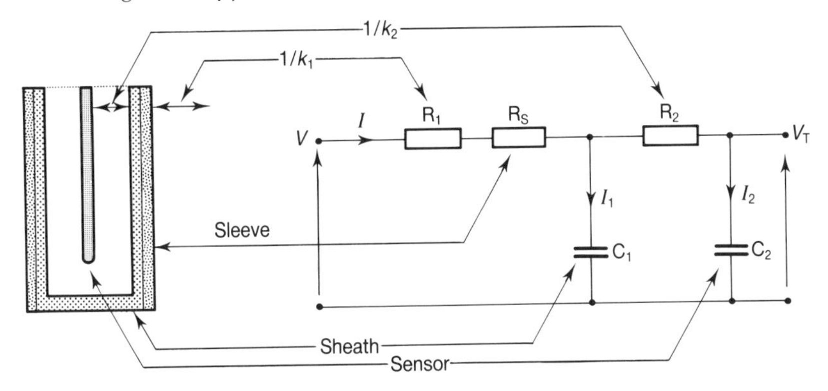

Figure 9.3 Lumped-parameter dynamic model for a laboratory-scale RTD.

multi-layer sensors to be modelled using lumped circuits consisting of capacitances, representing the storage of heat in the constructional materials of each layer, and resistances, which represent the thermal resistance at each boundary. Thirteen typical electrical temperature sensor structures (McGhee *et al.*, 1992a,b; Michalski *et al.*, 1991) may be represented by one of the five most commonly used equivalent thermal transfer functions. These take the form of a single or double thermal lag with or without pure time delay.

The sensor considered in this chapter is a small laboratory-grade Pt-100 Ω RTD mounted inside a stainless steel sheath. The sheath, which gives the probe an operating temperature range of up to 300°C, has a diameter of 2 mm and an active length of 15.9 mm as shown in Figure 9.3. A plastic sleeve with a diameter of 3 mm was used to form an auxiliary sheath. The sleeve, modelled by the thermal resistance, R_s, was assumed to be a good representation of any increase in sheath resistance due to the accumulation of impurity layers. The steel sheath, modelled by a thermal capacitance, C_1, encloses the RTD sensor with its thermal capacitance, C_2. A resistance, R_2, models the sheath. Newton cooling at the boundary between the sleeve/sheath and surface fluid film is represented by the resistor R_1.

9.3 Design and application of multi-frequency binary sequences

9.3.1 Status and trends in MBS signal design

McGhee *et al.* (1991b) have proposed a composite definition of identification which takes account of many different requirements, and is stated as:

Identification is a determination of an optimal model for a system from a certain class of permissible models using specific criteria, which are based upon a combination of *a priori* knowledge and measurement data from an identification channel.

Non-parametric modelling is a sub-division of identification which obtains non-parametric models in either the time or the frequency domains. Passive identification uses normal system-operating records as the basis of model evaluation whereas active identification interrogates the system under consideration. Analysis of the stimulation resulting from this interrogation reveals the model. The question then arises as to what form of test signal will satisfy the optimality criterion in the definition given above for identification. As there is no universally optimal signal, this question does not have a concise answer. However, it appears that satisfactory identification signals will provide suitable models from experiments which are as short as possible. This important point should be borne in mind for later purposes.

Multi-frequency binary sequences are candidate signals which can satisfy the definition for optimal test signals in frequency-domain identification. A multi-frequency binary sequence may be regarded as a hard-limited binary approximation to a non-binary multi-frequency signal. However, this approximating binary signal will generally exhibit large differences between its harmonic amplitudes and those of the non-binary signal it purports to approximate. This occurs even when the original non-binary signal has an even distribution of its harmonic power. Initial designs of MBS signals due to Jensen (1959) correspond to hard-limiting the non-binary multi-frequency signals of Levin (1959). These codes were used in Ager-Hansen (1962).

MBS signals with approximately equal power distribution were designed by Van den Bos (1967, 1974). Because of the important results obtained it will be beneficial to give a brief review of the design procedure. The problem to be solved in tailoring the design of MBS signals is to find the signal $u(t) = u^0(t)$ which maximizes the design index:

$$I_1 = \sum_{k=1}^{m} (\lambda_k |a_k| + \mu_k |b_k|) \tag{9.17a}$$

where λ_k and μ_k are weighting factors (>0) and a_k and b_k are the Fourier coefficients of the kth harmonic of $u(t)$, given by

$$a_k = \frac{2}{T}\int_0^T u(t)\cos\left(2\pi\frac{kt}{T}\right)\mathrm{d}t \tag{9.17b}$$

$$b_k = \frac{2}{T}\int_0^T u(t)\sin\left(2\pi\frac{kt}{T}\right)\mathrm{d}t \tag{9.17c}$$

It has been shown (Van den Bos, 1967) that the family of signals maximizing the power in a binary signal is given by

$$u^s(t) = \mathrm{signum}[f(t)] \tag{9.18}$$

where $f(t)$ is the multi-frequency signal:

$$f(t) = \sum_{k=1}^{m} \left[\pm \lambda_k \cos\left(2\pi k \frac{t}{T} \right) \pm \mu_k \sin\left(2\pi k \frac{t}{T} \right) \right] \tag{9.19}$$

In general, it is desirable to space the harmonics of a signal for frequency-response measurement evenly on a logarithmic frequency scale. For this reason, Van den Bos (1967) utilized his theoretical results to construct signals with octave spacing in the frequency domain. Selection of appropriate MBS codes used an algorithm which conducted a random search on an existing sequence. This involved the evaluation of the power in each harmonic of interest. A comparison of the actual power in each harmonic against a desired, prespecified value for each harmonic was performed. Such a comparison searched for those amplitudes which minimized the optimizing criterion

$$I_2 = \sum_{k=1}^{m} | P_{k\mathrm{d}} - P_{k\mathrm{a}} |^2 \tag{9.20}$$

where $P_{k\mathrm{d}}$ is the desired power and $P_{k\mathrm{a}}$ is the actual power of harmonic k for the trial sequence. More complete details of the algorithm are given in Van den Bos and Krol (1979).

Anti-symmetric MBS signals were designed at the Oak Ridge National Laboratory, USA. These signals are better for measuring frequency response in the presence of non-linearities (Buckner, 1970; Buckner and Kerlin, 1972) because their second half-period is the negative of the first half-period. It has been shown (Buckner, 1970) that anti-symmetric signals remove the even-numbered terms from a Volterra functional expansion for non-linear system representation. In other respects, the design approach did not differ significantly from the approach of Van den Bos (1967, 1974).

Octave linearly spaced harmonics are not always the best choice for a multi-harmonic test signal. Many situations occur in the non-parametric identification of higher-order systems, systems with resonances and controlled closed-loop processes, where a closer clustering of the harmonics of the test signal would be beneficial. Design of other MBS signals, showing this closer clustering of harmonics, was stimulated by this requirement (Harris and Mellichamp, 1980). The resulting signals are referred to as Harris MBS.

Another type of clustering is exhibited by sequences described in Chapter 7. They are generated using the principles of phase-shift keying to give what are called Strathclyde PSKMBS.

An important consideration in the choice of a test signal is the length of the experiment which is required to give results of a specified accuracy. Sequences have been designed to achieve the shortest possible experiment time for a specified error (Paehlike, 1980; Paehlike and Rake, 1979). Ströbel (1975) defines the RMS relative error, $\varepsilon_{\mathrm{C}}(\omega_k)$, as

$$\varepsilon_{\mathrm{C}}(\omega_k) = \left\{ E\left[\frac{| [\hat{G}_{\mathrm{N}}(\mathrm{j}\omega_k) - \hat{\hat{G}}_{\mathrm{NC}}(\mathrm{j}\omega_k)] |^2}{| \hat{G}_{\mathrm{N}}(\mathrm{j}\omega_k) |^2} \right] \right\}^{1/2} \tag{9.21}$$

where $\hat{G}_N(j\omega_k)$ is a finite record estimate of the frequency response and $\hat{\hat{G}}_{NC}(j\omega_k)$ a measured correlation analysis estimate of $\hat{G}_N(j\omega_k)$. With noise spectrum, $S(\omega_k)$, it is known (Paehlike, 1980; Paehlike and Rake, 1979) that the measuring time, T_m, for the response to harmonic A_{kr} is

$$T_m(A_{kr}) = \frac{2^2}{|A_{kr}|^2} \cdot \frac{S(\omega_k)}{[\,|\hat{G}_N(j\omega_k)|\,\varepsilon_C(\omega_k)]^2} \tag{9.22}$$

As all the points of the frequency response are obtained from the same experiment, the measuring time is selected as the largest of all measuring times. A library of some 40 sequences resulted from this work (Paehlike, 1980).

All the sequences previously described were designed as a result of what may be called 'bespoke' or 'made-to-measure' approaches. That is, the code of the sequence is tailored to meet the specification of harmonic content. Another approach could examine the characteristics of MBS codes which actually exist, so that suitable 'off-the-peg' sequences could be used. This examination of MBS sequences started in the Industrial Control Centre, University of Strathclyde, in the mid-1980s. The proposed design approach commenced from an adaptation of the source- and channel-coding theorems from the theory of information due to Shannon and Weaver (1972). As more details appear in Chapter 7 this design approach will not be considered further.

9.3.2 Applications of MBS test signals

The sequences developed by Jensen (1959) were applied in the dynamic testing of nuclear reactors at the Halden Heavy Water Boiling Reactor (Ager-Hansen, 1962). Standard control rods were used to introduce MBS reactivity inputs. Dynamic testing at the Molten Salt Reactor Experiment (Buckner and Kerlin, 1972), the High-Flux Isotope Reactor (Chen *et al*., 1972) and the Fast Flux Test Facility (Harris *et al*., 1989) used MBS test signals designed by Buckner (1970). Kerlin (1974) reviews reactor frequency-response testing.

Experimental measurement of the frequency response of process systems has been obtained by Harris and Mellichamp (1980). A stirred tank system was the first system studied. This consisted of two cylindrical constant-volume tanks connected in series by a transfer line which introduced a large delay into the process. The tanks also had water flowrate and temperature controllers fitted. For the purposes of identification, the transfer function between heat input and water temperature was examined. Evaluation tests were also performed on a liquid-level system. Polymer process extruders have also been identified (Berardino, 1983; Bezanson and Harris, 1986).

Successful applications of MBS in electromechanical systems have been reported. Henderson and McGhee (1989a, 1990a) have used Strathclyde PSKMBS to identify the closed-loop magnification of a d.c. servomechanism. PSKMBS are signals which give very precise control of the clustering of their harmonic content using a signal design analogy with optical zooming. Although they meet some of the requirements for close clustering referred to above, they belong to the class of modulated MBS. Electrical drives form an important group of electromechanical systems which have been studied by Paehlike (1980) and Paehlike and Rake (1979).

Identification of temperature and thermal systems is important. Thermal processes in the form of a cross-flow heat exchanger (Franck and Rake, 1985), heating and ventilating systems (Haghighat, 1988) and a warm-air system (Henderson and McGhee, 1989b) were identified. Electrical resistance furnaces used in the semiconductor and porcelain industries have been extensively treated by Sankowski (1989a–d). A review of thermal systems by McGhee *et al.* (1991b) also includes references. Another important application of MBS test signals to temperature sensor identification is the theme of this chapter.

9.4 Data measurement for temperature sensor testing and identification

9.4.1 Theoretical basis for MBS identification of temperature sensors

LEAST SQUARES MODELLING OF TEMPERATURE SENSORS

Measuring the step response or frequency response of real electrical contact temperature sensors is non-parametric identification. This method, which may entail obtaining estimates of behaviour in either the time or frequency domains, depends upon least squares methods for its theoretical basis.

Consider the situation illustrated in Figure 9.4. A real sensor, $\mathscr{S}$, is to be modelled by a linear constant coefficient model sensor, $\mathscr{M}$. Test signal generation and signal processing are included in the information-handling sub-system, $\mathscr{I}$. It is only realistic to process information from finite-duration records. For minimum $\overline{e^2}$ between the response of the real sensor and its linear model the theory of least squares estimation (Jenkins and Watts, 1969) shows that the sensor, $\mathscr{S}$, may be modelled by a linear impulse-response function, $\hat{h}(t)$, which characterizes $\mathscr{S}$. This impulse response is obtained from the finite record Wiener–Hopf equation as

$$\hat{R}_{\vartheta v}(\tau) = \int_0^{\infty} \hat{R}_{\vartheta\vartheta}(\tau - \lambda)\hat{h}(\lambda)\, \mathrm{d}\lambda, (\lambda > 0) \tag{9.23a}$$

where $\hat{R}_{\vartheta\vartheta}$ and $\hat{R}_{\vartheta v}$ are, respectively, finite record estimates of the auto- and cross-covariance functions between the input temperature signal, $\vartheta(t)$, and its evoked electrical response, $v(t)$.

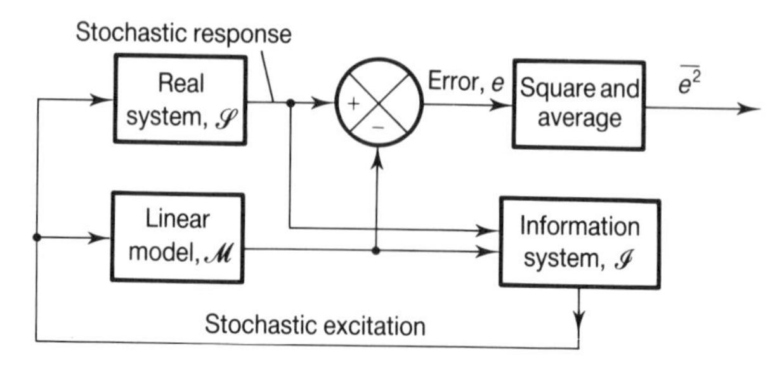

Figure 9.4 Least squares modelling of temperature sensors.

If the auto-covariance function (or ACF) of the input temperature, $\hat{R}_{\vartheta\vartheta}(\tau - \lambda)$, of equation (9.23a) is an impulse, $\delta(\tau - \lambda)$, expressed as

$$R_{\vartheta\vartheta}(\tau - \lambda) = \delta(\tau - \lambda) \tag{9.23b}$$

then equation (9.23a) becomes

$$\hat{R}_{\vartheta v}(\tau) = \hat{h}(\tau) \tag{9.24}$$

Note that impulses of thermal energy are not used in temperature sensor testing. Applying the sifting properties of the delta function in equation (9.23a) shows that $\hat{h}(t)$ may be obtained merely by acquiring the cross-covariance function (CCF), $\hat{R}_{\vartheta v}(\tau)$, between the input signal, $\vartheta(t)$, and response, $v(t)$.

The above approach, which corresponds to a time-domain solution of the finite-record Wiener–Hopf equation, requires a search for signals which exhibit impulsive ACFs. Stationary stochastic signals possess this property. As they cannot be generated in thermal systems, more easily generated deterministic test signals possessing as near to an impulsive ACF as possible are of significant interest. It is for this reason that the PRBS signal, with an ACF approximating an impulse, is so important and has been so popular. Since the sensor is described by its time-domain impulse response the technique may be referred to as non-parametric time-domain identification by the correlation method.

The one-to-one correspondence between impulse response in the time domain and frequency response means that the measurement of frequency response by data measurement is an alternative method for solving the finite-record Wiener–Hopf equation. In the past, single frequency (or mono-frequent) sinusoidal signals were exclusively used for system testing to obtain frequency-response estimates (Rake, 1980). This method also uses correlation techniques (Godfrey, 1980). The sensor under test reaches a steady thermal state in response to a sinusoidal excitation after any switch-on transient. When transients have died away, the effect of any noise is averaged out by using correlation analysis. Hence, this method is called non-parametric frequency-domain analysis by the correlation method.

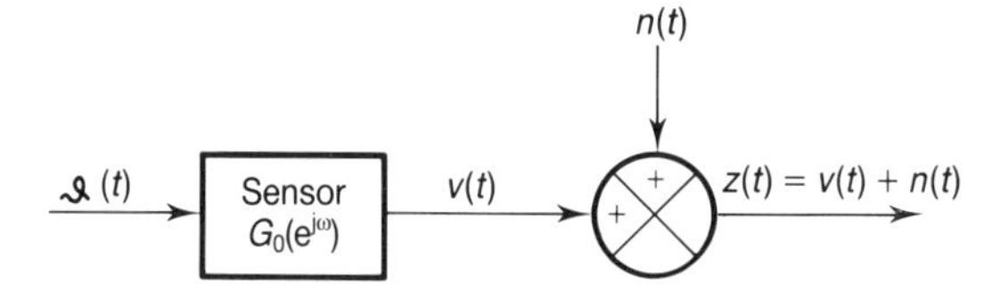

Figure 9.5 Signal flow diagram for empirical estimation of temperature sensor frequency response.

9.4.2 Empirical estimation of sensor frequency response

Consider the real-time-invariant electrical temperature sensor of Figure 9.5. The measured signal $z(t)$ is the output, $v(t)$, contaminated by the noise, $n(t)$. Assuming discrete time for simplicity, the signal is

$$z(t) = \sum_{k=1}^{\infty} g_0(k)\vartheta(t-k) + n(t)$$

$$\text{(with discrete } t = 1, 2, 3, \ldots, \text{etc.)} \tag{9.25}$$

This sensor has a frequency response, G_0, given by

$$G_0(\mathrm{e}^{\mathrm{j}\omega}) = \sum_{k=1}^{\infty} g_0(k)\,\mathrm{e}^{\mathrm{j}\omega k} \tag{9.26}$$

The noise, $n(t)$, is assumed to be zero mean, pure Gaussian and with the power-spectral density

$$S_\mathrm{n}(f) = \mathscr{F}[R_\mathrm{n}(\tau)] = \mathscr{F}\{E[n(t+\tau)n(t)]\} \tag{9.27}$$

where $R_\mathrm{n}(\tau)$ is the auto-covariance of $n(t)$, $\mathscr{F}$ is the operation of taking Fourier transforms and E is the mathematical expectation.

Estimating the frequency response, G_0, of the sensor by interrogation with periodic signals is the aim of empirical frequency-response estimation. This estimation problem is summarized in the following notation. Generate a totally deterministic periodic temperature signal, $\vartheta^N : \vartheta(k)\ k = 1, 2, 3, \ldots$, etc. Apply it to the sensor of Figure 9.5 and observe the evoked output signal $z^N : z(k)\ \ k = 1, 2, 3, \ldots$, etc. From these N completely known values of ϑ and the N observed values of z, form an estimate of the frequency-response function

$$\hat{G}_\mathrm{N}(\mathrm{e}^{\mathrm{j}\omega}) = G(\mathrm{e}^{\mathrm{j}\omega}; N, z^N; \vartheta^N) \tag{9.28}$$

The estimating experiment should be designed to ensure that $\hat{G}_\mathrm{N}$ is as close to the true transfer function, G_0, as possible.

FREQUENCY RESPONSE BY CORRELATION

One estimate for $\hat{G}_N(j\omega_p)$ can be formed by taking the ratio of the Fourier transforms of estimators of $v(t)$ and $\vartheta(t)$. This estimate, $\hat{\hat{G}}_{NC}(j\omega_p)$ (which can be called the empirical frequency-response estimate by correlation analysis) is given in terms of estimators $\hat{a}_{pv}$ and $\hat{B}_{pv}$ of the Fourier coefficients of the pth harmonic of $v(t)$ as

$$\hat{\hat{G}}_{NC}(j\omega_p) = \frac{\frac{1}{2}(\hat{a}_{pv} - j\hat{b}_{pv})}{\frac{1}{2}(a_{p\vartheta} - jb_{p\vartheta})} = \frac{Z_{NC}(\omega_p)}{\Theta_N(\omega_p)} \tag{9.29}$$

It has been assumed in equation (9.29) that the input temperature is an MBS signal with the equivalent multi-frequency representation

$$\vartheta(t) = \sum_{p=1}^{P} \left\{ a_{p\vartheta} \cos\left(2\pi p \frac{t}{T}\right) + b_{p\vartheta} \sin\left(2\pi p \frac{t}{T}\right) \right\} \tag{9.30}$$

in which some of the $a_{p\vartheta}$ and $b_{p\vartheta}$ may be zero. Since it is also assumed that the sensor is linear, the estimators $\hat{a}_{pv} = a_{pz}$ and $\hat{b}_{pv} = b_{pz}$ are

$$\left.\begin{matrix} a_{pz} = \hat{a}_{pv} \\ b_{pz} = \hat{b}_{pv} \end{matrix}\right\} = \frac{2}{T_{obs}} \int_0^{T_{obs}} z(t) \left\{ \begin{matrix} \cos\left(2\pi p \frac{t}{T}\right) \\ \sin\left(2\pi p \frac{t}{T}\right) \end{matrix} \right\} dt \tag{9.31}$$

The convergence and consistency conditions for estimates of periodic coefficients in noise which are derived in Chapter 8 apply equally well to the estimators in equation (9.31). Hence, the empirical frequency-response estimate, $\hat{\hat{G}}_{NC}(j\omega_p)$, obtained by the correlation method is an unbiased and consistent estimator of $\hat{G}_N(j\omega_p)$.

FREQUENCY RESPONSE BY PERIODOGRAM

Another empirical estimator, $\hat{\hat{G}}_{NP}(j\omega_p)$, for $\hat{G}_N(j\omega_p)$ is

$$\hat{\hat{G}}_{NP}(j\omega_p) = \frac{\mathscr{F}[z(k)]}{\mathscr{F}[\vartheta(k)]} = \frac{Z_{NP}(\omega_p)}{\Theta_{NP}(\omega_p)} \tag{9.32}$$

where Z_{NP} and Θ_{NP}, the Fourier transforms of $z(k)$ and $\vartheta(k)$, respectively, are defined as

$$\left.\begin{matrix} Z_{NP}(\omega_p) \\ \Theta_{NP}(\omega_p) \end{matrix}\right\} = \frac{1}{N^{1/2}} \sum_{t=1}^{N} \left\{ \begin{matrix} z(t) \\ \vartheta(t) \end{matrix} \right\} e^{-j\omega t} \tag{9.33}$$

The estimator, $\hat{\hat{G}}_{NP}(j\omega_p)$, will be recognized as the periodogram estimate of $\hat{G}_N(j\omega_p)$.

An important distinction should be made at this point. In general, when $\vartheta(t)$ is a realization of a stochastic process, the signal estimators are based upon spectral analysis (Jenkins and Watts, 1969). For these conditions the variance of the estimate does not decrease with increasing record length. In that case, $|\Theta_{NP}|^2$ has an erratic character whose average behaviour is like that of the input spectrum. What happens is that the frequency points become more and more tightly packed, with a constant variance which is equal to the noise-to-signal ratio at the frequency of interest. For this reason, the periodogram estimate is a poor estimator (Bendat and Piersol, 1971).

In contrast, $\vartheta(k)$, which is a bounded purely deterministic periodic temperature signal with a limited number of harmonics, is applied to a linear stable sensor. In addition, as it is also assumed that the noise, $n(t)$, contains only a finite amount of energy, this means that $S_n(f)$ asymptotically decreases with frequency. Under these conditions, it can be shown (Ljung, 1985, 1987) that the periodogram of the input temperature signal is

$$\Theta_{\mathrm{N}}(\omega_p) = \tfrac{1}{2}(a_{p\vartheta} - \mathrm{j}b_{p\vartheta}) = \Theta_{\mathrm{NP}}/N^{1/2} \tag{9.34}$$

Similarly, it can be shown that

$$Z_{\mathrm{NC}}(\omega_p) = \tfrac{1}{2}(a_{pz} - \mathrm{j}b_{pz}) = \tfrac{1}{2}(\hat{a}_{pv} - \mathrm{j}\hat{b}_{pv}) = Z_{\mathrm{NP}}(\omega_p)/N^{1/2} \tag{9.35}$$

Thus the non-parametric empirical frequency-response estimate, $\hat{\hat{G}}_{\mathrm{NC}}(\mathrm{j}\omega_p)$, of correlation analysis corresponds to the periodogram estimate, $\hat{\hat{G}}_{\mathrm{NP}}(\mathrm{j}\omega_p)$.

9.4.3 Experimentation for sensor testing

CLASSIFICATION OF SENSOR TESTING METHODS

Because of the difficulty in predicting sensor response by calculation it is important to devise methods for testing them. Tests may be for design-validation purposes or, in the case of working sensors (Michalski *et al.*, 1991), to ensure that they are still working with their specified dynamic properties. There is a great variety of methods for these purposes.

The main groups of sensor testing use either immersion or self-heat tests. In the most commonly applied approach to date, a step input of energy state or rate is generated by one of two methods (Chohan *et al.*, 1985). The first (or immersion) method plunges the sensor into a bath or pipeline where it is washed by a suitable fluid or gas. This plunging essentially applies a step input signal, which is the most commonly applied type of aperiodic signal for temperature sensor testing.

Temperature sensors may also be tested by applying an internally generated power input. These (called self-heating tests) are important because they allow dynamic response tests to be conducted *in situ*. A considerable benefit results as it is unnecessary to remove the sensor. The test is conducted under normal working conditions in the sensor's normal location.

Periodic test signals can be used to test temperature sensors. As periodic immersion or self-heating inputs give frequency-domain responses by correlation analysis, they are regarded as the most accurate methods. When these signals are used, the results are given as a Bode frequency-response function (Michalski *et al.*, 1981; Caldwell *et al.*, 1959). Periodic excitation of temperature sensors uses a very important group of test signals which may be either analog or binary. Using sinusoidal input signals in air-washed immersion tests is not very popular, because of the difficulty in precisely controlling the sinusoidal modulation of the air temperature. Binary-type squarewave temperature signals have been applied (Petit *et al.*, 1982). The application of multi-frequency binary sequences, described below, is important in sensor identification as they perform empirical multi-frequency response measurements using frequency analysis by the correlation method. Convective heat transfer dominates in immersion testing at lower temperatures (McGhee *et al.*, 1992a,b; Michalski *et al.*, 1991). Only a few contributions deal with temperatures above about 600°C where radiative heat transfer conditions apply (Michalski *et al.*, 1991).

It is of increasing importance to develop methods to identify temperature sensors *in situ*. The easiest way is to apply an internally generated power input in a self-heating test (Kerlin *et al.*, 1981). The simplest and most direct method of performing a self-heating test applies a step change in the bridge-conditioning supply for resistance temperature sensors (Kerlin *et al.*, 1978, 1982). Applying MBS signals in self-heating tests has been reported for RTDs (Jackowska-Strumillo *et al.*, 1992). Sinusoidal self-heating tests have been performed for thermocouples (Michalski and Eckersdorf, 1990).

DATA MEASUREMENT IMMERSION TESTING WITH MBS SIGNALS

An MBS signal for sensor testing is generated as a pattern of symbols. These are then used to stimulate the sensor under test in such a way that the binary symbols capture information about the sensor. Such a process of information capture may be referred to as data measurement. The capturing process is the dual of that in data communications, where information carrying, as opposed to capturing, is the aim.

The test rig for MBS testing of temperature sensors shown in Figure 9.6 (McGhee *et al.*, 1989, 1991a) consists of two sub-systems and SYstem Detection instrumentation (SYD). An environment management system is designed to control the temperatures and to maintain reasonably stable and equal flowrates. The temperature of each stream may be varied between 25°C and 100°C.

A test interface system provides signal conditioning for both the actuator and data acquisition from the sensors. For resistive-type modulating sensors, conditioning is by resistance deflection bridges, which are suitable for noble metal and thermistor sensors. An instrumentation amplifier is used to boost the signal at the deflection nodes of the bridge. In the case of thermocouple sensors it is also necessary to use a low-level amplifier.

Alternate immersion of the sensor under test into the managed air streams is accomplished using electromechanical solenoid actuators. The transverse mechanical motion is achieved by positive action on the field of each solenoid by a suitably conditioned and powered MBS signal. No difficulties associated with the transit time of the transverse displacement have arisen in the test conducted using the actuator.

A microprocessor development environment described in Dunlop (1984) was used as the basis for the SYD hardware. In addition to the system core, which consists of a bus-oriented modular instrument, an internally designed and manufactured data acquisition/distribution card is used. This card has a 12-bit ADC for analog signal conversion. Signal generation is achieved through a corresponding 12-bit DAC. The MBS generation and signal-processing routines, which are used for data-measurement

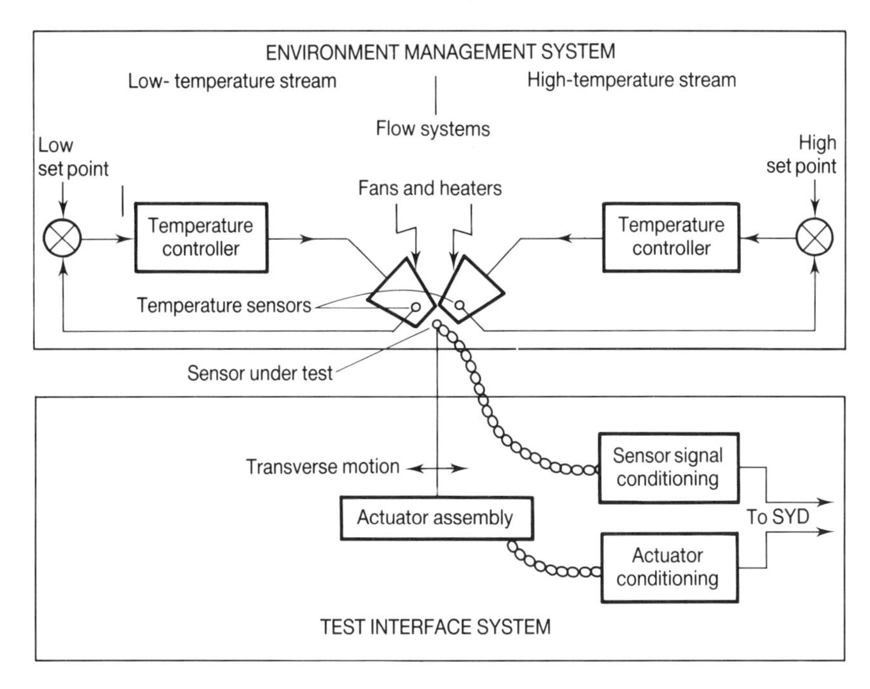

Figure 9.6 Block diagram of an immersion testing rig for temperature sensors.

purposes, are described in Henderson *et al.* (1987) and McGhee *et al.* (1987).

The probe used in the experiments is described in Section 9.2.2. Its equivalent lumped-parameter model is given in Figure 9.3. Immersion tests were conducted with and without the auxiliary PVC sleeve around the probe. The sleeve was assumed to be an adequate model of any increase in sheath thermal resistance due to accumulated impurity layers. Five bursts or periods of the Strathclyde compact octave MBS signal (see Chapter 7) were applied to the actuator. Data response during the first burst is discarded to eliminate errors due to any switch-on transient. Only data from the second to the fifth burst were used. The number of data points is

$$\begin{aligned} N_{\text{obs}} &= (\text{number of sequences}) \\ &\quad \times (\text{interrogations per sequence}) \times (\text{data points per interrogation}) \\ &= 4 \times 64 \times 16 = 4096 \end{aligned} \tag{9.36}$$

The frequency-response and calculated transfer function of the normal sensor for this immersion test are given in the Bode diagram shown in Figure 9.7. A two-part immersion test with MBS signal periods as shown in the Figure was then performed with the PVC sleeve in place. Overlapping measurements of the frequency response allow comparison of the frequency responses from different tests, as shown in Figure 9.7.

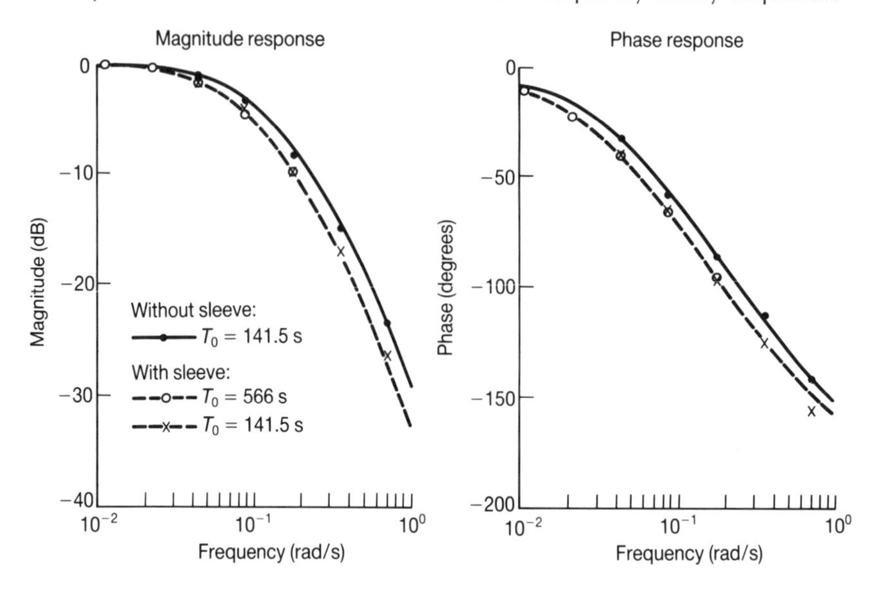

Figure 9.7 Temperature sensor frequency responses. Normalized responses: without sleeve $G_n = 1/(1 + j\omega 11.39)(1 + j\omega 2.25)$; with sleeve $G_p = 1/(1 + j\omega 14.39)(1 + j\omega 2.875)$.

As mentioned earlier, errors are caused by a switch-on transient. Figure 9.8 illustrates the transient which occurs during the first burst of the RTD response in the immersion test. The excitation and the response are overlaid with the display in normalized time.

Situations can occur in identification when more detailed frequency-response information is required. Measuring with this resolution, which is a legitimate aim of an identification experiment, was previously conducted using the classical sinusoidal method in a number of separate fixed-frequency experiments. The time-consuming repetitive nature of the measurement processes involved in this procedure can be considerably simplified by using the Strathclyde PSKMBS described in Chapter 7. These signals are constructed by modulating a squarewave carrier with a Strathclyde MBS. Frequency responses for immersion tests on the sensor with/without the sleeve are shown in Figure 9.9. Note the richness of the frequency-response information in this figure. The carrier signal was a squarewave temperature which was PSK modulated by the Strathclyde UC148 of Table 7.5.

9.4.4 Future trends in sensor testing

The above results for the identification of temperature sensors using MBS test signals have important ramifications. It is probable that the MBS technique will become the standard method for measuring the frequency response of temperature sensors.

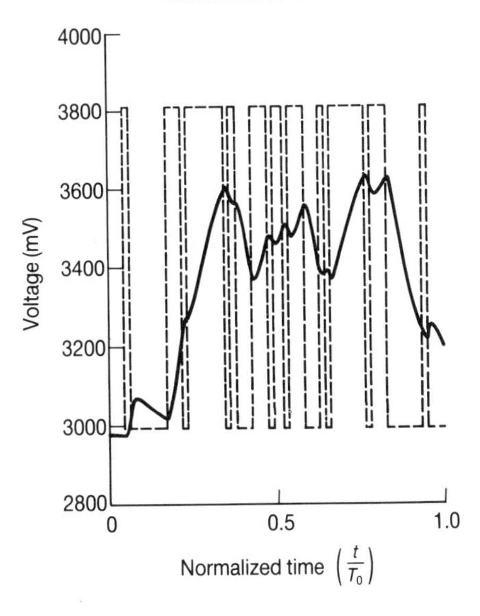

Figure 9.8 Response of RTD to first burst of Strathclyde octave MBS temperature excitation.

Development of the rig described in this chapter is possible in two principal areas. Verification of the external translational immersion technique will be allowed by modulating the temperature of one air stream with an MBS signal. This modulated air stream will then wash the tested sensor, whose position will remain stationary. A second important addition to the technology of temperature sensor testing applies MBS internal heating power for *in situ* testing. Preliminary work on this technique as applied to RTDs is described by Jackowska-Strumillo *et al.* (1992). A modification of the application of the internal technique to thermocouples using sinusoidal testing will incorporate MBS internal heating. Such a modification requires some practical difficulties to be overcome (Michalski and Eckersdorf, 1990).

Irrespective of whether the tests are by external or internal input or of test signal employed, it is to be expected that the use of computers for signal generation and data acquisition and analysis will continue with increasing momentum. The influence of MBS testing is likely to be significant. Another important addition will occur in extending the techniques of simulating MBS identification (McGhee *et al.*, 1990) to include non-linear heat-transfer conditions.

9.5 Conclusions

Testing of temperature sensors with MBS signals by the external method has been discussed in this chapter. The foundation for temperature sensor identification has

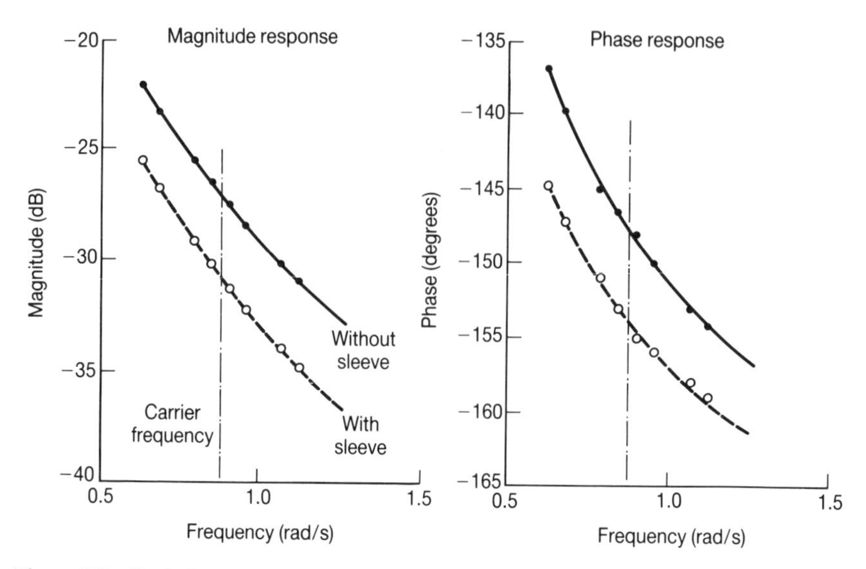

Figure 9.9 Bode frequency-response plots for an RTD with a temperature excitation in the form of a Strathclyde PSKMBS.

three main ingredients. Sensor modelling is important because it allows interpretation of the effects of the principal factors which contribute to their dynamic behaviour. An introduction to this modelling has been given. It can be concluded that theoretical modelling and calculative prediction can provide important insights into sensor behaviour. However, it is always difficult to give precise values for the parameters of materials used in construction. For this reason, performing real-time tests upon real components in realistic environments is an essential element for sensor identification by data measurement.

Data measurement, as described in this chapter, is that body of principles and technology which uses binary signals for data capture through system interrogation. Its dominant principles depend upon the use of computerized instrumentation for MBS signal generation and data acquisition and analysis. The design and application of MBS data-measurement signals are reviewed in this chapter. An appropriate application of these principles using microprocessor technology is embodied in instrumentation for system detection called SYD.

A sound theoretical basis for empirical frequency-response estimation by data measurement shows that the technique of MBS data measurement is frequency analysis by the correlation method. This approach leads to unbiased estimates whose variance decreases with experiment duration. Consequently, the estimates are also consistent. It is also shown that these frequency-response estimates by the correlation method correspond to the periodogram estimate. This correspondence arises from the deterministic nature of the MBS data-measurement interrogation.

Experimentation for sensor identification uses a test rig and instrumentation which have been described. The experimental results show that MBS data measurement provides the possibility of distinguishing differences in dynamic behaviour. Two types of test signal have been used in the experimental work. The first is a base-band Strathclyde MBS signal while the second uses the UC148 signal as the keying signal in the phase-shift-keyed modulation of a squarewave carrier. Low-frequency responses of a sensor with and without additional sheathing illustrate the possibility of sensor diagnosis. Strathclyde PSKMBS signals allow identification using signals which cluster the harmonic content of a data-measurement MBS signal around a suppressed carrier.

Data measurement by MBS signals is a significant means of obtaining frequency-response estimates of a wide range of other physical systems as well as temperature sensors.

Acknowledgements

L. M. Jackowska-Strumillo acknowledges the support of a scholarship from Strathclyde University in The Industrial Control Centre.

10

Design and Application of Test Signals for Helicopter Model Validation in the Frequency Domain

Ron Patton, Martin Miles and Paul Taylor

10.1 Introduction

Dynamic system testing in the frequency domain involves the design of suitable input test signals and the interpretation of input/output information obtained from a test. The coherence function is shown to be useful for both of the above stages of testing. A test signal may be selected on the basis of the accuracy of its results and the information contained in them. In this chapter it is shown that the coherence function allows, first, a comparison of the amount of contamination of results due to noise inside the loop and, second, an assessment of the degree of non-linearity present in the system under test. The latter can be approached by considering both the type of input signal and the shape of the coherence function obtained from tests using that signal. The reason for doing this research is to improve the accuracy of identification experiments, and to extract as much information as possible from any particular experiment. Some parameters can be optimized with the aid of the coherence function. For efficient estimation the input test signal length and peak factor, along with the type of signal which is best suited to the experiment, are also important factors.

As the length of test signal is increased then the accuracy of results is improved. However, to reduce mechanical fatigue and cut running costs, shorter signals are desired for tests. Furthermore, continual increases in signal length yield decreasing improvements. The variance of the coherence function (as defined in Section 10.3) shows this trade-off graphically and allows a suitable length to be chosen. The signal which is most appropriate for finding a linear representation of the system (the multi-frequency signal) may have its peak factor lowered in order to improve the accuracy of results. There is again a trade-off; finding a low peak factor is computationally intensive and the improvement gained in the accuracy of the results is reduced at lower peak factor levels, and therefore a further reduction in peak factor is less important. The coherence function again shows graphically this improvement.

There is growing interest in the use of frequency-domain analysis for identifying helicopter dynamics (Hamel, 1979; Kaletka, 1979; Black *et al*., 1986; Tischler *et al*., 1987; Young and Patton, 1988; Tischler and Kaletka, 1986; P. Young, 1989; Young and Patton, 1990). Recent control system development and evaluation work at Westland Helicopters has been concerned with the determination of helicopter frequency responses in order to validate the model used in a particular flight-dynamics simulator.

Recently, a frequency-response testing approach has been used by Westland as part of their flight-development programmes. This has been linked with research at York University in frequency-domain identification (Patton *et al*., 1990). Frequency responses derived from real flight tests allow identification of the aircraft transfer functions which are of interest without recourse to a postulated mathematical model. This focuses attention on, and facilitates an understanding of, the dominant modes of the aircraft's behaviour. The occurrence of non-linearities (hysteresis, backlash, saturation, etc.) can be isolated for further investigation. These flight-test results are compared with simulations from the helicopter and the flight-control system mathematical model. The comparison enables confidence in the simulation model to be established and highlights differences for use in future model development.

An automated test signal technique has been developed to give experimental repeatability and an adequate excitation of the dynamics of a helicopter while in flight; but, more importantly, it allows complex test signals to be put to use in genuine helicopter tests. Measures have been taken to ensure that the automated signal generation is safe, well controlled and repeatable. The safety issues dictate that the pilot should be free to observe, monitor and remove the test signal input.

This chapter focuses on the test signals and their properties, as applied to the identification of parameters of simple dynamic system examples. The work accompanies an ongoing research programme at York University into helicopter parameter identification.

The chapter examines non-parametric stages of identification. Evaluation of the frequency response and coherence function is dealt with, while curve fitting and parameter estimation are left as further topics. The chapter describes relevant aspects of frequency-domain identification, test signal design and spectral analysis.

The need for carefully chosen experimental conditions together with a suitable test input signal design is illustrated at the beginning of the chapter through a case study of model validation based on real flight test data from a prototype Westland helicopter. The data are processed to generate frequency-response gain and phase estimates and coherence plots. A comparison is then made between the responses derived from a simulator model with those derived by estimation. On the basis of this the simulator model is shown to be valid over a range of frequencies corresponding to the rigid-body dynamics of the helicopter.

The application example is based on the use of one test signal, the logarithmic swept sine wave. However, the study goes on to show how the coherence function, obtained during spectral analysis, is used to make a more appropriate choice of input signal required for a test. Further specific uses of the coherence function are examined in the latter part of the chapter, with a view to choosing a test signal, or selecting parameters for one.

10.2 Frequency-domain testing

10.2.1 Introduction

When performed correctly, frequency-response testing can give clear engineering insight into the dynamic structure of a system and the results can be readily interpreted in graphical form. Dominant modes and features such as resonances can often be observed immediately and related to known physical features. Information bandwidth and required spectral resolution need to be established before a test is carried out. This, together with the need to tailor a test signal spectrum to the required application, means that some *a priori* information about the system under test is required.

Four main stages are encountered in the process of performing an identification experiment in the frequency domain:

1. The design of an input test signal.
2. Application of that signal, in a test run on the plant, noting fatigue damage and non-linear effects.
3. Spectral analysis, using FFT and window-based averaging as standard.
4. Interpretation of results, relating the data to information required in, for example, model validation or control system design.

When carrying out an identification experiment on a helicopter we begin with a linear model corresponding to that helicopter. It is then necessary to analyze the real system in the following ways:

1. The model, which has been derived from physical principles and assumptions, will have some degree of inaccuracy, and the identification procedure will be used to establish any deficiencies of such a model or to account for modelling errors. Usually the transfer function of the system will be affected perhaps at higher frequencies by some extra, unmodelled, factors (for example, a body resonance, the existence of which should be accounted for in the data).
2. The achievement of high bandwidth, high authority control is an important aim for helicopter control. Unfortunately, such a system cannot easily be designed

without considering the effects of high-frequency dynamic modes. Existing flight dynamics models of helicopters contain selected blade modes. Clearly, for high bandwidth control further dynamic effects need to be modelled accurately, and this leads to the requirement of reliable identification over a larger bandwidth than with the rigid-body system in order that a robust control law may be developed.

3. The linear model is derived from the full non-linear, coupled system. As the complete system has high-frequency dynamic effects, an associated disturbance model is often also used to a simple model at high frequencies. The linear model information (taken to any complexity, i.e. with or without a high-frequency model) can then be used for the control system design. The complete system model is then useful for assessing the disturbance rejection of the controller and the robustness of the system to flight-parameter variations.

The frequency-domain approach to system identification offers a distinct advantage over time-domain methods in that non-parametric results may easily be obtained from the test data. Any major errors in the identification procedure may be detected by a graphical comparison of model and measured frequency responses, before a parametric transfer function is fitted.

10.2.2 Input signal design

The information required for successful system testing and identification will vary. The characteristics of the input signal will be influenced by the required frequency range of the information; the spectral shape of any noise in the system; non-linearities, the effects of which should be kept to a minimum; the linear dynamics of the system (poles, resonances, etc.).

For frequency-domain identification it is necessary to look at test signals from the point of view of their frequency composition. The length of time which a signal can be applied is important, and the magnitude of the spectral density of each signal is directly proportional to this. For any test the information required in the results leads to a need for specifying the characteristics of the test signal for qualities such as:

1. *Bandwidth*: The maximum and the minimum frequencies at which data are wanted must first be specified. For a result with good accuracy at all required frequencies there should be sufficient power in the signal over this band.
2. *Amplitude (system disturbance) criterion*: The maximum disturbance to the system caused by the test signal should be minimized. The test signal will be designed to have a low disturbance on the plant in question, while still enabling the required knowledge to be obtained from the measurements. A small amplitude is advantageous in terms of both obtaining a linear model and of keeping wear on the mechanical parts low. Too small an amplitude, however, will mean that results may be affected by low amplitude non-linearities, such as backlash, stiction, etc. In fact, in this situation it is advisable to repeat the tests for a range of amplitudes.

3. *Harmonic content*: For non-linearities (the source of additional harmonics in the output) to be rejected from interfering with the linear description of the system, the minimum amount of information should be lost when removing harmonic effects from the experiment. The controllability of the harmonic content of the input is a desirable property of a test signal.

One of the difficulties often encountered with the frequency-domain approach is that of inadequate persistency of input signal. It is important that the signal has a power-spectral density whose shape can be tailored to provide the best signal for the application under consideration. With most types of test signal this feature is of limited flexibility.

Once the input signal information criteria, such as bandwidth, frequency resolution, maximum signal amplitude, etc., have been established, the accuracy of results will be determined by the type of input signal used. It is therefore up to the designer of the experiment to make the appropriate decision. Many experiments in the past have required a simple form of accuracy suitable mainly for model validation. However, the demands upon identification methods for helicopter systems are increasing and accurate parameter estimation needs a reliable identification methodology.

The 3-2-1-1 (7-bit binary m-sequence) and linear swept sine wave are particularly simple and hence convenient forms of test input signals. Both of these signals have been used extensively for multi-frequency testing in aircraft systems (Kaletka, 1979; Tischler and Kaletka, 1986). Automated test inputs to the helicopter (see later under Section 10.4.1) means that more complex signals can be used in experiments. The test signals considered in this chapter are listed in the next section, with a brief explanation of their structure.

10.2.3 Test signal types

THE SWEPT SINE WAVE

As its name suggests, the swept sine wave is a sinusoid which varies in frequency over time. It may be sub-divided into two types (referring to the rate of change of frequency with time): the linear and the logarithmic. The use of a logarithmic rate is an attempt to compensate for the low spectral density at low frequencies, which is a particular drawback of the linear swept sine wave test signal.

Results of a test using swept sines will be affected at certain frequencies by events occurring at specific times. That is, they will be affected at frequencies corresponding to the particular state of the swept sine wave at the time of the event (e.g. transient noise). Results are also affected at those frequencies which correspond to an event occurring while the swept sine wave is at a certain amplitude, i.e. amplitude-dependent non-linearities.

BINARY M-SEQUENCE SIGNALS

As shown in Chapter 1, Section 1.4, of this book, the pseudo-random binary sequence (PRBS) or binary maximum-length sequence (m-sequence) can be generated by a shift register circuit arrangement or its software equivalent and leads to a signal of length $2^n - 1$ where n is the number of shift register stages. To compare the signal with the other chosen signals the bandwidth was set to a similar range (i.e. $f_{max} \approx 150 f_{min}$ for the other signals, f_{max} and f_{min} being the maximum and minimum frequency components in the signal). Therefore a signal of length 127 bits was chosen, with the clock rate $\Delta T = 1/f_{max}$ selected such that there are 127 frequency components below the maximum frequency (f_{max}). Setting the clock rate to a value which allowed more frequency components below f_{max} would mean that the test contained frequency ranges of very poor identification, where very low spectral components exist. As the signal stands, it gives very low power components at frequencies where the $\sin^2(\pi k/N)/(\pi k/N)^2$ curve is zero ($k = N = 2^n - 1$). The logarithmic spacing of the output compensates this to some extent.

THE MULTI-HARMONIC SINUSOID OR MULTI-FREQUENCY SIGNAL (MFS)

Comprising (as its name suggests) a number of sinusoids at chosen frequencies, this signal has proved useful in identification due to its versatility (for further details of these signals see Chapter 2, Section 2.4.2). The sequence is composed of harmonics of a base frequency, so that the signal has a fixed period. Design of the signal involves considering the following criteria:

1. The harmonics used determine first, the bandwidth of the identification and second, the effect which non-linearities will have on the results. By using, for example, prime harmonics (i.e. those harmonics that are multiples of the base frequency, such that the multiple is a prime number, not including 1 or 2) the effect of a non-linearity on the resulting frequency responses is suppressed. In Chapter 12, Rees and Jones compare such a prime harmonic signal with a pseudo-random binary signal and state that the rejection of non-linear effects is better with the prime sequence.
2. Some *a priori* knowledge of the power required at each frequency enables the specification of the spectral density of the signal, which leads to control over the effect of noise on the frequency-response estimates. The effect of the noise on the system will be modified according to its initial spectral shape and to the transfer function of the system to which the noise is added. It can be arranged, by choice of the signal power density function, that a disturbance affects the results equally across the frequency band of interest. Transfer function estimation should then be equally accurate over the whole of that frequency band. Alternatively, the spectral shape of the noise may be measured using knowledge of the system's transfer spectrum and the effect of the noise over the frequency band. The spectral density of the multi-frequency signal can be tuned to the user's wishes.

3. The relative phase of each harmonic is a very important feature, such that selective choice of each phase allows the signal to be generated with a low peak factor (Schroeder, 1970; Flower *et al.*, 1978b).

The Schroeder-phased signal is a special case of a multi-frequency wave signal with the phasing of the harmonics selected to provide a low peak factor. This type of signal was first described by Schroeder (1970), who wanted to minimize the peak-to-peak amplitude, or peak factor, of a multi-frequency signal. For identification of parameters of a dynamic system a low-peak-factor test signal will prevent large perturbations being introduced into the system under test, while giving an adequate signal over the period in which the test is applied. These are clearly desirable characteristics of a test signal.

The design of the Schroeder-phased signal is quite straightforward, as the specification of the phase angles of the harmonics depends on the relative magnitudes of the different frequency components. The objective is to select the phase angles so that the peak factor or maximum envelope excursion is small. The standard derivation given by Schroeder (1970) does not solve the minimum peak-factor problem; it simply yields low peak factors (in most cases).

10.3 Spectral analysis

10.3.1 Introduction

The third stage in frequency-domain identification is spectral analysis, the result of which is a non-parametric model of system behaviour, obtained from input–output data over the frequency range of interest. The procedures involved with finding this model (i.e. auto- and cross-spectral methods) facilitate the calculation of the magnitude-squared coherence function for the data (Wellstead, 1978). The coherence function has been used in this work as a mechanism for comparing the performance of the various test input signals.

Direst spectral estimation with overlapped Fast Fourier Transform (FFT) is used. The sets of input–output data are divided into segments, which overlap by an arbitrarily chosen amount (60% is used here since it is close to the optimum as suggested by Carter *et al.*, 1973). Each input–output pair is windowed (i.e. multiplied by a weighting function, in this case Hanning), and processed in turn; the power-spectral densities are thus obtained by averaging. The auto- and cross-power spectral estimates are found by the Fourier transform of the auto- and cross-correlation functions, respectively (see Wellstead, 1978, or Priestley, 1981).

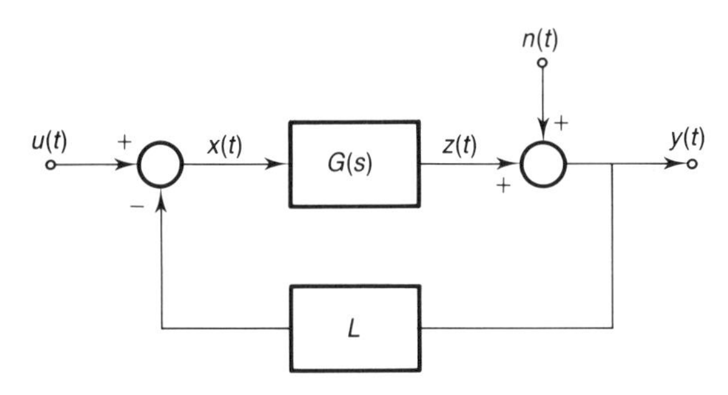

Figure 10.1 Closed-loop system, with noise in loop.

Figure 10.1 shows a general feedback system with the open-loop plant or process represented by the transfer function $G(s)$. The noise-free output(s) is (are) given by $z(t)$ ($\mathbf{z}(t)$ in the vector case). The term $n(t)$ is used to denote the measurement noise ($\mathbf{n}(t)$ in the vector case) for the multi-output case. Hence, $y(t)$ (or $\mathbf{y}(t)$) is (are) the noisy output(s) of the process used for output feedback, with L the (constant) gain output feedback parameter (matrix); $u(t)$ is the test input signal used to excite the system.

Assuming unbiased spectral estimation (i.e. no correlation between the noise and the system inputs), the frequency response can then be defined as

$$G(\mathrm{j}\omega) = \frac{[S_{xy}(\mathrm{j}\omega)]}{[S_x(\mathrm{j}\omega)]} \tag{10.1}$$

where S_x is the auto-spectral density of a single input signal and S_{xy} is the cross-spectral density between a pair of input and output signals. Strictly, this relation is true only when the measurements are uncorrelated with the actuation inputs – a situation which can break down in some closed-loop instances when the feedback L is applied between the measurements and the actuation signals. In the ideal case, the frequency response is found using equation (10.1). The variance of spectral estimates is reduced by the averaging of estimates from successive data segments.

10.3.2 Use of the coherence function

The magnitude-squared coherence function (commonly called the coherence function) serves as an indicator of how well the fundamental or first harmonic component of the power spectrum can be used to derive an accurate model of the input–output dynamics, i.e. it is a measure of the linear dependence of the output on the input defined in spectral terms. The coherence function $\Gamma^2_{xy}(\omega)$ is given by

$$\Gamma^2_{xy}(\omega) = \frac{|S_{xy}(\mathrm{j}\omega)|^2}{S_x(\omega)S_y(\omega)} \leqslant 1 \tag{10.2}$$

where S_y is the auto-spectral density of the output. By definition, the coherence function lies between 0 and 1. A totally noise-free linear system would yield $\Gamma^2_{xy}(\omega) = 1$. Three effects can cause the coherence function to take a value less than unity: non-linearities in the system under test, input and output noise and secondary inputs such as external disturbances. When a system is noisy or non-linear, the coherence function indicates the accuracy of a linear identification as a function of frequency. The closer it is to unity, the more reliance can be placed on an accompanying frequency-response estimate, at a given frequency. For a real application, which will be non-linear and affected to some extent by noise, a plot of coherence against frequency will indicate the way in which the disturbances change across the frequency band. However, for a noise-free linear system the coherence function gives an excellent comparison of the efficiency of a number of test signals in exciting the dynamic modes of a typical system.

The former of the above points regarding disturbances across the frequency band is reflected in the fact that the variance of a frequency-response estimate obtained during spectral analysis is related to the coherence function thus (Wellstead, 1981):

$$\frac{\mathrm{Var}\{|G(\mathrm{j}\omega)|\}}{|G(\mathrm{j}\omega)|} = \mathrm{Var}\{\arg G(\mathrm{j}\omega)\} = \frac{1}{2N}\left[\frac{1}{\Gamma^2_{xy}(\omega)} - 1\right] \tag{10.3}$$

where N is the number of averaged segments.

10.3.3 Effects of measurement noise

When measurement noise is added to the system at its output and fed back through the control loop (see Figure 10.1) the spectral estimates become biased (Bendat and Piersol, 1971). If the open-loop dynamics are required from the closed-loop test, then the analysis by Tischler (1987) is appropriate.

In the closed-loop situation shown in Figure 10.1 the actual system frequency response $G(\mathrm{j}\omega) = S_{xz}(\mathrm{j}\omega)/S_x(\mathrm{j}\omega)$, while the measured estimate is $\hat{G}(\mathrm{j}\omega) = S_{xy}(\mathrm{j}\omega)/S_x(\mathrm{j}\omega)$. Since

$$\frac{S_{xy}(\mathrm{j}\omega)}{S_x(\mathrm{j}\omega)} = \frac{S_{xz}(\mathrm{j}\omega)}{S_x(\mathrm{j}\omega)} + \frac{S_{xn}(\mathrm{j}\omega)}{S_x(\mathrm{j}\omega)} \tag{10.4}$$

the estimate consists of the true response together with a bias term $S_{xn}(\mathrm{j}\omega)/S_x(\mathrm{j}\omega)$. From the closed-loop system relationships,

$$\hat{G}(\mathrm{j}\omega) = G(\mathrm{j}\omega) - \frac{L}{(S^2 + L^2)\{1/[1 + LG(\mathrm{j}\omega)]\}} \tag{10.5}$$

where S^2 is the (power) signal-to-noise ratio $S_u(\mathrm{j}\omega)/S_n(\mathrm{j}\omega)$. Thus when the noisy output $y(t)$ is fed back to close the loop, equations (10.1) and (10.2) are not fully applicable since the input to the system $x(t)$ is correlated with the noise. However, if the feedback gain L is low, they become reasonable approximations and have been used in this chapter.

10.3.4 Design of test signal length

Signal length is an important parameter of a test signal when carrying out experiments on aircraft. Flying time is costly and mechanical parts have a finite life so that fatigue damage accumulated during frequency-domain testing must be minimized. Also, the time which the aircraft may be kept at one linear operating point is limited. One may select certain test signals that have less mechanically wearing characteristics and which use experimental procedures that minimize total application time. For all cases, however, each single test signal should be applied for the shortest time possible to yield accurate results.

Carter *et al.* (1973) used the coherence function to facilitate the choice of some parameters for spectral analysis. Specifically, the variance of the coherence function (based on the coherence function data over the frequency range considered) was used as a measure of the accuracy of the spectral analysis for different amounts of segment overlap (referring to segmented spectral analysis). The variance of the coherence function becomes lower as more averages are taken, and a smoother transfer function estimate is formed. The results obtained by Carter *et al.* demonstrate that as the overlap is increased the rate of decrease of variance slows down. A suitable value for overlap can be seen to be around 60%, such that a further increase in overlap yields little improvement in the accuracy of the results.

Using this idea, the variance of the coherence function was used to measure the performance of an input signal as a function of its length.

Using a multi-frequency sine wave of period 50 s and containing 100 harmonic frequencies from 0.02 Hz to 2 Hz, noise was added to the signal and the coherence function was calculated between the noisy and clean data sets, i.e. simulating the identification of a noisy piece of wire. Different lengths of signals were used and the variance of the coherence function in each case was calculated, as shown in

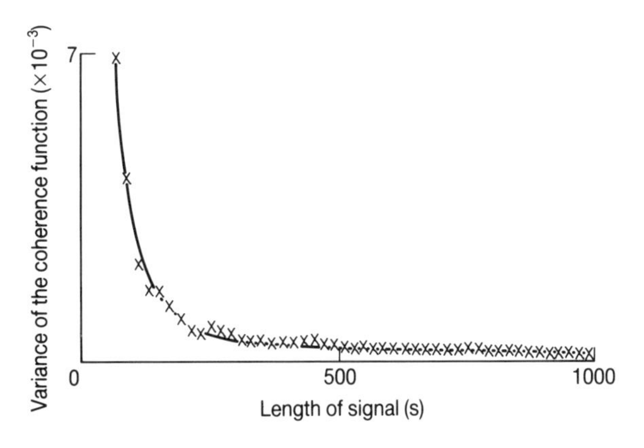

Figure 10.2 The improvement (or decreasing variance) given by increasing the length of a test signal.

Figure 10.2. On increasing the length of the signal, the variance correspondingly decreases. With a view to shortening the length of the input signal the graph shows that beyond about 300 s of data there is little improvement in the variance. This corresponds to approximately 15 segments of data.

In general (for any signal), if one requires a spectral resolution of f_{min} (thus fixing the window size to $1/f_{min}$ s), then (for spectral analysis using the optimum 60% overlap) the length of the test signal should be $6/f_{min}$.

10.3.5 Test signal peak factor and relationship with coherence function

An essential feature is the need to control the amplitude of the test signal; an accurate linear, small signal model is required, while the signal should be of sufficient amplitude so as not to be adversely affected by low level non-linearities (e.g. stiction, etc.).

The relative peak factor is defined in terms of the maximum, minimum and RMS values of the periodic signal $V(t)$ as (Schroeder, 1970):

$$P_{\mathrm{f}} = \frac{V_{max} - V_{min}}{2\sqrt{2}V_{rms}} \tag{10.6}$$

The lower it is, the higher the RMS value of the signal for a given maximum perturbation. Results of an experiment will then be better in the presence of noise, when using a low-peak-factor signal.

A swept sine signal has a peak factor of 1, while a PRBS with levels $\pm V$ has a peak factor of $1/\sqrt{2}$. However, as explained in Chapter 2 (Section 2.4.2), the peak factor of a PRBS with its d.c. level removed is

$$\frac{N}{\sqrt{N^2 - 1}} \cdot \frac{1}{\sqrt{2}}$$

which is equal to 0.7144 for $N = 7$ and 0.7071 for $N = 127$. If the frequency content of a PRBS above the first (main) lobe is removed, the effective peak factor (in the sense of this chapter) increases further, to 0.9914 for $N = 7$ and to 0.9896 for $N = 127$. This demonstrates that, when using a signal with a very low peak factor in a practical system, the effective peak factor will not be quite as low as that designed.

The MFS may have its peak factor altered by changing the relative phases of the sinusoids. The usefulness of a low peak factor may be shown graphically as follows.

The coherence function is also very useful as a measure of uncertainty (noise and non-linearities) across the test frequency band of interest. Its relevance to the noise can be observed when expressing the function in terms of the auto-spectra of the signal and of the noise for a linear system (Wellstead, 1981), thus:

$$\Gamma_{xy}^2(\omega) = \left[1 + \frac{S_n(\omega)}{S_z(\omega)}\right]^{-1} \tag{10.7}$$

where S_n and S_z are the auto-spectra of the noise and the noise-free output signal, respectively.

This property means that the shape of the coherence function in the frequency domain will be directly related to the shape of the noise and disturbances. Not only may a linearized model of the system be constructed from the data, but a more accurate model of the disturbance shape can also be obtained.

The final shape of the noise function will be affected by two considerations:

1. The original shape of the disturbances which affect the plant. That is, the average spectral density of a real interference source (for example, a gust) will be shaped in the frequency domain.
2. The frequency response of the system, and of the controller, will lead to a filtered noise source, and hence the transfer function estimate will not be equally affected by inaccuracies.

Using a signal with a finite bandwidth, a short application time is desired, yielding less fatigue damage to mechanical parts. We also require as many valid frequency points in the results as possible, hence utilizing as much information as is available. A high-average, flat coherence function is therefore an important specification of the experiment.

In a set of results which contains a coherence function that varies over a large range of values it is possible to consider information corresponding to those points with high coherence, or in some cases (Tischler, 1987) one may weight the accuracy of the results using the coherence values. In both cases it is obvious that a consistent level of coherence would be better so that minimal information is lost.

One may also compensate for the closed-loop noise-shaping effect by giving the input signal an appropriate spectral density function, the result of which is that the noise will have an even effect on the results, and so appear uncorrelated with frequency. Regions of large disturbances may be compensated for by increasing the power density in that area, and thus flattening the coherence function. Again, using the idea of Carter *et al.* (1973), the variance of the coherence function was used as a measure of performance of a signal but this time based on changing the peak factor of the signal.

Figure 10.2 shows that by using a variable signal length we can observe the improvement in the coherence function. A change in peak factor simply shifts the graph higher or lower. A lower peak factor will mean a higher signal-to-noise ratio and therefore a lower overall variance curve will be observed.

This trend, showing changes in the variance of the coherence against change in peak factor is illustrated in Figure 10.3. As the peak factor of the signal is lowered, the variance of the coherence function decreases. A noticeable upward trend leads to a strong argument for selecting signals with low peak factors in order to reduce the effects of noise on an experiment.

Relating equations (10.6) and (10.7) to each other leads us to determine the following relationship between the coherence function and the peak factor of the test signal:

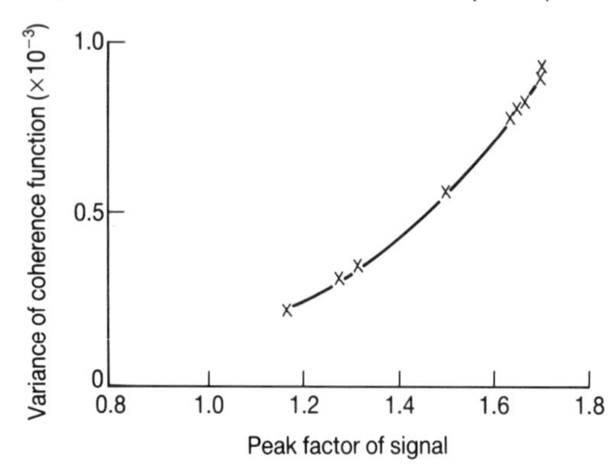

Figure 10.3 Relationship between signal peak factor and coherence function.

$$\overline{\Gamma_{xy}^2(\omega)} = \frac{\alpha}{\alpha + (P_f)^2} \tag{10.8}$$

where P_f is the peak factor of the signal as defined in equation (10.6) and α depends upon the averages of the noise and signal power spectra, and the peak-to-peak value of the signal. The latter two may be kept constant by specifying a flat-spectrum input signal with a constant maximum amplitude.

As the peak factor decreases, the coherence will approach unity, but at a slower rate. This rate will depend upon the noise level as well as the peak factor. Figure 10.4 shows the experimental variation of coherence mean with peak factor and the way that the measured data points (plotted with X) relate to the curve

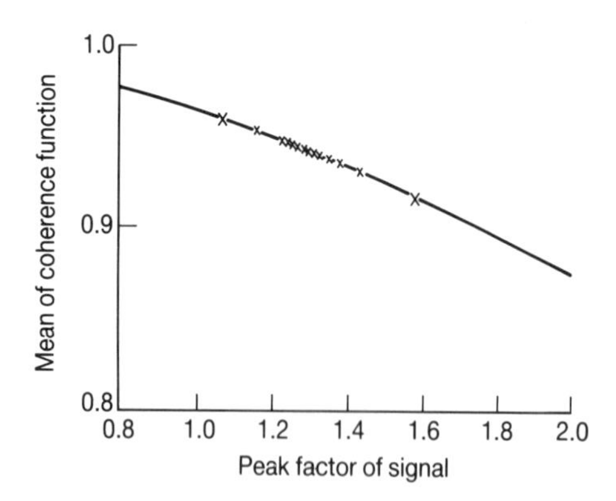

Figure 10.4 Variation of coherence mean against the input signal peak factor.

generated by equation (10.8) (the line). For a given system it is therefore advisable to find the lowest peak factor which will yield meaningful results. The smaller the peak factor, the smaller the amplitude range for a given power spectrum, hence the more linear the system, the higher the coherence.

10.4 Frequency-domain testing of a helicopter in flight

This section considers the application of one type of test signal, namely the logarithmic swept sine wave, to a real full-size helicopter in flight. The objective of this study has been to perform a validation of a non-linear dynamic model used in a Westland flight simulator. The recorded flight data (in response to the test signal sequence) were processed off-line using the spectral analysis procedure described in Section 10.3 to provide estimates of gain and phase of normally measured aerodynamic quantities over suitable frequency bands. These gain and phase estimates were compared with gain and phase data generated from a linearization of the simulator dynamic model, the main aim being to determine the validity of the simulator model against the real data and to assess over what frequency band the model is valid. The test signals were applied to the helicopter in flight using an automated system, based on special signal design and signal pre-recording.

10.4.1 Automated test signal application

A conflict would arise if the test signals were to be applied manually by the pilot; a simple signal sequence is all that is feasible in this situation and this is insufficient to ensure adequacy of spectral content and repeatability. The automation of the test signal application is thus an obvious way forward. The automation enables the experiment to be sufficiently controlled and repeatable and also guarantees the desired spectral content. There are other important advantages: limiting of the inevitable fatigue damage that this type of testing incurs and leaving the pilot free to observe and monitor the experiment and to intervene if necessary. Test signal automation allows us, therefore, to use more complex test signals to meet the above requirements.

The input signal, aircraft response and other outputs of interest are recorded on a system known as MODAS (Multiplexed On-board Data Acquisition System). The design of Automatic Stabilization Equipment (ASE) fitted to Westland helicopters is such that the series actuators (which are fast acting) facilitate the implementation of automatic test signal inputs. The signal is summed with the ASE signal at the series actuator input (single-lane). This implementation also conveniently avoids 'switching' of the control laws which may occur on some ASEs if the test input were applied through the cyclic control stick.

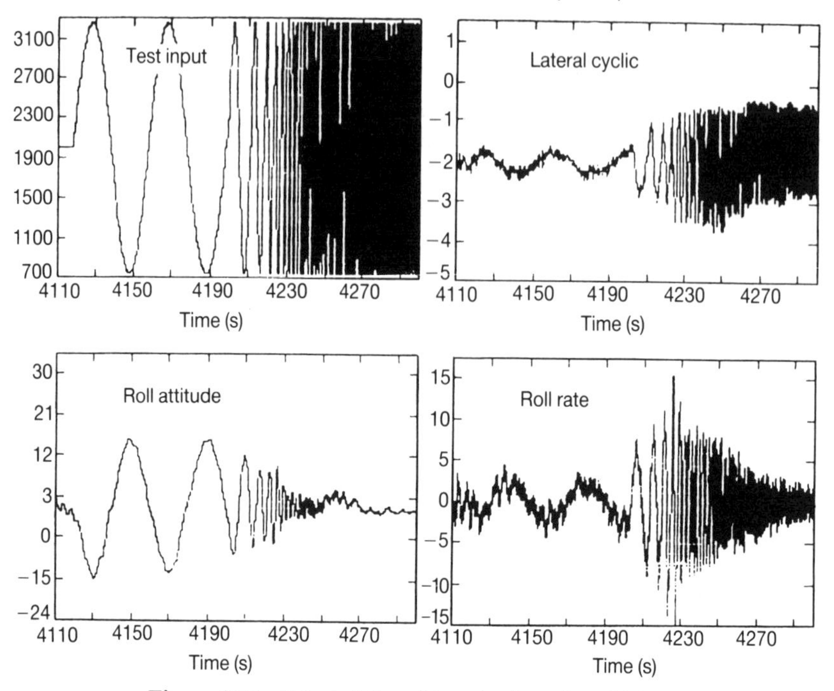

Figure 10.5 Selected time histories in roll at 80 knots.

In general, the series actuator amplitude (single-lane) cannot exceed 5% of the blade pitch range. The test inputs are introduced into one lane only for safety reasons. The swept sine wave is pre-recorded on tape and played back in flight. A change in any of the test signal parameters is easily accommodated by re-recording the signal.

The amplitude is selectable in flight, 25%, 50% and 75% approximately of actuator travel, allowing shakedown and safe progression to attaining a reasonable attitude response. For overall confidence and improvement in the quality of the Gain and Phase information at low frequencies, a signal at a selected constant frequency is introduced, typical values being 0.1 Hz, 0.25 Hz and 1 Hz.

Prior to flight, the hardware was bench tested and the techniques assessed in the flight dynamics simulator. Results were analyzed for possible fatigue damage.

The flight-test procedure involves trimming the aircraft to the required condition with the ASE engaged, i.e. the aircraft is in equilibrium. The auto-trim, if fitted, is switched to Manual to prevent the slow-acting parallel actuator opposing the low-frequency band of the test signal. The lane of the channel of interest, which does not include the test signal, is disengaged (this ensures adequate response by preventing the stabilization from opposing the test signal); the amplitude of the test signal is selected. The tape of the test signal is played. The pilot monitors the aircraft response and trim condition, applying manual control as necessary.

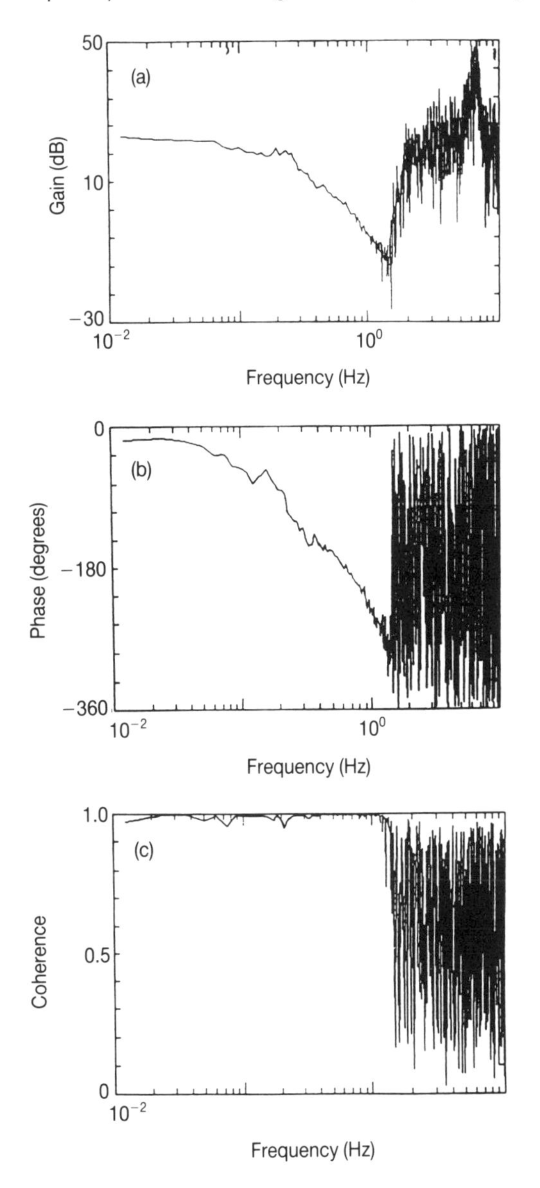

Figure 10.6 Aircraft roll attitude: (a) gain, (b) phase and (c) coherence at 80 knots.

10.4.2 Flight-test results

Typical flight results from frequency-domain testing obtained by the method described are shown in Figures 10.5 and 10.6. Figure 10.5 illustrates typical time histories of

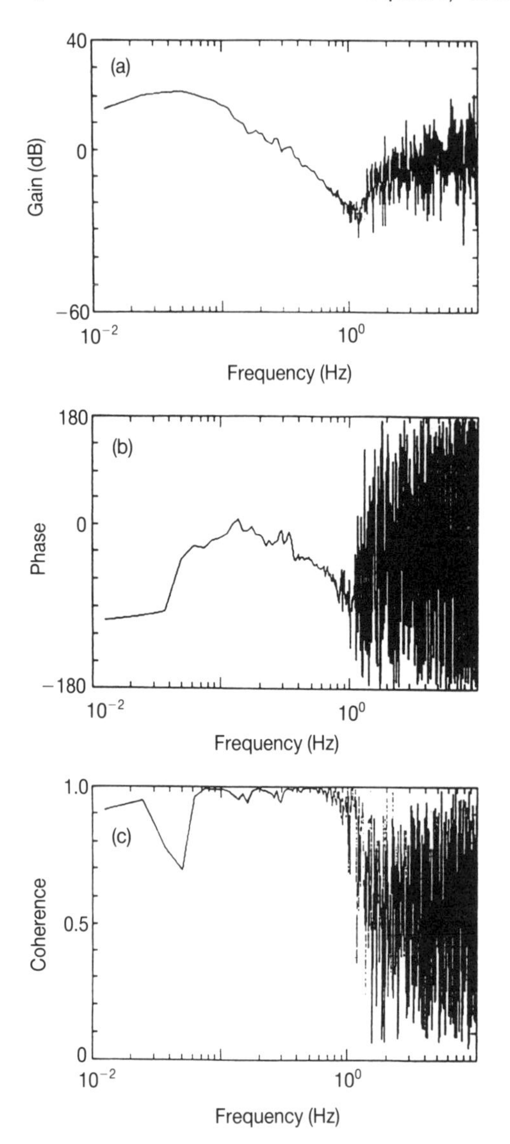

Figure 10.7 Aircraft pitch attitude: (a) gain, (b) phase and (c) coherence at 80 knots.

the variables of interest; the input test signal; lateral cyclic blade pitch angle; aircraft body-roll attitude; and roll rate. Particularly noticeable are the increased attenuation and phase lag of the estimated aircraft response at high frequencies. Here the test signal applied to the helicopter is a swept sine wave of constant amplitude, beginning at 25 mHz, ending at 2 Hz, and lasting for 180 s.

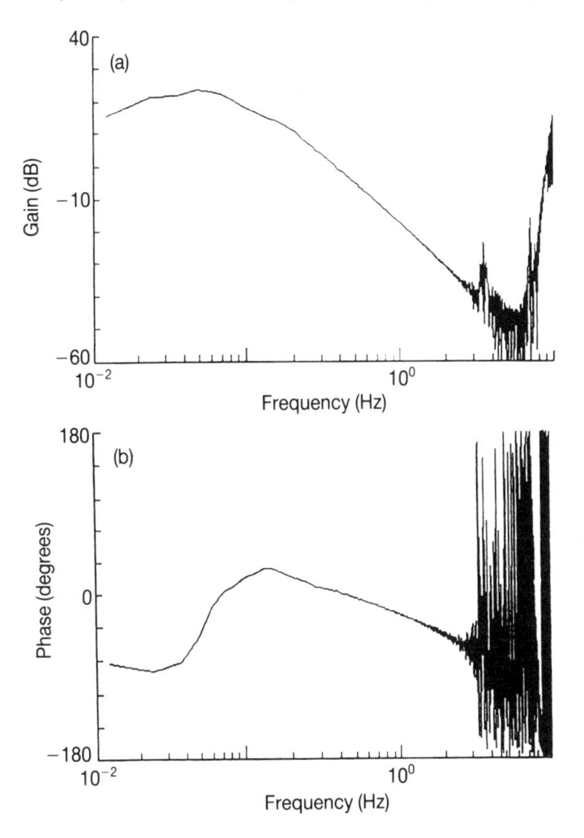

Figure 10.8 Simulated pitch attitude: (a) gain and (b) phase.

Figures 10.6(a) and 10.6(b) illustrate a conventional Bode diagram of Gain and Phase for aircraft roll attitude. The sign convention adopted is a positive lateral cyclic stick input (to the right) produces a positive lateral cyclic pitch change at the rotor and causes the aircraft to roll to the right (positive). The coherence plot (Figure 10.6(c)) shows that the information is good for frequencies up to 1 Hz, i.e. a bandwidth containing the dominant rigid modes.

Figure 10.7 shows corresponding Bode and coherence plots for aircraft body pitch attitude to longitudinal cyclic blade angle. Again the coherence shows that the information is good for frequencies up to 1 Hz.

Figure 10.8 illustrates the equivalent Bode plots to Figures 10.7(a) and 10.7(b) produced from a non-linear simulation model. Overlaying Figures 10.7(a) and 10.8(a) illustrates excellent agreement for the Gain over the frequency range of interest. Overlaying Figures 10.7(b) and 10.8(b) shows a generally good agreement between the phase plots. At higher frequencies there appears to be a tendency for the aircraft phase to roll off more rapidly than the simulation. Extending the frequency range would confirm this. The cause is either missing additional degrees of freedom or too large a time constant in the modelled degrees of freedom; and non-linearities present in mechanical control runs and power servo. The results demonstrate the adequacy of the logarithmic swept sine wave test input for flight applications.

In the above cases the output (body attitude) is correlated with the total signal at the power servo outputs resolved into blade lateral or longitudinal cyclic pitch applied to the rotor. This allows direct comparison of the stabilized mathematical model with the stabilized aircraft but without the presence of the ASE and mechanical control runs. In principle, Bode diagrams relating any two instrumented stations of the aircraft can be produced. A picture of the characteristics of the components within the closed-loop (ASE + aircraft) can be constructed. The comparison of these Bode diagrams with those derived from classical linear analysis identifies non-linearities which can subsequently be modelled using a 'describing function' approach, while giving a good physical understanding of the system characteristics.

The ability to analyze physical phenomena in this way demonstrates the power of the frequency-domain approach. This is considered an important prelude to reliable parameter identification; the estimation of system parameters should only be attempted once the dynamic system structure is well understood. This work shows that frequency-domain testing can facilitate this process, if the experiments are carried out with due care. The swept frequency type of test signal has been used exclusively in this application study, but the benefits of using other test signals for helicopter model validation are still being considered. Having considered an application study, we are now in a position to describe more fully the requirements of a frequency-domain testing procedure and, in particular, the requirements for the design of the most suitable test input signals to avoid problems such as fatigue damage (for the helicopter application) while also providing the best opportunity to derive the true linear model of the system at the point of operation.

10.5 The scaled helicopter model

The linear mathematical model used in this study is based on the work of Hendricks (1980) and is derived from a non-linear model of a 1/7 scale remotely piloted helicopter using standard aerodynamic equations, such as those given by Bramwell (1976). It has been further assumed that the coupling between the longitudinal and lateral motions can be considered negligible in this case. Furthermore, the effect of rotor dynamics and other high-frequency dynamic effects have not been considered. The four-state open-loop unstable longitudinal-motion model used is as follows.

LONGITUDINAL MOTION

The linearized longitudinal model of the helicopter can be written in state-space form as

$$\dot{\mathbf{m}} = A\mathbf{m} + B\omega \tag{10.9}$$

where the control inputs and the states are

$$\boldsymbol{\omega}(t) = (\beta_1, \theta_0)^{\mathrm{T}}, \qquad \mathbf{m}(t) = (u, w, \theta, q)^{\mathrm{T}} \tag{10.10}$$

where β_1 is the longitudinal cyclic pitch input while θ_0 is the collective pitch. The states are: horizontal velocity u, vertical velocity w, pitch angle θ, and pitch rate q. For the flight condition with the tip-speed ratio μ of 0.1 and zero angle of attack, the open-loop stability and control matrices are

$$A = \begin{bmatrix} -0.1 & 0.01 & -9.81 & 0 \\ -3 & -1.27 & 0 & 7.12 \\ 0 & 0 & 0 & 1 \\ 0.41 & -0.37 & 0 & -0.7 \end{bmatrix} \tag{10.11}$$

$$B = \begin{bmatrix} 9.74 & 0.03 \\ 9.03 & -64.75 \\ 0 & 0 \\ -32.15 & -0.11 \end{bmatrix} \qquad C = [0 \quad 0 \quad 0 \quad 1] \tag{10.12}$$

The output $q = C\mathbf{m}$ has been selected for the purpose of the frequency-response investigation. Note that all four of the model state variables have been fed back to stabilize the system. This has been done for convenience only in this example.

The state feedback controller gain matrix was computed using a pole-placement algorithm as

$$K = \begin{bmatrix} 1.7 & 7.4 & -360 & -220 \\ 6.1 & -3.0 & 1.1 & 15 \end{bmatrix} \times 10^{-3} \tag{10.13}$$

Note that the matrix K is not the same as the general feedback gain L shown in Figure 10.1.

The corresponding system closed-loop matrix is then given by

$$F = A - BK \tag{10.14}$$

The transfer function matrix giving the relationship between the two inputs and the one output signal is given by

$$H(s) = C(sI - F)^{-1}B = [q/\beta_1 \mid q/\theta_0] \tag{10.15}$$

leading to the collective pitch-to-pitch rate transfer function used in the simulation experiments of

$$\frac{q}{\theta_0} = \frac{s(s - 76.77)(s + 0.34)}{s^4 + 9.42s^3 + 24.73s^2 + 20.75s + 7.01} \tag{10.16}$$

which has poles of

$$-5.66 \quad -0.57 \pm \mathrm{j}0.39, \quad -2.63$$

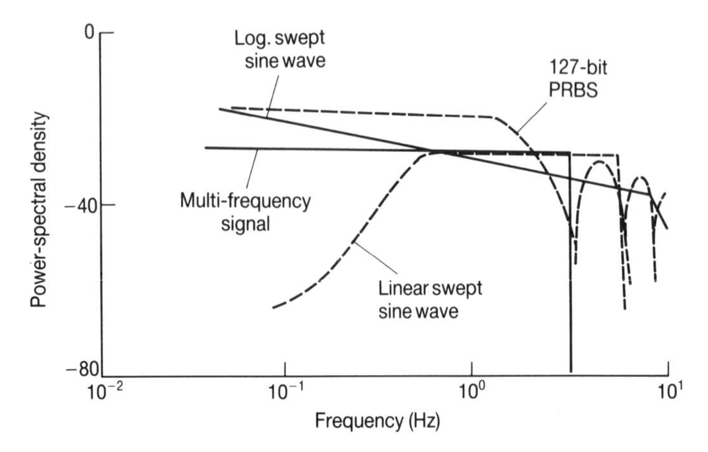

Figure 10.9 The power-spectral density of some input signals.

The comparison of the application of different test signals was performed on the above simulation model of the longitudinal dynamics of the scaled helicopter. The next section describes the comparison study and results in some detail.

10.6 Test signal comparison

Figure 10.9 shows a comparison of the power-spectral densities of the four signals used in this study, based on a simulation of the longitudinal dynamics of the scaled helicopter mathematical model. The signals used have all been described in Section 10.3 above.

The amplitudes of the sine wave, the PRBS and the multi-frequency signal were all equal. The low peak factor of the last means that the RMS values are of similar magnitude, so that the different signals will perturb the system under test to approximately the same extent. The test signals have the following parameters:

1. The logarithmic and linear swept sine waves cover a frequency band ranging from 0.02 Hz to 3 Hz, and have maximum amplitudes ± 1.
2. The multi-frequency sine wave (MFS) has consecutive, Schroeder-phased harmonics from 0.02 Hz to 3 Hz, with maximum amplitudes ± 1. The period of the wave was 50 s and 150 harmonics were included within the bandwidth.
3. The PRBS was of sequence length 127 bits, with the first side-lobe minimum at 3 Hz, and amplitude of ± 1.

The following observations for each test signal can be made which are generally in favour of the MFS, by referring to the spectral density plots shown in Figure 10.9:

1. The multi-frequency signal is flat (or of very specific spectral density) and therefore useful for accurate measurements in the frequency domain.
2. In this signal the lack of power existing outside the band of interest means that the test will be less affected by aliasing errors, and that it is more efficient in its use of the spectrum. Aliasing errors will occur due to high-frequency components of the signal being reflected to frequencies within the test band. The usual way to overcome aliasing is to use filters to reduce the frequency content above a maximum frequency. With this signal there are very few frequency components above the set maximum, and therefore low aliasing errors will be produced.
3. Due to its lower peak factor, the PRBS signal has a higher power at much of the relevant frequencies; but the spectral density function is of fixed shape: low power at higher frequencies, and spectral components outside the band of interest.

The following test was carried out in order to demonstrate how the coherence function can be used to analyze qualitatively the type of results that can be expected from different test signals.

The signals were applied, in turn, to the linear simulation of the fourth-order longitudinal helicopter model described in Section 10.5, with a Gaussian white-noise source of fixed RMS (0.05 rad/s) added to the signal at the system output before the feedback (as shown in Figure 10.1). The input signals were all of a fixed sample frequency and fixed peak-to-peak amplitude, to facilitate the comparison; and also, of necessity, of the same length. The output signals were each sampled at the same rate as the input signals.

Due to the random nature of the noise, the test on each signal was carried out twenty times, and the final coherence function was taken as the average of each result. Also, 60% overlap was used for each test. Figure 10.10 shows the results of this test.

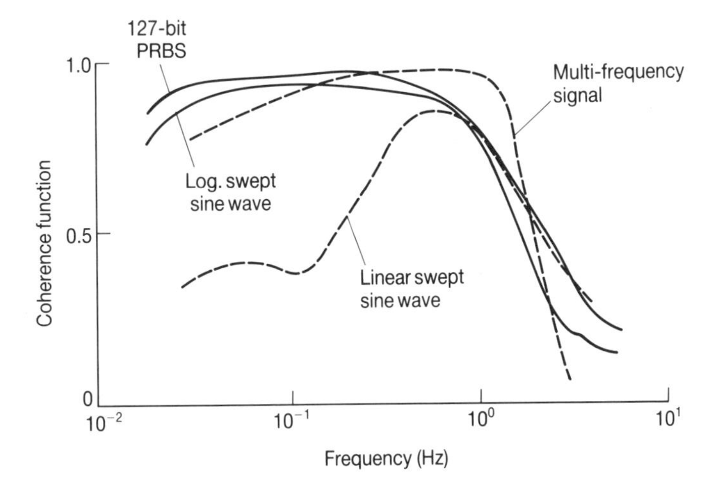

Figure 10.10 Coherence function, using the signals of Figure 10.9 and the system illustrated by Figure 10.1.

A comparison of Figures 10.9 and 10.10 illustrates the combined effect of signal-to-noise ratio and system transfer function.

A more appropriate signal (i.e. that with flatter coherence over the specific bandwidth) is the multi-harmonic non-binary signal. Although the signal is of a slightly lower spectral density than the m-sequence signal, and therefore the average value of the coherence function is lower (at low frequencies), the overall frequency range is affected more evenly by noise. In the case of the multi-frequency wave the information is valid throughout the whole frequency range of the signal.

10.7 Conclusions

The frequency-domain approach gives a powerful way of identifying the physical characteristics of mechanical systems. The results have shown the validation of the flight simulator model structure.

It is considered that the approach taken is a realistic and thorough way of validating dynamic structure for helicopters, based on flight testing and simulation data. It has also been demonstrated that the coherence function is a powerful tool for use with frequency-domain testing and model validation. In comparing test signals the coherence function depends upon the power-spectral density of the input, the shape of the system transfer function and the shape of the noise affecting the system.

Using the variance of the coherence function, the improvement in accuracy attained by increasing the length of test signal is displayed graphically. As a result, it is easy to choose an ideal test signal length, approximately $6/f_{min}$.

Using the mean as well as the variance of the coherence function, the improvement in accuracy attained by reducing the peak factor of the signal is, again, shown graphically. The trade-off between these two factors is visible, and the optimum peak factor may be found by trading off the improvements gained against the computational effort involved in finding low peak factors for multi-frequency sinusoids.

11

Design of Multi-level Pseudo-random Signals for System Identification

H. Anthony Barker

11.1 Introduction

Pseudo-random signals are the most popular choice for the persistently exciting perturbation signals required in system identification. Some simple examples of these signals have been given in Chapter 1. In this chapter the general class of pseudo-random signals will be described, and the properties of the class that are important for system identification will be developed.

When pseudo-random signals were first introduced, some thirty years ago, the binary form was preferred for two important reasons; simplicity of generation and simplicity of transduction. These no longer apply when generation is accomplished by a digital computer and digital-to-analog conversion to any reasonable number of levels is readily available.

Interest therefore turns naturally to multi-level pseudo-random signals for system identification. Such signals have been considered briefly in Sections 1.6 and 5.3 of this book. These signals test a system at many operating levels, and provide the possibility of identifying non-linear behaviour, or of identifying linear behaviour in the presence of non-linearities. The limitations of binary signals for this purpose is graphically illustrated by the second-order non-linear system $y = u^2$, which, when tested by a binary signal u with levels 1 and -1, gives an output y which is merely a constant level 1.

The principal factor which militates against the use of multi-level pseudo-random signals for system identification is the complexity of the underlying theory, which is based on the algebra of finite fields (Albert, 1956). To some extent, this complexity can be circumvented by using sum and product tables to define the field algebra (Barker, 1986), and this approach will be adopted in the next section.

11.2 Finite fields

A finite, or Galois, field is simply a collection of elements which together satisfy the requirements of the relevant finite field theory. The simplest finite fields are the prime fields, in each of which the number of elements is a prime number. If p is a prime number, then the corresponding prime field is GF(p), with p field elements a_0, a_1, $\ldots$, a_{p-1}.

The requirements of a prime field are satisfied if the field elements are taken to be the integers modulo-p, that is,

$$a_i = i \qquad i = 0, 1, \ldots, p-1 \qquad p \text{ prime} \tag{11.1}$$

and the field operations of addition and multiplication are accomplished by modulo-p arithmetic, that is,

$$a_i + a_j = i + j \qquad \text{mod } p \tag{11.2}$$

and

$$a_i a_j = ij \qquad \text{mod } p \tag{11.3}$$

These addition and multiplication operations may be implemented by means of sum and product tables which are constructed according to equations (11.2) and (11.3). Examples are given in Table 11.1 for GF(2), GF(3), GF(5) and GF(7).

In extensions of a prime field, the number of elements is a power of the prime, given by

$$q = p^k \qquad k \text{ a positive integer} \tag{11.4}$$

and the extension field is GF(q), with q field elements b_0, b_1, $\ldots$, b_{q-1}.

The requirements of an extension field are not as easily satisfied as those of a prime field. The field elements are actually polynomials of degree $k-1$ in the prime field GF(p), with coefficients which are the elements of GF(p) taken in all possible combinations, that is,

$$b_i(x) = \sum_{r=0}^{k-1} a_{ir} x^r \qquad i = 0, 1, \ldots, q-1 \tag{11.5}$$

Table 11.1 Sum and product tables for GF(2), GF(3), GF(5) and GF(7)

GF(2)

+	0	1
0	0	1
1	1	0

×	0	1
0	0	0
1	0	1

GF(3)

+	0	1	2
0	0	1	2
1	1	2	0
2	2	0	1

×	0	1	2
0	0	0	0
1	0	1	2
2	0	2	1

GF(5)

+	0	1	2	3	4
0	0	1	2	3	4
1	1	2	3	4	0
2	2	3	4	0	1
3	3	4	0	1	2
4	4	0	1	2	3

×	0	1	2	3	4
0	0	0	0	0	0
1	0	1	2	3	4
2	0	2	4	1	3
3	0	3	1	4	2
4	0	4	3	2	1

GF(7)

+	0	1	2	3	4	5	6
0	0	1	2	3	4	5	6
1	1	2	3	4	5	6	0
2	2	3	4	5	6	0	1
3	3	4	5	6	0	1	2
4	4	5	6	0	1	2	3
5	5	6	0	1	2	3	4
6	6	0	1	2	3	4	5

×	0	1	2	3	4	5	6
0	0	0	0	0	0	0	0
1	0	1	2	3	4	5	6
2	0	2	4	6	1	3	5
3	0	3	6	2	5	1	4
4	0	4	1	5	2	6	3
5	0	5	3	1	6	4	2
6	0	6	5	4	3	2	1

For example, the field elements of GF(4), GF(8) and GF(9) are as shown in Table 11.2. A compact form for these elements is obtained by substituting p for x in $b_i(x)$, in which case the elements are the integers modulo-q, as shown in Table 11.2.

Table 11.2 Field elements of GF(4),, GF(8) and GF(9) in polynomial and compact form

GF(4)	
Polynomial	Compact
$0 + 0x$	0
$1 + 0x$	1
$0 + 1x$	2
$1 + 1x$	3

GF(8)	
Polynomial	Compact
$0 + 0x + 0x^2$	0
$1 + 0x + 0x^2$	1
$0 + 1x + 0x^2$	2
$1 + 1x + 0x^2$	3
$0 + 0x + 1x^2$	4
$1 + 0x + 1x^2$	5
$0 + 1x + 1x^2$	6
$1 + 1x + 1x^2$	7

GF(9)	
Polynomial	Compact
$0 + 0x$	0
$1 + 0x$	1
$2 + 0x$	2
$0 + 1x$	3
$1 + 1x$	4
$2 + 1x$	5
$0 + 2x$	6
$1 + 2x$	7
$2 + 2x$	8

The operation of addition in the extension field GF(q) is accomplished by the addition modulo-p of the coefficients of corresponding powers of x, that is,

$$b_i(x) + b_j(x) = \sum_{r=0}^{k-1} (a_{ir} + a_{jr})x^r \tag{11.6}$$

For example, in GF(9)

$$(2 + 1x) + (2 + 2x) = 1 + 0x \tag{11.7}$$

This addition operation may be implemented with the elements in compact form by means of a sum table which is constructed according to equation (11.6). Examples are given in Table 11.3 for GF(4), GF(8) and GF(9).

The operation of multiplication in the extension field GF(q) is not as easily accomplished, because multiplication of elements $b_i(x)$ and $b_j(x)$ results in the polynomial $b_i(x)b_j(x)$ of degree $2(k - 1)$, which is not in the form of a field element as defined in equation (11.5). The field element in its proper form is obtained as the remainder when the polynomial $b_i(x)b_j(x)$ is divided by $f(x)$, a primitive, or indexing, polynomial of degree k in the prime field GF(p).

For example, in GF(9)

$$(2 + 1x)(2 + 2x) = 1 + 0x + 2x^2 \tag{11.8}$$

which is not in the form of a field element of GF(9), but when $1 + 0x + 2x^2$ is divided by the primitive polynomial $1 + x + 2x^2$ in GF(3), the remainder $0 + 2x$ is obtained and this is the required product of $2 + 1x$ and $2 + 2x$ in GF(9).

The multiplication operation may be implemented with the elements in compact form by means of a product table which is constructed according to this procedure. Examples are given in Table 11.3 for GF(4), obtained with the indexing polynomial $1 + x + x^2$ in GF(2), for GF(8), obtained with the indexing polynomial $1 + x^2 + x^3$ in GF(2), and for GF(9), obtained with the indexing polynomial $1 + x + 2x^2$ in GF(3).

Primitive polynomials are involved not only in the generation of extension fields but also in the generation of pseudo-random signals. A primitive polynomial $f(x)$ of

Table 11.3 Sum and product tables for GF(4), GF(8) and GF(9)

GF(4)

+	0	1	2	3
0	0	1	2	3
1	1	0	3	2
2	2	3	0	1
3	3	2	1	0

×	0	1	2	3
0	0	0	0	0
1	0	1	2	3
2	0	2	3	1
3	0	3	1	2

GF(8)

+	0	1	2	3	4	5	6	7
0	0	1	2	3	4	5	6	7
1	1	0	3	2	5	4	7	6
2	2	3	0	1	6	7	4	5
3	3	2	1	0	7	6	5	4
4	4	5	6	7	0	1	2	3
5	5	4	7	6	1	0	3	2
6	6	7	4	5	2	3	0	1
7	7	6	5	4	3	2	1	0

×	0	1	2	3	4	5	6	7
0	0	0	0	0	0	0	0	0
1	0	1	2	3	4	5	6	7
2	0	2	4	6	5	7	1	3
3	0	3	6	5	1	2	7	4
4	0	4	5	1	7	3	2	6
5	0	5	7	2	3	6	4	1
6	0	6	1	7	2	4	3	5
7	0	7	3	4	6	1	5	2

GF(9)

+	0	1	2	3	4	5	6	7	8
0	0	1	2	3	4	5	6	7	8
1	1	2	0	4	5	3	7	8	6
2	2	0	1	5	3	4	8	6	7
3	3	4	5	6	7	8	0	1	2
4	4	5	3	7	8	6	1	2	0
5	5	3	4	8	6	7	2	0	1
6	6	7	8	0	1	2	3	4	5
7	7	8	6	1	2	0	4	5	3
8	8	6	7	2	0	1	5	3	4

×	0	1	2	3	4	5	6	7	8
0	0	0	0	0	0	0	0	0	0
1	0	1	2	3	4	5	6	7	8
2	0	2	1	6	8	7	3	5	4
3	0	3	6	4	7	1	8	2	5
4	0	4	8	7	2	3	5	6	1
5	0	5	7	1	3	8	2	4	6
6	0	6	3	8	5	2	4	1	7
7	0	7	5	2	6	4	1	8	3
8	0	8	4	5	1	6	7	3	2

degree n in GF(q) is given by

$$f(x) = \sum_{r=0}^{n} c_r x^r \qquad c_0 = 1 \quad c_n \neq 0 \tag{11.9}$$

where $f(x)$ is irreducible, that is, it cannot be factored, and is not a factor of $1 + x^r$ for any $r < q^n - 1$. There are $n^{-1}\phi(q^n - 1)$ primitive polynomials of degree n in GF(q), where $\phi(r)$ is Euler's function, the number of positive integers less than r and prime to r.

For example, if $q = 4$ and $n = 2$, then the number of primitive polynomials of degree 2 in GF(4) is $2^{-1}\phi(15)$; the integers, 1, 2, 4, 7, 8, 11, 13 and 14 are all less than 15 and prime to 15, so $\phi(15) = 8$ and the number of primitive polynomials in this case is therefore 4.

Tables of primitive polynomials in prime fields are given in the literature (Church, 1935; Peterson, 1961; Watson, 1962; Alanen and Knuth, 1964; Everett, 1966) and a selection of these polynomials appears in Table 11.4.

Table 11.4 Selection of primitive polynomials in prime fields

GF(2)
$1 + x + x^2$
$1 + x^2 + x^3$
$1 + x^3 + x^4$
$1 + x^3 + x^5$
$1 + x^5 + x^6$
$1 + x^4 + x^7$
$1 + x^4 + x^5 + x^6 + x^8$
$1 + x^5 + x^9$
$1 + x^7 + x^{10}$
$1 + x^9 + x^{11}$
$1 + x^6 + x^8 + x^{11} + x^{12}$
$1 + x^9 + x^{10} + x^{12} + x^{13}$
$1 + x^4 + x^8 + x^{13} + x^{14}$

GF(3)
$1 + x + 2x^2$
$1 + 2x^2 + x^3$
$1 + x^3 + 2x^4$
$1 + 2x^4 + x^5$
$1 + x^5 + 2x^6$
$1 + 2x^5 + x^7$
$1 + x^5 + 2x^8$

GF(5)
$1 + x + 2x^2$
$1 + 3x^2 + 2x^3$
$1 + x^2 + 2x^3 + 2x^4$
$1 + 4x^4 + 2x^5$
$1 + x^5 + 2x^6$

GF(7)
$1 + x + 3x^2$
$1 + 3x^2 + 2x^3$
$1 + x^2 + 3x^3 + 5x^4$

GF(11)
$1 + x + 7x^2$
$1 + x^2 + 4x^3$
$1 + x^3 + 2x^4$

GF(13)
$1 + x + 2x^2$
$1 + x^2 + 6x^3$

GF(17)
$1 + x + 3x^2$
$1 + x^2 + 3x^3$

GF(19)
$1 + x + 3x^2$
$1 + x^2 + 4x^3$

GF(23)
$1 + x + 7x^2$
$1 + x^2 + 3x^3$

GF(29)
$1 + x + 3x^2$
$1 + x^2 + 11x^3$

GF(31)
$1 + x + 12x^2$
$1 + x^2 + 14x^3$

A consequence of this multiplicity of primitive polynomials is that there is a corresponding multiplicity of product tables for the extension field GF(q), one for each of the $k^{-1}\phi(p^k - 1)$ primitive polynomials of degree k in the prime field GF(p). These may appear to generate different extension fields, but as all of the fields are isomorphic any one of them may be taken as GF(q) without any loss of generality. When tabulating primitive polynomials in extension fields, it is necessary to give the indexing polynomials used to generate the fields in order to avoid any ambiguity. A selection of primitive polynomials in extension fields is given in Table 11.5.

Table 11.5 Selection of primitive polynomials in extension fields

	GF(4)
index	$1 + x + x^2$
	$1 + x + 2x^2$
	$1 + x + x^2 + 2x^3$
	$1 + x + x^2 + 2x^4$
	$1 + x^4 + 2x^5$
	$1 + x^4 + x^5 + 2x^6$
	$1 + x^5 + 2x^6 + 3x^7$

	GF(8)
index	$1 + x^2 + x^3$
	$1 + x + 2x^2$
	$1 + x + 3x^3$
	$1 + x + 5x^4$

	GF(9)
index	$1 + x + 2x^2$
	$1 + x + 3x^2$
	$1 + x^2 + 3x^3$
	$1 + x + 6x^4$

	GF(16)
index	$1 + x^3 + x^4$
	$1 + x + 2x^2$
	$1 + x^2 + 2x^3$

	GF(25)
index	$1 + x + 2x^2$
	$1 + x + 8x^2$

	GF(27)
index	$1 + 2x^2 + x^3$
	$1 + x + 5x^2$

	GF(32)
index	$1 + x^3 + x^5$
	$1 + x + 3x^2$

Although finite field operations are defined in terms of addition and multiplication, these operations may be extended to include subtraction and division by means of inverse and reciprocal elements. For the field element c, the inverse is the field element $-c$, for which

$$c + (-c) = 0 \tag{11.10}$$

and the reciprocal is the field element c^{-1}, for which

$$c \times (c^{-1}) = 1 \tag{11.11}$$

where 0^{-1} is taken as 0.

Table 11.6 Inverse and reciprocal elements of GF(9)

GF(9)		
Element	Inverse	Reciprocal
0	0	0
1	2	1
2	1	2
3	6	5
4	8	8
5	7	3
6	3	7
7	5	6
8	4	4

For example, the inverse and reciprocal elements of the field elements of GF(9), obtained from Table 11.3, are given in Table 11.6.

Finite field elements of particular interest are the primitive elements. A field element is primitive if its powers generate all the non-zero elements of the field. If g is a primitive element of GF(q), then the field elements are

$$0, 1, g, g^2, \ldots, g^{q-2}$$

In an extension field, the primitive elements include p, $-p$, p^{-1} and $-\mathrm{p}^{-1}$.

For example, the primitive elements of GF(9) are 3, 5, 6 and 7, and Table 11.7 shows how the powers of each of these primitive elements generate all the non-zero elements of the field.

In a primitive polynomial $f(x)$ of degree n in GF(q), the coefficient c_n is always the product of $(-1)^n$ and a primitive element of GF(q).

Table 11.7 Elements of GF(9) generated as powers of primitive elements

GF(9)								
0	g^0	g^1	g^2	g^3	g^4	g^5	g^6	g^7
0	1	3	4	7	2	6	8	5
0	1	5	8	6	2	7	4	3
0	1	6	4	5	2	3	8	7
0	1	7	8	3	2	5	4	6

11.3 Pseudo-random signals

11.3.1 Generation

Figure 11.1 illustrates the basic method for generating pseudo-random signals. A clocked shift register with feedback generates the maximum-length or m-sequence $\{s_i\}$ with members that are field elements of GF(q), according to the equation

$$s_i = \sum_{r=1}^{n} -c_r s_{i-r} \qquad \text{all } i \tag{11.12}$$

in which the coefficients c_r are those of the primitive polynomial $f(x)$ of degree n in GF(q), given in equation (11.9).

By mapping each finite field element c into a corresponding real number $u(c)$, the m-sequence $\{s_i\}$ is converted into the pseudo-random sequence $\{u_i\}$ of real numbers, according to the equation

$$u_i = u(s_i) \qquad \text{all } i \tag{11.13}$$

The pseudo-random sequence $\{u_i\}$ is converted by a zero-order-hold into the step-wise continuous pseudo-random signal $u(t)$, according to the equation

$$u(t) = u_i \qquad i\Delta t \leqslant t < (i+1)\Delta t \tag{11.14}$$

The clock interval Δt determines only the temporal nature of the pseudo-random signal $u(t)$. Its fundamental characteristics are shaped by those of the pseudo-random sequence $\{u_i\}$ on which it is based. These, in turn, depend on the nature of the m-sequence $\{s_i\}$ and on the way in which the field elements of GF(q) are mapped into real numbers.

11.3.2 m-sequences

Many of the characteristics of m-sequences given in this section were first described by Zierler (1959). They follow from the fact that equation (11.12) defines a recurrence relationship between $n+1$ successive members of the m-sequence $\{s_i\}$, which may be written

$$\sum_{r=0}^{n} c_r s_{i-r} = \sum_{r=0}^{n} c_r D^r s_i = f(D) s_i = 0 \qquad \text{all } i \tag{11.15}$$

in which D is the delay operator. Equation (11.15) is the characteristic equation of the sequence $\{s_i\}$, and it is because $f(D)$ is a primitive polynomial in GF(q) that $\{s_i\}$ is an m-sequence in GF(q), with the greatest possible period N, which is given by

$$N = q^n - 1 \tag{11.16}$$

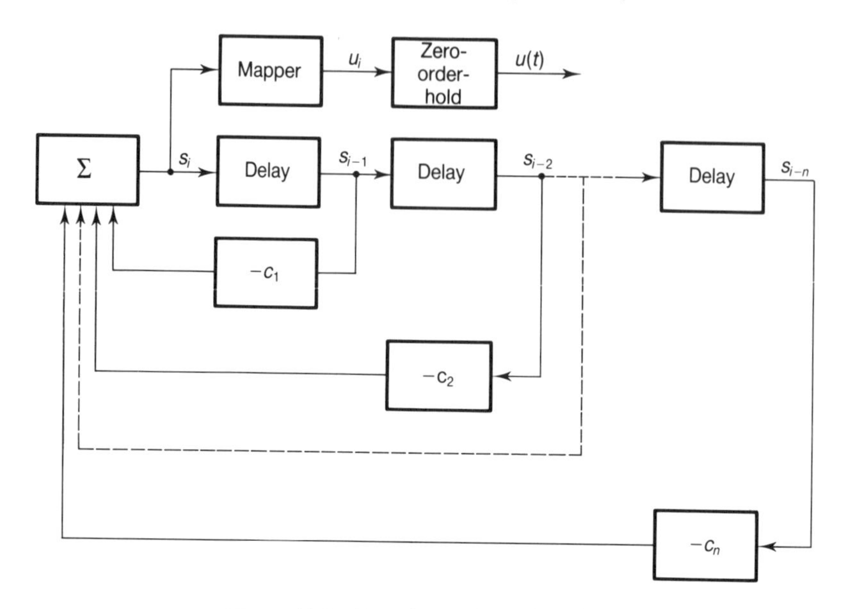

Figure 11.1 Pseudo-random signal generator.

This period of N sequence members may be sub-divided into $q-1$ sub-periods, each of $N/(q-1)$ members, and any sub-period may be formed by multiplying the members of the previous sub-period by a primitive element, g, given by

$$g = (-1)^n c_n \tag{11.17}$$

so that

$$s_{i+N/(q-1)} = g s_i \qquad \text{all } i \tag{11.18}$$

When the first sequence member is generated, the shift register in Figure 11.1 can contain any one of the q^n different combinations of the q field elements of GF(q), except for 00 ... 0, which would generate a null sequence. The m-sequence $\{s_i\}$ can therefore have N different phases, from among which it is convenient to select a reference phase with the property

$$s_{qi} = s_i \qquad \text{all } i \tag{11.19}$$

Because of the sub-periodicity of the sequence, there are in fact $q-1$ different phases with this property, from among which the actual reference phase is chosen to be the one for which the first non-zero member, starting from s_0, is 1.

Table 11.8 shows a period of the reference phase of the m-sequence in GF(7) generated with the primitive polynomial $1 + x + 3x^2$. The period of 48 members may be sub-divided into 6 sub-periods of 8 members. A member of a sub-period may be formed by multiplying the corresponding member of the previous sub-period by the primitive element 3; for example,

Table 11.8 Period of m-sequence in GF(7)

s_0

1	3	1	4	0	2	5	3	3	2	3	5	0	6	1	2	2	6	2	1	0	4	3	6
6	4	6	3	0	5	2	4	4	5	4	2	0	1	6	5	5	1	5	6	0	3	4	1

Table 11.9 Period of m-sequence in GF(9)

s_0

1	1	8	1	6	0	4	8	6	1	3	3	5	3	8	0	7	5	8	3
4	4	1	4	5	0	2	1	5	4	7	7	3	7	1	0	6	3	1	7
2	2	4	2	3	0	8	4	3	2	6	6	7	6	4	0	5	7	4	6
8	8	2	8	7	0	1	2	7	8	5	5	6	5	2	0	3	6	2	5

$$s_{11} = s_3 = 5 \tag{11.20}$$

The reference phase property gives, for example,

$$s_{21} = s_3 = 4 \tag{11.21}$$

Table 11.9 shows a period of the reference phase of the m-sequence in GF(9) generated with the primitive polynomial $1 + x + 3x^2$, with corresponding properties.

11.3.3 Pseudo-random sequences

If the q field elements of GF(q), in the form

$$0, g^0, g^1, g^2, \ldots, g^{q-2}$$

are mapped into q real numbers

$$u(0), u(g^0), u(g^1), u(g^2), \ldots, u(g^{q-2})$$

then the m-sequence $\{s_i\}$ is converted into the pseudo-random sequence $\{u_i\}$ according to equation (11.13).

For example, one of the primitive elements of GF(7) is 3, and the mapping

$$u(0) = 0$$

$$u(3^0) = u(1) = 1 \qquad u(3^1) = u(3) = 3 \qquad u(3^2) = u(2) = 2 \tag{11.22}$$

$$u(3^3) = u(6) = -1 \quad u(3^4) = u(4) = -3 \quad u(3^5) = u(5) = -2$$

converts the m-sequence in Table 11.8 into the pseudo-random sequence in Table 11.10. The pseudo-random sequence $\{u_i\}$ has period N, and the period may be sub-divided into $q - 1$ sub-periods, each of $N/(q - 1)$ members, but the relationships between the members of successive sub-periods depend not only on equation (11.18) but also on the mapping.

Table 11.10 Period of pseudo-random sequence based on m-sequence in GF(7)

u_0

1 3 1 −3 0 2 −2 3	3 2 3 −2 0 −1 1 2	2 −1 2 1 0 −3 3 −1
−1 −3 −1 3 0 −2 2 −3	−3 −2 −3 2 0 1 −1 −2	−2 1 −2 −1 0 3 −3 1

11.3.4 Pseudo-random signals

When the pseudo-random sequence $\{u_i\}$ is applied to a zero-order-hold it is converted into the step-wise continuous pseudo-random signal $u(t)$ with period $N\Delta t$ according to equation (11.14). For example, the pseudo-random sequence in Table 11.10 is converted into the pseudo-random signal in Figure 11.2.

11.4 Autocorrelation functions

As the pseudo-random signal $u(t)$ has period $N\Delta t$ its autocorrelation function $R_{uu}(\tau)$ is given by

$$R_{uu}(\tau) = \frac{1}{N\Delta t}\int_0^{N\Delta t} u(t)u(t + \tau)\,\mathrm{d}t \tag{11.23}$$

where $R_{uu}(\tau)$ also has period $N\Delta t$. Because $u(t)$ has the constant values u_i in the intervals $i\Delta t \leqslant t < (i + 1)\Delta t$ for all i, $R_{uu}(\tau)$ is linear between its values at integer multiples of Δt. These values are given by $R_{uu}(k)$, the autocorrelation function of the pseudo-random sequence $\{u_i\}$, given by

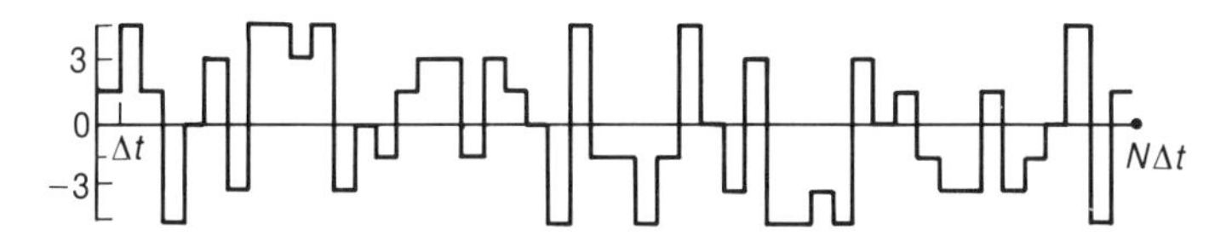

Figure 11.2 Period of pseudo-random signal.

$$R_{uu}(k) = \frac{1}{N}\sum_{i=0}^{N-1} u_i u_{i+k} = \frac{1}{N}\sum_{i=0}^{N-1} u(s_i)u(s_{i+k}) \qquad (11.24)$$

The autocorrelation function $R_{uu}(k)$ has period N, and is an even function, so that

$$R_{uu}(k) = R_{uu}(N + k) = R_{uu}(-k) = R_{uu}(N - k) \qquad (11.25)$$

For example, the autocorrelation function of the pseudo-random sequence in Table 11.10 is given in Table 11.11, and the autocorrelation function of the pseudo-random signal in Figure 11.2 is given in Figure 11.3.

Table 11.11 Period of autocorrelation function of pseudo-random sequence in Table 11.10

$R_{uu}(0)$

4 0 0 0 0 0 0 0	2 0 0 0 0 0 0 0	−2 0 0 0 0 0 0 0
−4 0 0 0 0 0 0 0	−2 0 0 0 0 0 0 0	2 0 0 0 0 0 0 0

The characteristics of the autocorrelation functions $R_{uu}(\tau)$ and $R_{uu}(k)$ follow from the ways in which the terms $u(s_i)u(s_{i+k})$ occur in the summation in equation (11.24). If k is not a multiple of $N/(q-1)$, then the terms $u(d)u(e)$ occur q^{n-2} times for all combinations of the elements d and e in GF(q), except for the term $u(0)u(0)$, which occurs only $q^{n-2} - 1$ times. Therefore

$$R_{uu}(k) = \frac{1}{N}\left[q^{n-2}\left\{u(0) + \sum_{h=0}^{q-2} u(g^h)\right\}^2 - u^2(0)\right] \qquad k \neq N/(q-1) \qquad (11.26)$$

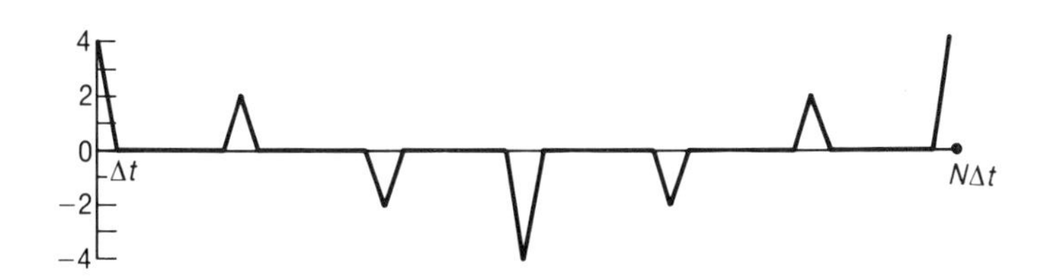

Figure 11.3 Autocorrelation function of pseudo-random signal.

If k is a multiple of $N/(q-1)$, then the ways in which the terms occur in the summation in equation (11.24) are restricted by equation (11.18), so that the terms $u(d)u(g^r d)$ occur q^{n-1} times for all non-zero elements d in GF(q), while $u(0)u(0)$ occurs only $q^{n-1}-1$ times. Therefore

$$R_{uu}(rN/\langle q-1\rangle) = \frac{1}{N}\left[q^{n-1}\left\{u^2(0) + \sum_{h=0}^{q-2} u(g^h)u(g^{h+r})\right\} - u^2(0)\right] \tag{11.27}$$

The process of designing a pseudo-random signal for system identification by cross-correlation involves choosing the real numbers into which the field elements of GF(q) are mapped so that the autocorrelation function $R_{uu}(k)$ is zero except at a few chosen values of k within a period N. As the majority of values of k are those for which $k \neq N/(q-1)$, it is desirable to choose the mapping so that $R_{uu}(k)$ is zero for all such values. Equation (11.26) shows that this can be accomplished by mapping the zero element of GF(q) into zero, that is

$$u(0) = 0 \tag{11.28}$$

and mapping the non-zero elements of GF(q) so as to satisfy the requirement

$$\sum_{h=0}^{q-2} u(g^h) = 0 \tag{11.29}$$

For example, for a pseudo-random sequence based on an m-sequence in GF(7) in which one of the primitive elements is 3, this requires that

$$u(0) = 0$$

$$u(g^0) + u(g^1) + u(g^2) + u(g^3) + u(g^4) + u(g^5)$$
$$= u(1) + u(3) + u(2) + u(6) + u(4) + u(5) = 0 \tag{11.30}$$

The mapping in equation (11.22) is one of many that satisfies this requirement.

By placing further restrictions on the mapping, it is possible to ensure that the autocorrelation function $R_{uu}(k)$ is zero for all except two values of k within a period N, in which case the autocorrelation function is primitive (Barker, 1969). If $p > 2$ and the non-zero elements of GF(q) are mapped to satisfy the requirement

$$u(g^h) + u(g^{h+(q-1)/2}) = 0 \qquad h = 0, 1, \ldots, (q-1)/2 - 1 \tag{11.31}$$

then from equation (11.18), the pseudo-random sequence $\{u_i\}$ has the property

$$u_i + u_{i+N/2} = 0 \qquad \text{all } i \tag{11.32}$$

so that the sequence is inverse-repeat, or anti-symmetric. In this case, from equation (11.24), the autocorrelation function $R_{uu}(k)$ has the property

$$R_{uu}(k) + R_{uu}(k + N/2) = 0 \qquad \text{all } k \tag{11.33}$$

so that it is also inverse-repeat.

Any mapping which satisfies the requirement in equation (11.31) automatically satisfies the requirement in equation (11.29).

Equation (11.33) shows that the two values of k within a period N for which the autocorrelation function $R_{uu}(k)$ is non-zero are $k = 0$ and $k = N/2$ and if $R_{uu}(k)$ is to be zero for all other values of k then equations (11.25), (11.27), (11.28) and (11.31) give the requirement

$$\sum_{h=0}^{(q-3)/2} u(g^h)u(g^{h+r}) = 0 \quad \begin{array}{l} r = 1, 2, \ldots, (q-3)/4 \text{ if } (q-3)/2 \text{ is even} \\ r = 1, 2, \ldots, (q-5)/4 \text{ if } (q-5)/2 \text{ is even} \end{array} \tag{11.34}$$

For example, for a pseudo-random sequence based on an m-sequence in GF(7), in which one of the primitive elements is 3, the inverse-repeat requirement in equation (11.31) gives

$$\begin{aligned} u(3^0) + u(3^3) &= u(1) + u(6) = 0 \\ u(3^1) + u(3^4) &= u(3) + u(4) = 0 \\ u(3^2) + u(3^5) &= u(2) + u(5) = 0 \end{aligned} \tag{11.35}$$

The mapping in equation (11.22) satisfies this requirement, so both the pseudo-random sequence in Table 11.10 and its autocorrelation function in Table 11.11 are inverse-repeat.

The primitive autocorrelation function requirement in equation (11.34) gives

$$u(3^0)u(3^1) + u(3^1)u(3^2) + u(3^2)u(3^3) = u(1)u(3) + u(3)u(2) + u(2)u(6) = 0 \tag{11.36}$$

The mapping in equation (11.22) does not satisfy this requirement, so the autocorrelation function in Table 11.11 is not primitive.

A mapping which satisfies all the requirements in equations (11.28), (11.35) and (11.36) is

$$\begin{aligned} u(0) &= 0 \\ u(1) &= -u(6) = 6 \\ u(2) &= -u(5) = 3 \\ u(3) &= -u(4) = 2 \end{aligned} \tag{11.37}$$

The pseudo-random sequence obtained with this mapping is shown in Table 11.12 and its primitive autocorrelation function in Table 11.13.

Table 11.12 Period of pseudo-random sequence with primitive autocorrelation function

u_0

6 2 6 −2 0 3 −3 2	2 3 2 −3 0 −6 6 3	3 −6 3 6 0 −2 2 −6
−6 −2 −6 2 0 −3 3 −2	−2 −3 −2 3 0 6 −6 −3	−3 6 −3 −6 0 2 −2 6

Table 11.13 Period of primitive autocorrelation function of pseudo-random sequence in Table 11.12

$R_{uu}(0)$

14.3 0 0 0 0 0 0 0	0 0 0 0 0 0 0 0	0 0 0 0 0 0 0 0
−14.3 0 0 0 0 0 0 0	0 0 0 0 0 0 0 0	0 0 0 0 0 0 0 0

11.5 Harmonic content

The pseudo-random signal $u(t)$ has period $N\Delta t$, and may therefore be represented by a Fourier series, so that

$$u(t) = \sum_{k=-\infty}^{\infty} C_k \exp(2\pi jkt/N\Delta t) \tag{11.38}$$

in which the coefficient C_k defines the magnitude and phase of the kth harmonic of $u(t)$, which has frequency $k/N\Delta t$ Hz.

The frequency response $H(\omega)$ of the zero-order-hold which converts the pseudo-random sequence $\{u_i\}$ into the pseudo-random signal $u(t)$ is given by

$$H(\omega) = \Delta t \frac{\sin(\omega\Delta t/2)}{\omega\Delta t/2} \exp(-j\omega\Delta t/2) \tag{11.39}$$

and the coefficient C_k is therefore given by

$$C_k = H(2\pi k/N\Delta t)\frac{1}{\Delta t} U_k = \frac{\sin(\pi k/N)}{\pi k/N} \exp(-j\pi k/N) U_k \tag{11.40}$$

where $\{U_k\}$ is the discrete Fourier transform of the pseudo-random sequence $\{u_i\}$, given by

$$U_k = \frac{1}{N} \sum_{i=0}^{N-1} u_i \exp(-2\pi jik/N) \qquad \text{all } k \tag{11.41}$$

Equation (11.40) shows that the magnitude and phase of the kth harmonic of the pseudo-random signal $u(t)$ are the magnitude phase of U_k, modulated by

$$\left|\frac{\sin(\pi k/N)}{\pi k/N}\right|$$

and shifted by $-180k/N^0$, respectively. These factors are shown in Figure 11.4.

The magnitude factor shows that, for system identification, only the first $N/2$ harmonics of the pseudo-random signal $u(t)$ are likely to be of interest. They contain about 77% of the signal power, and in any case the bandwidth of the system under test must not extend beyond the $N/2$-th harmonic if aliasing is to be avoided when sampling the output of the system at the same frequency as that of the pseudo-random signal generator clock, as is normally the case. Corrections for both the magnitude and phase factors may be applied to results obtained with the first $N/2$ harmonics. Alternatively, the phase factor can be eliminated entirely by sampling the output of the system under test halfway between the epochs of the pseudo-random signal generator clock.

Attention may therefore be focused on the discrete Fourier transform sequence $\{U_k\}$ for determining the harmonic content of the pseudo-random signal $u(t)$ and the pseudo-random sequence $\{u_i\}$. As in equation (11.41), the period N is not normally a power of 2, the popular radix-2 Fast Fourier Transform cannot normally be used

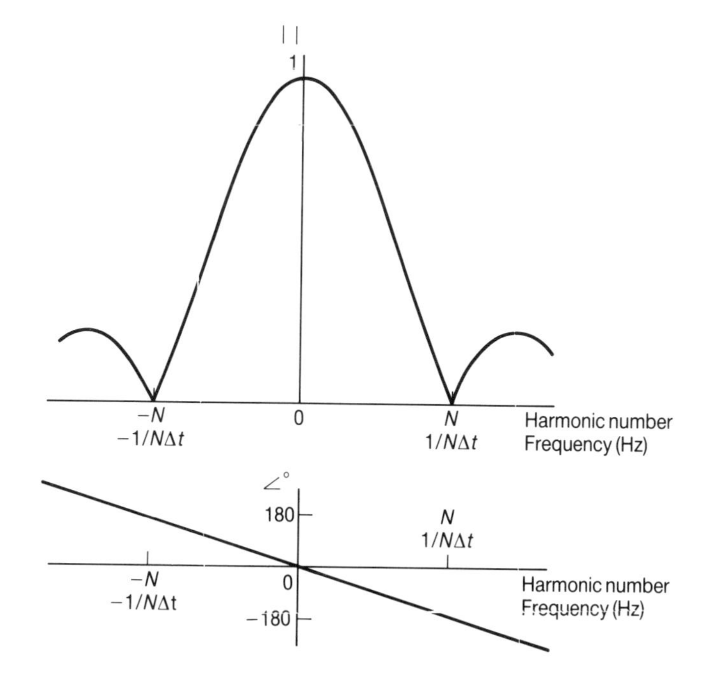

Figure 11.4 Magnitude and phase of correction factors.

to compute $\{U_k\}$ from $\{u_i\}$, but other efficient algorithms are available for this purpose (Rabiner *et al.*, 1969; Singleton, 1969; Winograd, 1976; Silverman, 1977).

From equation (11.41) the transform sequence $\{U_k\}$ has period N and is an odd function, so that

$$U_k = U_{N+k} = -U_{-k} = -U_{N-k} \qquad \text{all } k \tag{11.42}$$

Because the transform sequence $\{U_k\}$ is derived from a reference phase of the m-sequence $\{s_i\}$, it has a property similar to that in equation (11.19), which is

$$U_{qk} = U_k \qquad \text{all } k \tag{11.43}$$

Moreover, a consequence of the sub-periodicity of the m-sequence $\{s_i\}$ is that a period of N members of the transform sequence $\{U_k\}$ may also be sub-divided into sub-periods. In this case, however, there are $N/(q-1)$ sub-periods, each of $q-1$ members. The magnitudes of corresponding members in the sub-periods are equal, that is,

$$|U_{k+q-1}| = |U_k| \qquad \text{all } k \tag{11.44}$$

and only the phases are different.

Table 11.14 shows a period of the transform sequence obtained from the pseudo-random sequence in Table 11.12. The period of 48 members may be sub-divided into 8 sub-periods of 6 members. In Table 11.14 all harmonics which are

Table 11.14 Period of transform of pseudo-random sequence in Table 11.12

	U_0											
$\lvert\,\rvert$	0	0.77	0	0.77	0	0.77	0	0.77	0	0.77	0	0.77
∠°	0	36	0	24	0	−23	0	36	0	−99	0	100
$\lvert\,\rvert$	0	0.77	0	0.77	0	0.77	0	0.77	0	0.77	0	0.77
∠°	0	23	0	−99	0	−111	0	−100	0	24	0	−111
$\lvert\,\rvert$	0	0.77	0	0.77	0	0.77	0	0.77	0	0.77	0	0.77
∠°	0	111	0	−24	0	100	0	111	0	99	0	23
$\lvert\,\rvert$	0	0.77	0	0.77	0	0.77	0	0.77	0	0.77	0	0.77
∠°	0	100	0	−99	0	36	0	−23	0	24	0	36

multiples of 2 are absent. This will be shown to be a consequence of the inverse-repeat mapping in equation (11.37). In general, the process of designing a pseudo-random signal for system identification by harmonic analysis involves choosing the real numbers into which the field elements of GF(q) are mapped so that all harmonics which are multiples of a prime number v are absent, that is,

$$U_{iv} = 0 \qquad \text{all } i \tag{11.45}$$

To determine such a mapping, it is necessary to consider the magnitude of the kth harmonic, which is given by

$$|U_k| = \begin{cases} |M_0 q^{n-1} + u(0)(q^{n-1} - 1)/(q-1)|(q-1)/N & k = rN \\ |M_0 - u(0)|\, q^{(n-2)/2}(q-1)/N & k = s(q-1) \neq rN \\ |M_k|\, q^{(n-1)/2}(q-1)/N & k \neq s(q-1) \end{cases} \tag{11.46}$$

where $\{M_k\}$ is the discrete Fourier transform of the sequence of non-zero elements of GF(q) generated as powers of a primitive element g, that is, of

$$u(1),\ u(g),\ u(g^2),\ \ldots,\ u(g^{q-2})$$

which is given by

$$M_k = \frac{1}{q-1} \sum_{i=0}^{q-2} u(g^i)\exp(-2\pi jik/\langle q-1\rangle) \qquad \text{all } k \tag{11.47}$$

Equation (11.46) shows that if equation (11.45) is to be satisfied, so that all harmonics which are multiples of v are absent, then when $v \neq q - 1$ it is necessary that

$$M_{iv} = 0 \qquad \text{all } i \tag{11.48}$$

This equation can always be satisfied if v is a factor of $q - 1$, because equation (11.47) can then be rearranged as $(q - 1)/v$ groups of v terms, so that equation (11.48) becomes

$$M_{iv} = \frac{1}{q-1} \sum_{h=0}^{(q-1)/v-1} \sum_{r=0}^{v-1} u(g^{h+r(q-1)/v})\exp(-2\pi jhiv/\langle q-1\rangle) \quad \text{all } i \tag{11.49}$$

and this equation is satisfied when

$$\sum_{r=0}^{v-1} u(g^{h+r(q-1)/v}) = 0 \qquad h = 0, 1, \ldots, (q-1)/v - 1 \tag{11.50}$$

Interestingly, this important condition involves only those real numbers mapped from the non-zero elements of GF(q), and not $u(0)$.

Because $q - 1$ and its multiples are also multiples of v, it is also necessary that in equation (11.46)

$$|M_0 - u(0)| = 0 \tag{11.51}$$

and since M_0 is zero when equation (11.49) is satisfied, this requires that

$$u(0) = 0 \tag{11.52}$$

The remaining part of equation (11.46), relating to harmonics which are multiples of N, and therefore of v, is also zero when equations (11.50) and (11.52) are satisfied.

It follows from equation (11.18) that when the mapping satisfies equations (11.50) and (11.52) the pseudo-random sequence has the distinctive property

$$\sum_{r=0}^{v-1} u_{i+rN/v} = 0 \tag{11.53}$$

Equations (11.50) and (11.52), together with the requirement that v must be a factor of $q-1$, provide the basis for designing pseudo-random signals with all harmonics which are multiples of v absent.

For example, if all harmonics which are multiples of 2 are to be absent, then $v = 2$ and equations (11.50), (11.52) and (11.53) become equations (11.31), (11.28) and (11.32), respectively, showing that the pseudo-random sequence must be inverse-repeat. In order to satisfy the requirement that 2 is a factor of $q-1$ the pseudo-random sequence must be based on an m-sequence in a finite field GF(q) for which q is odd. A typical case is $q = 7$, for which an inverse-repeat pseudo-random sequence, obtained with the mapping in equation (11.37) which satisfies the necessary requirements, is shown in Table 11.12. The transform of this sequence is shown in Table 11.14, from which it can be seen that the harmonics which are multiples of 2 are absent as required.

In general, to design a pseudo-random signal so that all harmonics which are multiples of any one of several different prime numbers $v_1, v_2, \ldots, v_i$ are absent, the finite field GF(q) in which the m-sequence is generated must be chosen so that $q-1$ is a multiple of each of the prime numbers $v_1, v_2, \ldots, v_i$. In mapping the field elements of GF(q) into real numbers to convert the m-sequence into a pseudo-random sequence, the zero element of GF(q) must be mapped into zero and the non-zero elements of GF(q) must be mapped into numbers which satisfy equation (11.50) for each of the prime numbers $v_1, v_2, \ldots, v_i$.

For example, suppose that it is required to design a pseudo-random signal in which all harmonics which are multiples of 2 or 3 are to be absent. Then the smallest finite field GF(q) in which the m-sequence can be generated is such that $q - 1 = 2 \times 3$, so $q = 7$. One of the primitive elements of GF(7) is 3, so equations (11.50) and (11.52) give the requirements for mapping the field elements of GF(q) into real numbers as

$$\begin{aligned}
&x(0) = 0 \\
&x(3^0) + x(3^3) = x(1) + x(6) = 0 \\
&x(3^1) + x(3^4) = x(3) + x(4) = 0 \\
&x(3^2) + x(3^5) = x(2) + x(5) = 0 \\
&x(3^0) + x(3^2) + x(3^4) = x(1) + x(2) + x(4) = 0 \\
&x(3^1) + x(3^3) + x(3^5) = x(3) + x(6) + x(5) = 0
\end{aligned} \tag{11.54}$$

The mapping in equation (11.22) satisfies equation (11.54), so the pseudo-random sequence in Table 11.10 is one which satisfies the requirements. Note that in addition to the inverse-repeat characteristic

$$u_i + u_{i+24} = 0 \tag{11.55}$$

the sequence also has the characteristic expected from equation (11.53) when $v = 3$:

$$u_i + u_{i+16} + u_{i+32} = 0 \tag{11.56}$$

The transform of the pseudo-random sequence in Table 11.10 is shown in Table 11.15, from which it can be seen that the harmonics which are multiples of 2 or 3 are absent. Although Table 11.15 shows a period of the transform sequence, only the 24 harmonics in the first half of the period are useful, and 16 of these are absent so about 77% of the power of the signal is concentrated in the 8 useful harmonics 1, 5, 7, 11, 13, 17, 19 and 23. If more harmonics are required, then a longer pseudo-random sequence must be used.

Table 11.16 shows the first two of the six sub-periods of a pseudo-random sequence with period 342 based on the m-sequence in GF(7) generated with the primitive polynomial $1 + 3x^2 + 2x^3$ and using the mapping in equation (11.22) as before. Table 11.17 shows the first eight of the 57 sub-periods of the transform of the pseudo-random sequence in Table 11.16.

Table 11.15 Period of transform of pseudo-random sequence in Table 11.10

	U_0											
$\lvert\ \rvert$	0	0.51	0	0	0	0.51	0	0.51	0	0	0	0.51
$\angle^\circ$	0	69	0	0	0	−56	0	69	0	0	0	67
	0	0.51	0	0	0	0.51	0	0.51	0	0	0	0.51
	0	56	0	0	0	−144	0	−67	0	0	0	−144
	0	0.51	0	0	0	0.51	0	0.51	0	0	0	0.51
	0	144	0	0	0	67	0	144	0	0	0	−56
	0	0.51	0	0	0	0.51	0	0.51	0	0	0	0.51
	0	−67	0	0	0	−69	0	56	0	0	0	−69

Table 11.16 Sub-periods of pseudo-random sequence based on m-sequence in GF(7)

S_0																		
1	0	−2	−2	−1	3	0	0	1	0	−3	−2	2	−2	−2	2	3	−2	1
0	1	−2	−3	−3	−1	1	2	−1	−1	−1	−2	−2	1	3	1	3	−2	3
0	2	1	1	0	2	−2	1	2	1	−1	0	1	2	−3	−1	−2	2	1
−2	0	−3	−3	2	1	0	0	−2	0	−1	−3	3	−3	−3	3	1	−3	−2
0	−2	−3	−1	−1	2	−2	3	2	2	2	−3	−3	−2	1	−2	1	−3	1
0	3	−2	−2	0	3	−3	−2	3	−2	2	0	−2	3	−1	2	−3	3	−2

Table 11.17 Period of transform of pseudo-random sequence in Table 11.16

	U_0											
||	0	0.19	0	0	0	0.19	0	0.19	0	0	0	0.19
∠°	0	−67	0	0	0	−136	0	−67	0	0	0	143
	0	0.19	0	0	0	0.19	0	0.19	0	0	0	0.19
	0	123	0	0	0	12	0	−71	0	0	0	142
	0	0.19	0	0	0	0.19	0	0.19	0	0	0	0.19
	0	61	0	0	0	20	0	−144	0	0	0	−136
	0	0.19	0	0	0	0.19	0	0.19	0	0	0	0.19
	0	−61	0	0	0	9	0	−9	0	0	0	−123

Comparing Table 11.17 with Table 11.15 shows that the harmonic structure of the longer sequence is similar to that of the shorter, and in particular harmonics which are multiples of 2 or 3 are again absent. In this case, however, 114 of the 171 harmonics in the first half of the period are absent, so about 77% of the power of the signal is concentrated in the 57 useful harmonics 1, 5, 7, 11, 13, 17, 19, 23, 25, 29, 31, 35, 37, 41, 43, 47, ..., 169.

Table 11.18 shows the harmonics which are removable from pseudo-random sequences based on m-sequences in finite fields with less than 32 elements.

Table 11.18 Harmonics removable from pseudo-random sequences based on m-sequences in GF(q), for $q < 32$

Finite field (number of elements)	Removable harmonics (multiples of)
2	–
3	2
4	3
5	2
7	2,3
8	7
9	2
11	2,5
13	2,3
16	3,5
17	2
19	2,3
23	2,11
25	2,3
27	2,13
29	2,7
31	2,3,5

11.6 Application

Multi-level pseudo-random signals are used for the identification of both linear and non-linear systems. Detailed consideration of the latter is beyond the scope of this chapter because much of the work involves high-order autocorrelation functions of the signals (Barker and Pradisthayon, 1970). Correlation methods for non-linear system identification using pseudo-random signals have been described by Barker *et al.* (1972), Barker and Davy (1978) and Billings and Fakhouri (1980). An excellent survey has been given by Billings (1980). Harmonic analysis methods for non-linear system identification using pseudo-random signals have been described by Barker and Davy (1979) and Barker and Al-Hilal (1985).

For linear system identification, the greatest strength of multi-level pseudo-random signals lies not in correlation methods but in harmonic analysis methods. Indeed, it may be argued that the way in which the technology has evolved, together with the label 'pseudo-random' attached to these signals, has meant that their true nature as deterministic, multi-frequency periodic signals has tended to be obscured or ignored. The crude averaging process of correlation, so essential for truly random signals, simply destroys a good deal of useful information when applied to pseudo-random signals.

Kavanagh (1969) first described the harmonic properties of pseudo-random binary sequences. Lamb and Rees (1973) attempted to apply these sequences to the identification of linear systems by adding a member to increase the sequence period to a power of 2 so that the radix-2 Fast Fourier Transform could be used. Barker and Davy (1975) extended the work of Kavanagh to multi-level pseudo-random signals, and showed how they should properly be applied to the identification of linear systems.

The basis of the harmonic analysis method for system identification using multi-level pseudo-random signals is shown in Figure 11.5. The estimate of the system harmonic response at the kth harmonic, with frequency $k/N\Delta t$ Hz is

$$\hat{W}(2\pi k/N\Delta t) = Y_k/U_k \qquad 0 < k < N/2 \tag{11.57}$$

where $\{Y_k\}$ is the discrete Fourier transform of the sequence $\{y_i\}$ which is obtained by synchronously sampling the output $y(t)$ of the system under test.

Complete analyses of both the systematic and random errors which may be expected with this method, together with some corrections that may be applied, are given by Barker and Davy (1975) for binary, inverse-repeat binary and inverse-repeat ternary pseudo-random signals. The most common systematic errors for which corrections cannot be applied are those due to non-linearities, and it is therefore desirable to eliminate these errors from the estimates entirely. It is possible to achieve this using correlation methods, but only under quite severe restrictions (Barker and Obidegwu, 1973), and harmonic analysis methods are far better for this purpose because of the power of the spectral decomposition approach.

If even harmonics, multiples of 2, are absent from the pseudo-random signal $u(t)$, so that it is composed of odd harmonics only, then even-order non-linearities present in the system under test generate components in the output signal $y(t)$ that are composed of even harmonics, multiples of 2, only. These even harmonics are

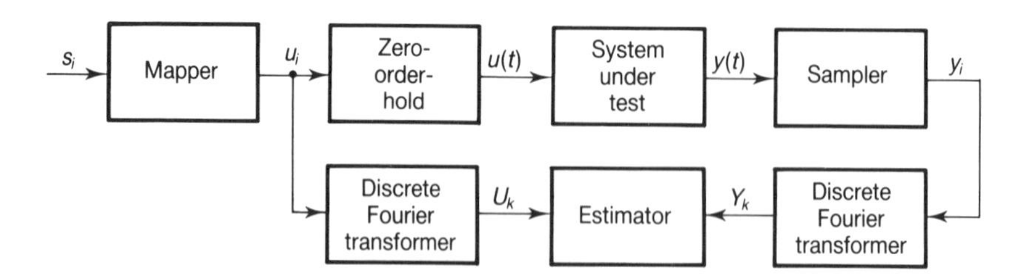

Figure 11.5 System-identification method.

completely independent of the odd harmonics which compose the components in the output signal $y(t)$ that are generated by the linear part of the system under test, or any odd-order non-linearities present in it.

The output sequence transform $\{Y_k\}$ may therefore be decomposed into two independent sequences, one of members for which k is even and the other of members for which k is odd. The former is either discarded or used to identify the second-order non-linearity (Barker and Davy, 1978). The latter is used to obtain, through equation (11.57), estimates of the harmonic response of the linear part of the system that are completely unbiased by the presence of even-order non-linearities in the system under test.

An extension of this approach is used when odd-order non-linearities are also present in the system under test, although in this case it is not possible to achieve complete independence. For example, if harmonics that are multiples of 3 are also absent from the pseudo-random signal $u(t)$, and a third-order non-linearity is present in the system under test, then that non-linearity generates a component in the output signal $y(t)$ that is composed of odd harmonics only. These harmonics include not only those that are multiples of 3 and are therefore absent from the pseudo-random signal $u(t)$, but also those harmonics that are present in the pseudo-random signal $u(t)$. The latter are identical to the harmonics which compose the component of the output signal $y(t)$ that is generated by the linear part of the system under test, except for a gain change in the magnitudes which reflects the equivalent linear gain of the third-order non-linearity. The remainder of the output sequence transform $\{Y_k\}$ is therefore decomposed into two further sequences, one of members for which k is odd and a multiple of 3 and the other of members for which k is odd and not a multiple of 3. The former is either discarded or used to identify the third-order non-linearity (Barker and Al-Hilal, 1985). The latter is used to obtain, through equation (11.57), estimates of the harmonic response of the linear part of the system that are completely unbiased by the presence of even-order non-linearities, or the third-order non-linearity, in the system under test.

The method is most easily demonstrated with an instantaneous non-linearity for which

$$y = u + u^2 + u^3 \tag{11.58}$$

A pseudo-random sequence $\{u_i\}$ with all harmonics which are multiples of 2 or 3 absent is given in Table 11.10, and its transform in Table 11.15. Using this pseudo-random sequence gives the output sequence $\{y_i\}$ shown in Table 11.19 for which the output sequence transform $\{Y_k\}$ is presented in Table 11.20.

Table 11.19 Period of output sequence

y_0

3 39 3 −21 0 14 −6 39	39 14 39 −6 0 −1 3 14	14 −1 14 3 0 −21 39 −1
−1 −21 −1 39 0 −6 14 −21	−21 −6 −21 14 0 3 −1 −6	−6 3 −6 −1 0 39 −21 3

Table 11.20 Period of transform of output sequence

	Y_0											
$\lvert\,\rvert$	4.08	4.04	0.77	1.98	0.77	4.04	0.58	4.04	0.77	1.98	0.77	4.04
$\langle^\circ$	0	69	74	−156	−101	−56	0	69	−177	81	150	67
	0.58	4.04	0.77	1.98	0.77	4.04	0.58	4.04	0.77	1.98	0.77	4.04
	180	56	74	81	79	−144	0	−67	101	−156	150	−144
	0.58	4.04	0.77	1.98	0.77	4.04	0.58	4.04	0.77	1.98	0.77	4.04
	180	144	−150	156	−101	67	0	144	−79	−81	−74	−56
	0.58	4.04	0.77	1.98	0.77	4.04	0.58	4.04	0.77	1.98	0.77	4.04
	180	−67	−150	−81	177	−69	0	56	101	156	−74	−69

Table 11.21 Period of decomposed transform of output sequence

	Y_0											
$\lvert\,\rvert$	0	4.04	0	0	0	4.04	0	4.04	0	0	0	4.04
$\langle^\circ$	0	69	0	0	0	−56	0	69	0	0	0	67
	0	4.04	0	0	0	4.04	0	4.04	0	0	0	4.04
	0	56	0	0	0	−144	0	−67	0	0	0	−144
	0	4.04	0	0	0	4.04	0	4.04	0	0	0	4.04
	0	144	0	0	0	67	0	144	0	0	0	−56
	0	4.04	0	0	0	4.04	0	4.04	0	0	0	4.04
	0	−67	0	0	0	−69	0	56	0	0	0	−69

The output sequence $\{Y_k\}$ is decomposed by removing all members for which k is a multiple of 2 or 3, to give the sequence in Table 11.21. Comparison of Tables 11.15 and 11.21 show that equation (11.57) will give the correct estimates of $W(2\pi k/N\Delta t)$ for values of k which are not multiples of 2 or 3. The gain of 7.92 is the sum of a linear gain of 1 from the linearity and an equivalent linear gain of 6.92 from the cubic non-linearity, but these are not separable in the result, which reflects the linearization of the non-linear system with the pseudo-random signal levels chosen.

Although the method has been demonstrated here with an instantaneous non-linearity, it is of course applicable to non-linear systems with dynamics, for which reliable estimates of linear behaviour may be obtained.

12

Design and Application of Non-binary Low-peak-factor Signals for System-dynamic Measurement

David Rees and David L. Jones

12.1 Introduction

A fundamental component of any control system design is the identification of a model of the dynamic characteristics of the system, which is often expressed in transfer-function terms. Generally, the transfer function of a system may be directly identified from a knowledge of the frequency response of a system, which in turn may be obtained by measurement of the system's input and output signals. The choice of input signal is critical, since it can affect the accuracy and the speed at which the system response is obtained. In some applications the choice of input stimulus is restricted by the operating environment and it is impossible to excite the system with an arbitrary input signal. However, in the main, the only restriction on the excitation signal is its maximum allowed peak-to-peak amplitude.

The conventional method for measuring the frequency response of a system employs a sinusoid as the input stimulus. Developments in signal-processing algorithms and hardware have made it possible to utilize more complex input test signals which have a broadband spectrum, termed multi-frequency test signals. These signals allow several spectral estimates to be measured simultaneously and this results in an important reduction in the overall measurement time. The mechanization of such multi-frequency waveform test procedures relies on the use of digital-to-analog conversion techniques for test signal generation, and analog-to-digital conversion for the capture of the system output. These techniques must be applied carefully, since an undesirable loss of accuracy can result unless special precautions are taken.

In general, a successful identification method must incorporate an accurate measurement of the system frequency response which should be obtained rapidly, with little disturbance to the process. Furthermore, identification is often undertaken on systems which are noisy and exhibit slight non-linear behaviour. It is desirable for the measurement procedure to provide both noise and harmonic rejection capability.

In order to fully exploit the advantages of multi-frequency testing the test waveform should possess the following attributes:

1. The waveform should have a rectangular magnitude envelope in the frequency domain, i.e. the waveform should contain components of equal energy. This avoids weak components which may be susceptible to noise in the system under test.
2. The waveform should have a large RMS value with respect to its time domain peak-to-peak value, i.e. a low peak factor (Schroeder, 1970), as described by:

$$\text{Peak factor} = \frac{x_{\text{max}} - x_{\text{min}}}{2\sqrt{2}x_{\text{rms}}} \tag{12.1}$$

A low peak factor allows system testing at high energy levels for a given dynamic range from the system under test (see Chapter 2, Section 2.4.2).

A number of multi-frequency waveforms have been suggested as suitable test signals and these include impulse and square waveforms, random noise, swept sine (Van Brussel, 1975), multi-frequency binary signals (Paehlike and Rake, 1979; Van den Bos, 1987), maximum-length pseudo-random binary signals (Godfrey, 1969a; Barker and Davy, 1975), and multi-frequency signals (Lamb and Rees, 1973; Rees, 1977). The relative merits of these signals when used to test real systems have been outlined elsewhere (Van der Ouderaa *et al.*, 1988a,b). It is therefore the aim of this chapter to focus on the design and development of one class of signals: non-binary multi-frequency waveforms.

12.2 Non-binary multi-frequency test waveforms

The general expression for a non-binary, multi-frequency test signal is given by

$$x(t) = \sum_{i=1}^{N} A_i \sin(\omega_i t + \theta_i) \tag{12.2}$$

The waveforms are defined by choosing the required values of N, A_i and ω_i. From this the signal magnitude envelope in the frequency domain is derived. The peak factor of such signals is dependent upon the choice of θ_i. The main problem in designing such waveforms is that of optimizing the choice of θ_i to minimize the peak factor of the test waveform.

Since all real systems exhibit some degree of non-linear behaviour it is a constraint that any multi-frequency waveform should be capable of producing meaningful frequency estimates when testing systems exhibiting non-linear effects. It is this fact which largely curtails the use of multi-frequency test signals, since spectral estimates obtained by these test signals can be heavily corrupted by harmonic intermodulation components.

12.2.1 Prime multi-frequency signal

It was shown by analyzing the harmonics generated by a general polynomial-type non-linearity to a multi-frequency signal that the number of spurious harmonics which fall on component frequencies of the original test signal are substantially reduced, provided the test signal consists of an assemblage of sinusoids of frequencies which are prime number multiples of some fundamental which is itself excluded from the signal (Rees, 1976, 1977). The test signal is given by

$$x(t) = \sum_{\substack{i \neq 1 \\ i \neq 2 \\ i = \text{numbers}(3,5,7,11,,\ldots,\text{prime})}}^{N} A_i \sin(\omega_i t + \theta_i) \qquad (12.3)$$

The amplitudes of the sine waves are chosen to be equal and the number of sinusoids included in the signal is determined by the frequency range that is to be tested.

This signal provides total immunity from harmonic distortion due to even-power non-linearities. Also, in the case of odd-power non-linearities a signal composed of prime frequencies can be expected to give a substantial reduction in the number of spurious harmonics that fall on the component frequencies (prime harmonics), compared with PRBS and inverse-repeat PRBS, respectively.

This is demonstrated in Table 12.1, which gives the percentage reduction in harmonic distortion using the prime signal compared with PRBS when testing an open-loop second-order system with saturation at the output. The results were obtained from a MATLAB simulation. The signals used had equal peak-to-peak amplitude (the dynamic range was the same). The prime sinusoid contained 15 harmonics and the PRBS signal (sequence length = 127) was chosen so that its −3 dB bandwidth covered an equivalent frequency range. Note that the prime signal gives greater power at the spectral lines of interest than the PRBS signal.

The percentage reduction in harmonic distortion is defined as follows:

% reduction in harmonic distortion

$$= \frac{\{[(Pp_j - Pps_j)/Pp_j] - [(Pm_j - Pms_j)/Pm_j]\}}{(Pm_j - Pms_j)/Pm_j} \times 100$$

where

Table 12.1 Percentage reduction in harmonic distortion using the prime sinusoid signal compared with the PRBS

Spectral line number j	% Reduction in harmonic distortion
3	103
5	160
7	26
11	61
13	74
17	24
19	51
23	115
29	18
31	123
37	43
41	310
43	493
47	30
53	52

Pp_j = power spectrum of system output at spectral line j with PRBS input signal and no saturation,
Pps_j = power spectrum of system output at spectral line j with PRBS input signal and saturation,
Pm_j = power spectrum of system output at spectral line j with prime input signal and no saturation, and
Pms_j = power spectrum of system output at spectral line j with prime input signal and saturation.

It can be seen from the results obtained that there is a significant benefit in using the prime multi-sine signal when testing non-linear systems, as it gives better harmonic rejection.

In order to minimize the number and power of the harmonics generated the test signal should contain few sinusoids. However, the number of sinusoids chosen and their frequency content is governed by the number of points required for satisfactory description of the frequency-response characteristic and this determines the value of N in equation (12.2).

12.2.2 Phase selection

The choice of phases in equation (12.3) is critical in that they determine the dynamic range of the signal. The peak factor provides a quantitative measure of the time history of the signal. For a single sinusoid the relative peak factor is 1, and for PRBS it is 0.707. Low-peak-factor signals are always desirable in system testing as their use tends to avoid large signal perturbations. This makes the transduction of the signal somewhat easier and also ensures the reduction of harmonics generated due to non-linear behaviour. Unfortunately, prime sinusoid signals with zero phases do possess large peak factors. This is mainly due to spikes or glitches occurring at the beginning and end of each half-cycle (Figure 12.1). This is further illustrated in Table 12.2, which gives peak factor values for a prime sinusoid signal composed of 10, 15, 20, 30 and 40 harmonics. Also included in Table 12.2 for comparison purposes are the peak factors for a multi-frequency signal made up of consecutive harmonics. From these results it can be seen that the peak factor of multi-frequency signals composed of consecutive harmonics with zero phases increases as the number of harmonics increase. This is not the case with the prime harmonic signal.

To address this issue of unacceptable peak factors, a number of alternative phase-generation algorithms were explored. These have been detailed elsewhere (Rees, 1986), and so only the two most significant will be considered here.

SCHROEDER PHASES

In a paper published in 1970 Schroeder developed formulas for phase generation which are based on a simple intuitive concept of the asymptotic relationship between the power spectra of frequency-modulated signals and their instantaneous frequencies. The observation was made that phase-modulated signals are known to have low peak factors, and on this basis he proceeded to construct such a signal with the desired harmonic power content which resulted in low-peak-factor multi-frequency signals. He showed that in the case of a spectrum containing N consecutive harmonics of equal power content the phase of the nth harmonic is given by

$$\theta_n = \pi \frac{n^2}{N} \tag{12.4}$$

This formula was then used to adjust the phases of a cosine representation of the prime sinusoid signal given in equation (12.2). The results of using this formula are shown in Table 12.2 for multi-frequency signals made up of prime and consecutive harmonics and the time history of both signals with 20 harmonics is given in Figure 12.2. The resulting reduction in peak factor is substantial in both cases, but as would be expected, the results are better in consecutive harmonics.

OPTIMUM PHASES

Peak-factor minimization involves the non-linear optimization of a function with a large number of local minima. A routine based on a Quasi-Newton algorithm (Gill and Murray, 1972) was used to attempt minimization of this function $f(\theta_1, \theta_2, \ldots, \theta_N)$ of the N independent phases $\theta_1, \theta_2, \ldots, \theta_N$. The phases generated by the routine

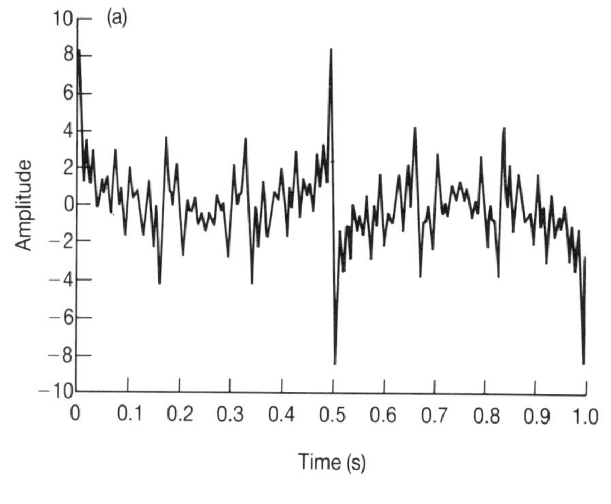

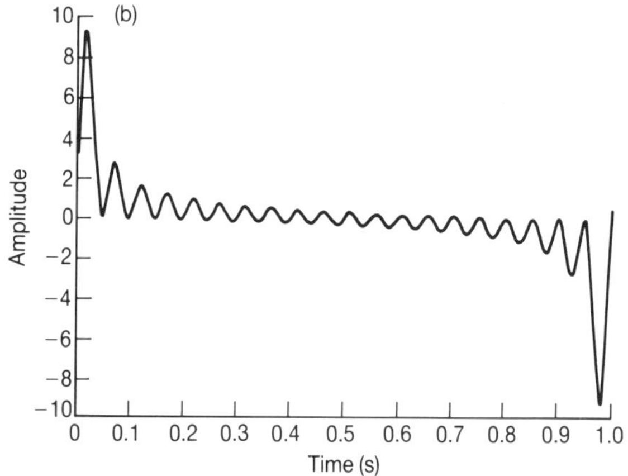

Figure 12.1 One period of a multi-frequency signal composed of 20 harmonics with all zero phases. (a) Prime harmonics; (b) consecutive harmonics.

gave signals with considerably lower peak factors than those with zero or Schroeder phases. These are sub-optimal results, however, as lower peak factors have been achieved (Van der Ouderaa *et al.*, 1988a,b) and an optimum solution to the problem has yet to be found. The routine required a considerable amount of processing time and was very sensitive to initial phase values. Schroeder phases are suggested as possible starting values. Figure 12.3 shows the time history over one period for a 20 harmonic prime signal and a signal composed of 20 consecutive harmonics. A comparison of the signals in Figure 12.1–12.3, which are of equal power, illustrates how much the peak-to-peak excursions can be reduced by phase optimization.

The peak factors for prime and consecutive harmonic signals are shown in Table 12.2. The phases used to generate these signals are given in the Appendix to this

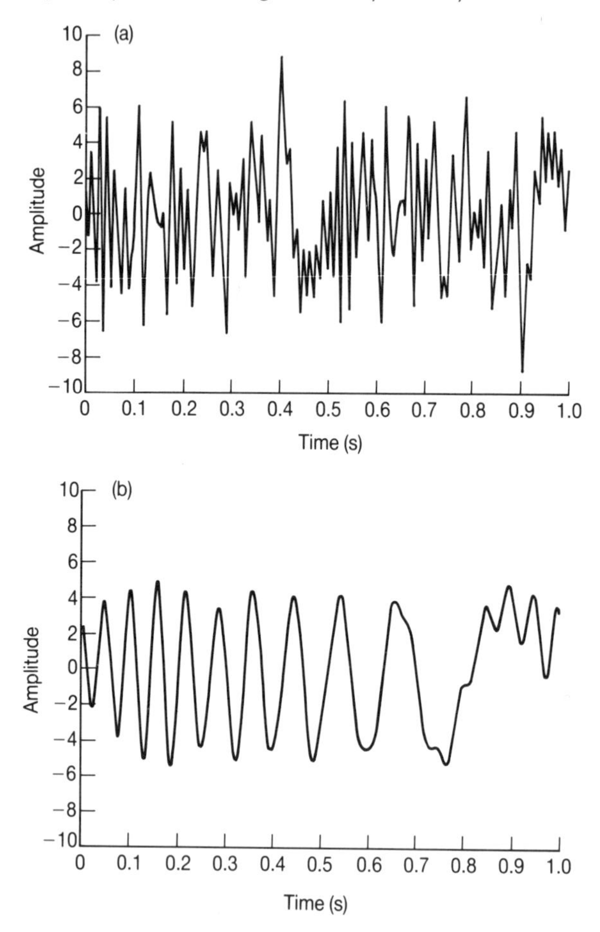

Figure 12.2 One period of a multi-frequency signal composed of 20 harmonics with Schroeder phases. (a) Prime harmonics; (b) consecutive harmonics.

chapter. Note that the 15 consecutive harmonic signal has a peak factor of less than one, a possibility dismissed by some recent authors (Boyd, 1986).

12.3 Testing systems using the prime multi-frequency signal

A frequency-response analyzer has been developed based upon the use of a 20-harmonic prime multi-frequency test waveform. Two hundred and fifty-six samples of the waveform were taken, digitized and stored within the Read Only Memory (ROM) of the analyzer. During a frequency-response measurement, these samples ($x^*(t)$) are output to the System Under Test (SUT) using a zero-order-hold digital-

Table 12.2 A comparison of peak factors for multi-frequency signals using different phase-generation algorithms

	Prime harmonics ($i = 3, 5, 7, \ldots$)			Consecutive harmonics ($i = 1, 2, 3, \ldots$)		
No. of sinusoids	Zero phases ($\Theta_i = 0$)	Schroeder phases	Optimized phases	Zero phases	Schroeder phases	Optimized phases
5	1.754	1.37	1.021	1.771	1.287	1.02
10	2.258	1.491	1.112	2.402	1.217	1.022
15	2.649	1.834	1.109	2.896	1.177	0.9972
20	2.909	1.939	1.147	3.298	1.165	1.098
30	3.444	1.814	1.105	4.028	1.134	1.082
40	2.264	2.18	1.065	4.526	1.158	1.097

to-analog converter (DAC). At the same time, the output of the SUT is captured, digitized and stored for further processing. A 256-point Fast Fourier Transform (FFT) algorithm is subsequently used to evaluate the spectral content of the captured signal. By comparing this with the spectral content of the injected test signal the transfer function of the system is simultaneously evaluated at 20 spectral points. Figure 12.4 is a functional block diagram of the measurement process.

12.3.1 Aliasing effects in the measurement process

In the complex frequency domain the test waveform is described by

$$X(\mathrm{j}\omega) = \sum_{n=\pm 3, \pm 5, \ldots, \text{prime number}}^{73} \delta\left(\omega - \frac{2\pi n}{N\,\Delta t}\right) \mathrm{e}^{\mathrm{j}\theta n} \tag{12.5}$$

This composite waveform is sampled, digitized and stored within the ROM of the instrument and this is effected by multiplying the waveform by a sequence of impulse functions $d_0(t)$, where

$$d_0(t) = \sum_{r=-\infty}^{\infty} \delta(t - r\,\Delta t) \tag{12.6}$$

$$D_0(\mathrm{j}\omega) = \frac{1}{\Delta t} \sum_{r=-\infty}^{\infty} \delta\left(\omega - \frac{2\pi r}{\Delta t}\right) \tag{12.7}$$

The sampling rate Δt was chosen so that the number of samples was 256 over the fundamental period of the waveform. This ensures that the spectrum of the samples stored in ROM are unaffected by aliasing.

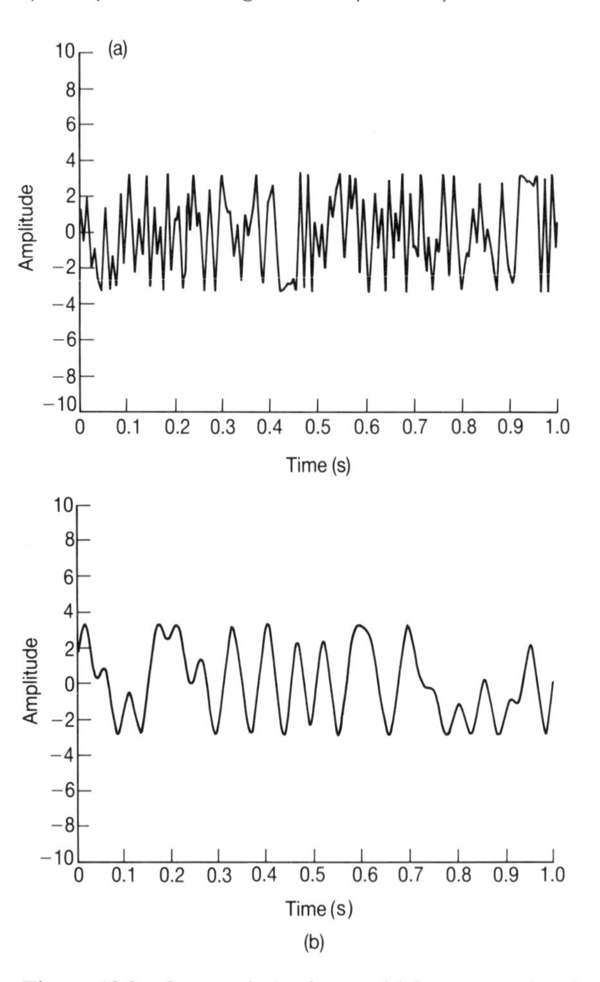

Figure 12.3 One period of a multi-frequency signal composed of 20 harmonics, with optimized phases. (a) Prime harmonics; (b) consecutive harmonics.

From equation (12.5) it can be seen that the frequency of the highest harmonic component is $73/N\,\Delta t$ or $73/256\Delta t$. The spectrum of the resulting waveform is given by

$$X^*(\mathrm{j}\omega) = X(\mathrm{j}\omega)^* D_0(\mathrm{j}\omega) \tag{12.8}$$

$$X^*(\mathrm{j}\omega) = \frac{1}{\Delta t}\sum_{r=-\infty}^{\infty} X\left[\mathrm{j}\left(\omega - \frac{2\pi r}{\Delta t}\right)\right] \tag{12.9}$$

When performing a frequency-response measurement this ROMed waveform is continuously read out of memory at a rate of one sample every Δt s and applied to the SUT through a zero-order-hold DAC.

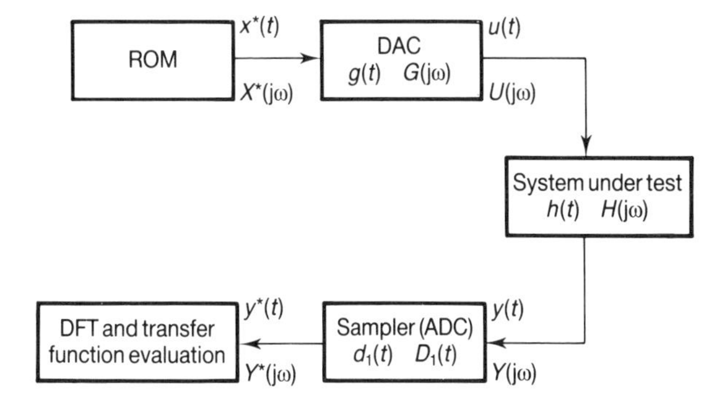

Figure 12.4 Instrument test procedure.

The time- and frequency-domain representation of the DAC are given as

$$g(t) = 0 \ldots |t| > \frac{\Delta t}{2} \tag{12.10}$$

$$= 1 \ldots |t| < \frac{\Delta t}{2}$$

$$= \frac{1}{2} \ldots |t| = \frac{\Delta t}{2}$$

and

$$|G(j\omega)| = \Delta t \frac{\sin(\omega \Delta t/2)}{(\omega \Delta t/2)} \tag{12.11}$$

The reconstituted signal applied to the SUT is therefore given by

$$u(t) = x^*(t)^*g(t) \tag{12.12}$$

$$U(j\omega) = X^*(j\omega) \cdot G(j\omega) \tag{12.13}$$

$$U(j\omega) = \frac{1}{\Delta t} \sum_{r=-\infty}^{\infty} X\left[j\left(\omega - \frac{2\pi r}{\Delta t}\right)\right] \cdot G\left[j\left(\omega - \frac{2\pi r}{\Delta t}\right)\right] \tag{12.14}$$

From equation (12.5)

$$U(j\omega) = \frac{1}{\Delta t} \sum_{r=-\infty}^{\infty} \sum_{n=\pm 3, \pm 5, \text{prime}}^{73} \delta\left[\omega - 2\pi\left(\frac{rN+n}{N\,\Delta t}\right)\right] e^{j\theta n} \cdot G\left[j\left(\omega - \frac{2\pi r}{\Delta t}\right)\right] \tag{12.15}$$

It can be seen from the above equation that the reconstituted signal varies markedly from the original prime composite signal given in equation (12.5). The signal is no longer band-limited and the original component harmonics have been shaped by the DAC ($G(\mathrm{j}\omega)$). It is this reconstituted signal which is applied to the SUT, whose frequency response $H(\mathrm{j}\omega)$ is to be measured.

The output of the system $Y(\mathrm{j}\omega)$ is therefore described in the time and frequency domains by

$$y(t) = u(t)^{*}h(t) \tag{12.16}$$

$$Y(\mathrm{j}\omega) = U(\mathrm{j}\omega)\cdot H(\mathrm{j}\omega) \tag{12.17}$$

$$Y(\mathrm{j}\omega) = \sum_{r=-\infty}^{\infty} X\left[\mathrm{j}\left(\omega + \frac{2\pi r}{\Delta t}\right)\right]\cdot G\left[\mathrm{j}\left(\omega - \frac{2\pi r}{\Delta t}\right)\right]\cdot H\left[\mathrm{j}\left(\omega - \frac{2\pi r}{\Delta t}\right)\right] \tag{12.18}$$

This output signal is synchronously sampled by the analog-to-digital converter (ADC) at the same rate (Δt) at which the input signal to the SUT is generated. This sampling function is given by

$$d_1(t) = \sum_{k=-\infty}^{\infty} \delta(t - k\Delta t) \tag{12.19}$$

$$D_1(\mathrm{j}\omega) = \frac{1}{\Delta t}\sum_{k=-\infty}^{\infty} \delta\left(\omega - \frac{2\pi k}{\Delta t}\right) \tag{12.20}$$

The resulting sampled waveform $y^{*}(t)$ is therefore given by

$$y^{*}(t) = y(t)\cdot d_1(t) \tag{12.21}$$

and

$$Y^{*}(\mathrm{j}\omega) = Y(\mathrm{j}\omega)^{*}D_1(\mathrm{j}\omega) \tag{12.22}$$

From equations (12.20) and (12.22),

$$Y^{*}(\mathrm{j}\omega) = \frac{1}{\Delta t}\sum_{k-\infty}^{\infty} \delta\left(\omega - \frac{2\pi k}{\Delta t}\right)^{*} Y(\mathrm{j}\omega) \tag{12.23}$$

or

$$Y^{*}(\mathrm{j}\omega) = \frac{1}{\Delta t}\sum_{k=-\infty}^{\infty} Y\left[\mathrm{j}\left(\omega + \frac{2\pi k}{\Delta t}\right)\right] \tag{12.24}$$

From equation (12.17) this may be rewritten as

$$Y^{*}(\mathrm{j}\omega) = \frac{1}{\Delta t}\sum_{k=-\infty}^{\infty} \frac{1}{\Delta t}\sum_{r=-\infty}^{\infty} X\left[\mathrm{j}\left(\omega + \frac{2\pi r}{\Delta t} + \frac{2\pi k}{\Delta t}\right)\right]\cdot G\left[\mathrm{j}\left(\omega + \frac{2\pi r}{\Delta t}\right)\right]\cdot H\left[\mathrm{j}\left(\omega + \frac{2\pi r}{\Delta t}\right)\right] \tag{12.25}$$

These time-domain samples of the system output are transformed using a FFT algorithm to yield the Discrete Fourier Transform (DFT) of the output signal of the system.

By comparing the calculated spectral content of this output signal with the injected signal spectrum for $n = 3, 5, \ldots, 73$ the frequency response of the system can be evaluated at 20 spectral points, since

$$Y(\mathrm{j}\omega) = U(\mathrm{j}\omega) \cdot H(\mathrm{j}\omega)$$

$$H(\mathrm{j}\omega) = \frac{Y(\mathrm{j}\omega)}{U(\mathrm{j}\omega)} \tag{12.26}$$

In order that the DFT correctly evaluates the spectrum of $y(t)$ it is necessary that the sampling rate is at least twice the frequency of the highest frequency component of the signal. Equation (12.15) shows that the signal applied to the system under test is not band-limited and therefore this condition cannot be guaranteed. As a result, the calculated spectrum $Y^*(\mathrm{j}\omega)$ may be markedly different from $Y(\mathrm{j}\omega)$ in the range $n = 3$ to $n = 73$, due to the effects of frequency folding or aliasing.

Rearranging equation (12.25), the calculated spectrum at any measurement harmonic $(2\pi m/N\delta t)$:

$$Y^*\left(\mathrm{j}\frac{2\pi m}{N\Delta t}\right) = \frac{1}{\Delta t^2} X\left(\mathrm{j}\frac{2\pi m}{N\Delta t}\right) \cdot G\left(\mathrm{j}\frac{2\pi m}{N\,\Delta t}\right) H\left(\mathrm{j}\frac{2\pi m}{N\,\Delta t}\right)$$

$$+ \frac{1}{\Delta t^2} \sum_{\substack{r=-\infty \\ r \neq 0}}^{\infty} X\left(\mathrm{j}\frac{2\pi m}{N\,\Delta t}\right) \cdot G\left(\mathrm{j}\frac{2\pi(rN+m)}{N\,\Delta t}\right) \cdot H\left(\mathrm{j}\frac{2\pi(rN+m)}{N\,\Delta t}\right) \tag{12.27}$$

From the above equation it can be seen that the spectrum of the sampled system output is affected by an infinite number of aliasing terms. This means that the calculated spectrum of the system output at any measured frequency differs from the actual spectrum due to the extra summation terms on the right-hand side of equation (12.27).

In order to reduce aliasing on the measurements the effect of these terms must be minimized. This may be done in a number of ways:

1. *Increase the number of samples per period (N) of the original test waveform stored in ROM*. It can be easily shown from equation (12.11) that the size of the $G(\mathrm{j}(2\pi(rN + M)/N\delta t))$ terms in the above equation decrease as the ratio M/N decreases. The main disadvantage with this approach lies in the extra memory required for storage of the test waveform. For a SUT with a constant gain and zero phase response it may be shown that for the prime multi-frequency test waveform the number of samples per period of the prime multi-frequency test waveform stored in ROM must be increased from 256 to 10 496 in order to reduce the effects of aliasing to less than −72 dB. Since the highest frequency component of the waveform is the 73rd harmonic, this sampling rate is over forty times the theoretical (Nyquist) rate. The drawback of this approach, therefore, is the fact that the storage requirement in the test waveform generation system must be significantly increased.

2. *Increase the sampling rate at the ADC.* The effect of this measure is to reduce the number of aliasing terms which affect the measurement at any harmonic test frequency. Again, for a system with a constant gain and zero phase response the sampling rate at the ADC must be increased by a factor of 40 in order to reduce the effect of aliasing to less than -72 dB. The effect of this approach therefore is to increase the number of samples handled by the waveform capture system. This in turn means that the storage required to capture the system output must be significantly increased. To overcome this, a decimation-in-time digital filter may be employed to reduce the sample numbers prior to storage. These filters, however, have a distorting effect on the spectrum of the captured system output. Therefore care must be taken to evaluate the characteristics of the filters so that their effect on measurements may be compensated for.

   The limiting factor in applying a digital filter is the extra processing capability required to perform the filter calculations in real time. This may require elaborate and expensive processing capability if measurement of a high-frequency response is required.
3. *Use a variable cut-off frequency analog filter to remove the high-frequency aliasing components.* A major requirement of modern frequency-response analyzers is that the frequency band over which measurements are made must be programmable from a few millihertz to hundreds of kilohertz. Thus the cut-off frequency of any anti-aliasing filter must be programmable over several decades. To realize this using analog techniques is very difficult due to the effects of component tolerances, drift and leakage. As with digital filters, the gain and phase characteristics of the analog filter must be carefully evaluated and compensated for in any frequency-response calculation. One approach which may prove to be effective in practice is to use a number of switchable fixed cut-off frequency filters in conjunction with a digital filter.
4. *Use of an anti-aliasing compensation algorithm.* A novel approach to the aliasing problem is to employ an anti-aliasing compensation algorithm. By rearranging equation (12.27) it can be shown that the response of the system under test at any test harmonic frequency $(2\pi m/N\delta t)$ is given by

$$H\left(\mathrm{j}\frac{2\pi m}{N\,\Delta t}\right)$$

$$= \left[Y^*[\mathrm{j}(2\pi m/N\,\Delta t)] - (1/\Delta t^2) \sum_{\substack{r=-\infty \\ r\neq 0}}^{\infty} X[\mathrm{j}(2\pi m/N\,\Delta t)] . G\{\mathrm{j}2\pi[(rN+m)/N\Delta t]\} \right.$$

$$\left. \cdot H\{\mathrm{j}2\pi[(rN+m)/N\Delta t]\} \right] (1/\Delta t^2) X[\mathrm{j}(2\pi m/N\Delta t)] \cdot G[\mathrm{j}(2\pi m/N\,\Delta t)] \quad (12.28)$$

In order to accurately calculate the system response the right-hand side of the above equation must be evaluated:

$Y^*(\text{j}2\pi m/N\,\Delta t)$ is the spectral estimate of the sampled system output as calculated by the DFT and is therefore known.
$X(\text{j}2\pi m/N\,\Delta t)$ is the discrete spectrum of the ROMed waveform and is known for all values of m.
$G(\text{j}2\pi m/N\,\Delta t)$, $G(\text{j}2\pi[rN+m]/N\,\Delta t]$ is the transfer function of the zero-order-hold DAC and is known at all frequencies.

Therefore, in order to accurately evaluate the system response at any frequency, only values for $H[\text{j}(2\pi(rN+m)/N\,\Delta t]$ must be obtained. That is, the frequency response of the system at the higher 'aliasing' frequencies must be measured. In theory, this involves the calculation of an infinite number of terms in order to obtain one spectral estimate. However, in practice only a small number of terms need be evaluated in order to obtain a high degree of accuracy.

It has been shown (Jones, 1990) that the values of $H[\text{j}2\pi(rN+m)/N\,\Delta t]$ need only be known for $|r| < 20$ in order to accurately calculate the system response at $H(\text{j}2\pi m/N\,\Delta t)$. For a worst-case system which exhibits no attenuation of the aliasing frequencies it was shown that the error introduced is less than -72 dB. Furthermore, the response of the system at these higher frequencies can be obtained by a single additional frequency-response measurement where the fundamental frequency of the test signal is set to $(2\pi 73/N\,\Delta t)$ rad s^{-1}, since the response of the system is then measured at frequencies up to $(2\pi 5329/N\,\Delta t)$ rad s^{-1}, whereas the system response is only required up to $(2\pi 5193/N\,\Delta t)$ rad s^{-1}.

However, this additional frequency-response measurement may in turn also be corrupted by the effect of aliasing, and, consequently, yet another frequency-response measurement may be required in order to correct for aliasing effects. This problem continues until a set of measurements are made which can be guaranteed to be negligibly affected by aliasing – for example, a measurement just below the maximum frequency range of the instrument. (Here it is assumed that any system tested will have no significant response at any frequency above the maximum range of the instrument. If this cannot be guaranteed, a simple fixed cut-off frequency filter may be employed.)

The maximum number of repeat tests required is, in practice, very small, since when using the prime multi-frequency waveform the response of the system is known up to the 73rd harmonic of the fundamental test frequency. Therefore two repeat tests will guarantee an accurate frequency-response analysis from 50 kHz down to 0.12 Hz and three repeat tests give 50 kHz down to 1.7×10^{-3} Hz.

The above algorithm represents a procedure by which the response of a system may be accurately measured without using an elaborate anti-aliasing filter. Furthermore, the algorithm may be implemented largely in software and offers significant advantages in terms of reduced circuit complexity and cost. The effectiveness of this procedure has been comprehensively demonstrated by Jones (1990).

12.4 Testing systems with saturation-type non-linearity

In order to illustrate the performance of the prime multi-frequency signal, tests were carried out on a third-order system with a controlled level of non-linear behaviour. The results of these tests were compared with those obtained from the same system when using a short sequence length pseudo-random binary signal and a monotonic frequency-response analyzer.

The system was arranged in both open- and closed-loop configuration as indicated in Figures 12.5(a) and 12.5(b). As can be seen, the non-linearity is of the saturation kind, which generates harmonics at odd multiples of the component test frequencies. The amplitude of these harmonics will be determined by the degree of non-linearity introduced. In order to quantify the degree of system non-linearity the following saturation figure is defined:

$$\text{Saturation figure } (\beta) = \frac{(A_{\mathrm{m}} - A_{\mathrm{s}})}{A_{\mathrm{m}}} \times 100\% \qquad (12.29)$$

where A_{m} is the peak amplitude of the input test signal and A_{s} is the voltage amplitude at which saturation occurs.

From the above definition it can be seen that there will be an increase in the degree of saturation as β increases. In the case of the closed-loop system, not only will there be harmonics generated at odd component frequencies due to saturation, there will also be sum and difference frequencies produced as the result of negative feedback. Therefore, one would expect the spectral estimates to have a greater harmonic content over the frequencies of test. However, these effects may be partially moderated as negative feedback will tend to linearize the effect of the non-linearity. A further expectation would be that the effective forward path gain will decrease with an increase in saturation, and this would reflect itself in reducing the magnitude plots for increasing values of β.

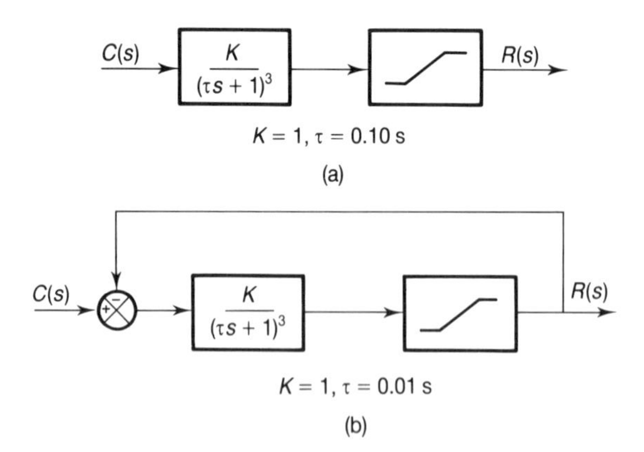

Figure 12.5 Third-order system with saturation. (a) Open-loop; (b) closed-loop.

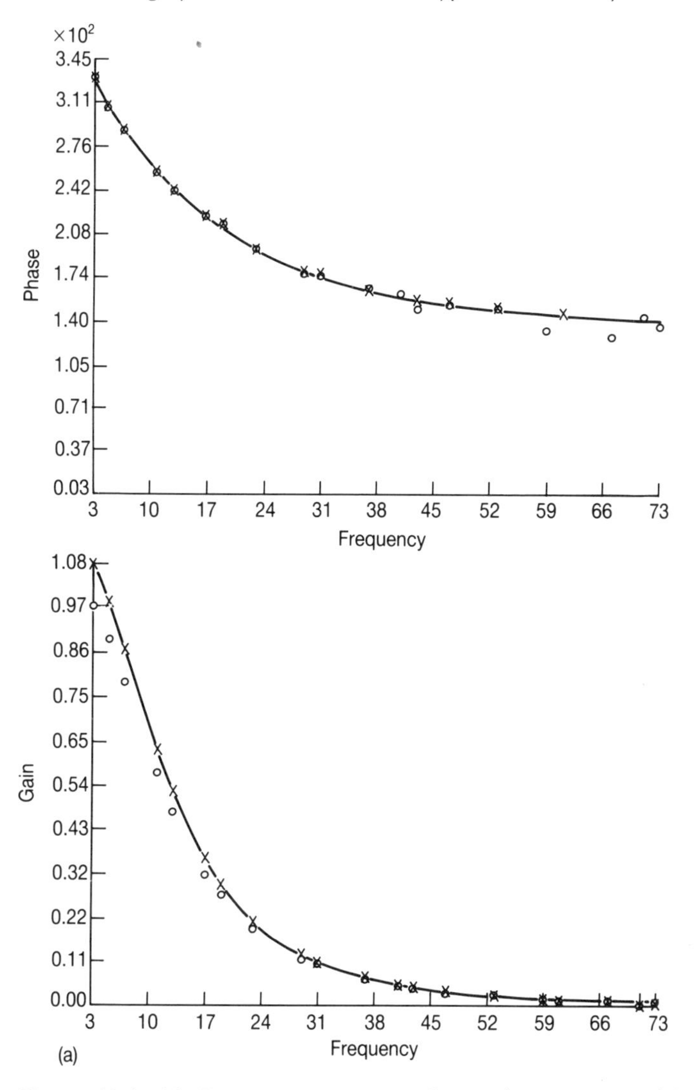

Figure 12.6 (a) Frequency response of open-loop system with saturation non-linearity using prime multi-frequency test signal; $\beta = 30\%$.

In the case of monotonic testing, the prediction of reduced gain can be accurately estimated using the describing function. However, the phenomenon is very much dependent on the amplitude probability distribution of the test signal, and for the signals considered in this investigation the reduction in gain for different β values will vary from those obtained using a single sinusoid test signal. The results obtained from the instrument for the system configurations shown in Figure 12.5, which includes a saturation non-linearity, are given in Figures 12.6–12.9.

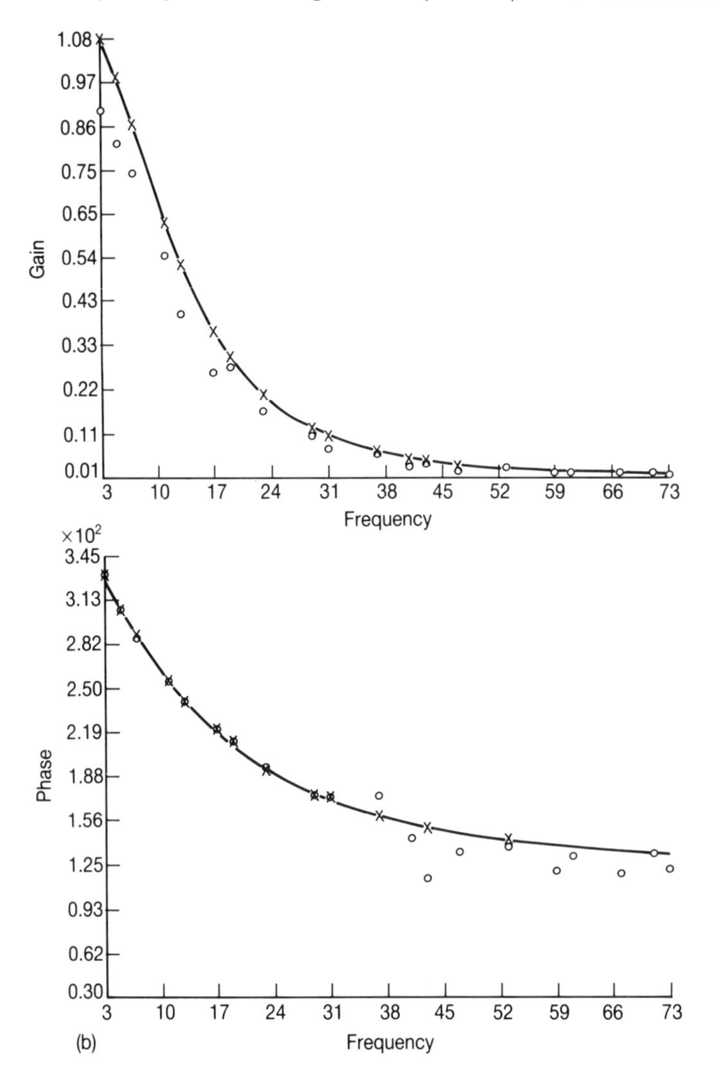

Figure 12.6 (cont.) (b) Frequency response of open-loop system with saturation non-linearity using prime multi-frequency test signal; $\beta = 50\%$.

Figures 12.6 and 12.7 show the results for the open-loop system, with different levels of saturation. The trends already discussed in this section are readily observable:

1. With the introduction of saturation there is a reduction in gain.
2. The prime multi-frequency test signal gives results that have a substantially reduced level of scatter when compared with those obtained using the PRBS waveform.

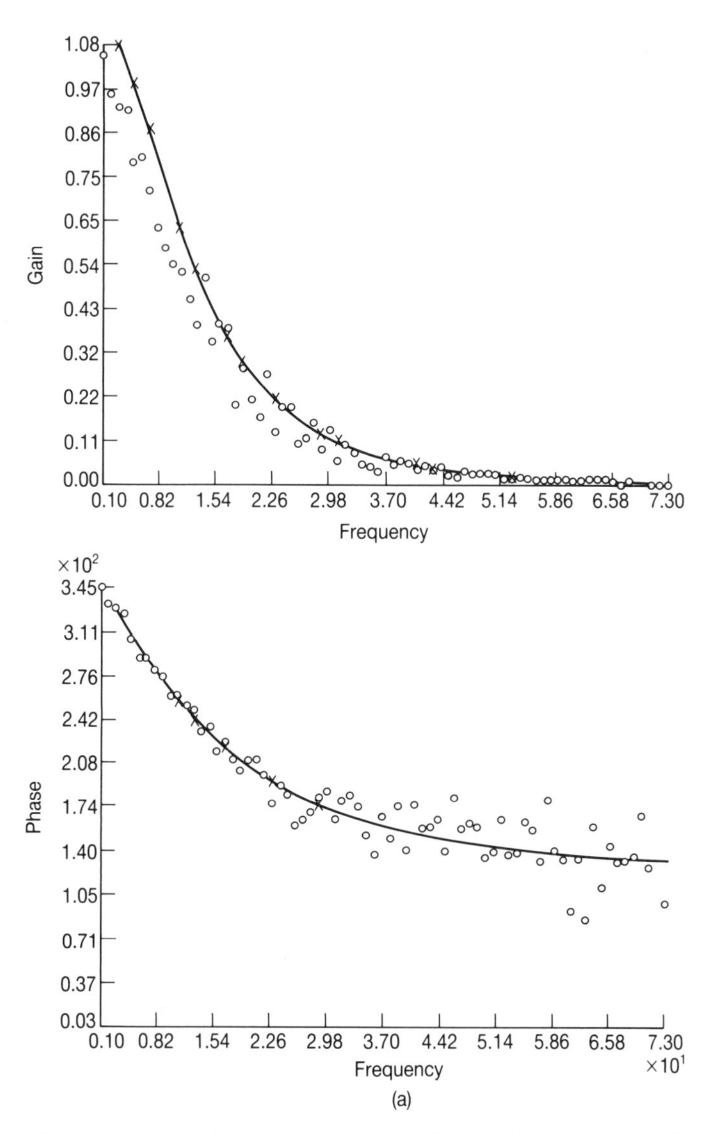

Figure 12.7 (a) Frequency response of open-loop system with saturation non-linearity using PRBS test signal; $\beta = 30\%$.

3. The degree of degradation of the spectral estimates is directly related to the amount of saturation introduced, i.e. as β increases so does the degree of scatter. However, in the case of the prime sinusoid this degradation is minimal compared to the PRBS, for the reasons that have been discussed previously.
4. The phase estimates have a greater susceptibility to scatter, but this can be partly attributed to a lower signal-to-noise ratio at the higher test frequencies.

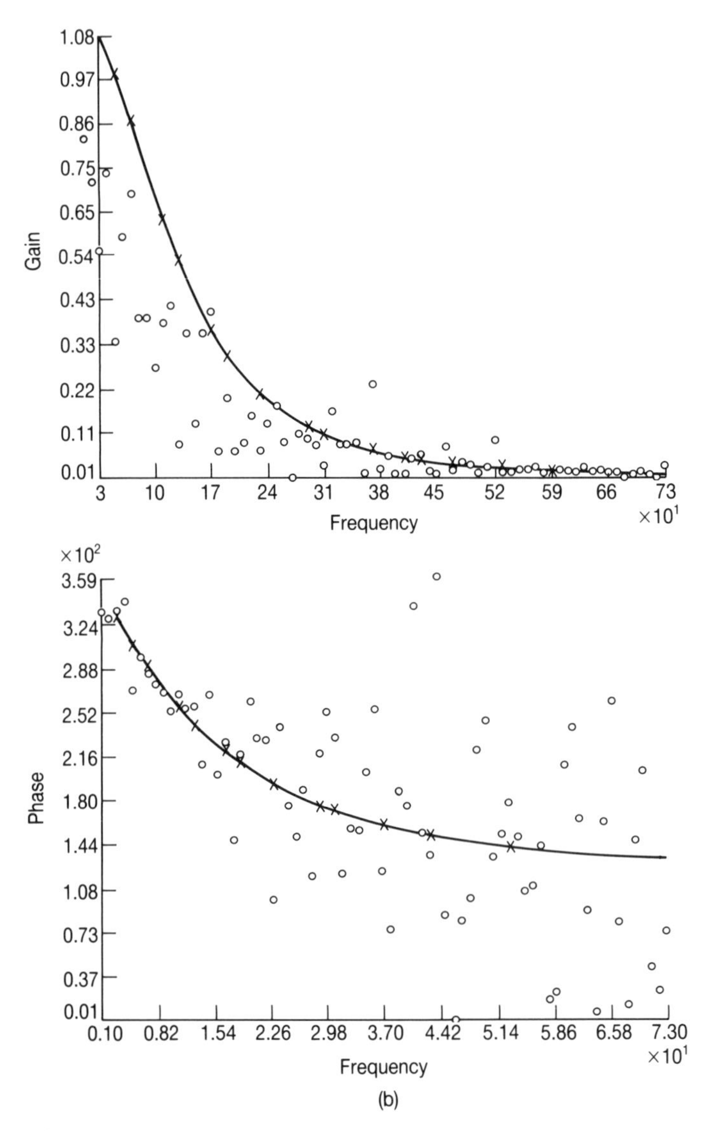

Figure 12.7 (cont.) (b) Frequency response of open-loop system with saturation non-linearity using PRBS test signal; $\beta = 50\%$.

The closed-loop results are shown in Figures 12.8 and 12.9. Again, the trends are consistent with those measured in the open-loop case, with the additional observations:

1. The result of negative feedback serves to linearize the behaviour of the system, so that the results obtained using a prime signal when the saturation level is 33.3% ($\beta = 33.3\%$) gives estimates that are comparable with a linear system ($\beta = 0$).

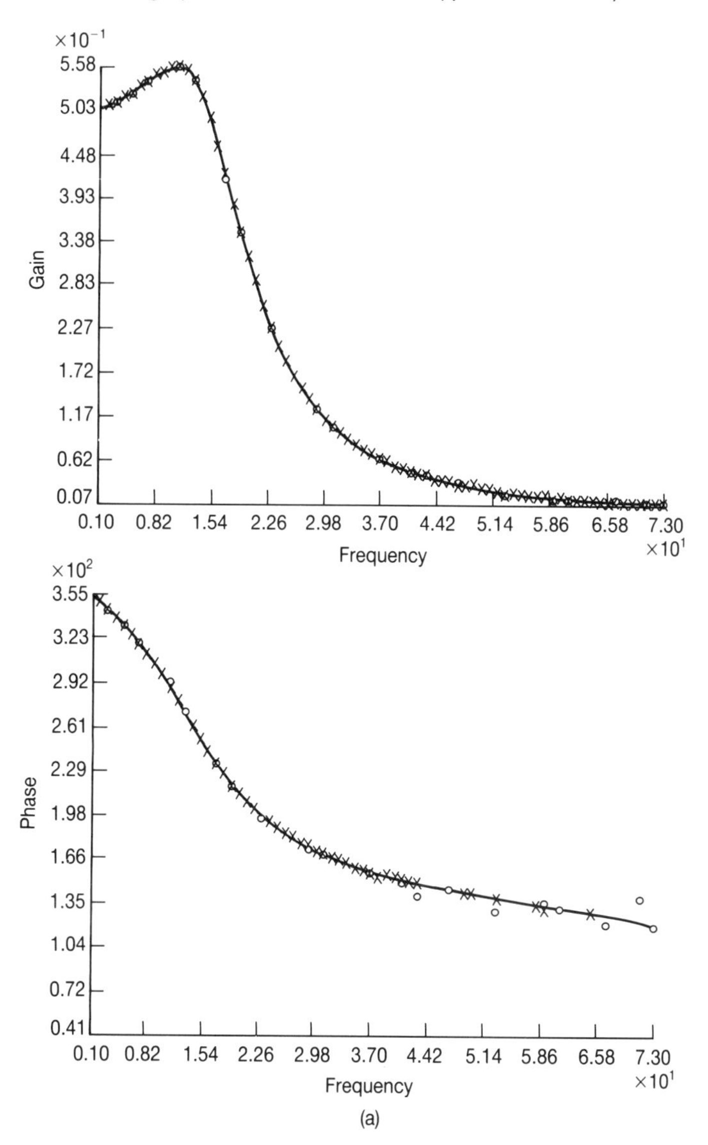

Figure 12.8 (a) Frequency response of closed-loop system with saturation non-linearity using prime multi-frequency test signal; $\beta = 30\%$.

2. As β increases, there is an observable reduction in the gain of the system. However, this reduction in relative terms is less than that observed in the open-loop case.
3. In all cases the prime sinusoid gives estimates with less scatter for both magnitude and phase measurement.
4. The estimates obtained for the closed-loop system with both signals show less scatter than the corresponding open-loop case.

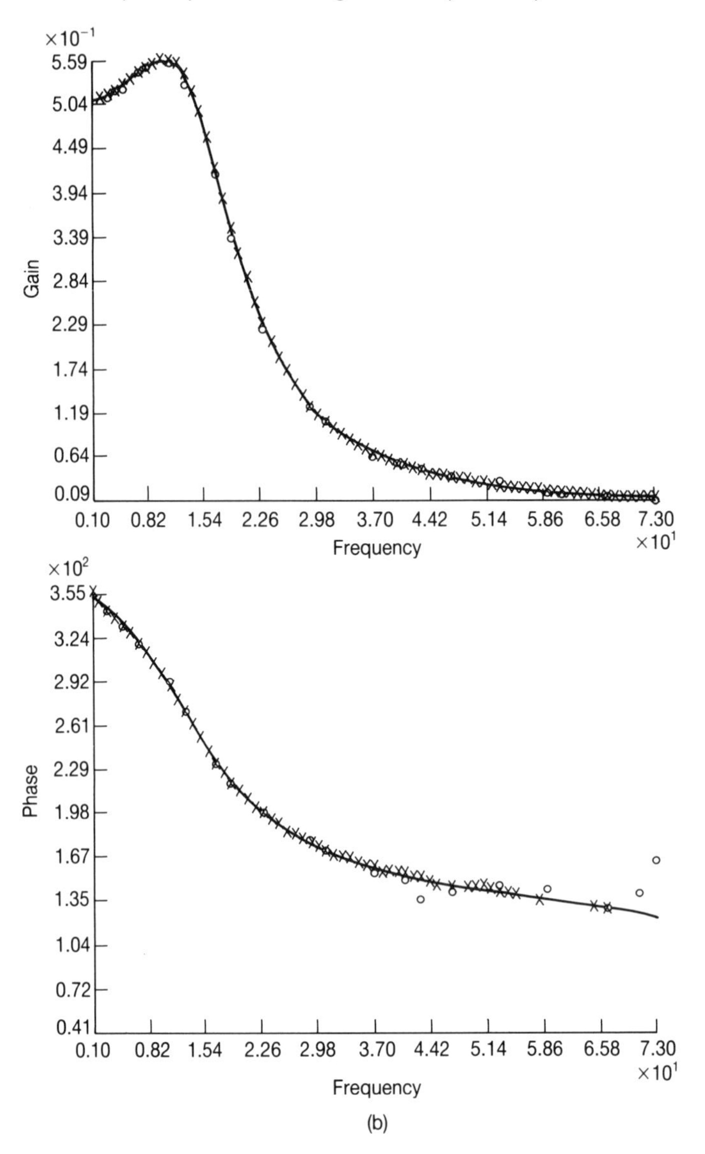

Figure 12.8 (cont.) (b) Frequency response of closed-loop system with saturation non-linearity using prime multi-frequency test signal; $\beta = 50\%$.

12.5 Conclusions

This chapter has examined the role of non-binary multi-frequency signals in frequency-domain non-parametric system identification. The main test stimulus considered was the prime multi-frequency signal, due to its superior harmonic rejection

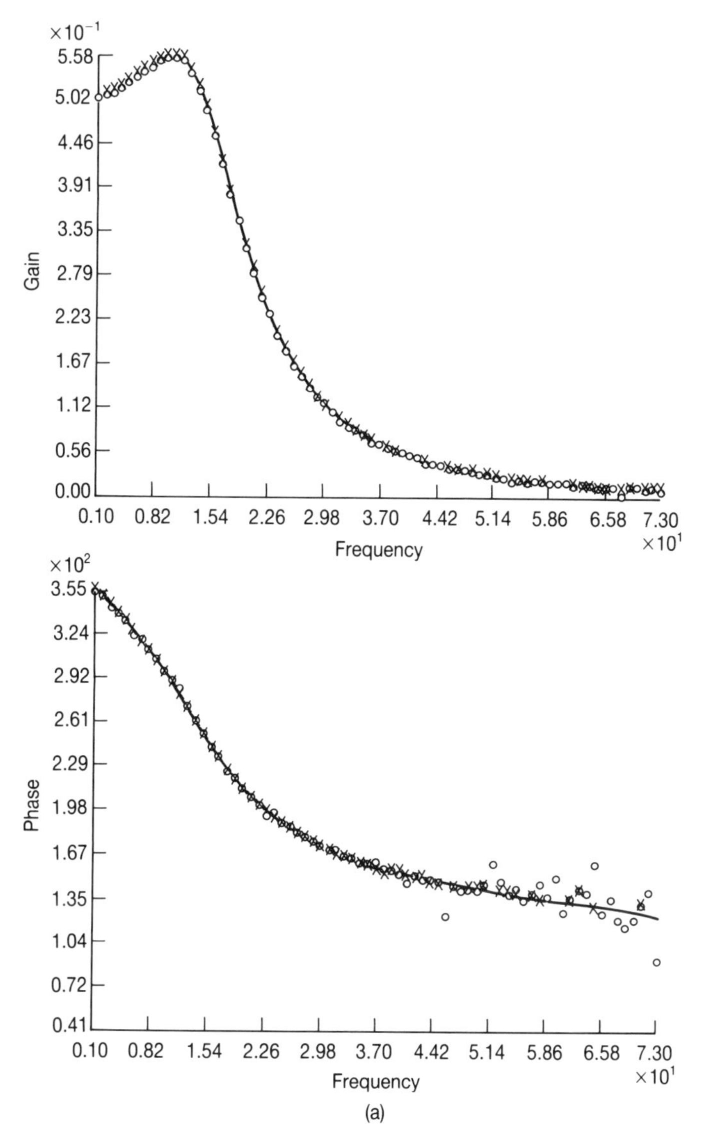

Figure 12.9 (a) Frequency response of closed-loop system with saturation non-linearity using PRBS test signal; $\beta = 30\%$.

properties compared with other multi-frequency signals. The chapter explored various algorithms for generating low peak factors for such a signal and presented a set of optimum phases.

The implications of mechanizing such a signal within an instrument were considered and it was shown that the aliasing problems associated with digital techniques used for the storage and generation of multi-frequency waveforms can be

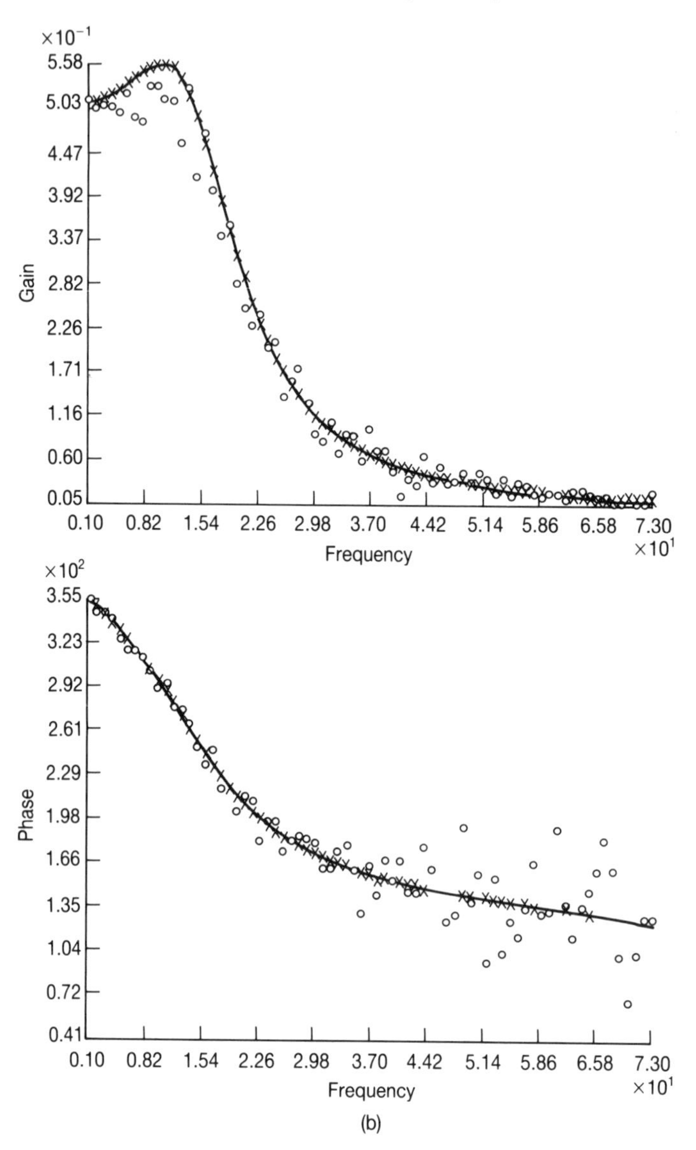

Figure 12.9 (cont.) (b) Frequency response of closed-loop system with saturation non-linearity using PRBS test signal; $\beta = 50\%$.

overcome in a number of ways, including the use of a novel software compensation algorithm. The theoretical basis for this algorithm was presented along with a method by which the algorithm can easily be incorporated with a multi-frequency response analyzer. The chapter then demonstrated the clear superiority of the prime sinusoid waveform when used for frequency-domain non-parametric identification on systems exhibiting some non-linearity.

Appendix

Optimized phases incorporated within prime sinusoid signal (radians)

10 Harmonics
5.5716 4.2064 1.4156 5.6458 1.1278 1.9900 1.4536
2.0784 5.9526 1.1567

15 Harmonics
3.9905 3.8804 5.1961 5.4526 4.7758 5.7851 1.7319
2.1795 3.1147 1.5086 0.0392 3.6960 0.2089 4.0436
1.8467

20 Harmonics
2.8017 3.2761 3.3538 6.1196 0.9324 1.1850 2.7593
5.6758 2.9598 4.7758 1.4462 5.4801 2.9448 3.5067
3.6925 0.0297 5.6201 0.2981 1.5402 6.2757

30 Harmonics
3.0413 2.4468 3.1949 3.9835 4.2950 3.0583 2.0867
1.5008 1.2969 4.1257 2.8612 2.3013 0.9771 2.1246
4.6795 2.4802 4.3835 1.1634 2.3215 1.2156 4.1648
1.3048 0.2414 2.9768 1.1840 1.0509 3.0888 2.9172
3.9864 4.5863

40 Harmonics
1.6244 2.9266 3.0395 4.0217 3.2199 3.6003 1.5192
1.9627 3.5807 1.8153 3.0943 4.6539 2.2617 4.0544
3.8800 2.6461 4.3996 2.3673 2.3366 3.4977 3.1340
0.5750 1.4078 3.4022 3.0598 2.0211 0.0975 1.8848
2.7045 3.5278 4.2948 5.0208 3.9311 2.7793 4.3954
2.9804 2.1767 1.9237 2.9439 4.4258

Optimized phases incorporated within composite sinusoid signal using consecutive harmonics

10 Harmonics
5.3672 5.9735 2.0660 1.7821 5.9459 3.1851 1.5868
2.8458 0.7851 1.4487

15 Harmonics
2.5056 4.7763 3.9403 4.6601 2.1013 0.0253 3.2886
2.9879 3.7726 4.4374 1.4438 3.8390 6.2387 6.2411
4.9591

20 Harmonics

5.6538	6.2417	2.3755	2.6367	1.6073	0.1789	0.2085
3.7323	5.0540	0.7510	4.2876	0.2041	0.1866	4.5226
1.6337	5.7635	2.8349	6.1688	0.8497	0.0750	

30 Harmonics

1.6503	0.7408	2.6323	5.6254	3.4973	5.6780	1.2257
2.6894	2.9795	2.8900	1.0830	1.3741	0.8813	2.4541
2.7467	2.2030	1.8456	4.5505	5.0845	0.8922	3.7390
1.3678	5.9397	5.6848	2.0323	5.7820	4.5476	2.1552
3.9818	0.0767					

40 Harmonics

4.8009	1.3915	5.5238	2.3963	5.6810	3.7118	2.3321
1.5870	0.2777	2.3456	0.3409	4.1726	4.1633	5.8255
3.3560	0.2746	5.0378	2.5447	1.4162	1.4213	1.2990
4.0722	5.3158	0.3432	2.4730	1.2470	2.0857	1.4657
2.0976	5.2141	0.6696	2.6350	0.4945	4.9201	5.5075
6.0702	0.6203	4.4694	5.5115	1.9307		

Acknowledgements

The authors acknowledge the help of Ceri Evans and Llyr Roberts in obtaining the results published in Table 12.2 and those shown in Figure 12.1–12.3.

13

Multi-frequency Signals for Plant Identification

Walter Ditmar and Ray Pettitt

13.1 Introduction

Results obtained from plant identification trials are affected by both the properties of the plant under test and the chosen method of identification. System properties include plant topology, non-linearity and non-stationarity, noise statistics, interface problems such as limited access to important plant variables and transducer errors. The method of identification involves the choice of system model to best represent the plant under study, and considerations concerning compatibility with the statistical properties of plant variables.

To obtain a mathematical model of plant dynamics based on an observation of input and output signals requires some form of measurable disturbance within the plant structure. Naturally occurring plant noise may be adequate for the purpose of identification, and indeed in some applications this is the only allowable or available disturbance.

The unpredictable behaviour of system signals and weak stationarity of plant dynamics are the most common problems in the on-line identification of system components. Both these problems may be greatly reduced by the insertion of a suitable perturbation signal into an accessible point of the system under study to increase statistical activity. Advantages include improved stability of parametric estimates, the avoidance of bias in experimental results plus a reduction of computational overhead, and a simplification of measurement procedure.

Following an overview of least squares frequency-response estimation principles, the properties of perturbation signals for system identification are presented and discussed. A novel pseudo-random signal which is based on properties of the Gaussian model is introduced and compared with the performance of some presently available perturbation signals.

13.2 Signal processing

13.2.1 Least squares frequency-response regression

In least squares spectral modelling a regression is performed between two complex valued phasors $X(k)_n$ and $Y(k)_n$, where k is an integer which indicates the harmonic frequency $k\omega_0$ and n is the nth phasor in a set of M phasors. In simplified notation, the regression model is defined by

$$E_n^y = Y_n - G_{xy} X_n \tag{13.1}$$

The least squares coefficient between the phasors is obtained by computing weighted averages of phasor products:

$$\hat{G}_{xy} = (\overline{X_n^* Y_n}) . (\overline{|X_n|^2})^{-1} \tag{13.2a}$$

The normalized magnitude squared regression coefficient between the phasor sets $\{X\}$ and $\{Y\}$ is known as the 'squared coherence' between the phasors:

$$\hat{\gamma}_{xy}^2 = (|\overline{X_n^* . Y_n}|^2) . (\overline{|X_n|^2} . \overline{|Y_n|^2})^{-1} \tag{13.2b}$$

Taking expected values of the stochastic estimates above, the familiar expressions for frequency response and squared coherence result (see, for example, Chapter 2, Section 2.5):

$$G_{xy} = \frac{S_{xy}}{S_{xx}} \tag{13.3a}$$

$$\gamma_{xy}^2 = \frac{|S_{xy}|^2}{S_{xx} . S_{yy}} \tag{13.3b}$$

The effective bandwidth of the spectral window through which the signal spectrum is viewed by the phasor estimators $B_e = \omega_0$. In stochastic analysis the spectral width of this window is not affected by averaging of successive records. The dashed curve in Figure 13.1 shows the composite spectral response of the stochastic estimators to coherent energy components in the signal spectrum.

Squared coherence represents the coherent signal power in an output y contributed by an input x through a linear process G_{xy}, divided by the total signal power in y:

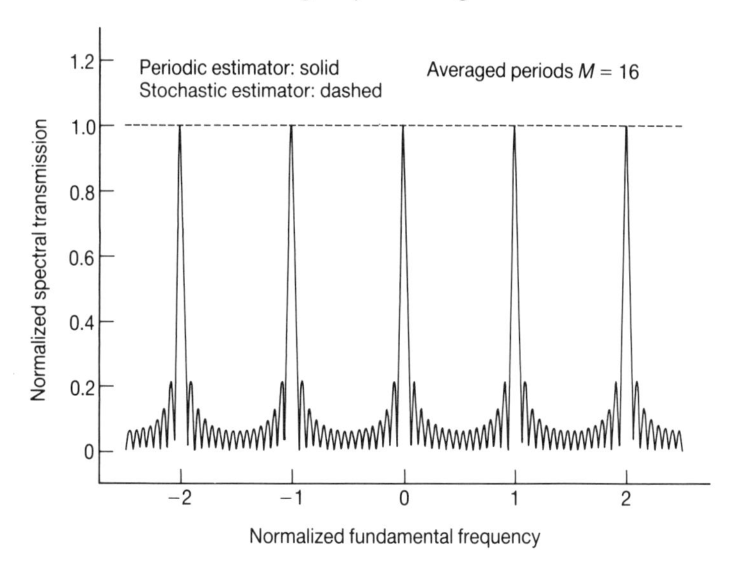

Figure 13.1 Response of stochastic and periodic spectral estimators.

$$\gamma_{xy}^2 = \frac{S_{xx} | G_{xy} |^2}{S_{yy}} \tag{13.3c}$$

13.2.2 Regression using external perturbation

An external perturbation signal, U, may be applied in cases of excessive variability to improve the frequency-response estimates of equation (13.2a), or when the desired frequency response is contained within a feedback structure (see Wellstead, 1970). The system equation (13.1) is now augmented with a corresponding expression which connects the perturbation signal, U, to the system:

$$E_n^x = X_n - G_{ux} U_n \tag{13.4a}$$

Substitution of equation (13.4a) into equation (13.1) results in a new regression model which is driven by the perturbation source U:

$$E_n^y + G_{xy} E_n^x = Y_n - G_{uy} U_n \tag{13.4b}$$

The frequency response between nodes x and y can then be defined as:

$$G_{xy \,.\, u} = \frac{G_{uy}}{G_{ux}} \tag{13.5}$$

Substitution of the least squares error solutions of equations (13.4a) and (13.4b) into equation (13.5) gives the frequency-response estimate in the form:

$$\hat{G}_{xy.u} = (\overline{U_n^* . Y_n}).(\overline{U_n^* . X_n})^{-1} \qquad (13.6)$$

Returning equation (13.6) into equation (13.1) results in an expression for the associated squared coherence:

$$\hat{\gamma}^2_{xy.u} = \frac{\overline{|X_n|^2}}{\overline{|Y_n|^2}} \cdot \frac{|\overline{U_n^* Y_n}|^2}{|\overline{U_n^* X_n}|^2} \qquad (13.7)$$

Taking expected values provides insight into the behaviour of this estimator:

$$G_{xy.u} = \frac{S_{uy}}{S_{ux}} \qquad (13.8a)$$

$$\gamma^2_{xy.u} = 1 - \frac{S_e^y}{S_{yy}} \qquad (13.8b)$$

Expression (13.8b) shows that only the incoherent error process E_n^y contributes to the variability of the estimator in equation (13.6). This condition only holds when the data for equations (13.4a) and (13.4b) are obtained simultaneously during a plant trial.

13.2.3 Periodic regression

Choosing $u(t)$ to be a periodic signal, it follows that

$$U_n = U_0, \qquad n = 1, 2, 3 \ldots \qquad (13.9)$$

Equations (13.6) and (13.7) now reduce to much simpler expressions. The frequency-response estimator and its associated squared coherence function are obtained, respectively, as

$$\hat{G}_{xy.u_0} = \frac{\overline{Y_n}}{\overline{X_n}} \qquad (13.10a)$$

$$\hat{\gamma}_{xy.u_0} = \frac{\overline{|X_n|^2}}{\overline{|Y_n|^2}} \cdot \frac{|\overline{Y_n}|^2}{|\overline{X_n}|^2} \qquad (13.10b)$$

The unbiased frequency-response estimator of equation (13.10a) consists of the ratio of two periodic spectral estimators. The advantage of equation (13.10a) over equation (13.6) lies in the fact that the perturbation signal U is not required in explicit form; only the duration of its period needs to be known. Use of the periodic frequency-response estimator also leads to a great reduction in computing effort since the complex multiplications in equation (13.6) are not required. Furthermore, the spectral smoothing in equation (13.10) may be replaced by simple periodic averaging in the time domain prior to Fourier transformation (see Appendix, equations (13.A1) and (13.A2)).

In contrast to the fixed all-pass spectral window associated with stochastic estimators, use of the periodic spectral estimator invokes a window-closing procedure round the discrete spectral lines of periodic signal components (see Figure 13.1 and Appendix, equation (13.A3)). Averaging M successive components results in an effective window bandwidth $B_e = \omega_0/M$, which will progressively eliminate unwanted incoherent signal components as M increases. The solid curve in Figure 13.1 shows the composite spectral response of the periodic estimators to coherent energy components in the signal spectrum.

13.3 Test-signal properties

An important constraint which must be placed on any perturbation signal for use in practical system tests is that its peak excursion should not exceed a given maximum value while maximizing its dynamic information content. This restriction avoids excessive stress and possibly damage to the system's hardware. In this context, two parameters of $x(t)$ can be optimized:

1. The signal's peak-to-peak value over the duration of the test
2. The distribution of spectral components in the power-spectral density $S_{xx}(f)$.

13.3.1 Power (P)

The probability density function which will maximize signal power subject to an amplitude constraint $\pm c$ is obtained from

Maximize:
$$P_x = \int_{-c}^{c} x^2 p(x)\,\mathrm{d}x \qquad (13.11\text{a})$$

Subject to:
$$1 = \int_{-c}^{c} p(x)\,\mathrm{d}x \qquad (13.11\text{b})$$

Variational calculus gives the optimum probability density function $p(x)$ in the form of two impulses:

$$p(x) = a\,.\,\delta(x-c) + b\,.\,\delta(x+c) \quad [a+b=1] \qquad (13.12)$$

This result defines a two-level waveform and the 'best' signal will therefore be a type of squarewave.

13.3.2 Relative peak factor (RPF)

A measure of amplitude deviation versus signal RMS power is given by

$$\mathrm{RPF} = \frac{x(t)_{\max} - x(t)_{\min}}{2\sqrt{2}\, x_{\mathrm{rms}}} \tag{13.13}$$

As shown in Chapter 2, Section 2.4.2, this factor should be as small as possible so that for a given RMS power the range of amplitude excursions is also as small as possible. Squarewaves and single sine waves have RPFs of 0.707 and 1, respectively. The pseudo-random binary noise (PRBN) waveform has an RPF = 0.707 and is therefore more suitable than a single sine wave for the testing of equipment which is sensitive to maximum signal excursion. In the case of normalized Gaussian random noise, approximately 99% of all values of $x(t)$ are contained within an interval of ± 2.57. The approximate value of RPF for such a signal is then obtained as 1.82. This is considerably larger than the value achieved for any two-level waveforms due to the shape of the probability density function which is clustered around the mean value, in contrast to the widest possible discrete distributions associated with binary signals.

13.3.3 Spectral entropy (H)

It is shown above that two-level signals contain maximum power for a given constraint in magnitude. The frequency content of the signal remains as yet undefined. For example, a simple 50% duty cycle squarewave will have an optimum RPF with its spectral components inversely proportional to frequency. For system-testing purposes, a more useful signal would contain equal energy in all its frequency components.

A performance indicator for the distribution of spectral components is the spectral entropy factor H (see Ditmar, 1977). Here the signal's normalized power-spectral density $S_{xx}(f)/R_{xx}(0)$ is regarded as a probabilistic statement describing the presence of mutually independent sinusoidal messages. An optimum signal would maximize the information content or entropy, subject to a constrained bandwidth of $\pm B$ Hz:

Given that:
$$p(f_x) = \frac{S_{xx}(f)}{R_{xx}(0)} \tag{13.14a}$$

Maximize:
$$H = -\int_{-B}^{B} p(f_x)\,.\ln[p(f_x)]\,.\mathrm{d}f_x \tag{13.14b}$$

Subject to:
$$1 = \int_{-B}^{B} p(f_x)\,\mathrm{d}f_x \tag{13.14c}$$

Variational calculus gives the optimum probability density function as:

$$p(f_x) = \begin{cases} \dfrac{1}{2B} & f_x \in \pm B \\ 0 & \text{otherwise} \end{cases} \qquad (13.15)$$

Thus H becomes a maximum when the probability of obtaining any frequency component is the same for all frequencies f_x. $S_{xx}(f)$ is therefore constant and for a signal constrained over a bandwidth $\pm B$, $S_{xx}(f) = R_{xx}(0) \,.\, p(f_x) = R_{xx}(0)/2B$, giving a rectangular power-spectral density over a frequency interval of $\pm B$.

An example of a signal with optimum H is Gaussian distributed noise with a uniform truncated spectral envelope. The PRBN signal has a lower entropy since its power-spectral density is a sine–cosine squared function extending over all frequencies. In practice, it is difficult to obtain idealized band-limited noise by corrective filtering, due to the sharp cut-off required and the constant power over the bandwidth of the signal.

Consideration of the requirements for minimum RPF and maximum H reveals the contradiction that the maximum entropy condition imposes a finite bandwidth on the signal which limits its rise time. Therefore, the signal can no longer be two-level as is required from minimum RPF considerations. Relaxing the two-level condition allows the power-spectral density to be shaped optimally as in the case of the Gaussian distributed random signal. Thus a trade-off results between the minimum RPF condition and the required spectral envelope of the signal.

13.3.4 Degrees of freedom (DF) and equivalent bandwidth (EB)

DF and EB are useful quantities for the analysis of pseudo-random or stochastic signals (see Bendat and Piersol, 1986). Consider the autocorrelation function $R_{xx}(\tau)$ and its Fourier transform, the power-spectral density $S_{xx}(f)$. The total signal power is given by $P_x = R_{xx}(0)$. Whenever $R_{xx}(\tau) = 0$ then the signal and its delayed version $x(t - \tau)$ will be uncorrelated. When $R_{xx}(k\Delta t) = 0$, k integer, then a string of independent samples $x(t - k\Delta t)$ can be taken from the signal. It follows that the DF of such a signal with $T_r = N\Delta t$ s duration will be N. The autocorrelation function of a signal with optimum H and bandwidth of B Hz is obtained by taking the inverse Fourier transform of equation (13.15):

$$R_{xx}(\tau) = P_x \frac{\sin(2B\tau\pi)}{2B\tau\pi} \qquad (13.16)$$

This autocorrelation function is zero at multiples of $1/2B$, so that the samples spaced by $1/2B$ will be statistically independent. The DF of a record of T_r s duration becomes

$$\mathrm{DF} = 2 \, . \, B \, . \, T_r \tag{13.17}$$

Practical signals $x(t)$ rarely have a rectangular power-spectral density. They may, however, be classified in terms of an 'equivalent' signal $x_e(t)$ with rectangular power-spectral density $S_{xx_e}(f)$:

$$S_{xx_e}(f) = \begin{cases} S_{xx}(0) & f \in \pm B_e \\ 0 & \text{otherwise} \end{cases} \tag{13.18}$$

The total power of $x(t)$ is obtained by integrating over the full range of $S_{xx}(f)$. Equating the power of both signals leads to the concept of EB associated with $x(t)$:

$$B_e = \frac{1}{2} \frac{\int S_{xx}(f) \, \mathrm{d}f}{S_{xx}(0)} = \frac{1}{2} \frac{R_{xx}(0)}{S_{xx}(0)} \tag{13.19}$$

In analogy with equation (13.17), the Equivalent DF (EDF) of the signal follows as

$$\mathrm{EDF} = 2 \, . \, B_e \, . \, T_r = \frac{R_{xx}(0)}{S_{xx}(0)} \, T_r \tag{13.20}$$

Application of equation (13.20) to the PRBN and Gaussian signals reveals that they contain identical EDFs.

The concepts of EB and EDF find wide application in the spectral analysis of stochastic signals. EDF is used to compute confidence intervals for spectral estimates. Alternatively, the EDF required to obtain a specified confidence level can be calculated and the result used to decide the minimum record length for a given analysis bandwidth.

13.3.5 Signal comparison

RPF and spectral entropy H are useful performance indices for the classification of signal performance. The sinusoidal wave has unit RPF, but it has an H value of only one bit/harmonic. Although the signal is obviously useful in single frequency applications, it is unsuited to multi-frequency analysis.

Two signals with superior performance indices are the PRBN and Gaussian noise signals with uniformly distributed spectra. Each is optimal in one index and performs well in the other: whereas PRBN features minimum RPF, the Gaussian signal has maximum H. Both perform well as test signals but have the disadvantage that their phase is uniformly distributed over an interval of 2π radians; this necessitates correlation-type analysis.

13.4 Synthesis of pseudo-random Gaussian noise (PRGN)

13.4.1 Test signal requirements

From the discussion in Sections 13.2 and 13.3 above, six specifications follow for a test signal:

1. The amplitude deviation must be a minimum for a given power level, i.e. the signal must have a low RPF.
2. The signal must be periodic in time to stabilize parametric estimates.
3. Compatibility with the radix-2 Fast Fourier Transform (FFT) algorithm is required to minimize computational effort.
4. Sufficient energy must be present in all spectral lines up to the Nyquist folding frequency to achieve a good signal-to-noise ratio of spectral estimates.
5. The spectral dimensions should be adjustable to match the response of the system under test.
6. No energy should be contained outside the Nyquist interval to eliminate the need for guard filters in the measurement hardware.

13.4.2 Magnitude and phase encoding

The above requirements can be met by performing an inverse FFT on suitably chosen magnitude and phase elements $A(k)$ and $\phi(k)$, respectively:

$$x(i) = \sum_{k=0}^{N-1} A(k)\, e^{j[ik(2\pi/N) + \phi(k)]} \qquad i \in 0, N-1 \tag{13.21}$$

This procedure results in a multi-frequency signal with the desired harmonic content. If all phases were made zero then the resulting signal would peak at index $i = 0$, resulting in a large RPF value. This effect can be reduced by adjusting the phases of individual harmonics using a suitable phase-encoding algorithm.

The method used to synthesize pseudo-random Gaussian noise (PRGN) attempts to simulate the magnitude and phase probability distribution of a band-limited Gaussian signal. Magnitude values are adjusted to provide a spectral envelope as required by the intended application of the signal. Phase values are independently assigned in pseudo-random order such that their distribution is uniform over the interval of 2π radians.

Transformation of an N-point data sequence results in two sets of independent magnitude and phase values of $N/2$ points each in the associated spectral domain. Having selected the required magnitude values, the phase values are obtained from a scrambling algorithm given by Pettitt (1983) and Ditmar (1977) as

$$\phi(k) = \begin{cases} 0 & k = 0 \\ s(k)\dfrac{2\pi}{N} & k = 1, 2, \ldots, N/2 - 1 \end{cases} \tag{13.22}$$

Here the integer N is the length of the FFT array. The factors $s(k)$ are pseudo-randomly ordered integer values in the range 1, 2, ..., $N/2 - 1$ taken from the register outputs of a v-bit chain code generator, where $v = \log_2(N/2)$. The phase distribution over the full Nyquist range is

$$p[\phi] = \frac{1}{N}\sum_{k=-(N/2)}^{(N/2)-1} \delta\left(\phi - \frac{2\pi k}{N}\right) \tag{13.23}$$

The resulting time sequence has Gaussian properties combined with a low RPF. The spectral envelope is programmable and can be matched to system requirements.

13.4.3 Spectral composition

Due to the frequency-domain correspondence and the bidirectional properties of the discrete Fourier transform, the signal's amplitude probability density function, $p(x)$, can be expected to resemble the normal distribution. The PRGN signal derived from a PRBS may therefore be viewed as a form of periodic band-limited Gaussian distributed perturbation. This signal finds the same application in periodic-based, direct spectral estimation procedures using the periodic transfer-function estimator of equation (13.10a), as the application of Gaussian noise in cross-spectral estimation procedures using the stochastic estimator of equation (13.2).

With a uniformly selected magnitude spectrum, the signal's autocorrelation function $R_{xx}(\tau)$ becomes a sine–cosine function which is periodic with the same period T as the time sequence $x(t, T)$, and with zero crossings at intervals of $T/N = \Delta t$. Samples taken at intervals Δt are therefore uncorrelated and a total of N uncorrelated samples can be obtained from the signal. The DF of this signal are the same as for a signal in EB format for which EDF $= 2B_e T = N$, and where $N/2$ complex estimates are obtained of the harmonic spectrum, placed at $N/2$ discrete frequency ordinates. The signal's EB is therefore $N/(2T)$. The entropy, H, is obtained from a discrete version of equation (13.14b), where the relative probability of each of the N equal harmonics $p(f_k)$ is $1/N$. For a v-bit chain code generator:

$$H = \sum_{k=1}^{N} \frac{1}{N}\log_2(N) = \log_2(N) = v + 1 \tag{13.24}$$

Therefore for this type of signal, EDF and H are related by

$$H = \log_2(\text{EDF}) \tag{13.25}$$

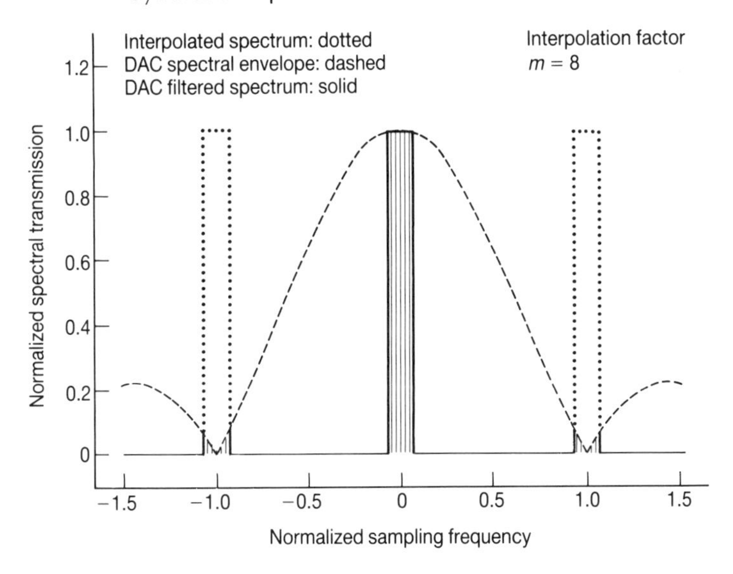

Figure 13.2 Test signal spectrum after interpolation and D/A conversion.

13.4.4 Filtering

To satisfy condition (3) in Section 13.4.1 above, the desired complex conjugate spectrum in the frequency domain of length N can be padded with zeros to a length mN (m integer) and then inverse-transformed using the FFT algorithm. This results in an $(m - 1)$ point interpolation between the original N data points. The resulting test signal data points are applied to a D/A converter which is then oversampled at a rate of m times the rate at which the measured system response signals are sampled. The first alias frequencies are now shifted to $(1 - 0.5/m)$ times the sampling frequency of the interpolated signal. This is illustrated in Figure 13.2 with the response of the zero-order-hold interpolator superimposed and shown dashed on the graph. The combined effect is to greatly reduce the amplitude of the first alias. By this method, the need for sophisticated guard filters and the attendant differential measurement errors are eliminated. If required, the interpolation can be adjusted such that the amount of first alias is below the minimum resolution of A/D converters which are used in the measurement system.

13.4.5 Optimization

The sequence in which PRGN phase values can be assigned to the FFT array is not unique. Slight changes in RPF will be experienced when signals are synthesized using different phase orders. To obtain the lowest RPF for a given spectral envelope, it

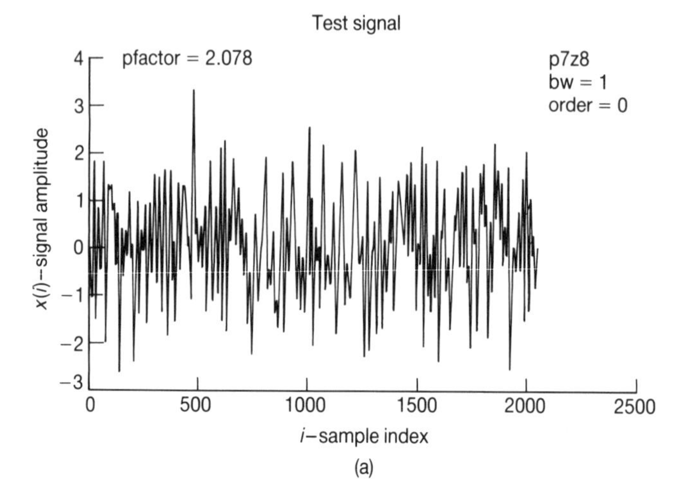

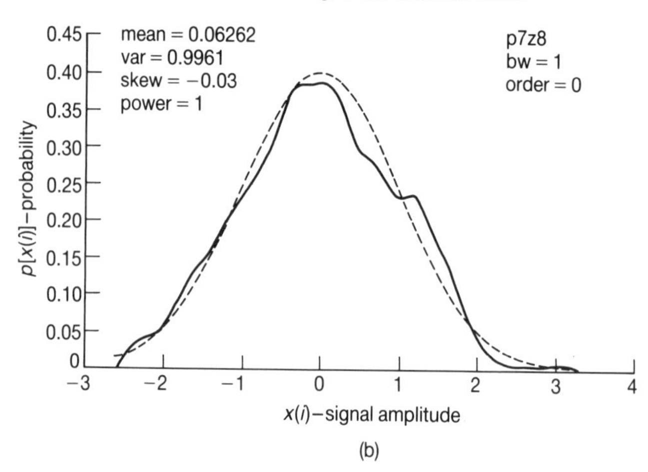

Figure 13.3 PRGN signal with $N/2 = 128$ harmonics and interpolation factor $m = 8$. (a) unfiltered time signal; (b) probability density function of unfiltered time signal.

will be necessary to search all possible $\{(N/2) - 1\}!$ phase combinations for this parameter.

A reduced set of $N/2$ combinations is obtained by successively rotating the pseudo-random phase set in steps of $2\pi/N$ radians. Although this is not a complete search, a sub-optimum RPF value can be found for the synthesized signal (see Khoshlajeh-Motamed and Pettitt, 1991). The searches are readily implemented by changing the initial pseudo-random integer sequence $s(k)$ in equation (13.22) to $s(k + r)$, for $r = 0, 1, \ldots, N/2 - 1$.

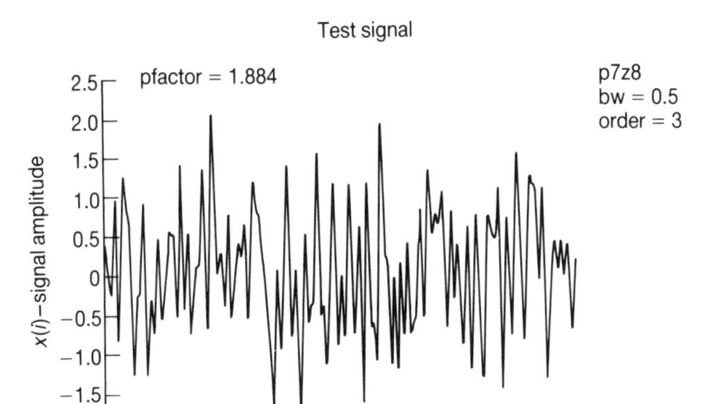

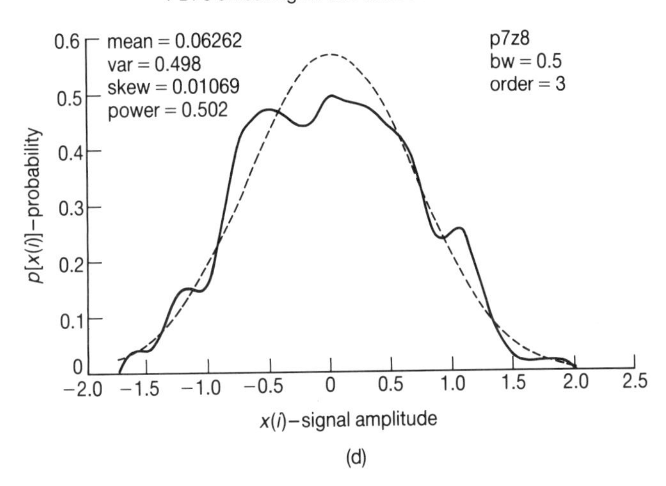

Figure 13.3 (cont.) (c) Filtered time signal; (d) probability density function of filtered time signal.

13.5 Results

13.5.1 Practical synthesis of PRGN

The signals shown in Figures 13.3 and 13.4 were synthesized using $N/2 = 128$ discrete frequencies. The phasor components are equal in magnitude and have phases ordered according to the algorithm of equation (13.22) above.

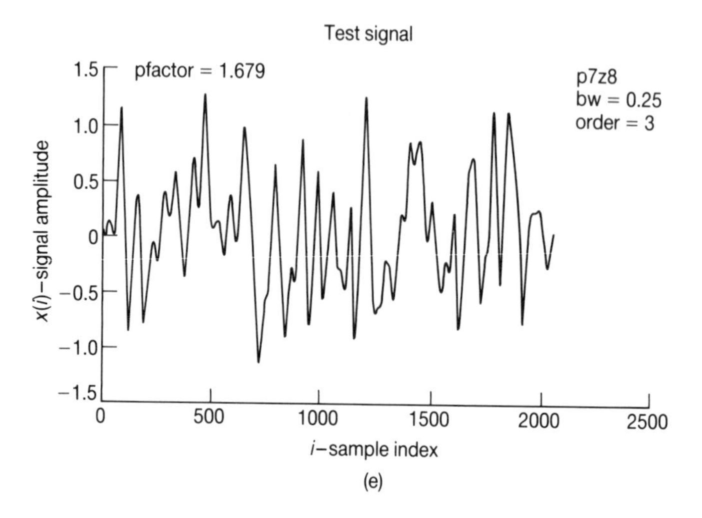

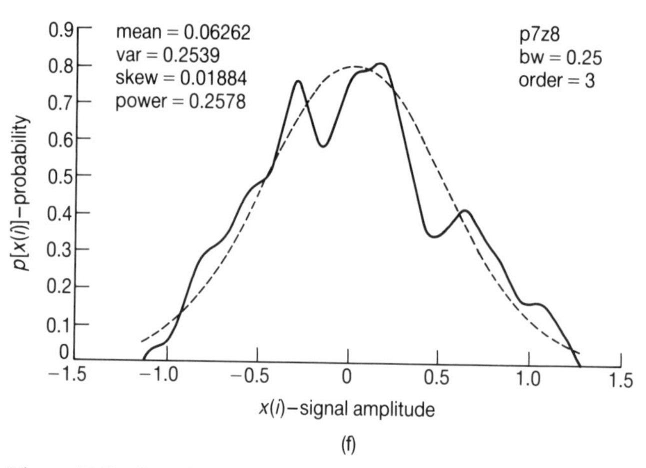

Figure 13.3 (cont.) (e) Filtered time signal; (f) probability density function of filtered time signal.

Figure 13.3(a) shows the PRGN signal before optimization with an initial shift register integer $s(k) = 127_{10}$, the corresponding RPF is 2.08. To obtain a sub-optimum signal the search process outlined in Section 13.4.5 was undertaken. Since the maximum-length sequence register can only generate $2^7 - 1 = 127$ values (0 not allowed), a total of 127 different shifted phase sets $\phi(k + r)$ with associated RPF_r's were obtained. An initial integer $s(k + r) = 101_{10}$ returned the minimum $\text{RPF}_r = 1.62$ for the signal shown in Figure 13.4(a).

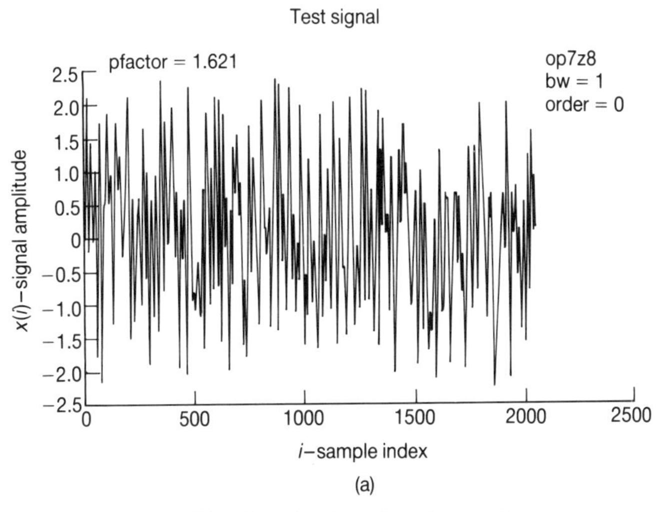

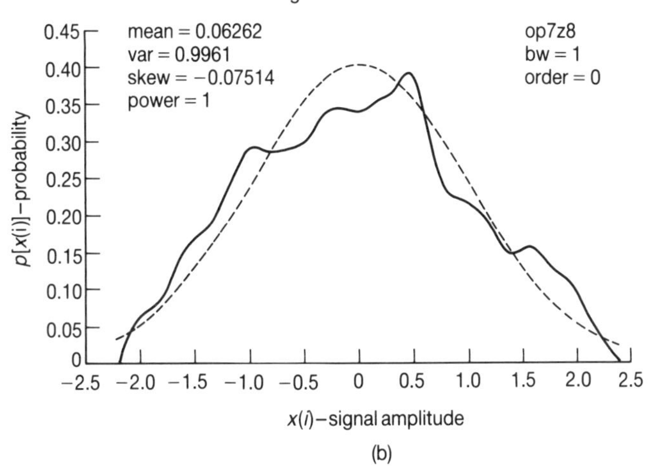

Figure 13.4 Sub-optimized PRGN signal with $N/2 = 128$ harmonics and interpolation factor $m = 8$. (a) unfiltered time signal; (b) probability density function of unfiltered time signal.

The synthesized signals were interpolated using the filtering procedure outlined in Section 13.4.4. An interpolation factor $m = 8$ was chosen which gives an FFT dimension of $mN = 2048$ data cells. Due to the $\sin^2 X/X^2$ shaped spectral envelope associated with the zero-order-hold element, the magnitude of the nearest aliasing component centred round the sampling frequency of the interpolator is reduced to 47 dB below that of the magnitude of the highest baseband component in the signal.

The probability density function $p(x)$ associated with the non-optimized and the

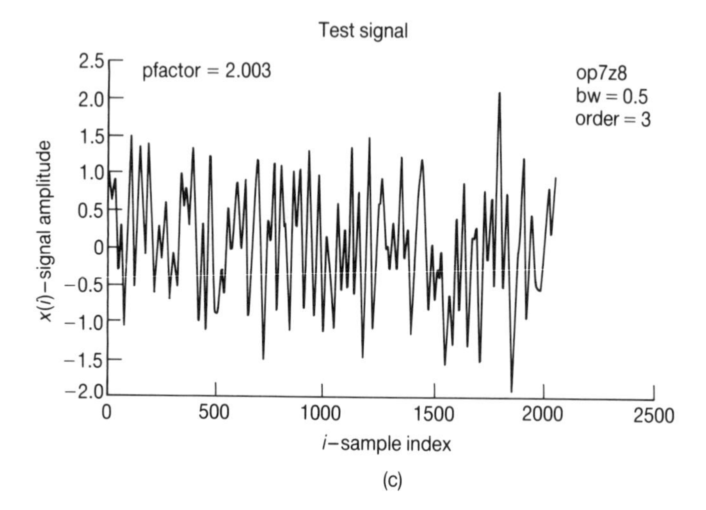

(c)

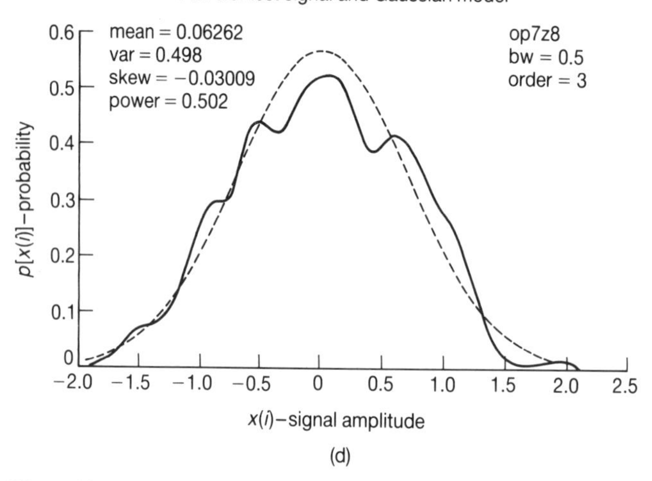

(d)

Figure 13.4 (cont.) (c) Filtered time signal; (d) probability density function of filtered time signal.

optimized PRGN signals are shown in Figures 13.3(b) and 13.4(b) together with a superimposed Gaussian distribution having the same first and second centralized moments as the signal shown. As might be expected, the distribution of the measured signal tends to approach the Gaussian form.

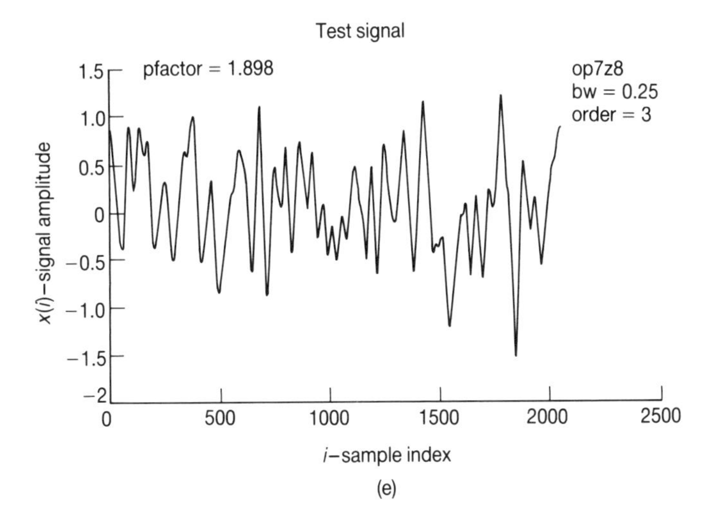

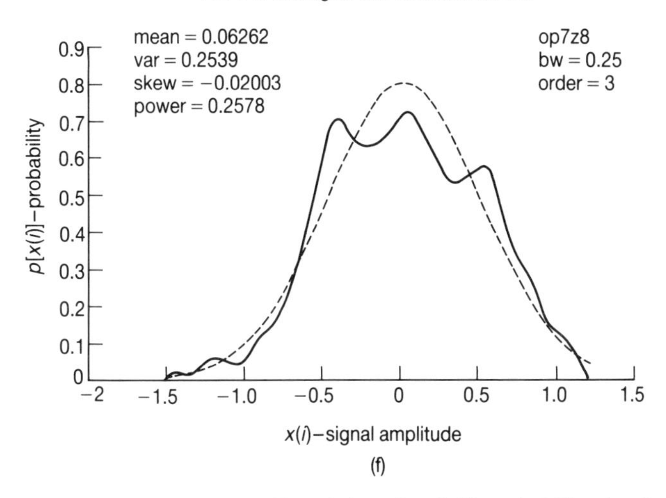

Figure 13.4 (cont.) (e) Filtered time signal; (f) probability density function of filtered time signal.

13.5.2 Comparison of PRGN with PRBS and the Schroeder-phased signal

The conventional PRBS signal is inferior to PRGN in three important respects:

1. There is no control over the signal's $\sin^2 X/X^2$ shaped spectral envelope.

2. The slow spectral roll-off inversely proportional to frequency causes aliasing problems in identification experiments.
3. Its length of $2^N - 1$ sample points is incompatible with analysis based on the radix-2 FFT algorithm.

The Schroeder algorithm (1970) is a deterministic method which obtains a low RPF by selecting phase angles dictated by an algorithm which is functionally dependent on the magnitude of the selected harmonic spectral components. The RPF obtained with this method is not necessarily the minimum achievable, since no explicit minimization routine is used. The resulting frequency spectrum has a similar format to the PRGN spectrum defined above and the filtering techniques of Section 13.4.4 can be used to reduce aliasing.

Since its power-spectral density contains no phase information, the Schroeder waveform has the same DF and H as the corresponding PRGN signal. For a uniform spectrum the Schroeder phase algorithm is given as

$$\phi(k) = \frac{\pi}{N}(k^2 + k) \equiv \frac{\pi}{N} k^2, \quad k = 0, 1, \ldots, \frac{N}{2} - 1 \tag{13.26}$$

The Schroeder probability density function is non-Gaussian with twin asymmetric peaks, similar to the probability density function of a single sine wave. This produces an RPF typically between 1.1 and 1.3. However, this result is not universal and for non-uniform spectra the Schroeder RPF usually increases.

A Schroeder signal having the same spectral composition and length as the signals shown in Figures 13.3 and 13.4 was synthesized and is shown in Figure 13.5(a). The RPF is lower than either of the PRGN signals in Figures 13.3(a) and 13.4(a) (see also Table 13.1). The probability density function in Figure 13.5(b) shows the characteristic 'U' shape associated with the distribution of sinusoidal waveforms. The signal achieves the low RPF value of 1.17 by approaching the 'ideal' impulsive distribution associated with a square waveform (see Section 13.3.1).

Table 13.1 Comparison of relative peak factors for three perturbation signals

	Unfiltered	Filtered BW = 0.5	Filtered BW = 0.25
PRGN	2.08	1.88	1.68
Sub-optimized PRGN	1.62	2.00	1.90
Schroeder-phased	1.17	1.42	1.94

13.5.3 Signal properties after low-pass filtering

Passive low-pass filtering will always reduce signal power, particularly when the ratio of equivalent bandwidth of the input signal to the filter's corner frequency $\lambda = B_e/f_{CO}$ is large, as might be expected.

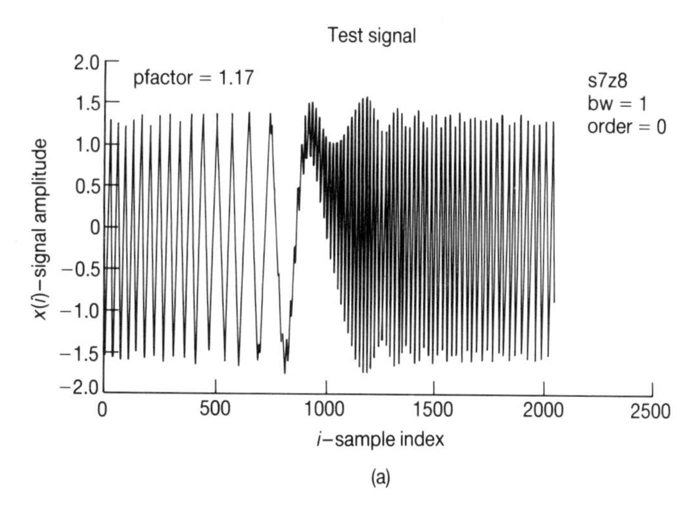

(a)

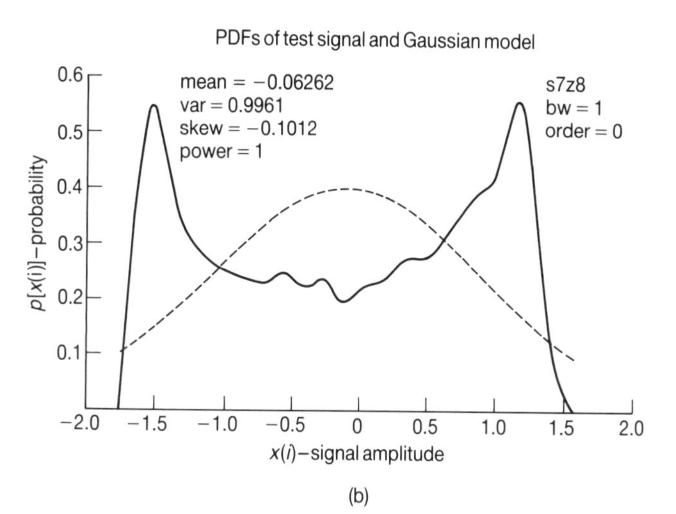

(b)

Figure 13.5 Schroeder-phased signal with $N/2 = 128$ harmonics and interpolation factor $m = 8$. (a) unfiltered time signal; (b) probability density function of unfiltered time signal.

The central limit theorem dictates that a summation of non-Gaussian processes will tend to the Gaussian form (see Bendat and Piersol, 1986). This happens when a signal with arbitrary probability density function is passed through a physical linear-time-invariant (LTI) system. Therefore the already nearly Gaussian distributed PRGN signal should remain relatively unaffected by passage through an LTI system. The Schroeder signal, however, has a non-Gaussian probability density function which will be changed by the filtering action of the LTI system.

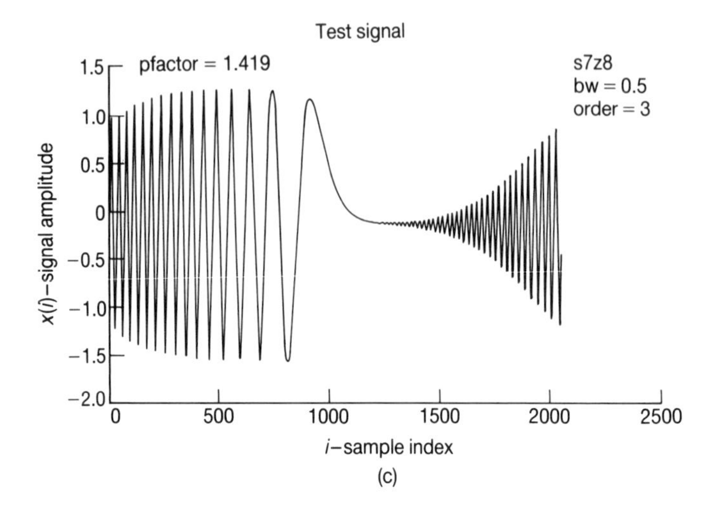

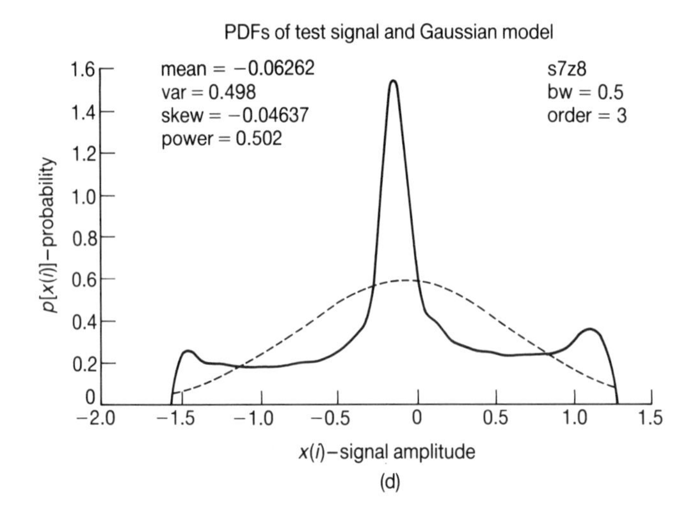

Figure 13.5 (cont.) (c) Filtered time signal; (d) probability density function of filtered time signal

The remaining graphs in Figures 13.3, 13.4 and 13.5(c)–(f) show the time responses and probability density functions of signals which result when a dynamic system is perturbed with each of the three synthesized signals in turn. Third-order low-pass Butterworth filter sections with relative bandwidths of 0.5 and 0.25 times the normalized Nyquist frequency were chosen to model the characteristics of a perturbed dynamic system.

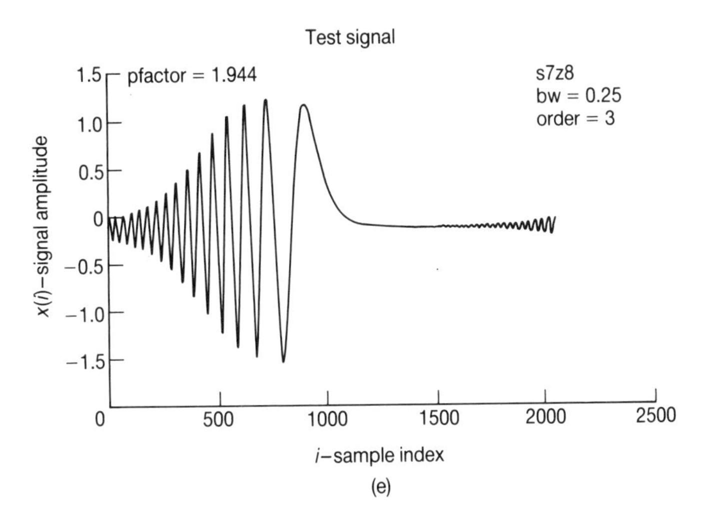

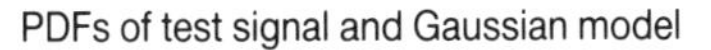

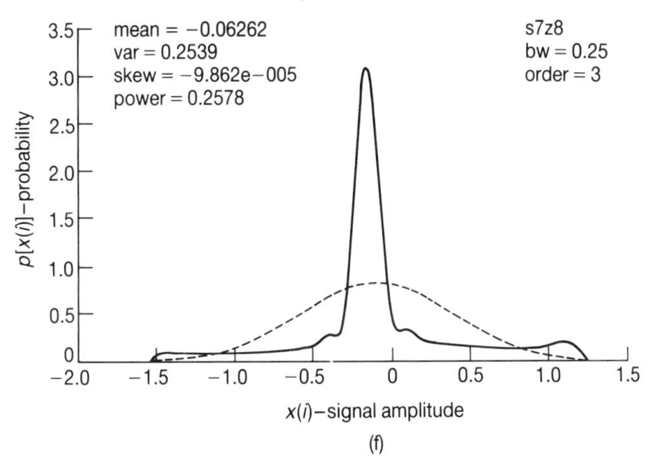

Figure 13.5 (cont.) (e) Filtered time signal; (f) probability density function of filtered time signal

Comparison of the time-domain waveforms shows that both PRGN in Figures 13.3(c) and 13.3(e) and optimized PRGN in Figures 13.4(c) and 13.4(e) retain their characteristic Gaussian appearance after filtering. In contrast, the Schroeder wave shows a change of form, indicating a localized loss of high-frequency energy in the centre of the waveshape. The unfiltered Schroeder waveform in Figure 13.5(a) with an RPF of 1.17 is superior to the RPF values for PRGN of 2.08 and 1.62 shown in

Figures 13.3(a) and 13.4(a). It is interesting to note that filtering reverses this position, with associated RPFs of 1.94 for the Schroeder waveform and RPFs of 1.68 and 1.89 for the PRGN waveforms, respectively. This effect is due to the way in which the harmonic frequencies are distributed in the time-domain waveform. Since the Schroeder waveform essentially consists of a constant-amplitude sinusoidal wave of increasing frequency (a 'chirp'), selective filtering renders part of the waveform inactive with a resulting loss of power and an associated increase of RPF. Frequency components are more uniformly distributed over the PRGN period, and filtering does not now incapacitate part of the waveshape. As a result, the PRGN RPF remains relatively resistant to filtering operations.

Probability density functions of the PRGN type signals in Figures 13.3(d) and 13.3(f), and 13.4(d) and 13.4(f) retain their overall approximate Gaussian shape after filtering as predicted by the central limit theorem. The probability density function of the Schroeder waveform, in contrast, changes from its characteristic 'U' shape to a probability density function, which indicates excessive concentration of signal components narrowly clustered around its mean value.

13.6 Conclusions

Several perturbation signals for application in plant-identification trials were studied. The relative peak factor (RPF), equivalent bandwidth (B_e), equivalent degrees of freedom (EDF) and spectral entropy H are all useful indicators of signal performance.

Simple pulse trains and single sinusoids perform badly in multi-frequency spectral estimation applications. Both pseudo-random binary noise (PRBN) and Gaussian random noise offer better performance but can still cause problems. The PRBN spectrum has a predetermined $\sin^2 X/X^2$ shape, is incompatible with radix-2 FFT analysis and causes aliasing errors in measurement results. The use of a non-periodic perturbation signal such as Gaussian noise increases the variance of parametric estimates and can cause bias errors in closed-loop system identification.

A pseudo-random Gaussian noise (PRGN) signal was synthesized which avoids the problems associated with the signals above. PRGN is based on properties of Gaussian noise; it features good RPF and H indices and is compatible with radix-2 FFT-based spectrum analysis procedures. Both stochastic and periodic estimators may be used in conjunction with this signal.

FFT-based interpolation was used to avoid the need for anti-alias pre-filtering of system-response signals. For systems with low-pass characteristics this technique avoids the need for guard filters on the systems' measurement channels and eliminates any associated differential filter errors.

PRGN was compared with a compatible synthesized Schroeder-phased signal. Both signals are programmed with a uniform spectral envelope and therefore feature identical H values. The unfiltered Schroeder wave has a non-Gaussian probability density function and a slightly lower RPF value than the corresponding PRGN waveform. However, when both signals are passed through a physical dynamic system the RPF of the Schroeder waveform becomes increasingly inferior to the RPF of the PRGN signal as the test system bandwidth is reduced. From this observation it can be argued that only limited returns can be expected from the use of RPF optimized signals, when these signals are filtered by cascaded sections of a dynamic system.

It would appear that the use of a Schroeder-phased synthesized signal as a test stimulus for the identification of a composite system, comprising cascaded low-pass sections, may offer little benefit over PRGN. In this case, sections which precede the measurement point will transform the original Schroeder probability density function to a distribution with excessive concentration around its mean value while returning higher RPFs.

Advantages of the new PRGN signal over the Schroeder wave include:

1. Magnitude and phase spectra are independent and can be separately programmed. This is an important feature when fast updating of the spectral envelope is required (in adaptive estimation procedures, for example).
2. The signal is based on the Gaussian model and is compatible with the characteristics of physical dynamic systems. When the signal is passed through a linear-time-invariant dynamic system it will remain approximately Gaussian distributed and only the signal variance will change.

Appendix

The unweighted average of M Fourier transformed N-point signal vectors x_n, $n = 0, 1, 2, \ldots, M-1$, at frequency index k can be expressed as

$$\overline{X(k)} = \frac{1}{M} \sum_{n=0}^{M-1} \left[\frac{1}{N} \sum_{i=0}^{N-1} x_{i+nN} \cdot \lambda^{-ik} \right] \tag{13.A1}$$

where λ represents the discrete Fourier kernel $e^{j2\pi/N}$. Swapping the order of summation gives

$$\overline{X(k)} = \frac{1}{N} \sum_{i=0}^{N-1} \left[\frac{1}{M} \sum_{n=0}^{M-1} x_{i+nN} \right] \lambda^{-ik} \tag{13.A2}$$

In words: 'The average of M, N-point Fourier transformed signal vectors equals the N-point Fourier transform of the M averaged signal vectors.'

Using $F^{+}\{\ \}$ to represent a Fourier transformation of a discrete time series x_i over an interval $i \in \pm\infty$, the response at frequency ω of the periodically averaged process is obtained as shown below. Assuming unit spacing of the signal samples:

$$F^{+}\left\{ \frac{1}{M} \sum_{n=0}^{M-1} x_{i+nN} \right\} = F^{+}\{x_i\} \cdot \frac{1}{M} \sum_{n=0}^{M-1} e^{j\omega nN}$$

$$= F^{+}\{x_i\} \cdot \frac{\sin(\omega MN/2)}{M \ \sin(\omega N/2)} \tag{13.A3}$$

Acknowledgements

The authors would like to acknowledge suport from the School of Systems Engineering, University of Portsmouth and the School of Electrical, Electronic and Information Engineering, South Bank University. Mr M. Khoshlahjeh-Motamed of South Bank University generated the original time sequences.

14

Application of Multi-frequency Test Signals to an Industrial Water Boiler

Haydn R. Porch

14.1 Introduction

In an attempt to improve the efficiency of process plant within the gas industry through improved plant control, modern control techniques have been investigated. Invariably all the techniques examined required process plant models in the form of transfer functions. As an alternative to model development through first principles of the physico-chemical laws and provided the process plant exists, system-identification techniques can be employed where the process plant input or excitation signal and the process plant output or response signal are analyzed to determine the transfer function of the plant. This immediately appears an attractive proposition to avoid the sometimes complicated and idealistic use of the physico-chemical laws and the process of determining the physical property values of the system. However, many process plants possess non-linearities to some degree, which complicate the process of linear system identification techniques to such an extent that they may be impractical to use. Typical non-linearities such as hysteresis and flow control valves (dependent upon the type of trim used) are common to many process plants within the gas industry, and these and other non-linearities must be investigated to assess their relative contribution to the system before system identification can be performed. Restricting the amplitude variation of the process plant excitation signal may allow the plant response to appear almost linear and therefore make the plant suitable for linear system-identification techniques.

The conventional method of performing system identification is by obtaining the frequency response of a plant using a sinusoid signal as the plant excitation and comparing this with the plant response to obtain the phase and gain values at that frequency of signal. A range of signal frequencies is required to obtain the complete frequency response of the plant. It can be seen clearly that the logical progression of this type of plant identification is to incorporate the frequency sweeping of the sinusoid into a single signal to perform the complete frequency response in a single plant test, reducing the testing time considerably. This progression has led to a whole range of multi-frequency type plant test signals with various properties. These are described in detail elsewhere in this book and only the major properties and their practical application will be discussed here.

In many cases the control engineer is understandably more concerned with plant control analysis than system identification. There appeared a need for a simple 'tool' to be applied to a plant such that a model can be achieved quickly with minimal effort and understanding of the sometimes complicated equations necessary to analyze the process plant data. Both Ljung (1987) and Söderström and Stoica (1989) indicate the necessary characteristics or requirements of such a tool and many of these requirements have been included in a software package called Plant System Identification (PSI), which was developed at British Gas, Midlands Research Station (See Note 1, p. 421).

The package is designed to run on any standard PC computer with a commercially available analog input/output card to perform the complete linear system-identification process through excitation signal generation, plant testing and data capture, data analysis in both frequency and time domain, transfer function fit in frequency domain and model validation. Consideration was given to the user interface to minimize the complication by using single-screen menus to prompt the user for the necessary information required in a particular module of identification.

Figure 14.1 shows the PSI master menu indicating the various modules within the package for complete system identification. The package builds on the work by Mercer and Mailey (1986) and uses many of the techniques described by Wellstead (1981). What follows is a description of the salient features of the developed PSI package for both frequency- and time-domain analysis, with emphasis on the practical use of multi-frequency type plant excitation signals followed by an example on the system identification of an industrial water boiler.

14.2 Excitation signal generation

The PSI software has the ability to generate several types of plant-excitation signal with various properties to suit the test application, two of which are of the multi-frequency type. To generate a suitable signal some *a priori* information must be known about the process plant. Figure 14.2 shows the layout of the excitation signal-generation menu which allows the user to configure a suitable test signal for the process plant under test.

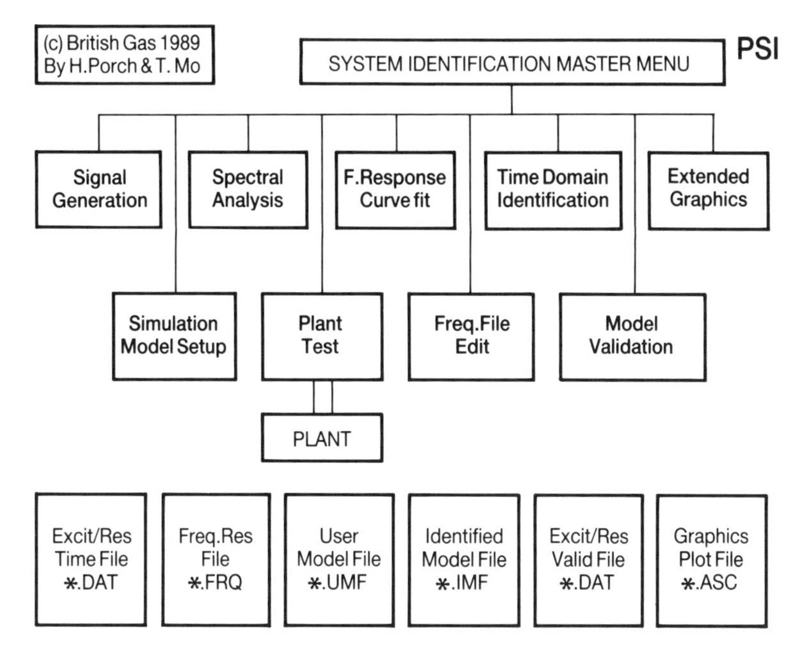

Figure 14.1 PSI Master Menu.

The third row in Figure 14.2 (indicated by $*$) shows the various test signals that can be generated while the remainder of the menu is used to configure the parameters of the signal and view the results. Figure 14.3 shows the generated plant test signal resulting from the configuration set-up in Figure 14.2.

When designing a suitable test signal it is important that the amplitude of the signal be as large as possible but within a linear operating range of the process plant under test. There are two other important properties of a suitable signal: signal power and signal bandwidth.

Let us first consider the property of bandwidth. For a test signal of period T s with N discrete data points, the lowest or fundamental frequency (ω_{f} rad/s) that can be examined is given by

$$\omega_{\mathrm{f}} = \frac{2\pi}{T} \tag{14.1}$$

If the Fast Fourier Transform (FFT) is used to analyze the data (as with the PSI package) then the maximum frequency (ω_{m}), or Nyquist frequency, examined is given by

$$\omega_{\mathrm{m}} = \frac{N\pi}{T} \tag{14.2}$$

MEM:329200	VERSION 3.0 SIGNAL GENERATION MENU

Title:[ ]

★ Square wave:[ ] Sine wave:[ ] Low Peak Factor:[ ] PRBS : [Y]

Load file : [ ] Save file : [ ] :

Plot signal:( ) Generate signal:( ) Points/PRBS sample : [4]

Display PSD:( )

Signal period : [1000.0]seconds Sample interval:[1.95312]seconds

Amplitude (p-p) : [2.000] volts Offset :[3.000] volts

No.of points:[512] Import PSD profile for LPF signal:[ ]

Fundamental frequency :[0.0010] Hz Maximum frequency: [0.256] Hz

[0.0063] Rad/s [1.6085] Rad/s

ESC to exit menu

Use cursor keys to select item to be changed.

Figure 14.2 PSI Signal Generation Menu.

The maximum frequency from an FFT analysis is always the Nyquist frequency, and, by definition, would indicate that there are only two samples per cycle. In practice, this is barely sufficient to reproduce the original signal and a much more practical maximum frequency (ω_p) would be half of the Nyquist frequency, as in

$$\omega_p = \frac{N\pi}{2T} \tag{14.3}$$

For total plant identification, the plant under test must have all its major dynamic modes within the range of ω_f to ω_p, ignoring subsequent FFT analysis results from ω_p to ω_m. This suggests that some idea of the frequency response of the plant under test must be known in advance before a total plant identification can be performed. A simple squarewave test signal is useful to establish an approximate time constant for the plant using simple slope constructs on the squarewave plant response to determine the dominant time constant τ. The resulting time constant can then be used to indicate a Bode plot corner frequency ω_c, given by

$$\omega_c = \frac{1}{\tau} \tag{14.4}$$

Thus the test signal frequency range must be chosen such that the inequality

$$\omega_f < \omega_c < \omega_p \tag{14.5}$$

is satisfied.

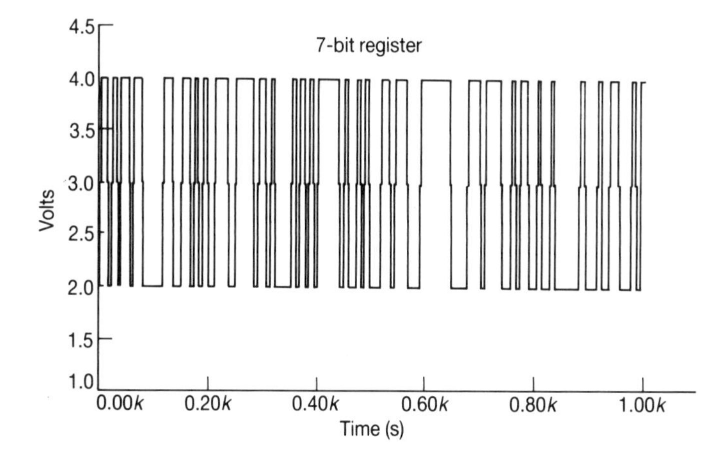

Figure 14.3 PSI-generated PRBS signal.

From a suitable selection of ω_{f} and ω_{p}, based upon inequality (14.5) the desired test signal period (T) and number of discrete points (N) can be found for the particular plant under test, having time constant τ, using equations (14.1) and (14.3). For example, if the experimental time constant τ was found to be approximately 20 s then $\omega_{\mathrm{c}} = 0.05$ rad/s. Suitable test signal parameters will be $T = 1000.0$ s and $N = 512$, yielding $\omega_{\mathrm{f}} = 0.0063$ rad/s and $\omega_{\mathrm{m}} = 1.6085$ rad/s, as indicated in Figure 14.2, from which equation (14.3) gives $\omega_{\mathrm{p}} = 0.8043$ rad/s.

Let us now consider the power in a test signal or, more precisely, the distribution of test signal power with frequency. This can be described using the power-spectral density function (S_{xx}), and the reader is referred to Chapter 2 of this book for a detailed explanation of this function. During a system-identification operation it is desirable to have as much power injected into a plant as possible. This can be performed simply by increasing the amplitude of the test signal. However, a compromise must be sought to prevent violating any plant-linearity constraints. With the test signal amplitude fixed by the process plant, the signal power spectrum will be determined by the signal type.

Consider a pseudo-random binary signal (PRBS) of amplitude $\pm a$, signal period T, with clock pulse interval of Δt, based on a maximum-length sequence (MLS) with N digits. Then the signal power spectrum $S_{xx}(n/N\,\Delta t)$ is, from Chapter 2, Example 2E:

$$S_{xx}\left(\frac{n}{N\,\Delta t}\right) = \frac{a^2(N+1)}{N^2} \cdot \frac{\sin^2[n\pi/N]}{[n\pi/N]^2}, \quad n = \pm 1, \pm 2, \pm 3, \ldots \tag{14.6}$$

The PRBS power spectrum envelope illustrated in Figure 14.4 shows that the effective frequency band for a PRBS signal is well below $1/\Delta t$ Hz. If such a signal is used to excite a plant and the resulting data, sampled at Δt intervals, are analyzed

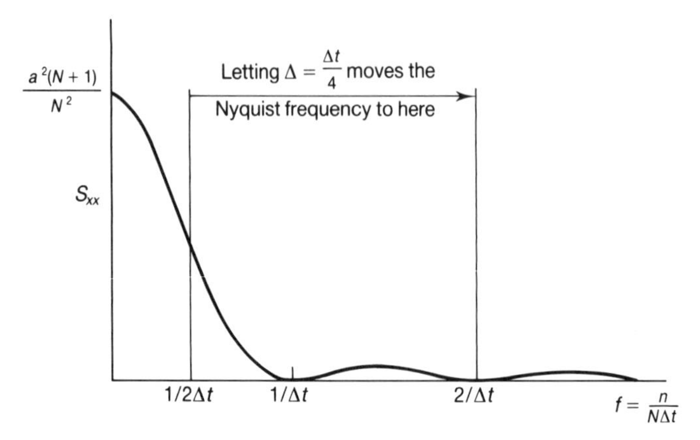

Figure 14.4 PRBS continuous power spectrum.

using FFT analysis, then the maximum analysis frequency will be $1/2\Delta t$ Hz. This indicates that there is still considerable power within the test signal above the Nyquist frequency and unless suitable low-pass filters are used when capturing the data, serious 'foldover' or aliasing errors will occur; this is also illustrated in Example 2F in Chapter 2.

An alternative solution is to reduce the sampling interval Δ from Δt to, say, $\Delta = \Delta t/4$. This will not alter the power spectrum of the PRBS signal, but the maximum analysis frequency will be increased to $2/\Delta t$ Hz. Now the signal power content above the Nyquist frequency is much less than the previous analysis and the foldover effects will be reduced. If we only consider data analysis results up to half of the Nyquist frequency (or ω_p) then we are still using the effective frequency band for the PRBS signal, although the power is reducing to a minimum at ω_p. If foldover frequencies are still a problem, the sampling interval can be reduced further by any interger power of 2. This ultimately yields more frequency-analysis data points, many of which have to be discarded due to the low power content at the higher frequencies.

The MLS PRBS signal does not lend itself readily to FFT analysis, due to the number of sequence intervals being $2^n - 1$, $n = 1, 2, 3, \ldots$; always one less than required for FFT analysis using the radix-2 algorithm. Increasing the number of samples does not improve matters. For example, having four samples per sequence interval gives four samples less than required for an FFT analysis. The obvious alternative is to use a mixed radix FFT algorithm similar to that proposed by Singleton (1969). These algorithms, although more flexible in the number of data points, generally require greater computation and are therefore slower to perform than the standard Cooley–Tukey algorithm described by Brigham (1988). To avoid this extra computational burden and to be compatible with other spectral analyses, the PSI software package uses the standard Cooley–Tukey algorithm and the PRBS signal is modified to suit. This modification is to corrupt the PRBS signal by 'stretching' it to the

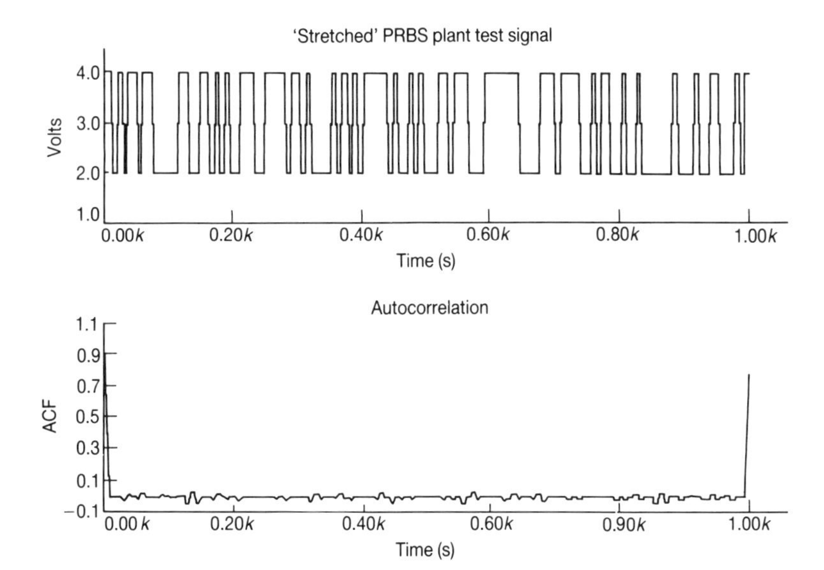

Figure 14.5 Stretched PRBS with its corresponding autocorrelation.

number of samples required to complete 2^n, $n = 1, 2, 3 \ldots$, samples, by adding the required number of points evenly spaced along the signal length. A true MLS, PRBS signal will produce an autocorrelation function of a triangular pulse which, from Chapter 1, Section 1.4, is described by

$$R_{xx}(\tau) = a^2\left[\frac{N+1}{N}\left(1 - \frac{|\tau|}{\Delta t}\right) - \frac{1}{N}\right], \qquad |\tau| < \Delta t \tag{14.7}$$

$$R_{xx}(\tau) = -\frac{a^2}{N}, \qquad \Delta t < \tau < (N-1)\Delta t \tag{14.8}$$

Using the radix-2 FFT algorithm to calculate the autocorrelation function of a 'stretched' PRBS signal produces the slightly corrupted ideal autocorrelation function shown in Figure 14.5. Whereas for the ideal autocorrelation function the steady-state bias is constant, the corrupted autocorrelation function has a non-steady bias. For large values of N this corruption reduces to practical insignificance.

There is a wide range of multi-frequency binary sequence (MBS) type signals each with its own particular property (see Chapter 2, Section 2.4.1 and Chapter 7). These have an advantage over the MLS PRBS in as much as they concentrate their signal power into predefined harmonics and these power levels are generally higher than that of an equivalent MLS PRBS. They also have one other major advantage over the MLS PRBS; they generally have 2^n, $n = 1, 2, 3, \ldots$, data points, which makes them more suitable for FFT analysis using the radix-2 algorithm.

To accommodate these and any other binary or non-binary user-defined signals, the facility has been provided within the PSI package to import user signals in the form of an ASCII list describing a normalized signal profile. A signal is simply defined by specifying the relative amplitude of each successive point within an ASCII list. The PSI software detects the ASCII file and converts the relative amplitude and signal period to that defined within the Signal Generation menu. Once a signal has been imported into the PSI package, the plant testing and data-analysis routines will be available as with the internally generated test signals, giving total flexibility in the type of test signal used.

The PRBS signal is classed as a binary type signal, that is, it exists in only one of two states ($\pm a$). As shown in Chapter 2, Section 2.4.2 and in Chapters 3, 12 and 13, there is also a class of analog signals that are multi-frequency in nature and have specific power spectra. If we consider a desired relative power spectrum (P_n) of a test signal ($x(kt)$, $k = 0, 1, 2, \ldots$) and perform the inverse discrete Fourier transform to yield the test signal, the following is obtained:

$$x(kt) = \sum_{n=1}^{N-1} \sqrt{2P_n} \cos\left[\frac{2\pi nk}{N} + \theta_n\right], \qquad k = 0, 1, 2, 3, \ldots, N \tag{14.9}$$

where P_n is the relative power at the nth harmonic and θ_n is the phase angle at the nth harmonic.

To obtain a suitable plant test signal from equation (14.9) the practical requirements such as the desired relative power P_n of a test signal need to be considered. In the practical case where the process plant frequency response is unknown, other than an indication of an approximate corner frequency, then it is desirable to have constant power across a frequency spectrum either side of the corner frequency, effectively providing signal power at all frequencies of the test signal. To generate subsequently a suitable signal, the desired signal phase angles θ_n need to be found. These can be any arbitrary values, but poor selection can lead to large peaks in the resulting test signal, which are undesirable in practice.

The concept of the Peak Factor is a measure of the size of variations in signal amplitude, the minimization of which is the design objective of any multi-frequency signal (for example and comparison, binary type signals with equal positive and negative levels possess a fixed peak factor of $1/\sqrt{2}$). Schroeder (1970) went some way towards solving the problem of finding suitable phase angles (θ_n) that minimize the peaking effect of a non-binary signal and still retain the signal power over the frequency range. The phase angle-generation analysis is based upon heuristic arguments which ultimately produced a simple formula to determine the phase variation at different signal harmonics:

$$\theta_n = 2\pi \sum_{i=1}^{n} iP_i \qquad n = 1, 2, 3, \ldots, N \tag{14.10}$$

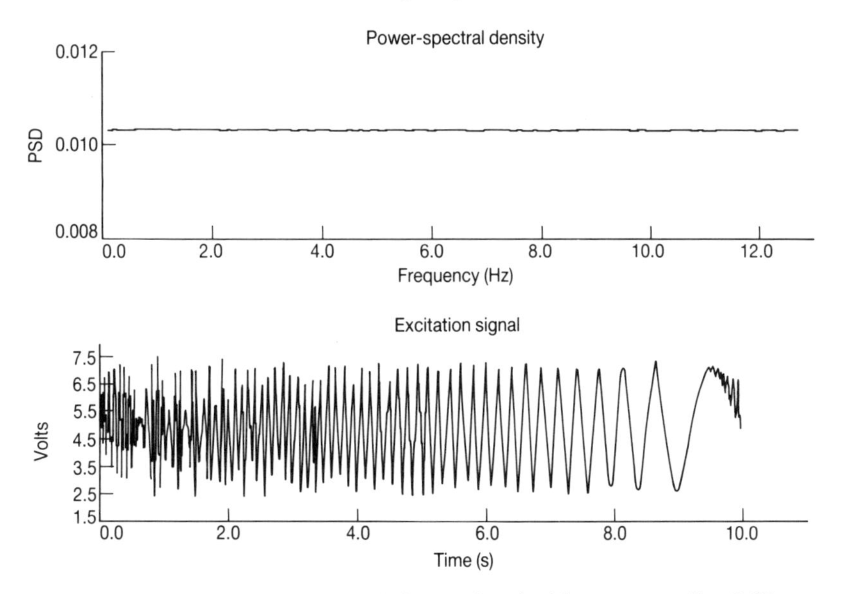

Figure 14.6 Schroeder low-peak-factor signal with corresponding PSD.

A multi-frequency plant test generated from equations (14.9) and (14.10) is known as a low-peak-factor (LPF) type signal and a typical constant-power is shown in Figure 14.6.

Having the ability to specify the desired power spectrum for the test signal leads to a whole family of Schroeder design type signals, a few examples of which are shown in Figures 14.7–14.9 with their associated predefined power spectra. The PSI software

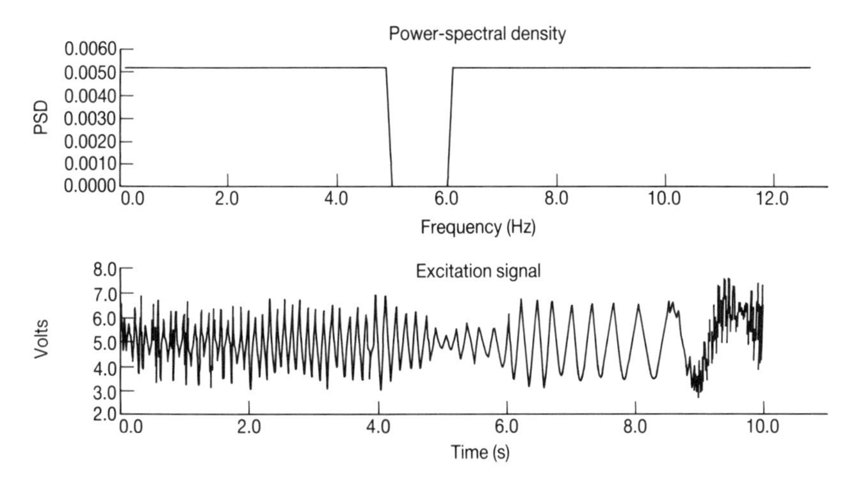

Figure 14.7 Schroeder signal with notched PSD.

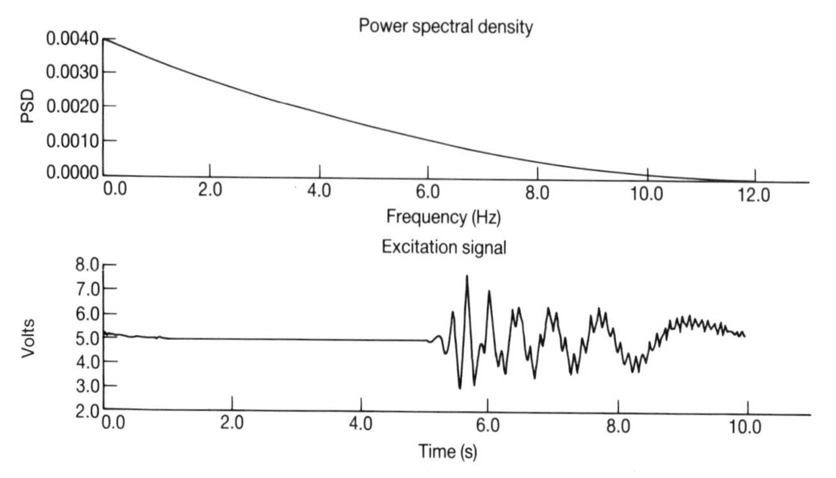

Figure 14.8 Schroeder signal with falling PSD.

provides the ability to generate a constant power LPF signal with user-configurable signal periods, amplitudes and number of points (must be given as 2^n, $n = 3, 4, 5 \ldots$ data points). Further flexibility is also included to import user-definable power spectra, to suit the test application, and generate a Schroeder LPF signal.

14.3 Plant testing

Designing and generating a suitable plant test signal is of major importance when

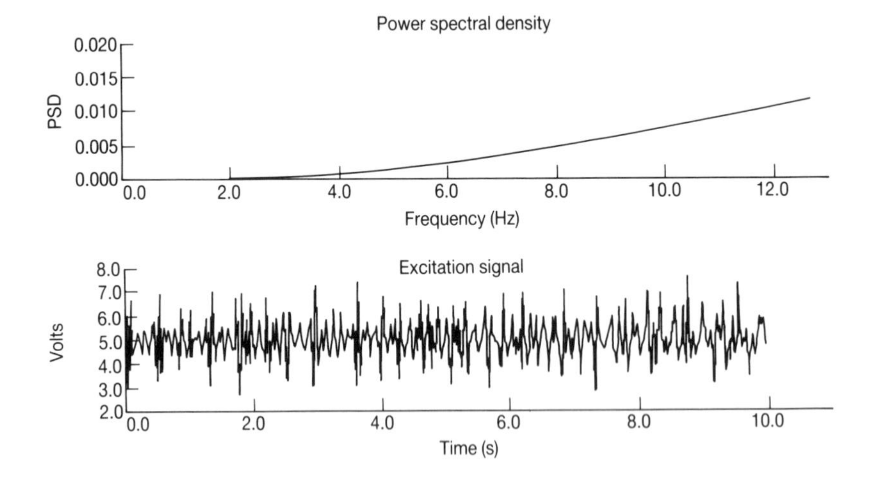

Figure 14.9 Schroeder signal with rising PSD.

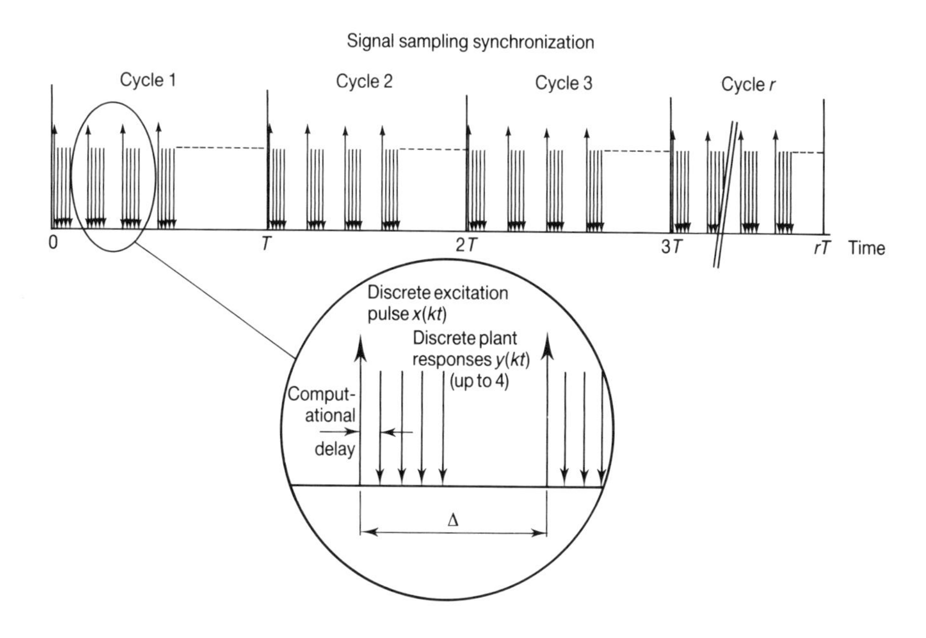

Figure 14.10 Signal timing arrangement for plant testing.

performing system identification. However, it is only one stage in the process of obtaining a plant model. Once a suitable plant test signal is generated, its discrete values must be output from a computer at real-time intervals and the subsequent plant responses captured. A schematic of the time arrangement for plant testing is shown in Figure 14.10, where T is the test signal period and Δ is the sampling interval.

Figure 14.10 highlights some important considerations when plant testing. First, repeated cycles of the test signal can be output to excite the plant, and the corresponding plant response captured; ultimately this reduces the final data-analysis error. This point will be discussed further below. The second consideration is the sampling delays: within every sampling interval (Δ) there is a value output to the plant and the corresponding plant response captured (up to four plant responses can be captured per Δ in the PSI software). This delay between the computer-generated output and the plant-response capture should be at a minimum. Practically, this is the computational delay required to perform the individual data output and capture operations. Any delay in capturing the plant responses, following the output signal, has the effect of subtracting the equivalent time delay from the final model or transfer function of the plant. Similarly, the data capture points must be co-ordinated to follow immediately each of the plant excitation output signals to prevent inconsistent time-delay effects from the final model result.

The effect of aliasing and how careful test signal selection can minimize foldover error have been discussed previously. Whenever possible, it is advisable that anti-aliasing filters should be used on both plant excitation and response signals.

There are many more factors that need to be considered when performing plant testing, which are dependent upon the particular plant under test and will not be discussed here. It must be considered that, in general, the interfacing of the computer to the plant instrumentation, and the subsequent calibration, can take considerably more time than the actual plant testing.

14.4 Frequency-response analysis

Once the plant excitation and response data are captured, then 'off-line' analysis can be performed to identify a transfer function of the plant. To obtain the frequency response of the plant it is necessary to find the ratio of the cross- and power-spectral densities. The power-spectral density equation for a signal $x(kt)$ and a similar equation for the cross-spectral density S_{xy} are as follows:

$$S_{xx}\left(\frac{n}{NT}\right) = \left[X\left(\frac{n}{NT}\right)X^*\left(\frac{n}{NT}\right)\right], \qquad n = 0, 1, 2, 3, \ldots, N/2 \tag{14.11}$$

$$S_{xy}\left(\frac{n}{NT}\right) = \left[X\left(\frac{n}{NT}\right)Y^*\left(\frac{n}{NT}\right)\right], \qquad n = 0, 1, 2, 3, \ldots, N/2 \tag{14.12}$$

where $X(n/NT)$ denotes the discrete Fourier transform of the plant excitation signal $x(kt)$, $Y(n/NT)$ the discrete Fourier transform of the plant response signal $y(kt)$ and an asterisk represents the complex conjugate.

However, to minimize errors in the signals due to disturbances while plant testing, the spectral estimates of the plant data are averaged over a sufficient number of signal cycles (r), and so the power- and cross-spectral densities are given by

$$\overline{S_{xx}}\left(\frac{n}{NT}\right) = \frac{1}{r}\sum_{i=1}^{r}\left[X\left(\frac{n}{NT}\right)_i X^*\left(\frac{n}{NT}\right)_i\right], \qquad n = 0, 1, 2, 3, \ldots, N/2 \tag{14.13}$$

$$\overline{S_{xy}}\left(\frac{n}{NT}\right) = \frac{1}{r}\sum_{i=1}^{r}\left[X\left(\frac{n}{NT}\right)_i Y^*\left(\frac{n}{NT}\right)_i\right], \qquad n = 0, 1, 2, 3, \ldots, N/2 \tag{14.14}$$

Once the averaged spectral estimates are calculated, the frequency response $G(n/NT)$ of the plant can be found from

$$G\left(\frac{n}{NT}\right) = \frac{\overline{S_{xy}}(n/NT)}{\overline{S_{xx}}(n/NT)}, \qquad n = 0, 1, 2, 3, \ldots, N/2 \tag{14.15}$$

The variance of the resulting frequency-response estimates calculated using equation (14.15) is given by

$$\frac{\mathrm{Var}\{|G(f)|\}}{|G(f)|^2} = \mathrm{Var}\{\mathrm{Arg}\, G(f)\} = \frac{1}{2r}\left(\frac{1}{\gamma_{xy}^2(f)} - 1\right) \tag{14.16}$$

where $f =$ frequency (n/NT), $\gamma_{xy}^2(f) =$ squared data coherence and $r =$ number of test signal cycles.

From equation (14.16) it can be seen that the variability of the frequency response is reduced by increasing the number of test signal cycles (r). It is, therefore, desirable to have data from as many test signal cycles as possible. However, for process plant with long time constants, the test signal period can be quite large and therefore the number of signal cycles chosen is a function of the practical limitations of the process plant.

The squared data coherence function, introduced in Section 2.5.1, is a measure of the interdependence of the two signals $x(kt)$ and $y(kt)$ and is thus a normalized frequency-domain representation of the signal-to-noise ratio in the process plant system. The squared coherence is defined as

$$\gamma_{xy}^2(f) = \frac{|S_{xy}(f)|^2}{S_{xx}(f)S_{yy}(f)} \tag{14.17}$$

In the noise-free case the coherence (γ) approaches unity and for large values of the noise spectrum the coherence approaches zero. A coherence value of 0.8 is generally regarded as the minimum acceptable value to reproduce a reasonable frequency-response estimate.

The frequency-response estimates of the process plant obtained from equation (14.14) include the predictable effects of the sample-and-hold device from the physical data-capturing hardware (computer input/output card). This sample-and-hold (or zero-order-hold) device contributes additional frequency components $F(\omega)$ to the calculated plant spectrum, where

$$F(\omega) = \mathrm{e}^{-\mathrm{j}(\omega\Delta/2)}\,\Delta\,\mathrm{sinc}\left(\frac{\omega\Delta}{2}\right) \tag{14.18}$$

where Δ is the sampling interval. The effect of the sample-and-hold device on the frequency response manifests itself mainly as being equivalent to a delay of half the sample period and therefore causes the phase response to increase considerably at higher frequencies. The effect on the magnitude response is practically insignificant up to the Nyquist frequency. To give the true continuous frequency-response estimates of the tested plant, the calculated frequency spectra must be modified to remove the frequency-response effects of this zero-order-hold device.

The resulting frequency response is useful, as it can give an indication of the dynamics of the plant. However, the data need to be parameterized to give a transfer function that can be used for further control analysis. A technique first proposed by Levy (1959), which used a complex least squares approach to estimate the unknown coefficients of a predefined transfer structure, has been extensively modified by Sanathan and Koerner (1963) to minimize the bias of the high-frequency data points. As such, this modified complex least squares fit is used in the PSI software.

14.5 Time-domain analysis

Time-domain system identification has generated much interest in recent years due mainly to increased industrial interest in self-tuning controllers, where some of the time-domain identification techniques are of the recursive type. When using these recursive techniques special precautions have to be taken to ensure that the algorithm is continually numerically robust and to avoid what is known as estimator wind-up. In general, batch methods of time-domain system identification do not suffer from many of the problems of recursive identification. However, although it is not obvious in time-domain identification as in frequency-response identification, the process plant excitation must be rich in frequency content to excite all the modes of the process. This frequency information is not readily available when performing recursive identification as the process excitation generally relies upon normal disturbances and noise within the process (for example, control inputs and plant responses).

Batch time-domain identification can be treated in a similar manner to frequency-response identification in that the process plant is excited with a suitable test signal and the responses captured; the results are analyzed using suitable algorithms.

The problem here is, what is a suitable algorithm? There are many algorithms available for the user, each with attributes that make them suitable for a limited type of process plant. From studies of many of these algorithms it has been found that one of the most general and computationally reliable methods was the multi-step instrumental variable (MIV), which uses a mixture of ordinary least square (OLS), generalized least squares (GLS) and instrumental variable (IV) techniques. For an explanation of these methods the reader is referred to a text on system identification (for example, Ljung, 1987, or Söderström and Stoica, 1989).

The form of plant model structure assumed within the PSI package is shown in Figure 14.11. This can be expressed as

$$y_t = \frac{B(q^{-1})}{1 + A(q^{-1})} q^{-k} u_t + \frac{1 + F(q^{-1})}{1 + G(q^{-1})} e_t \tag{14.19}$$

where y_t, u_t and e_t are the process output sequence (response), input sequence (excitation) and assumed zero mean white-noise sequence, respectively. k is the dead time expressed as an integer multiple of the sampling interval and must be greater than or equal to zero and q^{-1} is the backward shift operator.

Polynomials $A(q^{-1})$, $B(q^{-1})$, $F(q^{-1})$ and $G(q^{-1})$ are given by

$$A(q^{-1}) = a_1 q^{-1} + \ldots + a_{na} q^{-na} \tag{14.20}$$

$$B(q^{-1}) = b_1 q^{-1} + \ldots + b_{nb} q^{-nb} \text{ (with the assumed unit dead time)}$$

$$F(q^{-1}) = f_1 q^{-1} + \ldots + f_{nf} q^{-nf}$$

$$G(q^{-1}) = g_1 q^{-1} + \ldots + g_{ng} q^{-ng}$$

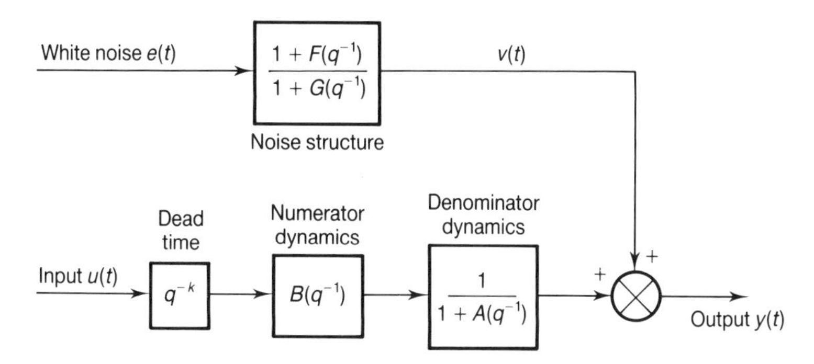

Figure 14.11 Plant model structure.

where na, nb, nf, ng are the number of parameters in each of the polynomials, respectively.

The basic IV algorithm is at least a two-stage process, with the two stages being repeated iteratively to reduce the model residuals. However, Söderström and his co-workers performed a detailed analysis of a four-step algorithm which, theoretically, will not require any further iterations provided the number of data points is large. This has the advantage of having relatively fixed computation time irrespective of the quality of the data. Examples of four-step IV techniques are given in Ljung (1987, Section 15.3) and Söderström and Stoica (1989, Section 8.2).

In the PSI package, the four steps are:

Step 1: OLS to obtain crude parameter estimates
Step 2: IV to get improved estimates
Step 3: OLS to estimate noise structure
Step 4: IV to obtain final estimates.

14.6 Model structure selection

In both frequency- and time-domain identification the model order or structure has to be predefined before parameterization can begin. An overparameterized model structure can lead to complicated use of the estimated model and possibly produce ill-conditioned parameters that could cause problems during control analysis. An underparameterized model may be very inaccurate. A method of determining a suitable model structure for a set of data is required.

In frequency-response analysis the magnitude and phase against frequency curves give the control engineer an indication of the desired model structure. For example, the fall-off rate of the magnitude is 20 dB/decade for every first-order component, the phase curve approaching $-90°$ for every first-order component, delay effect on the phase curve causing an increase in phase lag at higher frequencies, etc. These and other pointers give an indication of the true model structure. However, as the structure becomes more complicated the use of these pointers becomes more difficult to assess. To aid model structure selection in the frequency domain, two error criteria of magnitude (M_e) and phase (P_e) are used within the PSI software, based on the average absolute error over the number of frequency points (N_f) selected:

$$M_e = \frac{1}{N_f} \sum_{i=1}^{N_f} \frac{\text{abs}(|G(j\omega)_i| - |G_m(j\omega)_i|)}{G(j\omega)_i} \tag{14.21}$$

$$P_e = \frac{1}{N_f} \sum_{i=1}^{N_f} \text{abs}(\arg\{G(j\omega)_i\} - \arg\{G_m(j\omega)_i\}) \tag{14.22}$$

where $G_m(j\omega)$ = resulting analysis frequency response, $G(j\omega)$ = actual frequency-response data and N_f = number of frequency data points selected.

By repeated calculations a model structure can be found that ultimately minimizes the coefficients of fit from equations (14.21) and (14.22) but gives no indication on overparameterization or ill-conditioned parameters, and so general user common sense on model structures is still required.

A similar approach to model structure selection can be used within the time-domain identification by analyzing the prediction errors (more often called the residuals). These are the errors observed between the model response and the actual response of the plant to the same excitation. By repeated calculation a model structure can be found that minimizes the absolute sum of the residuals. By defining a function V as the sum of the residuals squared (equation (14.23)), a suitable model structure can be found that minimizes this function:

$$V \triangleq \sum_{t=1}^{N} (\text{actual } y_t - \text{estimated } y_t)^2$$

$$= \sum_{t=1}^{n} (y_t + a_1 y_{t-1} + \ldots + a_{na} y_{t-na} - b_1 u_{t-k-1} \ldots - b_{nb} u_{t-k-nb})^2 \tag{14.23}$$

However, to bring this model structure criterion into line with other criteria, the loss function criterion (LFC):

$$\text{LFC} = \ln(V) \tag{14.24}$$

which must be minimized to identify a suitable model structure. This criterion, as with the criterion for the frequency-domain structure selection, does not take into account any overparameterization and could lead to excessive model structures. An alternative criterion is the well-known Akaike information criterion (AIC):

$$\text{AIC} = N \log(\hat{\sigma}_e^2) + 2p \tag{14.25}$$

where $\hat{\sigma}_e^2$ is the estimated variance of the residuals, N is the number of data pairs and p is the number of parameters to be estimated. The first term of equation (14.25) represents a measure of how well the data fit the model and the second term provides a penalty on the complexity of the model. The minimization of the AIC criteria tends to seek a compromise between the degree of model fit and model complexity.

Another criterion of fit is described by P.C. Young (1989), which involves the use of two statistical measures which together define a suitable model structure that fits the data well and has well-defined parameter estimates. The criterion is called Young's information criterion (YIC):

$$\mathrm{YIC} = \ln(\mathrm{NSVR}) + \ln(\mathrm{NEVN}) \tag{14.26}$$

where

$$\mathrm{NSVR} \triangleq \frac{\sigma_e^2}{\sigma_y^2} \quad \text{and} \quad \mathrm{NEVN} \triangleq \left(\frac{\sigma_e^2}{N}\right) \sum_{i=1}^{p} \frac{P_{i,i}}{\hat{\theta}_i^2}$$

in which p is the number of parameters to be estimated, σ_e^2 is the residual variance, σ_y^2 is the actual plant output variance, N is the number of data pairs, $\hat{\theta}_i$ is the ith parameter identified and $P_{i,i}$ is the diagonal elements of the covariance matrix. The first term in equation (14.26) gives a normalized measure of how well the model explains the data and the second a normalized measure of how well the parameter estimates are defined for the model. As with the AIC, minimizing the YIC provides a compromise between model fit and parameter efficiency.

The three information criteria described in equations (14.24)–(14.26) have been discussed here as they are all included within the time-domain identification module of the PSI software. The following application example will indicate a comparison of the various model structures identified by each of the criteria.

14.7 Identification application

To satisfy the requirements of a control analysis project it was required that a transfer function of a 200 kW gas-fired industrial water boiler be found. A schematic of the boiler is shown in Figure 14.12, indicating that the internal temperature is the measured variable and the plant excitation signal drives a gas flow control valve which has a separate ratio controller linked to the air flow control valve. As the computer input/output card is only capable of taking voltage signals in and out, some signal conversions had to be performed on the plant instrumentation.

The first stage was to calibrate the computer voltage signals with actual plant parameters and to investigate the linearity of the total plant. From initial inspection, there are considerable non-linear components within the plant. The fuel valves are of the 'butterfly' type which exhibit a characteristic 'S' curve and have several mechanical linkages. The temperature measurement, from a linearized thermocouple signal, detects the radiative temperature of the gas combustion, which, from the physics of heat radiation, includes a 'temperature to the power of 4' component. Finally, the insulated box, which contains the gas burner and heated components,

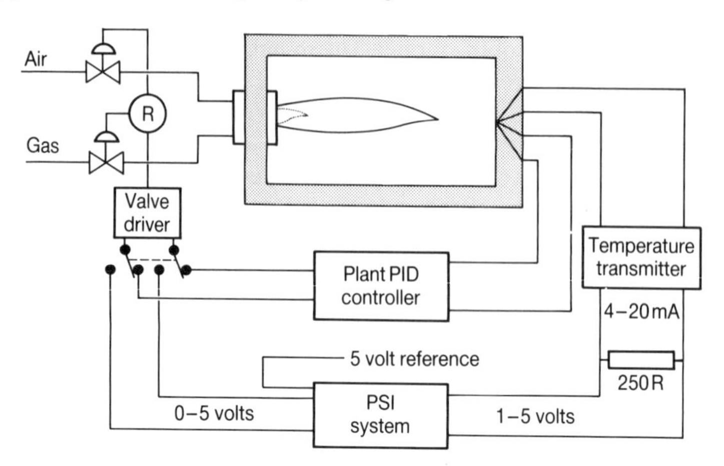

Figure 14.12 Schematic of water boiler and identification set-up.

reduces the amount of heat transferred to the atmosphere. This produces bi-directional dynamics; to reduce the temperature the fuel flow is reduced and the heat within the plant must be taken out with the water or through the insulation at a different rate from the heat input. Initial calibration tests on the boiler indicated that a test signal amplitude could be found that reflected a practical proportion of the working range of the plant, and did not emphasize the non-linear components to any great degree.

Initial squarewave excitation signal tests revealed that the plant had an approximate time constant of $\tau_c = 70$ s. This gave an expected Bode corner frequency $\omega_c = 0.0143$ rad/s from which the suitable plant multi-frequency test signal components can be determined. A constant-power Schroeder low-peak-factor signal was selected from the Excitation Signal Generation menu of the PSI software, with a signal period of $T = 6300$ s and number of discrete points of $N = 256$. This gives a frequency range of $\omega_f = 0.001$ to $\omega_p = 0.064$ rad/s (half of the Nyquist frequency) following FFT analysis.

Using the Plant Test menu of the PSI software, the generated test signal was injected into the gas control valve and the resulting temperature response signal was captured. To eliminate the effects of initial transients the boiler was subjected to one start-up cycle of the test signal where no temperature response was recorded, followed consecutively by two test cycles where the temperature response was recorded. Figure 14.13 shows the excitation signal used and the corresponding averaged plant response.

14.7.1 Frequency-domain analysis

These data were analyzed in the Spectral Analysis menu to yield the frequency response of the plant (with the effect of the zero-order-hold removed) shown in Figure 14.14, where the phase curve shows a 'dip' at the higher frequencies which could

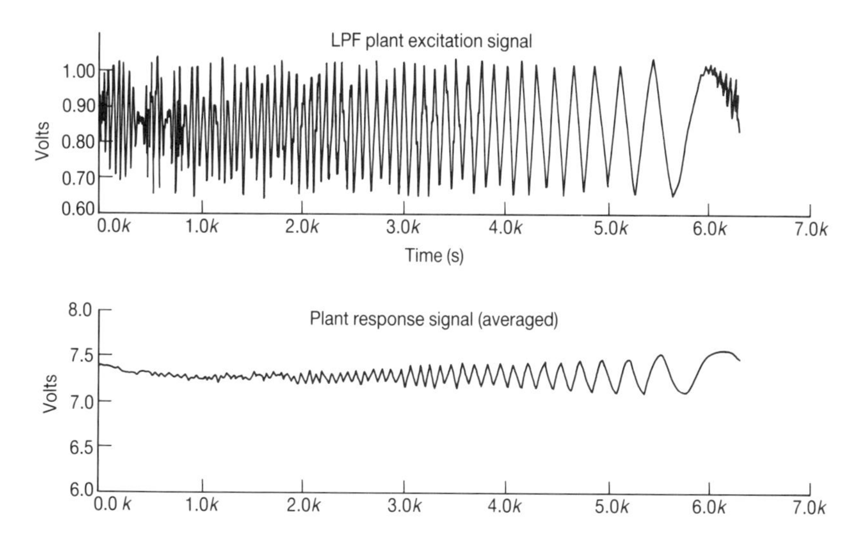

Figure 14.13 Boiler test signal and response.

indicate a transport delay within the plant (a small delay was evident from the early squarewave tests). The sharp drop in the phase component at low frequencies is just the phase foldover where, due to noise, the phase was slightly positive but has been interpreted as slightly greater than $-360°$. As the LPF test signal was designed to have constant power-spectral density over its frequency range, a complete frequency response was obtained of the plant under test over that frequency range. If the signal-to-noise ratio was poorer than that exhibited, then other more suitable plant-excitation signals could have been used to concentrate the signal power at specific frequency components. The resulting frequency response would then require editing to remove the unwanted harmonics at which minimal signal power was apparent. A module is provided within the PSI package to perform this frequency editing before attempting to perform a parametric fit to the frequency data. The coherence of the frequency-response data is shown in Figure 14.15, indicating good coherence up to approximately 0.015 Hz.

To obtain a parameterized transfer function, the frequency data were further analyzed in the Frequency Response Curve Fit menu where frequency data with a coherence value below $\gamma^2 = 0.8$ were edited out and the maximum frequency component limited to $\omega_p = 0.063$ rad/s (0.01 Hz). The structure of the resulting transfer function is determined partly by the characteristics of the frequency-response data and partly by the ultimate use of the transfer function, so a compromise is to be sought. A request for a high-order transfer function could yield ill-conditioned parameters which contribute dynamic effects outside the frequency range of the initial test. A low-order transfer function will be easier to use in further control analysis, but may be a poor representation of the process plant. To aid transfer structure

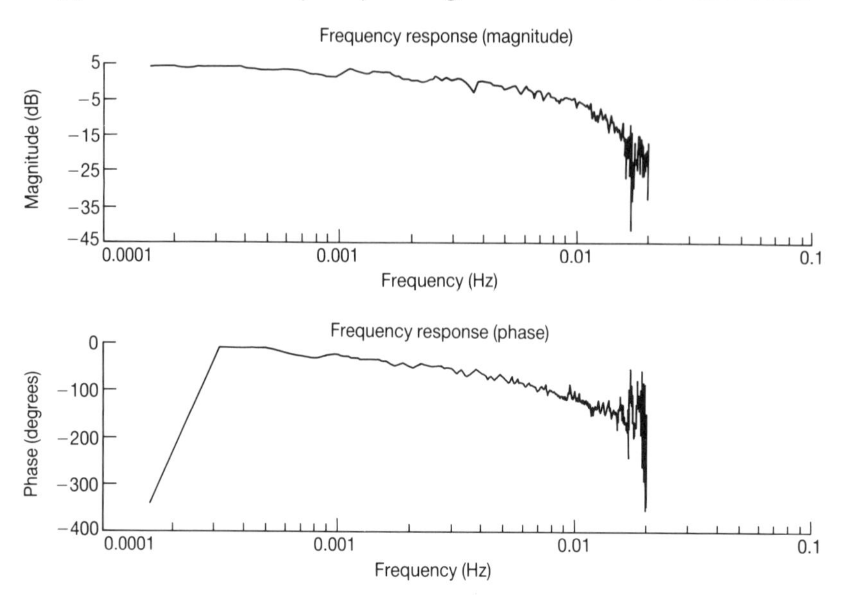

Figure 14.14 Frequency response of boiler.

selection, the two error criteria of magnitude (M_e) and phase (P_e) were used.

From the boiler frequency-response data a first-order transfer function with a 10 s delay was first requested. The least squares fit analysis produced the transfer function $G_1(s)$:

$$G_1(s) = \frac{1.5336}{1 + 44.495\,s} e^{-10\,s} \tag{14.27}$$

which yields error coefficient values of

$$M_e = 0.848 \qquad \text{and} \qquad P_e = 5.576$$

The comparison of the frequency response of $G_1(s)$ and the actual frequency-response data is shown in Figure 14.16.

A second-order transfer function was requested in an attempt to absorb the delay. The transfer function $G_2(s)$ was produced:

$$G_2(s) = \frac{1.4939}{1 + 49.616\,s + 518.91\,s^2} \tag{14.28}$$

which yields error coefficient values of

$$M_e = 0.999 \qquad P_e = 5.613$$

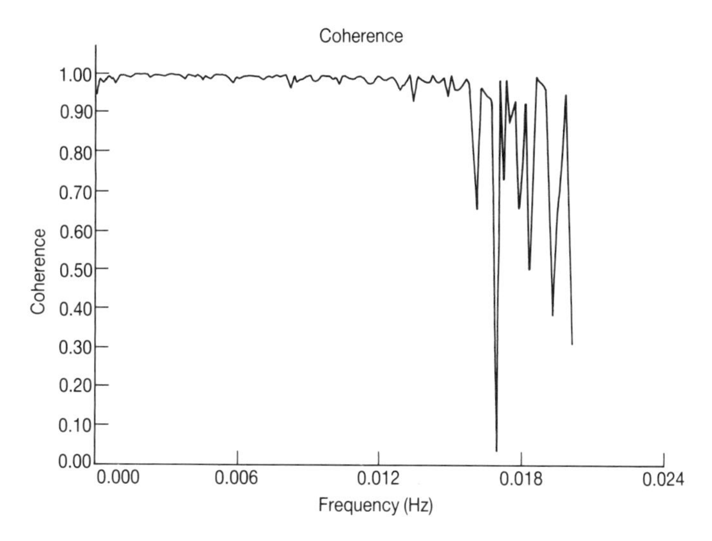

Figure 14.15 Coherence function of boiler test results.

indicating a slightly poorer fit than equation (14.26) and having ill-conditioned denominator coefficients. This poorer frequency-response comparison of $G_2(s)$ can be clearly seen in Figure 14.17.

A further fit was performed in which the second-order structure was preserved but a delay was added to give the desired transfer function $G_3(s)$ of the boiler.

$$G_3(s) = \frac{1.487}{1 + 42.9616\,s + 90.102\,s^2}\,e^{-10s} \tag{14.29}$$

which yields error coefficient values of $M_e = 0.765$ and $P_e = 5.453$. The frequency-response comparison is shown in Figure 14.18. This latter transfer function, $G_3(s)$, was of the desired complexity and accuracy for further control analysis.

14.7.2 Time-domain analysis

The plant excitation and response data shown in Figure 14.13 were used in the time-domain analysis. Rather than manually selecting the desired time-domain model structure, each of the information criteria of LFC, AIC and YIC were used to examine the various structures they selected. Each was instructed to check all the transfer function structure combinations up to a maximum of one delay interval, a numerator order of 2 and a denominator order of 2.

Table 14.1 summarizes the criterion results where both the minimum error (ERR) and AIC indicate that the best structure for the data is no delay and a numerator

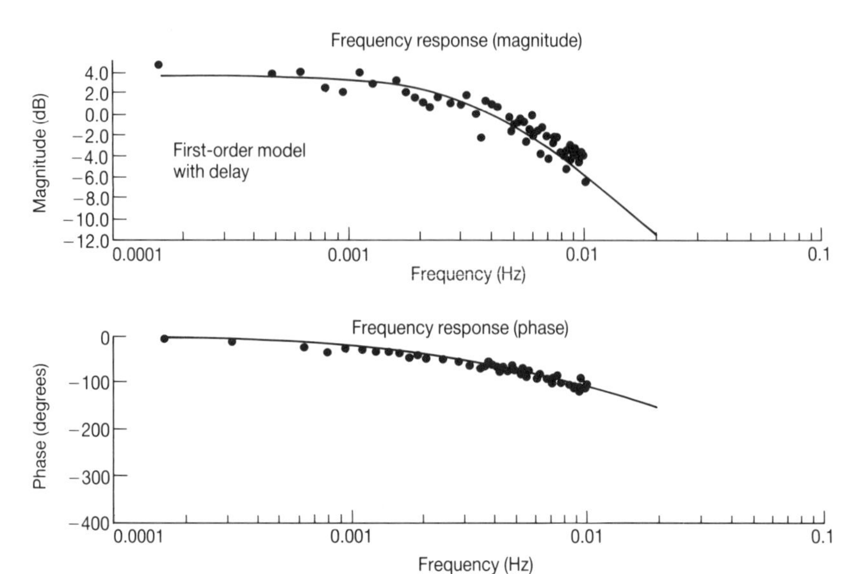

Figure 14.16 Frequency response of boiler and first-order model with delay.

and denominator order of 2. The YIC indicates that the best structure is slightly less at no delay, a numerator order of 2 and a denominator order of 1. In general, it has been found that the minimum error criteria target the higher-order models, the YIC tends to target slightly lower-order models which may be more suitable for control analysis, and the AIC targets model structure between the two.

Table 14.1 Values of AIC, YIC and ERR for various structures of the time-domain model of the industrial boiler

Delay	Numerator	Denominator	AIC	YIC	ERR
0	1	0	−2.10	−3.00	−2.11
0	1	1	−2.83	−9.60	−2.84
0	1	2	−3.02	−9.90	−3.03
0	2	0	−2.63	−8.20	−2.64
0	2	1	−3.33	−12.62	−3.34
0	2	2	−3.34	−6.29	−3.36
1	1	0	−2.33	−5.88	−2.34
1	1	1	−2.46	−5.86	−2.46
1	1	2	−2.46	0.29	−2.47
1	2	0	−2.48	−5.95	−2.48
1	2	1	−2.44	0.21	−2.45
1	2	2	−2.16	0.95	−2.18

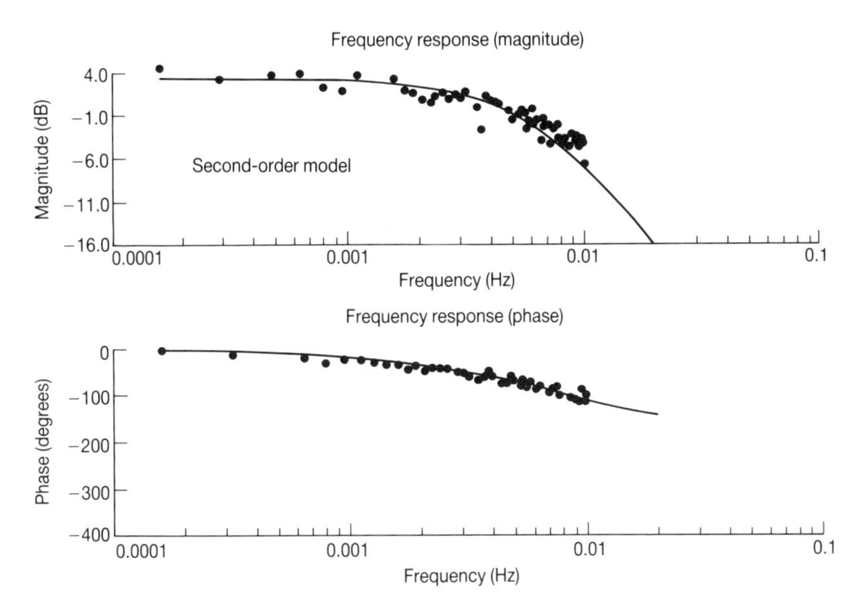

Figure 14.17 Frequency response of boiler and second-order model.

The corresponding difference equations for the two criterion selected model structures are

$$G_1(z) = \frac{0.391z^{-1} + 0.377z^{-2}}{1.0 - 0.473z^{-1} - 0.018z^{-2}} \tag{14.30}$$

$$G_2(z) = \frac{0.394z^{-1} + 0.375z^{-2}}{1.0 - 0.484z^{-1}} \tag{14.31}$$

The sample interval time for equations (14.30) and (14.31) was the same as the data-capture sample time of 24.61 s.

14.7.3 Model validation

After any system-identification process some form of model validation should be performed. Various statistical methods of model validation are provided within the Model Validation module of the PSI software. The method that appeals intuitively is to inject the plant-excitation signal into the resulting transfer function and visually compare the model response to that of the actual plant response. Major deficiencies in model structure and parameter estimates would give rise to obvious errors in the model output sequence. The excitation signal used in the comparison could be the same as was used to identify the model. In practice, it is desirable to obtain further

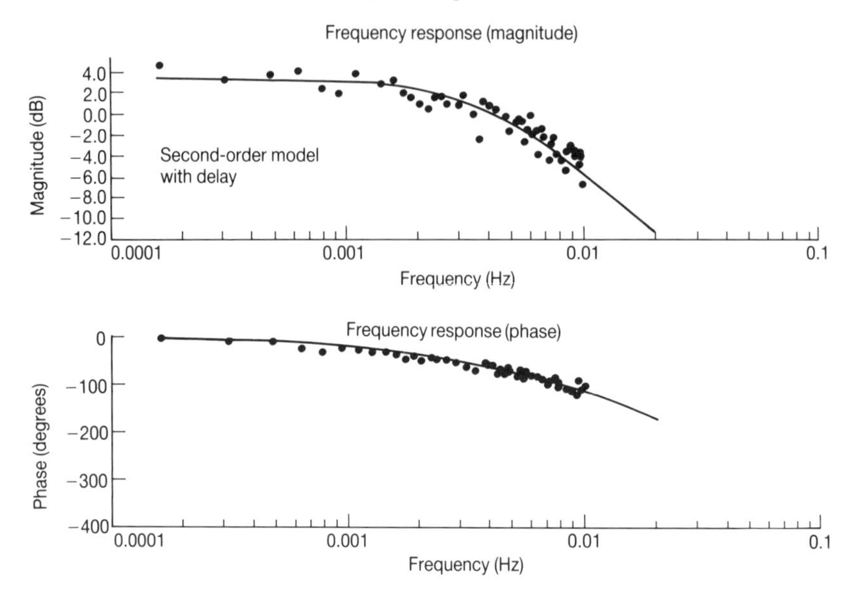

Figure 14.18 Frequency response of boiler and second-order model with delay.

plant responses to an excitation signal that had slightly different frequency components. A PRBS plant test signal was generated and used to inject into the plant and the plant responses captured. These data were used to compare the derived model responses to the PRBS signal with the actual plant responses.

Figure 14.19 shows the model and plant output comparison for the continuous model $G_3(s)$ (equation (14.29)). All the models identified (equations (14.27)–(14.31)) gave very similar comparisons to that shown in Figure 14.19, where the visual comparison of the responses does show some obvious errors which are due to the non-linearities within the plant. These non-linearities appear to manifest themselves in the inconsistent peaking of the plant response. Comparing the third peak in Figure 14.19 at approximately 1000 s with the maximum peak at approximately 4500 s shows that the plant gain has changed. To improve the model further would require an approach outside the conventional linear technique capabilities.

On the whole, the faster dynamics of the model do correspond well with the plant results but the slower more dominant dynamics do not respond too well. However, the resulting models are suitable for further control analysis. It is difficult to select the most suitable model structure from these comparisons. Instead, they just give the user confidence in knowing that the model does respond in a similar way to the actual plant.

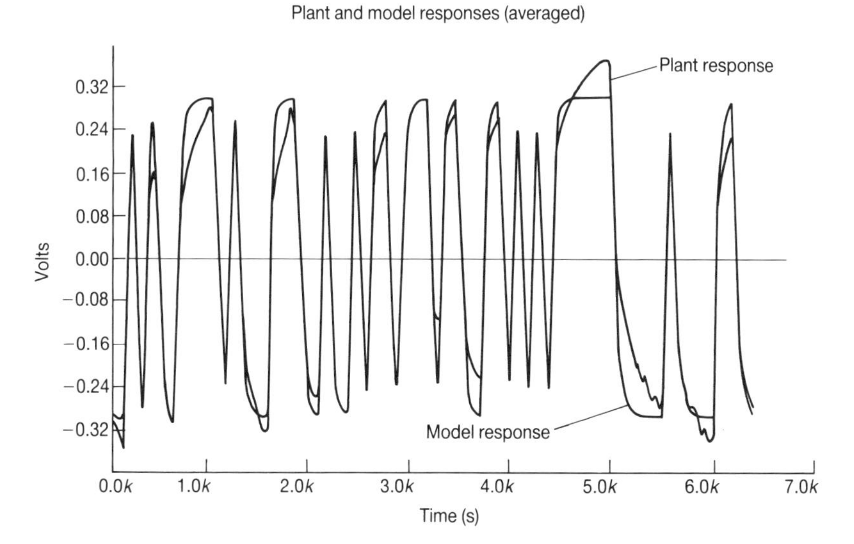

Figure 14.19 Comparison of plant and model responses to a PRBS signal.

14.8 Conclusions

The aim of this chapter has been to highlight some of the more practical implications of system identification and to describe the well-established algorithms used to perform frequency-response analysis. Emphasis has been given to the important topic of plant test signal selection, not only of type but also the various parameters that make up a specific signal. An attempt has also been made to highlight the potential non-linearities that appear in the type of plant described in the application and that linear system identification is generally limited to small regions of the total plant working range. This limitation must be considered when using the resulting transfer function for further control analysis. In many cases the non-linearities within process plant and the effects of non-measurable (or non-controllable) disturbances make the plant unsuitable for linear identification analysis.

In many control-orientated projects the exercise of plant identification takes far more time than the actual control analysis. It was for this reason that the PSI software was developed: to attempt to reduce the identification time by rationalizing the testing procedure, subsequent data analysis and model validation, leaving, it is hoped, more time for the plant control analysis.

Note

1. The PSI package is marketed and supported under licence by Industrial Systems and Control Ltd, 50 George St., Glasgow.

References

Ackroyd, M. H. (1970). 'The design of Huffman sequences', *IEEE Trans. Aerospace Electronic Systs,* **AES-6**, 790–96.

Ager-Hansen, H. (1962). 'Halden – Operating and research experiences', *Nucl. Power*, **7**, 57–65.

Alanen, J. D. and D. E. Knuth (1964). 'Tables of finite fields', *Sankhya,* **26**, 305–28.

Albert, A. A. (1956). *Fundamental Concepts of Higher Algebra*, University of Chicago Press: Chicago.

Al-Dabbagh, A. H. T. and M. Darnell (1991). 'New pseudorandom sequences derived from reflections in enclosures', *Proc. 1st Int. Symp. on Communication Theory and Application* (Hull-Warwick Comms Research Group), Crieff, Scotland, Paper 24.

Allemang, R. J. (1985). 'Frequency response function error considerations', *Proc. of the Tenth International Seminar on Modal Analysis*, KU Leuven, Belgium, 30 September to 4 October.

Åström, K. J. and T. Bohlin (1965). 'Numerical identification of linear dynamic systems from normal operating records', in P. H. Hammond (ed.), *Proc. IFAC Symposium on the Theory of Self-Adaptive Control*, Plenum Press: New York.

Baheti, R. S., R. R. Mohler and H. A. Spang (1980). 'Second-order correlation method for bilinear system identification', *IEEE Trans. Autom. Contr.*, **AC-25**, 1141–6.

Bailey, J. E. (1973). 'Periodic operation of chemical reactors: a review', *Chemical Engineering Communications*, **1**, 111–24.

Balmer, L. (1991). *Signals and Systems: An Introduction*, Prentice Hall: Englewood Cliffs, NJ.

Balza, C., A. Fromageot and M. Maniere (1967). 'Four-level pseudorandom sequences', *Electron. Lett.*, **3**, 313–15.

Barker, H. A. (1967). 'Choice of pseudorandom binary signals for system identification', *Electron. Lett.*, **3**, 524–6.

Barker, H. A. (1969). '*p*-level sequences with primitive autocorrelation functions', *Electron. Lett.*, **5**, 531–2.

Barker, H. A. (1986). 'Sum and product tables for Galois fields', *International Journal of Mathematical Education in Science and Technology*, **17**, 473–85.

Barker, H. A. (1991). 'Design of multilevel pseudorandom signals for specified harmonic content', *IEE International Conference Control 91*, Edinburgh, March 1991; IEE Conference Publication 332, pp. 556–61.

Barker, H. A. and M. H. Al-Hilal (1985). 'Nonlinear system identification using pseudorandom signals with partially orthogonal transforms', *Proceedings Seventh IFAC/IFORS Symposium on Identification and System Parameter Estimation*, York, pp. 415–20.

Barker, H. A. and R. W. Davy (1975). 'System identification using pseudorandom signals and the discrete Fourier transform', *Proceedings IEE,* **122**, 305–11.

Barker, H. A. and R. W. Davy (1978). 'Measurement of second-order Volterra kernels using pseudorandom ternary signals', *International Journal of Control*, **27**, 277–91.

Barker, H. A. and R. W. Davy (1979). 'Second-order Volterra kernel measurement using pseudorandom ternary signals and discrete Fourier transforms', *Proceedings IEE*, **126**, 457–60.

Barker, H. A. and S. N. Obidegwu (1973). 'Effects of nonlinearities on the measurement of weighting functions by cross correlation using pseudorandom signals', *Proceedings IEE*, **120**, 1293–300.

Barker, H. A. and T. Pradisthayon (1970). 'High-order autocorrelation functions of pseudorandom signals based on m sequences', *Proceedings IEE*, **117**, 1857–63.

Barker, H. A., S. N. Obidegwu and T. Pradisthayon (1972). 'Performance of antisymmetric pseudorandom signals in the measurement of second-order Volterra kernals by cross-correlation', *Proceedings IEE*, **119**, 353–62.

Barker, R. H. (1953). 'Group synchronising of binary digital systems', in Jackson, W. (ed.), *Communication Theory*, Butterworths: London, pp. 273–87.

Bendat, J. S. and A. G. Piersol (1966). *Measurement and Analysis of Random Data*, John Wiley: New York.

Bendat, J. S. and A. G. Piersol (1971). *Random Data: Analysis and Measurement Procedures*, Wiley-Interscience, London.

Bendat, J. S. and A. G. Piersol (1980). *Engineering Applications of Correlation and Spectral Analysis*, John Wiley: New York.

Bendat, J. S. and A. G. Piersol (1986). *Random Data – Analysis and Measurement Procedures*, 2nd edn, John Wiley: New York.

Berardino, R. M. (1983). *Identification of a Single Screw Extruder*, MS thesis, Clarkson University, NY.

Bezanson, L. W. and S. L. Harris (1986). 'Identification and control of an extruder using multivariable algorithms', *Proceedings IEE*, **133**, Part D, 145–52.

Billings, S. A. (1980). 'Identification of nonlinear systems – a survey', *Proceedings IEE*, **127**, 272–85.

Billings, S. A. (1986). 'Introduction to nonlinear systems analysis and identification', in Godfrey, K. R. and R. P. Jones (eds.), *Signal Processing for Control*, Springer Lecture Notes in Control and Information Sciences No. 79, pp. 263–94.

Billings, S. A. and S. Y. Fakhouri (1980). 'Identification of linear systems using correlation analysis and pseudorandom inputs', *International Journal of Systems Science*, **11**, 261–79.

Billings, S. A. and K. M. Tsang (1989). 'Spectral analysis for non-linear systems, Part 1: Parametric non-linear spectral analysis', *Mechanical Systems and Signal Processing*, **3**, 319–39.

Black, C. G., D. J. Murray-Smith and G. Padfield (1986). 'Experience with frequency-domain methods in helicopter system identification', *Proceedings of the 12th European Rotorcraft Forum*, Paper No. 76, Garmisch-Partenkirchen, FDR.

Boyd, S. (1986). 'Multitone signals with low crest factor', *IEEE Trans. on Circuits and Systems*, **CAS-33**, 1018–22.

Bramwell A. R. S. (1976). *Helicopter Dynamics*, Edward Arnold: London.

Briggs, P. A. N. and K. R. Godfrey (1966). 'Pseudorandom signals for the dynamic analysis of multivariable systems', *Proceedings IEE*, **133**, 1259–67.

Briggs, P. A. N. and K. R. Godfrey (1968). 'Autocorrelation function of a 4-level m-sequence', *Electron. Lett.*, **4**, 232–3.

Briggs, P. A. N. and K. R. Godfrey (1976). 'Design of uncorrelated signals', *Electron. Lett.*, **12**, 555–6.

Briggs, P. A. N., P. H. Hammond, M. T. G. Hughes and G. O. Plumb (1965). 'Correlation analysis of process dynamics using pseudo-random binary test perturbations', *Proc. Inst. Mech. Engrs*, **179**, Part 3H, 37–51.

Brigham, E. O. (1974). *The Fast Fourier Transform*, Prentice Hall: Englewood Cliffs, NJ.

Brigham, E. O. (1988). *The Fast Fourier Transform and its Applications*, Prentice Hall: Englewood Cliffs, NJ.

Brillinger, D. R. (1975). *Time Series: Data Analysis and Theory*, Holt, Rinehart and Winston: London.

Brillinger, D. R. (1981). *Time Series: Data Analysis and Theory* (expanded edition), Holden-Day: San Francisco, CA.

Brown, D., G. Carbon and K. Ramsey (1977). 'Survey of excitation techniques applicable to the testing of automotive structures', *International Automotive Eng. Congress and Exposition*, Cobo Hall, Detroit, 28 February to 4 March.

Buckner, M. R. (1970). *Optimum binary signals for frequency response testing*, ORNL-TM-3198, Oak Ridge National Laboratory.

Buckner, M. R. and T. W. Kerlin (1972). 'Optimum binary signals for reactor frequency response measurements', *Nucl. Sci. Eng.*, **49**, 255–62.

Caldwell, W. I., G. A. Coon and I. M. Zoss (1959). *Frequency Response for Process Control*, McGraw-Hill: New York.

Carter, C. G., C. H. Knapp and A. H. Nuttal (1973). 'Estimation of the magnitude-squared coherence function via overlapped Fast Fourier Transform Processing', *IEEE Trans. on Audio and Electroacoustics*, **AU-21**, 337–44.

Chang, J. A. (1966). 'Generation of 5-level maximal-length sequences', *Electron. Lett.*, **2**, 258.

Chang, J. A., A. G. Owen, M. Zaman and A. W. J. Griffin (1968). 'Dynamic modelling of a four-stand cold rolling mill using multi-level pseudo-random sequences', *Meas. and Control*, **1**, T80–84.

Chen, C. H., T. W. Kerlin and D. N. Fry (1972). 'Experiences with binary periodic signals for dynamic testing at HFIR', *IEEE Trans. on Nucl. Sci.*, **19**, 828–36.

Chesmore, E. D. (1987). *A Speech Message Transmission System for Low Capacity Channels*, DPhil thesis, University of York, pp. 6.13–6.14.

Chohan, R. K., F. Abdullah and L. Finkelstein (1985). 'Mathematical modelling of industrial thermometers', *Trans. Inst. MC*, **7**, 151–8.

Church, R. (1935). 'Tables of irreducible polynomials for the first four prime moduli', *Ann. Math.*, **36**, 198–209.

Clarke, D. W. (1967). 'Generalised least squares estimation of the parameters of a dynamic model', *1st IFAC Symposium on Identification in Automatic Control Systems*, Prague, Paper 3.17.

Clarke, D. W. and K. R. Godfrey (1966). 'Simultaneous estimation of the first and second derivatives of a cost function', *Electron. Lett.*, **2**, 338–9.

Clarke, D. W. and K. R. Godfrey (1967). 'Simulation study of a two-derivative hill climber', *Electron. Lett.*, **3**, 261–3.

Classen, T. A. C. M., W. F. G. Mecklengrauker, J. B. H. Peek and N. Van Hurck (1980). 'Signal processing methods for improving the dynamic range of A/D and D/A converters', *IEEE Trans. on Acoustics, Speech and Signal Processing*, **ASSP-28**, 529–37.

Czekai, D. A. (1984). *An Optimal Dual Controller with Active Binary Multifrequency Signal Perturbation*, MS thesis, Clarkson University, NY.

Darnell, M. (1968). *Multilevel Pseudorandom Signals for System Evaluation*, PhD thesis, University of Cambridge, pp. 52–6.

Darnell, M. (1989). 'The theory and generation of sets of uncorrelated digital sequences', in Beker, H. J. and F. C. Piper (eds.), *Cryptography and Coding*, Oxford University Press: Oxford, pp. 23–65.

Darnell, M. (1991a). 'Multi-level signals with good autocorrelation properties', *IEE Int. Conf. Control 91*, Edinburgh, March 1991, IEE Conf. Publ. 332, pp. 562–6.

Darnell, M. (1991b). 'Non-periodic binary signals with good autocorrelation properties', *IEE Int. Conf. Control 91*, Edinburgh, March 1991, IEE Conf. Publ. 332, pp. 178–83.

Darnell, M. and A. H. Kemp (1991). 'Multi-level complementary sequence sets: synthesis and applications', in Mitchell, C. (ed.), *Cryptography and Coding II*, Oxford University Press: Oxford, pp. 36–66.

Delbaen, F. (1990) 'Optimizing the determinant of a positive definite matrix', *Bulletin Société Mathématique de Belgique – Tijdschrift Belgisch Wiskundig Genootschap*, **42**, ser. B, 333–46.

Ditmar, W. P. A. (1977). *Identification of Power System Dynamics using Deterministic Disturbances*, PhD thesis, CNAA, London.

Dunlop, J. (1984). 'Multiuser microprocessor development system for undergraduate training', *Proceedings IEE, Part A*, **131**, 170–3.

Eykhoff, P. (1974). *System Identification*, John Wiley: New York.

Everett, D. (1966). 'Periodic digital sequences with pseudonoise properties', *GEC Jnl Sci. Technol.*, **33**, 115–26.

Fedorov, V. V. (1972). *Theory of Optimal Experiments*, Academic Press: New York.

Finkelstein, L. (1977). 'Introductory article: Instrument science and technology', *J. Phys. E.: Sc. Instr.*, **13**, 1041–5.

Finkelstein, L. (1985). 'State and advances of general principles of measurement and instrumentation science', *Measurement*, **3**, 2–6.

Flower, J. O., S. C. Forge, and G. F. Knott (1976). 'On-line computer determination of the unsteady lift characteristics of hydrofoils by a multifrequency technique', *International Shipbuilding Progress*, **23**, 313–16.

Flower, J. O., S. C. Forge, N. G. Ratcliffe and C. B. Roust (1978a). 'Dynamic measurements of a nuclear reactor using low peak factor excitation signals', *Nucl. Sci. Eng.*, **68**, 110–15.

Flower, J. O., G. F. Knott and S. C. Forge (1978b). 'Application of Schroeder phased harmonic signals to practical identification', *Measurement and Control*, **11**, 69–73.

Foster, D. M., R. L. Aamodt, R. I. Henkin and M. Berman (1979). 'Zinc metabolism in humans: a kinetic model', *Amer. Jnl Physiol.*, **237**, R340–49.

Franck, G. and H. Rake (1985). 'Identification of a large water-heated crossflow heat exchanger with binary multifrequency signals', *7th IFAC/IFORS Symposium on Identification and System Parameter Estimation*, York, July 1985; preprints, pp. 1859–64.

Frank, R. L. (1963). 'Polyphase codes with good nonperiodic correlation properties', *IEEE Trans. Inf. Theory*, **IT-9**, 43–5.

Gade, S. and H. Herlufsen (1987a). 'Use of weighting functions in DFT/FFT analysis (Part I), *Technical Review Brüel & Kjaer*, no. 3, Naerum, Denmark.

Gade, S. and H. Herlufsen (1987b). 'Use of weighting functions in DFT/FFT analysis (Part II), *Technical Review Brüel & Kjaer*, no. 4, Naerum, Denmark.

Gill, P. E. and W. Murray (1972). 'Quasi-Newton methods for unconstrained optimisation', *Journal of the Institute of Mathematics and its Applications*, **9**, 91–108.

Gillenwater, W. T. (1988). *Identification and Control of a Single Screw Extruder*, MS thesis, Clarkson University, NY.

Godfrey, K. R. (1969a). 'The theory of the correlation method of dynamic analysis and its application to industrial processes and nuclear power plant', *Meas. and Control*, **2**, T65–72.

Godfrey, K. R. (1969b). 'Dynamic analysis of an oil refinery unit under normal operating conditions', *Proceedings IEE*, **116**, 879–88.

Godfrey, K. R. (1975). 'Filtered binary sequences with no skewing', *Electron. Lett.*, **11**, 456–7.

Godfrey, K. R. (1980). 'Correlation methods', *Automatica*, **16**, 527–34.

Godfrey, K. R. (1983). *Compartmental Models and their Application*, Academic Press: New York.

Godfrey, K. R. and P. A. N. Briggs (1972). 'Identification of processes with direction-dependent dynamic responses', *Proceedings IEE*, **119**, 1733–9.

Godfrey, K. R. and D. J. Moore (1974). 'Identification of processes having direction-dependent dynamic responses, with gas turbine engine applications', *Automatica*, **10**, 469–81.

Godfrey, K. R. and B. Shackcloth (1970). 'Dynamic modelling of a steam reformer and the implementation of feedforward/feedback control', *Meas. and Control*, **3**, T65–72.

Golay, M. J. E. (1961). 'Complementary series', *IRE Trans. Inf. Theory*, **IT-7**, 82–7.

Goldberger, A. S. (1964). *Econometric Theory*, John Wiley: New York.

Golomb, S. W. (1967). *Shift Register Sequences*, Holden-Day: San Francisco, CA.

Golomb, S. W. and R. A. Scholtz (1965). 'Generalized Barker sequences', *IEEE Trans. Inf. Theory*, **IT-11**, 533–7.

Goodwin, G. C. (1971). 'Optimal input signals for nonlinear system identification', *Proceedings IEE*, **118**, 922–6.

Goodwin, G. C. (1987). 'Identification: experiment design', in Singh, M. G. (ed.), *Systems and Control Encyclopedia*, Pergamon Press: Oxford, pp. 2257–64.

Goodwin, G. C. and R. L. Payne (1977). *Dynamic System Identification: Experiment Design and Data Analysis*, Academic Press: New York.

Gradshteyn, I. S. and I. M. Ryzhik (1980). *Tables of Integrals, Series and Products*, Academic Press: New York.

Grenander, U. and G. Szegö (1958). *Toeplitz Forms and their Applications*, University of California Press: Berkeley, CA.

Guillaume, P. (1991). *Analytical Expression for the Bias of the Hlog Average*, Internal note 9-91 dep. ELEC, Vrije Universiteit Brussel.

Guillaume, P., J. Schoukens, R. Pintelon and I. Kollár (1991). 'Crest factor minimization using non-linear Tchebycheff Approximation methods', *IEEE Trans. Instrum. Meas.*, **IM-40**, 982–9.

Hackforth, H. L. (1960). *Infrared Radiation*, McGraw-Hill: New York.

Haghighat, F. (1988). 'Applications of system identification techniques to the determination of thermal response factors', *Proc. 8th IFAC Symp. Ident. Sys. Par. Estimation*, Beijing, PRC, pp. 511–15.

Halvorsen, W. G. and D. L. Brown (1977). 'Impulse technique for structural frequency response testing', *Sound and Vibration*, November, 8–21.

Hamel, P. G. (1979). 'Aircraft parameter identification methods and their applications – Survey and future aspects', *Proceedings of the AGARD FMP Symposium on Parameter Identification in Aerospace Systems*, Series LS-104, CP-1.

Harris, F. J. (1978). 'On the use of windows for harmonic analysis with the Discrete Fourier Transform', *IEEE Proceedings*, **66**, 51–83.

Harris, R. A., Yau Ming Chien, D. A. Clark, W. M. Ritter, R. A. Bennett, R. B. Rothrock, Jr and R. A. Sevenich (1989). 'Multifrequency binary sequence testing at the Fast Flux Test Facility', *Nucl. Sci. Eng.*, **103**, 294–301.

Harris, S. L. (1987). 'Generation of binary multifrequency signals for use in adaptive control algorithms', *AIChE Annual Conference*, New York.

Harris, S. L. and D. A. Mellichamp (1980). 'On-line identification of process dynamics: use of multifrequency binary sequences', *Industrial and Engineering Chemistry Process Design and*

Development, **19**, 166–74.

Harris, S. L. and D. A. Mellichamp (1981). 'Frequency domain adaptive controller', *Industrial and Engineering Chemistry Process Design and Development*, **20**, 188–96.

Hastings-James, R. and M. W. Sage (1969). 'Recursive generalised least-squares procedure for on-line identification of process parameters', *Proceedings IEE*, **116**, 2057–62.

Hazlerigg, A. D. G. and A. R. M. Noton (1965). 'Application of crosscorrelating equipment to linear system identification', *Proceedings IEE*, **112**, 2385–400.

Henderson, I. A. and J. McGhee (1988). 'PSK maxent MBS test signals for narrowband electro-thermal identification', *IMACS XII, 12th World Congress on Scientific Computation*, Paris, Vol. 3, pp. 351–3.

Henderson, I. A. and J. McGhee (1989a). 'Phase shift keyed technique for microcomputer narrowband identification of a control system', *ICCON 89, IEEE Int. Conf. on Control and Applications*, Jerusalem, WA-5-1, pp. 1–6.

Henderson, I. A. and J. McGhee (1989b). 'Maxent binary identification of a warm air system', *IMEKO TC-12 Sym. on Microprocessors in Temperature and Thermal Measurement*, Lodz, Poland, pp. 89–96.

Henderson, I. A and J. McGhee (1990a). 'A digital phase-shift-keyed technique for narrowband system identification', *Trans Inst., MC*, **12**, 147–55.

Henderson, I. A. and J. McGhee (1990b). 'Compact symmetrical binary codes for system identification', *Mathl Comput. Modelling*, **14**, 213–18.

Henderson, I. A. and J. McGhee (1990c). 'Minimisation of MBS switch-on transient by linear prediction', in *Multifrequency Testing for System Identification*, London, 8 June 1990, IEE Digest No. 1990/097.

Henderson, I. A. and J. McGhee (1990d). 'System identification instrumentation', in Rao, R. B. K., J. Au and B. Griffiths (eds.), *COMADEM 90, 2nd Int. Conf. on Condition Monitoring and Diagnostic Engineering Management*, Chapman and Hall: London, pp. 349–54.

Henderson, I. A. and J. McGhee (1990e). 'A taxonomy of temperature measuring instruments', *TEMPMEKO 90, 4th Symp. on Temp. and Therm. Mea. in Sc. and Ind.*, Helsinki, Finland, pp. 400–5.

Henderson, I. A. and J. McGhee (1991a). 'Multifrequency binary signal identification of control systems by simulation', *IEE Int. Conf. Control '91*, Edinburgh, March 1991, IEE Conf. Publ. 332, pp. 167–72.

Henderson, I. A. and J. McGhee (1991b). 'A digital frequency shift keyed technique for system identification', *9th IFAC/IFORS Sym. on Ident. and Sys. Par. Est.*, Budapest, Hungary, pp. 124–9.

Henderson, I. A. and J. McGhee (1991c). 'Compact multifrequency binary signals', UK Patent Application GB 2 243 978 A.

Henderson, I. A. and J. McGhee (1993). 'Classical taxonomy: a holistic perspective of temperature measuring instruments to be published in *Proceedings IEEE, Part A*.

Henderson, I. A., A. A. Ibrahim, J. McGhee and D. Sankowski (1987). 'Assembler generated binary sequences for process identification', in Adali, E. and F. Tunali (eds.), *Microcomputer Application in Process Control*, IFAC Proc. Series, No. 7, Pergamon Press: Oxford, pp. 77–82.

Henderson, I. A., J. McGhee, G. Smith and L. Jackowska-Strumillo (1991a). 'Eye patterns and active condition monitoring', in Rao, R. B. K. and A. D. Hope (eds.), *COMADEM 91, 3rd aInt. Conf. on Condition Monitoring and Diagnostic Engineering Management*, Adam Hilger: Bristol, pp. 305–9.

Henderson, I. A., J. McGhee, M. Al-Muhaisni and T. Arbuckle (1991b). 'Multifrequency measurement of R and C using an MBS ratio bridge', *12th IMEKO World Congress*, Beijing, PRC.

Hendricks, E. (1980). *Extended Kalman Filtering Applied to Helicopter Control*, MSc thesis Herlufsen, H. (1984). 'Dual channel F.F.T analysis (part I)', *Technical Review Brüel & Kjaer*, no. 1, Naerum, Denmark.

IMSOR, Technical University of Denmark, Lyngby, Denmark.

Hoffmann de Visme, G. (1971). *Binary Sequences*, English Universities Press: London.

Hofmann, D. (1976). *Dynamische Temperaturmessung*, VEB Verlag Technik: Berlin.

Holmes, J. K. (1982). *Coherent Spread Spectrum Systems*, John Wiley: New York.

Houston, S. S. (1989). 'Identification of a coupled body/coning/inflow model of Puma vertical response in the Hover', *Int. Journal of Rotorcraft & Powered Lift Aircraft* (Vertica), **13**, 229–49.

Huffman, D. A. (1962). 'The generation of impulse-equivalent pulse trains', *IRE Trans. Inf. Theory*, **IT-8**, S.10–16.

Isermann, R. and U. Baur (1974). 'Two-step process identification with correlation analysis and least squares parameter estimation', *J. Dynam. Syst. Meas. Cont.*, **96**, 426–32.

Isermann, R., U. Baur, W. Bamberger, P. Kneppo and H. Siebert (1974) 'Comparison of six on-line identification and parameter estimation methods', *Automatica*, **10**, 81–103.

Jackowska-Strumillo, L. M., J. McGhee and I. A. Henderson (1992). 'Instrumentation and experimentation for identification, monitoring and diagnosis of temperature sensors using multifrequency binary sequences', Part 2, 1043-8, in Scholey, J. F. (ed.), *Temperature: Its Measurement and Control in Science and Industry*, Vol. 6, American Institute of Physics.

Jakob, M. (1957). *Heat Transfer*, Vol. 2, John Wiley: New York.

Jakob, M. (1958). *Heat Transfer*, Vol. 1, 6th edn, John Wiley: New York.

Jenkins, G. M. and D. G. Watts (1969). *Spectral Analysis and its Applications*, Holden-Day: San Francisco, CA.

Jensen, J. R. (1959). *Notes on measurement of dynamic characteristics of linear systems, Part III*, Report Servoteknisk forskningslaboratorium, Danmarks Tekniske Højskole, Copenhagen, Denmark.

Johnson, P. C. and D. A. Mellichamp (1972). 'On-line digital computer determination of chemical process dynamics', *Industrial and Engineering Chemistry Process Design and Development*, **11**, 203–12.

Jones, D. L. (1990 . *Design and Development of a Composite Frequency Response Analyser*, PhD thesis, Polytechnic of Wales.

Jones, R. P. (1986). 'Signal analysis II', in Godfrey, K. R. and R. P. Jones (eds.), *Signal Processing for Control*, Springer: Berlin, pp. 143–54.

Kaletka, J. (1979). 'Rotorcraft identification experience', *Proceedings of the AGARD FMP Symposium on Parameter Identification in Aerospace Systems*, Series LS-1-4, CP-7, pp. 7-1–7-32.

Kavanagh, R. J. (1969). 'Fourier analysis of pseudorandom binary sequences', *Electron Lett.*, **5**, 173–4.

Kemp, A. H. and M. Darnell (1989). 'Synthesis of uncorrelated and nonsquare sets of multilevel complementary sequences', *Electron. Lett.*, **25**, 791–2.

Kendall, M. and A. Stuart (1979). *The Advanced Theory of Statistics*, Charles Griffin: London.

Kerlin, T. W. (1974). *Frequency Response Testing in Nuclear Reactors*, Academic Press: New York.

Kerlin, T. W., L. F. Miller and N. M. Hashemian (1978). '*In situ* response time testing of platinum resistance thermometers', *ISA Trans.*, **17**(4), 71–88.

Kerlin, T. W., N. M. Hashemian and K. M. Peterson (1981). 'Time response of temperature sensors', *ISA Trans.*, **20**(1), 65–7.

Kerlin, T. W., N. M. Hashemian and K. M. Peterson (1982). 'Response characteristics of temperature sensors installed in processes', *ACTA IMEKO III*, North-Holland: Amsterdam, pp. 95–103.

Khoshlahjeh-Motamed, M. and R. R. Pettitt (1991). 'Choice of multifrequency test signals for alternator identification', *26th Universities Power Engineering Conference*, Brighton Polytechnic, 18–20 September 1991.

Kirkpatrick, S., C. D. Gelatt and M. P. Vecchi (1983). 'Optimization by simulated annealing', *Science*, **220**, 671–80.

Komo, J. J. and Shyh-Chang Liu (1990). 'Maximal length sequences for frequency hopping', *IEEE Jnl on Selected Areas in Communications*, **8**, 819–22.

Kruger, M. (1983). 'Binary sequences II: Homogeneity and symmetry', *Information Sciences*, **31**, 15–31.

Lamb, J. D. and D. Rees (1973). 'Digital processing of system responses to pseudorandom binary sequences to obtain frequency response characteristics using the Fast Fourier Transform', in *The Use of Digital Computers in Measurement*, Conference Publication 103. IEE, London, p. 141–6.

Lee, Y. W. (1960). *Statistical Theory of Communication*, John Wiley: New York, pp. 341–8.

Lenstra, A. J. (1987). *Simulated Annealing Applied to the Synthesis of Low Peak Factor Signals*, Master of Technical Science Thesis, Department of Applied Physics, Delft University of Technology (in Dutch).

Levin, M. J. (1959). *Estimation of the Characteristics of Linear Systems in the Presence of Noise*, Doctor of Engineering Science Thesis, Department of Electrical Engineering, Columbia University, New York.

Levy, E. C. (1959). Complex curve fitting', *IEEE Trans. Autom. Contr.*, **AC-4**, 37–44.

Lidner, J. (1975). 'Binary sequences up to length 40 with best possible autocorrelation function', *Electron. Lett.*, **11**, 507.

Lieneweg, F. (1975). *Handbuch, Technische Temperaturmessung*, F. Vieweg: Braunschweig.

Ljung, L. (1985). 'On the estimation of transfer function', *Automatica*, **21**, 677–96.

Ljung, L. (1987). *System Identification: Theory for the User*, Prentice Hall: Englewood Cliffs, NJ.

Lobodzinski, W., M. Orzylowski, D. Sankowski and W. Szmanda (1983). 'Piec dyfuzyjny jako wielowymiarowy obiekt regulacji temperatury', *Prace PIE nr 88*, pp. 5–27, Warsaw, Poland.

MacWilliams, J. (1967). 'An example of two cyclically orthogonal sequences with maximum period', *IEEE Trans. Inf. Theory*, **IT-13**, 338–9.

Mayham, R. J. (1984). *Discrete-Time and Continuous-Time Linear Systems*, Addison-Wesley: Reading, MA.

McGhee, J. (1990). 'Holistic approaches for knowledge based process control', in McGhee, J., M. J. Grimble and P. Mowforth (eds.), *Knowledge Based Systems for Industrial Control*, Peter Peregrinus: Stevenage.

McGhee, J. and I. A. Henderson (1989). 'Holistic perception in measurement and control: applying keys adapted from classical taxonomy', *TRICMED 88, Trends in Control and Measurement Education*, IFAC Proc. Series, No. 5, pp. 31–6.

McGhee, J. and I. A. Henderson (1991a). 'The nature and scope of taxonomy in measurement education', *12th IMEKO World Congress*, Beijing, PRC.

McGhee, J. and I. A. Henderson (1991b). 'Classification science as a rational basis for sensor science', *Int. Workshop on Sensor Science*, Vrije Universiteit Brussel, Brussels.

McGhee, J., I. A. Henderson and D. Sankowski (1986a). 'Identification: systems, structures, signals and similarities', *Int. Conf. on Systems Science No. 9*, Wroclaw, Poland.

McGhee, J., I. A. Henderson and D. Sankowski (1986b). 'Functions and structures in measuring systems: A systems engineering context for instrumentation', *Measurement*, **4**, 111–19.

McGhee, J., G. Fisher and I. A. Henderson (1987). 'A fast DFT algorithm for on-line MBS process identification', *6th International Conference on Control Systems and Computer Science*, Bucharest, Romania, pp. 111–29.

McGhee, J., D. Sheppard and I. A. Henderson (1989). 'Immersion testing of temperature sensors using a microprocessor based MBS generator/analyser', *Proc. IMEKO TC-12 Symposium, Microprocessors in Temperature and Thermal Measurement*, Lodz, Poland, pp. 65–72.

McGhee, J., I. A. Henderson and S. Mackie (1990). 'Simulation of MBS identification of temperature sensors', *TEMPMEKO 90, 4th Symp on Temp. and Therm. in Sci. and Ind.*, Helsinki, Finland, pp. 201–6.

McGhee, J., I. A. Henderson and L. Jackowska-Strumillo (1991a). 'Identifying temperature sensors using Strathclyde compact multi-frequency binary sequences', *Proc. CSCS 8, 8th Int. Conf. Cont. Sys. Comp. Sc.*, Vol. 1, Politechnical Institute of Bucharest, pp. 144–52.

McGhee, J., I. A. Henderson and D. Sankowski (1991b), 'Electronics and computers in thermal systems: Reviewing current trends', *Trans. Inst. MC*, **13**, 75–90.

McGhee, J., I. A. Henderson, L. Michalski, K. Eckersdorf and D. Sankowski (1992a). 'Dynamic properties of contact temperature sensors, I: thermokinetic modelling and the idealized temperature sensor', to be published in Scholey, J. F. (ed.), *Temperature: Its Measurement and Control in Science and Industry*, Vol. 6, American Institute of Physics.

McGhee, J., I. A. Henderson, L. Michalski, K. Eckersdorf and D. Sankowski (1992b). 'Dynamic properties of contact temperature sensors, II: modelling, characterisation and testing of real temperature sensors', Part 2, 1163–8, in Scholey, J. F. (ed.), *Temperature: Its Measurement and Control in Science and Industry*, Vol. 6, American Institute of Physics.

McGreal, S. P. (1989). *Unsteady-state Reactor Simulation Using Binary Multifrequency Test Sequences*, MS thesis, Clarkson University, NY.

Mehra, R. K. (1974). 'Optimal input signals for parameter estimation in dynamic systems – survey and new results', *IEEE Trans, Autom. Contr.*, **AC-19**, 753–68.

Mehra, R. K. (1981). 'Choice of input signals', in Eykhoff, P. (ed.), *Trends and Progress in System Identification*, Pergamon Press: Oxford, pp. 305–66.

Mercer, R. W. and S. G. Mailey (1986). 'Computer aided identification of an air heating system', *Measurement and Control*, **19**, 293–300.

Michalski, L. (1966). 'Temperatur eines kammerofens', *Electrotechnek (Netherlands)*, **44**(20), 466–71.

Michalski, L. and K. Eckersdorf (1987). 'Temperature sensors in closed loop temperature control', *TEMPMEKO 87, Thermal and Temperature Measurement in Science and Industry*, Sheffield, pp. 127–33.

Michalski, L. and K. Eckersdorf (1990). '*In situ* determination of dynamics of temperature sensors', *TEMPMEKO 90, Proc 4th IMEKO Symposium Thermal and Temperature Measurement in Science and Industry*, Helsinki, pp. 193–200.

Michalski, L., K. Kuzminski and J. Sadowski (1981). *Temperature Control in Electroheat*, WNT: Warsaw (in Polish).

Michalski, L., D. Sankowski and J. Sadowski (1985). 'Hybrid simulation of temperature control in electric resistance furnaces', *11th IMACS World Congress*, Oslo, Vol. 5, pp. 61–6.

Michalski, L., K. Eckersdorf and J. McGhee (1991). *Temperature Measurement*, John Wiley: Chichester.

Moore, D. J. (1970). *Error Correction Applied to Dynamic Analysis*, Rolls Royce (1971) Ltd, Bristol Engine Division, Report EER/5033/70.

NAG (1990). *NAG Fortran Library Manual*, Mark 14, Numerical Algorithms Group, Oxford.

Newland, D. E. (1984). *An Introduction to Random Vibrations and Spectral Analysis*, 2nd edn,

Longman Scientific and Technical: Harlow.
Norden, R. H. (1972). 'A survey of maximum likelihood estimation, Part 1', *International Statistical Review*, **40**, 329–54.
Norden, R. H. (1973). 'A survey of maximum likelihood estimation, Part 2', *International Statistical Review*, **41**, 39–58.
Norton, J. P. (1986). *An Introduction to Identification*, Academic Press: New York.
Oppenheim, A. and R. Schafer (1975). *Digital Signal Processing*, Prentice Hall, Englewood Cliffs, NJ.
Overton, M. L. (1982) 'Algorithms for nonlinear l_1 and l_∞ fitting', in Powell, M. J. D. (ed.), *Nonlinear Optimization 1981*, Academic Press: London.
Paehlike, K. D. (1980). *Regelstreckenidentifikation mit binaren Mehrfrequenzsignalen*, PhD thesis, Rheinisch-Westfalischen Technischen Hochschule, Aachen, Germany.
Paehlike, K. D. and H. Rake (1979). 'Binary multifrequency signals – synthesis and application', *5th IFAC Symposium on Identification and System Parameter Estimation*, Darmstadt, September 1979, paper M8.2; preprints, pp. 589–96.
Papentin, F. (1983). 'Binary sequences. I. Complexity', *Information Sciences*, **31**, 1–14.
Parker, G. A. and E. L. Moore (1980). 'A modified Volterra series representation for a class of single-valued, continuous nonlinear systems', *J. Dynam. Syst. Meas. Contr.*, **102**, 163–7.
Parker, G. A. and E. L. Moore (1982). 'Practical nonlinear system identification using a modified Volterra series approach', *Automatica*, **18**, 85–91.
Patton, R. J., P. Taylor and P. Young (1990). 'Frequency domain testing of helicopter dynamics using automated test signals', *Proceedings of the 19th European Rotorcraft Forum*, Glasgow, 18–21 September, pp. III.3.1.1–14.
Peterson, W. W. (1961). *Error-Correcting Codes*, MIT Press: Cambridge, MA.
Petit, C., J. C. Gajen and P. Parantheon (1982). 'Frequency response of fine wire thermocouples', *J. Phys E: Sci. Instr.*, **15**, 760–64.
Pettitt, R. R. (1983). *On-line Modelling of Alternator Dynamics*, PhD thesis, CNAA, London.
Pintelon, R. and J. Schoukens (1990). 'Robust identification of transfer functions in the *s*- and *z*-domains', *IEEE Trans. Instrum. Meas.*, **IM-39**, No. 4, 565–73.
Plaskowski, A. and D. Sankowski (1984). 'The use of multifrequency binary sequences MBS to on-line system identification in electroheat', *IMACS European Simulation Meeting*, Eger, Hungary, pp. 285–96.
Poljak, B. T. and Ya. Z. Tsypkin (1980). 'Robust identification', *Automatica*, **16**, 53–63.
Porch, H. R. (1991). 'Application of multifrequency test signals to an industrial water boiler', *IEE Int. Conf. Control 91*, Edinburgh, March 1991, IEE Conference Publication No. 332, pp. 658–64.
Press, W. H., B. P. Flannery, S. A. Teukolsky and W. T. Vetterling (1986). *Numerical Recipes: The Art of Scientific Computing*, Cambridge University Press: Cambridge.
Priestley, M. B. (1981). *Spectral Analysis and Time Series*, Vol. 1, Academic Press: London, pp. 528–63.
Rabiner, L. R. and B. Gold (1975). *Theory and Application of Digital Signal Processing*, Prentice Hall: Englewood Cliffs, NJ.
Rabiner, L. R., R. W. Schafer and C. M. Rader (1969). 'The chirp-*z* transform algorithm and its application', *Bell System Technical Journal*, **48**, 1249–92.
Rake, H. (1980). 'Step response and frequency response methods', *Automatica*, **16**, 519–26.
Rakshit, A., S. N. Bhattacharyya and J. K. Choudhury (1985). 'A microprocessor-based multipoint signal averager for repetitive bioelectric signals', *Measurement*, **3**, No. 4, 169–74.
Rees, D. (1976). *Digital Processing of System Responses*, PhD thesis, Polytechnic of Wales.

Rees, D. (1977). 'Automatic testing of dynamic systems using multifrequency signals and the discrete Fourier transform', *Proc. IEE Conf. on New Developments in Automatic Testing*, IEE Conf. Pub. No. 158, pp. 24–7.

Rees, D. (1986). 'System identification using composite frequency signals with low peak factors', *Proc. Int. AMSE Conf. on Modelling and Simulation*, Sorrento, Italy.

Rees, D. (1990). 'Design of non-binary signals with low peak factors', *IEE Colloquium on Multifrequency Testing for System Identification*, London, 8 June 1990, IEE Colloquium Digest 1990/097, pp. 4.1–6.

Sakrison, D. J. (1966). 'Stochastic approximation: a recursive method for solving regression problems', *Advances in Communication Systems*, **2**, 51–106.

Sanathan, E. K. and J. Koerner (1963). 'Transfer function synthesis as a ratio of two complex polynomials', *IEEE Trans. Autom. Contr*, **AC-8**, 56–8.

Sankowski, D. (1983). 'Sposoby identyfikacji wlasnosci dynamicznych elektrycznych pieców komorowych', *Prace PIE nr 87*, pp. 21–45, Warsaw, Poland.

Sankowski, D. (1989a) 'Trend elimination by a microcomputer in temperature measurements for on-line identification of electro-heat systems', *Proc. IMEKO TC-12 Symposium, Microprocessors in Temperature and Thermal Measurement*, Lodz, Poland, pp. 97–104.

Sankowski, D. (1989b). *Wykorzystanie Wieloczestotliwosciowych sygnalów Binarnych (MBS) do Identyfikacji 'on-line' Rezystancyjnych Urzadzen Grzejnych*, DSc (habilitation) Thesis, Technical University of Lodz, Poland.

Sankowski, D. (1989c). 'Frequency domain multivariable model of an electric diffusion furnace', *Elektrowärme International*, **47**, B131–5.

Sankowski, D. (1989d). 'Closed loop identification of electric resistance furnaces', *Rozprawy Electrotechnizne*, **38**, 193–210.

Sankowski, D. (1990). 'Fast on-line identification of electroheat systems due to the shifted MBS', *International Symposium on System-Modelling-Control*, Vol. 3, Zakopane, Poland, pp. 78–84.

Sankowski, D. (1991). 'Stastistical characterisation of MBS frequency response estimation', *8th International Conference on Control Systems and Computer Science*, Vol. 1, Bucharest, pp. 177–84.

Sankowski, D. (1992). 'Removal of bias error of frequency estimates in on-line identification of electroheat systems', *Singapore International Conference on Intelligent Control and Instrumentation*, Singapore, February 1992.

Sankowski, D., J. McGhee, I. Henderson and P. Urbanek (1991). 'Compensation of switch-on transient in electric resistance furnaces using rotated multifrequency signals', *IEE Int. Conf. Control 91*, Edinburgh, March 1991, IEE Conf. Publ. No. 332, pp. 175–80.

Saridis, G. N. (1974). 'Comparison of six on-line identification algorithms', *Automatica*, **10**, 69–79.

Saridis, G. N. and G. Stein (1968). 'Stochastic approximation algorithms for linear system identification', *IEEE Trans. Autom. Contr.*, **AC-13**, 515–23 and 592–4.

Schoukens, J. and R. Pintelon (1990). 'Measurement of frequency response functions in noisy environments', *IEEE Trans. Instrum. Meas.*, **IM-39**, 905–9.

Schoukens, J. and R. Pintelon (1991). *Identification of Linear Systems. A Practical Guideline to Accurate Modeling*, Pergamon Press: Oxford.

Schoukens, J. and J. Renneboog (1986). 'Modeling the noise influence on the Fourier coefficients after a Discrete Fourier Transform', *IEEE Trans. Instrum. Meas.*, **IM-35**, 278–86.

Schoukens, J., R. Pintelon and J. Renneboog (1988a). 'A maximum likelihood method for linear and nonlinear systems – A practical application of estimation techniques in measurement problems', *IEEE Transactions Instrum. Meas.*, **IM-37**, 10–17.

Schoukens, J., R. Pintelon, E. Van der Ouderaa and J. Renneboog (1988b). 'Survey of excitation signals for FFT based signal analyzers', *IEEE Trans. Instrum. Meas.*, **IM-37**, 342–52.

Schroeder, M. R. (1970). 'Synthesis of low-peak factor signals and binary sequences with low autocorrelation', *IEEE Trans. Inf. Theory*, **IT16**, 85–9.

Schroeder, M. R. (1984). *Number Theory in Science and Communication*, Springer: Berlin.

Shannon, C. E. and W. Weaver (1972). *A Mathematical Theory of Communication*, University of Illinois Press.

Sharda, R., E. Wasil and B. L. Golden (1986). 'Mathematical programming software for microcomputers', in Gass, S. I., H. J. Greenberg, K. L. Hoffman and R. W. Langley (eds.), *Impacts of Microcomputers on Operations Research*, North-Holland: New York.

Silverman, H. F. (1977). 'An introduction to programming the Winograd fast Fourier transform algorithm (WFTA)', *IEEE Trans. on Acoustics, Speech and Signal Processing*, **ASSP-25**, 152–65.

Simpson, H. R. (1966). 'Statistical properties of a class of pseudorandom sequences', *Proceedings IEE*, **113**, 2075–81.

Simpson, R. J. and H. M. Power (1972). 'Correlation techniques for the identification of nonlinear systems', *Meas. and Control*, **5**, 316–21.

Singleton, R. C. (1969). 'An algorithm for computing the mixed radix fast Fourier transform', *IEEE Trans. on Audio and Electroacoustics*, **AU-17**, 93–103.

Sivaswamy, R. (1978). 'Multiphase complementary codes', *IEEE Trans. Inf. Theory*, **IT-24**, 546–52.

Söderström, T. (1977). 'On model structure testing in system identification', *International Journal of Control*, **26**, 1–18.

Söderström, T. and P. Stoica (1989). *System Identification*, Prentice Hall: Englewood Cliffs, NJ.

Ströbel, H. (1975). *Experimentelle Systemanalyse*, Akademie-Verlag: Berlin.

Stuart, A. and J. K. Ord (1991). *Kendall's Advanced Theory of Statistics*, Vol. 2, Edward Arnold: London.

Sydenham, P. H (1979). *Measuring Instruments Tools of Knowledge and Understanding*, Peter Peregrinus: Stevenage.

Tiefenthaler, C. (1980). 'Oversampling to increase S/N ratio of ADCs', *Electronic Product Design*, 59–62.

Tischler, M. B. (1987). *Frequency Response Identification of XV-15 Tilt Rotor Aircraft Dynamics*, PhD thesis, NASA, TM 89428.

Tischler, M. B. (1989). 'Advancements in frequency domain methods for rotorcraft system identification, *Int. Journal of Rotorcraft & Powered Lift Aircraft* (Vertica), **13**, 327–42.

Tischler, M. B. and J. Kaletka (1986). 'Modelling XV-15 tilt rotor aircraft dynamics by frequency and time-domain techniques', *Proceedings of the AGARD FMP Symposium on Rotorcraft Design for Operations*, CP-423, Amsterdam.

Tischler, M. B., J. W. Fletcher, V. L. Diekmann, R. A. Williams and R. W. Cusen (1987). *Demonstration of frequency-sweep testing technique using a Bell 214-ST helicopter*, NASA Technical Memorandum 89422.

Tomlinson, G. H. and P. Galvin (1974). 'Analysis of skewing in amplitude distributions of filtered m-sequences', *Proceedings IEE*, **121**, 1475–9.

Tseng, C. C. and C. L. Liu (1972). 'Complementary sets of sequences', *IEEE Trans. Inf. Theory*, **IT-18**, 644–52.

Turyn, R. (1968). 'Sequences with small correlation', in Mann, H. B. (ed.), *Error Correcting Codes*, John Wiley: New York, pp. 195–228.

Van Brussel, H. (1975). 'Comparative assessment of harmonic, random, swept sine and shock excitation methods for the identification of machine tool structures with rotating spindles', *Annals of the CIRP*, January, 291–6.

Van den Bos, A. (1967). 'Construction of binary multifrequency test signals', *1st IFAC Symposium on Identification in Automatic Control Systems*, Prague, June 1967, paper 4.6.

Van den Bos, A. (1970). 'Estimation of linear system coefficients from noisy responses to binary multifrequency test signals', *2nd IFAC Symposium on Identification and Process Parameter Estimation*, Prague, June 1970, paper 7.2.

Van den Bos, A. (1973). 'Selection of periodic test signals for estimation of linear system dynamics', *Third IFAC Symposium on Identification and System Parameter Estimation*, The Hague, pp. 1015–22.

Van den Bos, A. (1974). *Estimation of Parameters of Linear Systems using Periodic Test Signals*, Doctor of Technical Science Thesis, University of Technology, Delft University Press: Delft.

Van den Bos, A. (1987). 'A new method for synthesis of low-peak-factor signals', *IEEE Trans. Von Acoustics, Speech and Signal Processing*, **ASSP-35**, 120–22.

Van den Bos, A. (1989). 'Estimation of Fourier coefficients', *IEEE Trans. Instrum. Meas.*, **IM-38**, 1005–7.

Van den Bos, A (1991). 'Identification of continuous-time systems using multiharmonic test signals', in Sinha, N. K. and G. P. Rao (eds.), *Identification of Continuous-Time Systems*, Kluwer: Dordrecht.

Van den Bos, A. and R. G. Krol (1979). 'Synthesis of discrete-interval binary signals with specified Fourier amplitude spectra', *International Journal of Control*, **30**, 871–84.

Van den Eijnde, E. and J. Schoukens (1991). 'On the design of optimal excitation signals', *9th IFAC/IFORS Symposium on Identification and System Parameter Estimation*, Budapest, Hungary, pp. 827–32.

Van der Ouderaa, E. (1988). *Design of Optimal Input Signals with Minimal Crest Factor*, Doctor of Science Thesis, Faculty of Science, Free University of Brussels.

Van der Ouderaa, E., J. Schoukens and J. Renneboog (1988a). 'Peak factor minimization using a time-frequency domain swapping algorithm', *IEEE Trans. Instrum. Meas.*, **IM-37**, 145–7.

Van der Ouderaa, E., J. Schoukens and J. Renneboog (1988b). 'Peak factor minimization of input and output signals of linear systems', *IEEE Trans. on Instrum. Meas.*, **IM-37**, 207–12.

Watson, E. J. (1962). 'Primitive polynomials (modulo-2)', *Mathematics of Computing*, **16**, 368–9.

Wellstead, P. E. (1970). *Aspects of Real-Time Digital Spectral Analysis*, PhD thesis, University of Warwick.

Wellstead, P. E. (1977). 'Reference signals for closed-loop identification', *Int. Jnl Control*, **26**, 945–62.

Wellstead, P. E. (1978). *Spectral Analysis and Applications*, Control Systems Centre Report, No. 411, University of Manchester Institute of Science and Technology.

Wellstead, P. E. (1981). 'Non parametric methods of system identification', *Automatica*, **17**, 55–69.

Wellstead, P. E. (1986). 'Spectral analysis and applications', in Godfrey, K. R. and R. P. Jones (eds.), *Signal Processing for Control*, Springer Lecture Notes in Control and Information Sciences No. 79, pp. 210–44.

Wieslander, J. and B. Wittenmark (1971). 'An approach to adaptive control using real time identification', *Automatica*, **7**, 211–17.

Winograd, S. (1976). 'On computing the discrete Fourier transform', *Proceedings National Academy of Sciences (U.S.)*, **73**, 1005–6.

Yarmolik, V. N. and S. N. Demidenko (1988). *Generation and Application of Pseudorandom Sequences for Random Testing*, John Wiley: Chichester.

Young, P. (1989). *An Assessment of Techniques for Frequency Domain Identification of Helicopter Dynamics*, DPhil thesis, Department of Electronics, University of York.

Young, P. and R. J. Patton (1988), 'Frequency domain identification of remotely-piloted helicopter dynamics using frequency-sweep and Schroeder-phased test signals', *Proceedings of the AIAA Flight Mechanics Conference*, Minneapolis, USA, pp. 161–9 (AIAA paper No. 88-4349).

Young, P. and R. J. Patton (1990). 'Comparison of test signals for aircraft frequency domain identification', *Journal of Guidance, Control and Dynamics*, **13**, 430–38.

Young, P. C. (1970). 'An instrumental variable method for real-time identification of a noisy process', *Automatica*, **6**, 271–87.

Young, P. C. (1989). 'Recursive estimation, forecasting, and adaptive control', *Control and Dynamic Systems*, **30**, 119–65.

Zanakis, S. H. and J. S. Rustagi (eds.) (1982). *Optimization in Statistics*, North-Holland: Amsterdam.

Zarrop, M. (1979). *Optimal Experiment Design for Dynamic System Identification*, Springer: New York.

Zierler, N. (1959). 'Linear recurring sequences', *Jnl Soc. Ind. Appl. Math.*, **7**, 31–48.

Zypkin, Ja. S. (1987). *Grundlagen der informationellen Theorie der Identifikation*, VEB Verlag Technik, Berlin, Democratic Republic of Germany.

Index

adaptive control, 55, 219–21
Akaike Information Criterion (AIC), 412–13, 417–18
aliasing, 73–7, 133, 319, 355–61, 383, 394, 402, 407
autocorrelation function, 9–21, 31–3, 35–8, 45–9, 61–9, 71–2, 128, 177–81, 183–208, 332–6, 379–80
  higher-order, 45–6
  non-periodic, 55–6, 178–81, 187–208
  variance due to finite time averaging, 18
autocovariance function, 9–11, 287–8
averaging of data segments, 72–3, 108, 112–15, 306, 377

Barker sequences, 56, 188–9
  generalised, 195
Barker signals *see* Barker sequences
binary multi-frequency signals *see* multi-frequency binary signals
blast furnace 21–3
burst (in adaptive control) *see* adaptive control

CC package, 244–5, 248
chain code *see* maximum length sequences, binary
chirp, periodic *see* swept sine
chirp z-transform, 91–2, 113, 139
closed loop experiments, 21–3, 215, 264, 275–6, 304–6, 362–8
coherence function, 108–9, 112–15, 119–20, 298–9, 305–6, 308–11, 313–14, 319–20, 374–7, 408–9, 415–17
  variance of, 307–11, 320
compact multi-frequency binary signals, 96–7, 211, 215, 227–8, 248
complementary signals-binary, 191–2
  multi-level, 197–200
  recursive binary, 192–3
  uncorrelated binary, 194
  use in non-periodic system identification, 204–8
condition monitoring, 249–53
convolution integral, 2–4, 19–21, 31–2, 35, 78, 107, 205, 287–8
  graphical illustration of, 3
convolution sum, 2–3, 20, 289
Cramér-Rao bound, 144, 147, 168–70
crest factor, 131–42, 147–50
  minimisation of, 147–8, 155–8
crosscorrelation function, 10, 12, 17–23, 31–2, 35–8, 51–5, 107–8, 128, 287–8
  confidence interval of, 22–3
  graphical illustration of, 11
  non-periodic, 55, 178, 187–208
  variance due to finite data length, 18
cross-covariance function, 287–8
cross power spectral density, 108, 305–6
cross power spectrum (discrete), 108–9, 113–16, 374–7, 408–9

degrees of freedom (of a signal), 379–80, 382
delta function, 2, 7, 20
describing function, 316, 363
descriptive language technique, 239–40
diffusion furnace, 273–6
Dirac delta function *see* delta function
direction dependent dynamics, 50–3.
Discrete Fourier Transform *see* Fourier Transform, discrete
discrete interval random binary signal, 14–15, 20–1, 62–3
dispersion function, 144–6

distillation column, 7, 9, 32
drift, 33, 232, 256, 264–7, 270–2

electric resistance furnaces, 255–7, 268–72
equivalent bandwidth (of a signal), 379–80, 382
ergodic hypothesis, 9–10.
experimentation time, 33, 55, 240, 243, 256, 269, 286, 298–9, 301, 309, 409, 414
extruder, 216–18, 220
eye patterns, 232–3, 245–53

Fast Fourier Transform, 19, 73, 91–2, 95, 113, 126, 128, 212–13, 300, 304, 337–8, 344, 355, 381, 400, 402–3
feedback shift registers
  binary, 24–9, 136–9, 182, 382
  multi-level, 39–49, 182–3, 186, 189, 329–30
feedforward control, 4–5
finite fields, 321–8
  extension, 322, 324, 327
  field elements, 322, 324
  inverse element, 327–8
  prime, 322, 326
  primitive element, 328
  product table, 323, 325
  reciprocal element, 327–8
  sum table, 323, 325
five-level signals, 43–4, 54, 182–3
flat pass window, 81–92
folding frequency, 71, 400
Fourier coefficients
  estimation of, 167–70, 261–7
  use in discrete-time identification, 171–4.
Fourier series, 64–70, 336
Fourier Transform
  continuous, 61, 71, 78–81, 107–8, 233–4
  discrete, 70–8, 91–2, 108–9, 113, 126, 162–5, 336–9, 344, 395–6
four-level signals, 46–9
frequency response, 46, 107, 110–15, 117–20, 127–42, 153–5, 261–7, 269–76, 280, 289–96, 304–6, 343–7, 408–9, 414–20
frequency response estimation
  empirical estimates, 289–91, 296
  H_1 estimates, 128–30, 153–4
  H_2 estimates, 128–30, 154
  H_{log} estimates, 129–30, 154
  least squares regression, 374–5
  variance of estimate, 155, 240, 243, 290–1,

Galois fields, *see* finite fields.
gas turbine engine dynamics, 52–3
Gaussian signals (*see also* pseudo-random Gaussian noise), 18, 22, 39, 289, 319–20, 378–80, 394
  peak factor of, 378
generalised Barker sequences *see* Barker sequences, generalised
generalised least squares, 32, 58, 410

Hall sequences, 30
Hanning window, 80–91
helicopter model, 316–20
helicopter tests, 311–16
  aircraft body roll attitude, 313–14
  automatic stabilisation equipment, 311–12, 316
  lateral cyclic blade pitch angle, 314–16
  roll rate, 313–16
hill climbing, 55
Huffman signals, 196–7

identification channel coding theorem, 227
identification source coding theorem, 227
impact testing, 139–40
impulse response (*see also* weighting function), 1–2, 7, 46, 63, 107, 109, 287–8
information matrix, 144, 146–7
instrumental variable methods
  frequency domain, 172–3
  time domain, 58, 410–11
integer level transformation, 183–5
inverse repeat signals
  binary multi-frequency, 101
  pseudo-random binary, 36–8, 50–2, 92–3, 120
  pseudo-random non-binary, 334–6, 340
irreducible polynomial
  binary, 26–9
  non-binary, 41

leakage *see* spectral leakage
least squares estimator (frequency domain), 173–4

level transformation *see* pseudo-random signals (multi-level)
liquid level system, 214–16
loss function criterion, 412

MATLAB Toolset, 105, 107
maximum entropy signals
  definition, 225
maximum-length sequences (m-sequences)
  binary, 24–8, 33, 50–2, 136–9, 141, 161, 302–3, 382, 386
  multi-level, 39–49, 54, 181–7, 329–32
maximum likelihood estimator, 58, 143, 147, 169–70
model validation, 419–21
modulo–2 addition, definition, 24
multi-frequency binary signals, 95–101, 113–16, 139, 141, 164–5, 210–22, 225–54, 255–64, 268–71, 276, 280–1, 283–7, 292–6, 349, 403
multi-harmonic signals, 101–7, 110–13, 116–20, 121–4, 134–5, 141, 147–50, 155–60, 161–7, 303–4, 307–11, 318–20, 349–56, 362–72, 381–95, 404–6, 414–17
multi-input systems, 35–8, 217–18, 273–4
multi-sine signals *see* multi-harmonic signals

NARMA model, 124
non-linear systems, 49–55, 120–5, 222, 285, 298–9, 301–2, 311, 316, 343, 362–70, 397, 413, 420–1
non-periodic pseudo-random signals
  binary, 55–6, 187–94
  multi-level, 194–201
  use in identification, 201–8
normal operating records, 21–3
Nyquist frequency *see* folding frequency

optimal binary m-sequence (for drift elimination), 33
optimal starting point (for transient removal)
  pseudo random binary signal, 33
  multi-frequency binary signal, 260–1
optimal test signal design, 56, 58, 130–3, 213
optimal three-level *m*-sequence (for non-linear identification), 54
orthogonal signals *see* uncorrelated signals
overlapping data segments, 73, 304, 307
oversampling, 383

peak factor, 101–5, 121, 159, 164, 308–11, 349, 378, 383–95, 404–6
periodic noise, 135–6, 141
persistent excitation, 21, 58–9, 209, 302
phase shift keying, 232–3, 236–43, 247–53
Plant System Identification (PSI) package, 105, 398–401, 409, 411, 413–21
Poisson distribution, 16
polyphase signals, 195–6
power spectral density, 61–4, 107–8, 318
power spectrum
  continuous, 65–70, 74, 78, 318, 401–2
  discrete, 71–108, 123–4, 374–80, 408–9
prime harmonic signals, 105, 303, 350–6, 363–8, 371
primitive polynomial
  binary, 25–6, 28–9
  non-binary, 40–1, 324–8, 330–1
principle of superposition *see* superposition principle
pseudo-random binary signals
  non-periodic, 187–94
  periodic, 23–39, 68–70, 74–7, 87–95, 110–16, 181, 210, 221, 225, 302–3, 318–20, 349, 362–70, 378–80 389-90, 394, 401–3, 420
pseudo-random Gaussian noise, 381–95.
pseudo-random multi-level signals
  non-periodic, 194–201
  periodic, 39–49, 54–5, 181–7, 329–47
pulse signal, 139–140

quadratic residue codes, 29–30, 50–2

random binary signal, 16–17, 63–4
random burst, 140–1
random noise, 140–1
reconstruction filter, 160
rectangular window, 77–92
repeat frequency, 71
resistance temperature detector (*see also* temperature sensors), 278–83, 292–5
robust estimators, 170

Schroeder phased multi-harmonic signals,

102–5, 121–3, 134–5, 141, 165–6, 304, 352, 354–5, 389–95, 404–6
selection of harmonics, 222
settling time, 2, 34
shift registers *see* feedback shift registers
simulated annealing (for peak factor minimisation), 166–7
sine wave
  autocorrelation function, 12–14
  continuous power spectrum, 66
  discrete power spectrum, 82–7
sinusoidal level transformation, 184–8
skewing of probability density function, 39
spectral composition, 382
spectral entropy, 378–9, 382
spectral leakage, 77–92, 110–15, 133, 139–41
square wave
  autocorrelation function 13–14
  continuous power spectrum, 67–8
  use in testing, 221, 227, 291–2, 349, 400, 414
stationarity, 9–12
steam reformer, 4–7, 9
steelworks blast furnace *see* blast furnace
stepped sine signal, 133, 141
step response 4–9, 32, 209, 214, 287
stirred tank system, 214–16, 219–20
structure selection (of a model), 411–13, 416–17
sum of harmonics signals *see* multi-harmonic signals
superposition principle, 1, 280
swept sine signal, 133–4, 141, 300, 302, 311–16, 318–20, 349
switch-on transient, 232, 248–9, 256–61, 269, 271, 293–5
SYDLAB system detection laboratory, 245, 256–7, 292

temperature sensors,
  dynamic behaviour, 278–83
  frequency response, 278–80, 289–91
  mathematical models, 279–83
  testing methods, 291–5
three-level signals, 41–5, 54–5, 213–14
time factor, 127, 132–41
trajectory derived signals, 200–1
transfer function
  continuous (Laplace), 5–6, 8, 109, 117, 127 142–52, 215, 244, 248–53, 257, 274, 293, 416–17
  discrete (z-transform), 32, 58–9, 109, 167–74, 217–18, 410–11, 419
trend *see* drift
turn off (in adaptive control) *see* adaptive control
twin prime sequences, 30

ultra-compact multi-frequency binary signals, 227, 229–31, 241–2
uncorrelated signals, 35–8, 92–5, 101, 194, 217

Volterra kernel, 49–50, 53–4, 59, 120, 285
Volterra kernel transformation, 120, 122–4

warm air flow system, 114–17, 287
water boiler, 413–21
weighting function (*see also* impulse response), 2, 20, 31, 33, 49, 54
weighting sequence, 3, 21, 55
Westland flight simulator, 311
windows, 77–92, 140, 300

Young's Information Criterion, 413, 417–18

zero order hold, 160, 336–7, 409
zoom factor, 237–43